Lecture Notes in Control and Information Sciences 337

Editors: M. Thoma · M. Morari

Henk A. P. Blom · John Lygeros (Eds.)

Stochastic Hybrid Systems

Theory and Safety Critical Applications

With 88 Figures

Editors

Henk A.P. Blom
National Aerospace Laboratory
NLR
P.O. Box 09502
1006 BM Amsterdam
The Netherlands
blom@nlr.nl

John Lygeros
University of Patras
Department of Electrical and Computer Engineering
Systems and Measurements Laboratory
265 00 Patras
Greece
lygeros@ee.upatras.gr

This publication is a result of the HYBRIDGE project, a project within the 5th Framework Programme IST-2001-IV.2.1 (iii) (Distributed Control), funded by the European Commission under contract number IST-2001-32460. This publication does not represent the opinion of the Community, and the Community is not responsible for any use that might be made of data appearing therein.

ISSN 0170-8643

ISBN-10 3-540-33466-1 **Springer Berlin Heidelberg New York**
ISBN-13 978-3-540-33466-8 **Springer Berlin Heidelberg New York**

Library of Congress Control Number: 2006924574

Springer is a part of Springer Science+Business Media

springer.com

Typesetting: Data conversion by editors.
Final processing by PTP-Berlin Protago-TEX-Production GmbH, Germany (www.ptp-berlin.com)
Cover-Design: design & production GmbH, Heidelberg
Printed on acid-free paper 89/3141/Yu - 5 4 3 2 1 0

Preface

The first decade of the new millennium finds the global economy at an important juncture. The rapid technological advances of recent decades, coupled with economic pressure, are forcing together sectors of the economy that have evolved separately to date. Among these sectors are

- Industrial processes, an area of intense activity for more than a century.
- The information revolution, whose implications became apparent to the wider public in the 1990's, but whose foundations were being laid for decades.
- Service oriented society, which asks for an approach where humans remain responsible.

This rapprochement of "mind" and "matter" presents historic opportunities and challenges in many areas of economic and social activity. Some of the greatest challenges arise in the area of safety-critical embedded systems.

Embedded systems, i.e. systems where digital devices have to interact with a predominantly analog environment on the one hand, and with humans on the other, are the outcome of the merging of industrial and information processes. Many of these embedded systems are found in applications in which safety is a primary concern. Examples include automotive electronics, transportation systems and energy generation and distribution. The need to provide safety guarantees for the operation of these systems imposes particularly stringent requirements on the engineering design.

The design of safety-critical embedded systems is further complicated by the fact that their evolution often involves substantial levels of uncertainty, arising either from the physical process itself, or from the actions of human operators (e.g. the drivers, air traffic controllers, pilots, etc.). The theoretical development in handling uncertainty is facing a significant gap in how to incorporate the mind-setting of humans who are ultimately responsible for safety. This requires one to manage uncertainty in a predictable and safe way.

Air Traffic Management as example of distributed interactions in a safety critical system

Air Traffic Management (ATM) is one example of this class of systems that poses exceptional challenges. One of the defining features of the air traffic management process is the interplay between distributed decision making and safety criticality. Figure 1 highlights this point. Unlike other safety-critical industries, such as nuclear and chemical plants, decision making is carried out at many levels in the air traffic management process, and involves interactions between many stake holders: pilots, air traffic controllers, airline operation centers, airport authorities, government regulators and even the traveling public. The actions of all of these agents have an impact on both the safety and the economic efficiency of the system.

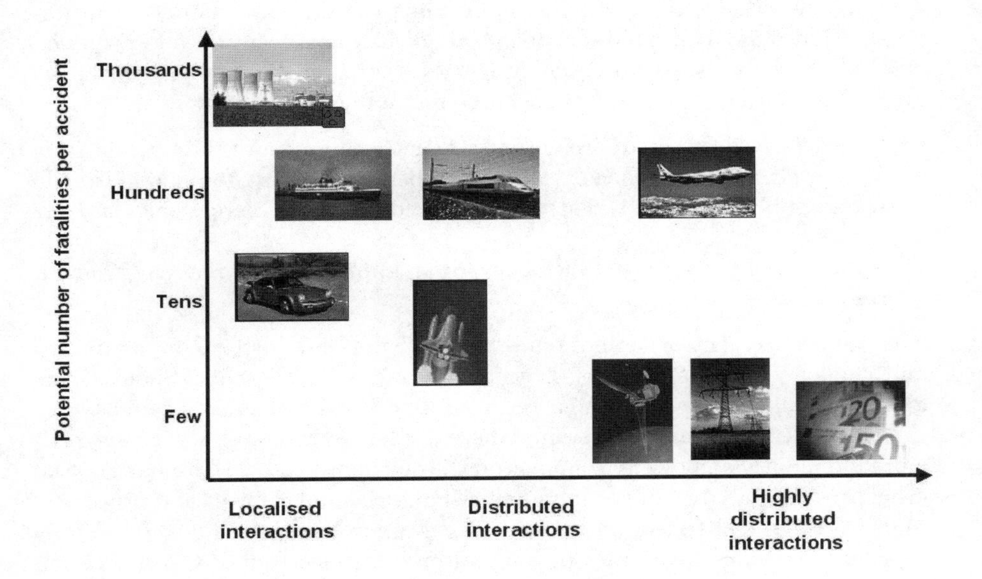

Fig. 1. Air traffic compared with other safety-critical processes in terms of potential number of fatalities per accident and the distribution of safety-critical interactions between human and system agents

Despite technological advances, including powerful on-board computers, advanced flight management and navigation systems, satellite positioning and communication systems, etc., air traffic management still is, to a large extent, built around a rigid airspace structure and a centralized, mostly human-operated system architecture. Despite this, the level of safety achieved in air traffic is very impressive, when one considers the volume of traffic and the relatively low number of accidents.

The increasing demand for air travel is stretching current air traffic management practices to their limits. Air-Traffic in Europe is projected to double every 10 to 15 years; even higher rates of growth are expected for the U.S., Asia and for trans-oceanic flights. The requirement is to improve current practice to be able to sustain this growth rate, without causing safety, or performance degradation, or placing an additional burden on the already overloaded human operators.

Research has shown that introducing automation of current controller tasks will not solve this problem alone. There is rather a need for fundamental changes in the human roles and tasks. One proposed advanced approach is to increase the role of pilots and airborne separation assistance systems in the air traffic management process. It is believed that in this way the safety and economy of air traffic can be improved and the tasks of ground controllers can be simplified, allowing them to handle the increased demand in air traffic without compromising the current high safety levels.

The main problem with introducing such changes to air traffic practices is that the system has evolved for a number of years in a rather ad hoc way. The current air traffic management system involves an uncomfortable mixture of rules, regulations, guidelines for the human operators, automated and semi-automated components, computer tools, etc. As a consequence, even though the current system delivers an admirable level of safety, it does so at the expense of complexity and conservativeness. Introducing any changes and assessing their impact on the safety of the system is therefore a very challenging task, which requires research in order to be built on solid foundations.

Stochastic Hybrid System Research Challenges

Stochastic hybrid system analysis can play a central role in restructuring complex safety critical processes such as air traffic management. In principle one can use stochastic analysis tools to investigate the safety of the current system, determine the impact of proposed changes, and suggest ways of improving the situation. This approach has had considerable success in the nuclear and chemical industries. Air traffic, however, poses a number of additional challenges for stochastic analysis methods.

- Complexity and distribution: The air traffic management system is highly distributed, involving the interaction of a large number of semi-autonomous agents (the aircraft) with centralized components (air traffic control). As discussed above, the complexity of the system increases further if one considers the impact of other stake holders, e.g. airlines, passengers and airports.
- Human in the loop: Current air traffic management is centered around the air traffic controllers and, to a lesser extent, the pilots. These human operators are likely to be an integral part of the system for many years to come. Therefore, assessing the impact of their actions (and potential errors) on the safety and performance of the system is crucial.

- Hybrid dynamics: When viewed as a dynamical system, air traffic management involves diverse types of dynamics:
  - Continuous dynamics, that arise from the physical movement of the aircraft, response times of the human operators, etc.
  - Discrete dynamics, that arise when aircraft take off or land, change cruising altitudes, etc., move from one airspace sector to another.
  - Stochastic dynamics, that arise due to weather uncertainty, errors of the human operators, the possibility of mechanical failure, etc.

The aim of this book is to provide an overview of recent research activity that addresses many of these challenges. The research contributions are organised in three parts:

Part 1. Stochastic Hybrid Processes
Part 2. Analytical Approaches
Part 3. Complexity and Randomization

Acknowledgment

Most of the research presented in this volume was funded by the European Commission, under the project HYBRIDGE, IST-2001-32460. This project brought together some fifty system theorists and mathematicians from seven universities (University of Cambridge, University of Twente, University of L'Aquila, National Technical University of Athens (NTUA), University of Brescia, University of Patras and Polytechnico of Milan) and three research institutes (National Aerospace Laboratory (NLR), Institut National de Recherche en Informatique et en Automatique (INRIA) and Centre d'Études de la Navigation Aérienne (CENA)) to develop innovative approaches for handling uncertainty in complex safety-critical systems through furthering state-of-the-art approaches developed in mathematics, control theory and computer science for dealing with uncertainty in automation, finance, robotics and transportation. In collaboration with experts from the Eurocontrol Experimental Centre, BAESystems, and AEA Technology these state-of-the-art approaches were then tailored to some specific and pressing problems in air traffic management.

The contents of this book reflects the authors views; the Community is not liable for any use that may be made of the information contained therein.

Amsterdam,
31th January 2006

Henk Blom
John Lygeros

List of Contributors

Henk A.P. Blom
National Aerospace Laboratory NLR
P.O. Box 90502, 1006 BM Amsterdam, The Netherlands
blom@nlr.nl

Manuela L. Bujorianu
University of Twente
Faculty of Computer Science
P.O. Box 217, 7500 AE Enschede, The Netherlands
manuela@cs.utwente.nl

Pierre Del Moral
Université de Nice
Sophia Antipolis-06108
Nice Cedex 02, France
delmoral@math.unice.fr

Elena De Santis
University of L'Aquila
Center of Excellence DEWS
Department of Electrical Engineering
Poggio di Roio, 67040
L'Aquila, Italy
desantis@ing.univaq.it

Maria D. Di Benedetto
University of L'Aquila
Center of Excellence DEWS
Department of Electrical Engineering
Poggio di Roio, 67040
L'Aquila, Italy
dibenede@ing.univaq.it

Stefano Di Gennaro
University of L'Aquila
Center of Excellence DEWS
Department of Electrical Engineering
Poggio di Roio, 67040
L'Aquila, Italy
digennar@ing.univaq.it

Dimos V. Dimarogonas
National Technical University of Athens
Control Systems Laboratory
9 Heroon Polytechniou Street
Zografou 15780, Athens, Greece
ddimar@mail.ntua.gr

Alessandro D'Innocenzo
University of L'Aquila
Center of Excellence DEWS
Department of Electrical Engineering
Poggio di Roio, 67040
L'Aquila, Italy
adinnoce@ing.univaq.it

Mariken H.C. Everdij
National Aerospace Laboratory NLR
P.O. Box 90502, 1006 BM Amsterdam, The Netherlands
everdij@nlr.nl

William Glover
University of Cambridge
Department of Engineering
Cambridge CB2 1PZ, U.K.
University of Cambridge, Cambridge, CB2 1PZ, UK
wg214@eng.cam.ac.uk

Jianghai Hu
Purdue University
School of Electrical and Computer Engineering
West Lafayette, IN 47906, USA
jianghai@purdue.edu

Bart Klein Obbink
National Aerospace Laboratory NLR
P.O. Box 90502, 1006 BM Amsterdam, The Netherlands
bklein@nlr.nl

Margriet B. Klompstra
National Aerospace Laboratory NLR
P.O. Box 90502, 1006 BM Amsterdam, The Netherlands
klompstr@nlr.nl

Kostas J. Kyriakopoulos
National Technical University of Athens
Control Systems Laboratory
9 Heroon Polytechniou Street
Zografou 15780, Athens, Greece
kkyria@central.ntua.gr

Andrea Lecchini
University of Cambridge
Department of Engineering
Cambridge CB2 1PZ, U.K.
al394@eng.cam.ac.uk

François LeGland
IRISA / INRIA
Campus de Beaulieu 35042 RENNES
Avenue du General Leclerc
Cédex, France
legland@irisa.fr

Pascal Lezaud
Centre d'Etudes de la Navigation Aérienne
31055 Toulouse Cedex, France
pascal.lezaud@recherche.enac.fr

Savvas G. Loizou
National Technical University of Athens
Control Systems Laboratory
9 Heroon Polytechniou Street
Zografou 15780, Athens, Greece
sloizou@central.ntua.gr

John Lygeros
University of Patras
Department of Electrical and Computer Engineering
Rio, Patras, GR-26500, Greece
lygeros@ee.upatras.gr

Jan Maciejowski
Department of Engineering
University of Cambridge, Cambridge, CB2 1PZ, UK
jmm@eng.cam.ac.uk

Nadia Oudjane
EDF, Division R&D
1 avenue du Géenéral de Gaulle
92141 CLAMART Cédex, France
nadia.oudjane@edf.fr

Giordano Pola
University of L'Aquila
Center of Excellence DEWS
Department of Electrical Engineering
Poggio di Roio, 67040
L'Aquila, Italy
pola@ing.univaq.it

Maria Prandini
Politecnico di Milano
Dipartimento di Elettronica e Informazione
Piazza Leonardo da Vinci 32, 20133
Milano, Italy
prandini@elet.polimi.it

Stefan Strubbe
University of Twente
Department of Applied Mathematics
P.O. Box 217, 7500 AE Enschede,
The Netherlands
s.n.strubbe@math.utwente.nl

Arjan Van der Schaft
University of Groningen
Institute for Mathematics and
Computer Science
P.O. Box 800, 9700 AV Groningen,
The Netherlands
A.J.van.der.Schaft@math.rug.nl

Contents

Part III Complexity and Randomization

Part I

Stochastic Hybrid Processes

Toward a General Theory of Stochastic Hybrid Systems

Manuela L. Bujorianu[1] and John Lygeros[2]

[1] Department of Engineering, University of Cambridge, Cambridge CB2 1PZ, U.K. lmb56@eng.cam.ac.uk
[2] Department of Electrical and Computer Engineering, University of Patras, Rio, Patras, GR-26500, Greece, lygeros@ee.upatras.gr

Summary. In this chapter we set up a mathematical structure, called Markov string, to obtaining a very general class of models for stochastic hybrid systems. Markov Strings are, in fact, a class of Markov processes, obtained by a mixing mechanism of stochastic processes, introduced by Meyer. We prove that Markov strings are strong Markov processes with the càdlàg property. We then show how a very general class of stochastic hybrid processes can be embedded in the framework of Markov strings. This class, which is referred to as the General Stochastic Hybrid Systems (GSHS), includes as special cases all the classes of stochastic hybrid processes, proposed in the literature.

1 Introduction

In the face of growing complexity of control systems, stochastic modeling has got a crucial role. Indeed, stochastic techniques for modeling control and hybrid systems have attracted attention of many researchers and constitute one of the hottest issues in contemporary high level research.

Hybrid systems have been extensively studied in the past decade, both concerning their theoretical framework, as well as relating to the increasing number of applications they are employed for. However, the subfield of stochastic hybrid systems is fairly young. There has been considerable current interest in stochastic hybrid systems due to their ability to represent such systems as maneuvering aircraft [18], switching communication networks [16]. Different issues related to stochastic hybrid systems have found applications to insurance pricing [12], capacity expansion models for the power industry [11], flexible manufacturing and fault tolerant control [13, 14], etc.

A considerable amount of research has been directed towards this topic, both in the direction of extending the theory of deterministic hybrid systems [17], as well as discovering new applications unique to the probabilistic framework.

H.A.P. Blom, J. Lygeros (Eds.): Stochastic Hybrid Systems, LNCIS 337, pp. 3–30, 2006.

1.1 Objectives of the Chapter

This chapter has three objectives:

1. Introduce a very general framework for modeling stochastic hybrid processes: General Stochastic Hybrid System, abbreviated with GSHS.
2. Develop a theoretical construction for mixing Markov processes which preserves the Markov property. The result of this mixing operation will be called *Markov string*.
3. Show how GSHS can be embedded in the Markov string constructions and hence deduce the basic properties of GSHS as Markov property, strong Markov property

A GSHS might be thought of a 'conventional' hybrid system enriched with three uncertainty characteristics:

1. the continuous-time dynamics are driven by stochastic differential equations (SDE) rather then classical ODE,
2. a jump takes place when the continuous state hits the mode boundary or according with a transition rate
3. the post jump locations are randomly chosen according with a stochastic kernel.

Intuitively, GSHS can be described as an interleaving between a finite or countable family of diffusion processes and a jump process. Our goal is to prove that GSHS is indeed a 'good model'. This means that we need to investigate the stochastic properties of this model. A natural property we were looking for is the Markov property. Analyzing the form of the GSHS executions (paths or trajectories), the first observation is that these are, in fact, 'concatenations' of the diffusion component paths. The continuity inherited from the diffusion trajectories is perturbed by the jumps between the diffusion components.

This observation leads to the investigation of a general mechanism for mixing Markov processes that preserves the Markov property. Given a finite or countable family of Markov processes with reasonably good properties, this machinery will allow us to get a new Markov process whose paths are obtained by 'sticking' together the component paths. Roughly speaking, Markov strings are sequences of Markov processes. The jump structure of a Markov string is completely described by a renewal kernel given a priori and a family of terminal times associated with the initial processes. We require that the Markov string have finitely many jumps in finite time. Under these assumptions we prove that the Markov strings, as stochastic processes, enjoy useful properties like the strong Markov property and the càdlàg property.

We then return to GSHS and show how GSHS can be embedded in the framework of Markov strings. The class of GSHS inherits the strong Markov and càdlàg properties from Markov strings.

Finally, we develop the expression of the infinitesimal generator associated to GSHS.

1.2 Related Work

A well-known and very powerful class of continuous time stochastic processes with stochastic jumps (for the discrete state and also for the continuous state) is the piecewise-deterministic Markov processes (PDMP), introduced in [10], and applied to hybrid system modeling in [8]. The other modeling approaches are those presented in [17] (stochastic hybrid systems abbreviated SHS), [2] (stochastic hybrid models abbreviated SHM), [14, 15] (switching diffusion processes, abbreviated SDP), [6] (general switching diffusion processes abbreviated GSDP), see, also, [24] for quick presentation and comparisons. A very general formal model for stochastic hybrid systems is proposed in [7], which extends the model from [17], where the deterministic differential equations for the continuous flow are replaced by their stochastic counterparts, and the reset maps are generalized to (state-dependent) distributions that define the probability density of the state after a discrete transition. In this model transitions are always triggered by deterministic conditions (guards) on the state.

GSHS generalize PDMP allowing a stochastic evolution (diffusion process) between two consecutive jumps, while for PDMP the inter-jump motion is deterministic, according to a vector field. As well, GSHS might be thought of as a kind of extended SHS for which the transitions between modes are triggered by some stochastic event (boundary hitting time and transition rate). Moreover, GSHS generalize SDP permitting that also the continuous state to have discontinuities when the process jumps from one diffusion to another.

Another model for stochastic hybrid processes with hybrid jumps, which allows switching diffusions with jumps both in the discrete state and the continuous state, is developed in [4]. It can be shown that the class of these models can be considered as a subclass of GSHS whose stochastic kernel, which gives the post jump locations, is chosen in an appropriate way such that the change of the discrete state at a jump depends on the pre-jump location (continuous and discrete) and the change of the continuous state depends on the pre-jump location and on the new discrete state.

1.3 ATM Motivation

The ultimate goal of our work (under the European Commission's HYBRIDGE project [19]) is to use theoretical tools developed for stochastic hybrid models as a basis for designing and analyzing advanced Air Traffic Management (ATM) concepts for the European airspace. The modeling of ATM systems is a stochastic hybrid process, since it involves the interaction of continuous dynamics (e.g. the movement of the aircraft), discrete dynamics (e.g. aircraft landing and taking off, moving from one air traffic control sector to another, etc.) and stochastic dynamics (e.g. due to wind, uncertainty about the actions of the human operators, malfunctions, etc.).

In the context of ATM we are interested in modeling and analyzing safety-critical situations. In [26], a number of such situations were identified. Each one appears to have different modeling needs. In the following, we highlight the stochastic hybrid issues that arise in two aspects of ATM modeling: aircraft and weather models. Different models developed in the literature for stochastic hybrid processes might be used to model different safety critical situations identified in ATM. The difference between these models consists in where the stochastic phenomena appear: in the discrete dynamics, in the continuous dynamics or in both. For different safety-critical situations identified in the ATM modeling different models might be appropriate depending where the randomness lies:

- In the modeling of *aircraft climbing* the most suitable models appear to be SHS [17].
- Uncertainty in the ATC *sector transition process* can be treated in the framework of PDMP [8].
- For *missed approaches*, an appropriate model seems to be the SDP model [14]. SDP can also model changes in the flight plan segment when the aircraft reaches a way point (by introducing rate functions with support in a neighborhood of the way point). For missed approaches due to runway incursions, a general stochastic hybrid systems model is needed to accurately model this case.
- For modeling *overtake maneuvers* in unmanaged airspace the most appropriate models are SDP [14].

For more details see [9]. The conclusions of the above discussion is that it is necessary to develop further a more general class of stochastic hybrid processes than those found in the literature. This is because

1. Different types of models seem to be needed to capture the different situations. This implies that a number of different techniques and tools must be mastered to be able to deal with all the cases of interest. If a GSHS framework were available the process would be more efficient, since a single set of results, simulation procedures, etc. could be used in all cases.
2. Certain situations, such as vertical crossings during descent and missed approaches due to runway incursions, would be more accurately modeled by a GSHS.

2 General Stochastic Hybrid Systems

2.1 Informal Discussion

General Stochastic Hybrid Systems (GSHS) are a class of non-linear stochastic continuous-time hybrid dynamical systems. GSHS are characterized by a hybrid state defined by two components: the continuous state and the discrete state. The continuous and the discrete parts of the state variable have their own natural dynamics, but the main point is to capture the interaction between them.

The time t is measured continuously. The state of the system is represented by a continuous variable x and a discrete variable i. The continuous variable evolves in some "cells" X^i (open sets in the Euclidean space) and the discrete variable belongs to a countable set Q. The intrinsic difference between the discrete and continuous variables, consists of the way that they evolve through time. The continuous state evolves according to an SDE whose vector field and drift factor depend on the hybrid state. The discrete dynamics produces transitions in both (continuous and discrete) state variables x, i. Switching between two discrete states is governed by a probability law or occurs when the continuous state hits the boundary of its state space. Whenever a switching occurs, the hybrid state is reset instantly to a new state according to a probability law which depends itself on the past hybrid state. Transitions, which occur when the continuous state hits the boundary of the state space are called forced transitions, and those which occur probabilistically according to a state dependent rate are called spontaneous transitions. Thus, a sample trajectory has the form $(q_t, x_t, t \geq 0)$, where $(x_t, t \geq 0)$ is piecewise continuous and $q_t \in Q$ is piecewise constant. Let $(0 \leq T_1 < T_2 < ... < T_i < T_{i+1} < ...)$ be the sequence of jump times.

It is easy to show that GSHS include, as special cases, many classes of stochastic hybrid processes found in the literature PDMP, SHS, etc.

In the following we make use of some standard notions from the Markov process theory as: underlying probability space, natural filtration, translation operator, Wiener probabilities, admissible filtration, stopping time, strong Markov property [5]. The basic definitions from the Markov process theory are summarized in the Appendix.

2.2 The Mathematical Model

If X is a Hausdorff topological space we use to denote by $\mathcal{B}(X)$ or $\mathcal{B}$ its Borel σ-algebra(the σ-algebra generated by all open sets). A topological space, which is homeomorphic to a Borel subset of a complete separable metric space is called Borel space. A topological space, which is is a homeomorphic with a Borel subset of a compact metric space is called Lusin space.

State space. Let Q be a countable set of discrete states, and let $d : Q \to \mathbb{N}$ and $\mathcal{X} : Q \to \mathbb{R}^{d(.)}$ be two maps assigning to each discrete state $i \in Q$ an open subset X^i of $\mathbb{R}^{d(i)}$. We call the set

$$X(Q, d, \mathcal{X}) = \bigcup_{i \in Q} \{i\} \times X^i$$

the hybrid state space of the GSHS and $x = (i, x^i) \in X(Q, d, \mathcal{X})$ the hybrid state. The closure of the hybrid state space will be

$$\overline{X} = X \cup \partial X$$

where

$$\partial X = \bigcup_{i \in Q} \{i\} \times \partial X^i.$$

It is clear that, for each $i \in Q$, the state space X^i is a Borel space. It is possible to define a metric ρ on X such that $\rho(x_n, x) \to 0$ as $n \to \infty$ with $x_n = (i_n, x_n^{i_n})$, $x = (i, x^i)$ if and only if there exists m such that $i_n = i$ for all $n \geq m$ and $x_{m+k}^i \to x^i$ as $k \to \infty$. The metric ρ restricted to any component X^i is equivalent to the usual Euclidean metric [10]. Each $\{i\} \times X^i$, being a Borel space, will be homeomorphic to a measurable subset of the Hilbert cube, $\mathcal{H}$ (Urysohn's theorem, Prop. 7.2 [3]). Recall that $\mathcal{H}$ is the product of countable many copies of $[0, 1]$. The definition of X shows that X is, as well, homeomorphic to a measurable subset of $\mathcal{H}$. Then $(X, \mathcal{B}(X))$ is a Borel space. Moreover, X is a Lusin space because it is a locally compact Hausdorff space with countable base (see [10] and the references therein).

Continuous and discrete dynamics. In each mode X^i, the continuous evolution is driven by the following stochastic differential equation (SDE)

$$dx(t) = b(i, x(t))dt + \sigma(i, x(t))dW_t, \tag{1}$$

where $(W_t, t \geq 0)$ is the m-dimensional standard Wiener process in a complete probability space.

Assumption 1 (Continuous evolution) *Suppose that* $b : Q \times X^{(\cdot)} \to \mathbb{R}^{d(\cdot)}$, $\sigma : Q \times X^{(\cdot)} \to \mathbb{R}^{d(\cdot) \times m}$, $m \in \mathbb{N}$, *are bounded and Lipschitz continuous in* x.

This assumption ensures, for any $i \in Q$, the existence and uniqueness (Theorem 6.2.2. in [1]) of the solution for the above SDE.

In this way, when i runs in Q, the equation (1) defines a family of diffusion processes $\mathbb{M}^i = (\Omega^i, \mathcal{F}^i, \mathcal{F}_t^i, x_t^i, \theta_t^i, P^i)$, $i \in Q$ with the state spaces $\mathbb{R}^{d(i)}$, $i \in Q$. For each $i \in Q$, the elements $\mathcal{F}^i$, $\mathcal{F}_t^i$, θ_t^i, P^i, $P_{x^i}^i$ have the usual meaning as in the Markov process theory (see Appendix).

The jump (switching) mechanism between the diffusions is governed by two functions: the jump rate λ and the transition measure R. The jump rate $\lambda : X \to \mathbb{R}_+$ is a measurable bounded function and the transition measure R maps X into the set $\mathcal{P}(X)$ of probability measures on $(X, \mathcal{B}(X))$. Alternatively, one can consider the transition measure $R : \overline{X} \times \mathcal{B}(X) \to [0, 1]$ as a reset probability kernel.

Assumption 2 (Discrete transitions) *(i) for all* $A \in \mathcal{B}(X)$, $R(\cdot, A)$ *is measurable;*
(ii) for all $x \in \overline{X}$ *the function* $R(x, \cdot)$ *is a probability measure.*
(iii) $\lambda : X \to \mathbb{R}_+$ *is a measurable function such that* $t \to \lambda(x_t^i(\omega^i))$ *is integrable on* $[0, \varepsilon(\omega^i))$, *for some* $\varepsilon(\omega^i) > 0$, *for each* $\omega^i \in \Omega^i$.

Since $\overline{X}$ is a Borel space, then $\overline{X}$ is homeomorphic to a subset of the Hilbert cube, $\mathcal{H}$. Therefore, its space of probabilities is homeomorphic to the space of probabilities of the corresponding subset of $\mathcal{H}$ (Lemma 7.10 [3]). There exists a measurable function $\digamma : \mathcal{H} \times \overline{X} \to X$ such that $R(x, A) = \mathfrak{p}\digamma^{-1}(A)$, $A \in \mathcal{B}(X)$, where $\mathfrak{p}$ is the probability measure on $\mathcal{H}$ associated to $R(x, \cdot)$ and $\digamma^{-1}(A) = \{\omega \in \mathcal{H} | \digamma(\omega, x) \in A\}$. The measurability of such a function is guaranteed by the measurability properties of the transition measure R.

Construction. We construct an GSHS as a *Markov 'sequence'* H, which admits $(\mathbb{M}^i)$ as subprocesses. The sample path of the stochastic process $(x_t)_{t>0}$ with values in X, starting from a fixed initial point $x_0 = (i_0, x_0^{i_0}) \in X$ is defined in a similar manner as PDMP [10].

Let ω^i be a trajectory which starts in (i, x^i). Let $t_*(\omega^i)$ be the first hitting time of ∂X^i of the process (x_t^i). Let us define the following right continuous multiplicative functional

$$F(t, \omega^i) = I_{(t<t_*(\omega^i))} \exp[-\int_0^t \lambda(i, x_s^i(\omega^i))ds]. \tag{2}$$

This function will be the survivor function for the stopping time S^i associated to the diffusion (x_t^i), which will be employed in the construction of our model. This means that "killing" of the process (x_t^i) is done according to the multiplicative functional $F(t, \cdot)$. The stopping time S^i can be thought of as the minimum of two other stopping times:

1. first hitting time of boundary, i.e. $t_*|_{\Omega^i}$;
2. the stopping time $S^{i\prime}$ given by the following continuous multiplicative functional (which plays the role of the survivor function)

$$M(t, \omega^i) = \exp(-\int_0^t \lambda(i, x_s^i(\omega^i)))ds.$$

The stopping time $S^{i\prime}$ can be defined as

$$S^{i\prime}(\omega^i) = \sup\{t | \Lambda_t^i(\omega^i) \leq m^i(\omega^i)\},$$

where Λ_t^i is the following additive functional associated to the diffusion (x_t^i)

$$\Lambda_t^i(\omega^i) = \int_0^t \lambda(i, x_s^i(\omega^i)))ds$$

and m^i is an $\mathbb{R}_+$-valued random variable on Ω^i, which is exponentially distributed with the survivor function $P_{x^i}^i[m^i > t] = e^{-t}$. Then

$$P_{x^i}^i[S^{i\prime} > t] = P_{x^i}^i[\Lambda_t^i \leq m^i]. \tag{3}$$

We set $\omega = \omega^{i_0}$ and the first jump time of the process is $T_1(\omega) = T_1(\omega^{i_0}) = S^{i_0}(\omega^{i_0})$. The sample path $x_t(\omega)$ up to the first jump time is now defined as follows:

$$\begin{aligned}&\text{if } T_1(\omega)=\infty : x_t(\omega)=(i_0,x_t^{i_0}(\omega^{i_0})),\ t\geq 0\\&\text{if } T_1(\omega)<\infty : x_t(\omega)=(i_0,x_t^{i_0}(\omega^{i_0})),\ 0\leq t<T_1(\omega)\\&\qquad\qquad x_{T_1}(\omega) \text{ is a r.v. w.r.t. } R((i_0,x_{T_1}^{i_0}(\omega^{i_0})),\cdot).\end{aligned}$$

The process restarts from $x_{T_1}(\omega) = (i_1, x_1^{i_1})$ according to the same recipe, using now the process $x_t^{i_1}$. Thus if $T_1(\omega) < \infty$ we define $\omega = (\omega^{i_0}, \omega^{i_1})$ and the next jump time

$$T_2(\omega) = T_2(\omega^{i_0}, \omega^{i_1}) = T_1(\omega^{i_0}) + S^{i_1}(\omega^{i_1})$$

The sample path $x_t(\omega)$ between the two jump times is now defined as follows:

$$\begin{aligned}&\text{if } T_2(\omega)=\infty : x_t(\omega)=(i_1,x_{t-T_1}^{i_1}(\omega)),\ t\geq T_1(\omega)\\&\text{if } T_2(\omega)<\infty : x_t(\omega)=(i_1,x_t^{i_1}(\omega)),\ 0\leq T_1(\omega)\leq t<T_2(\omega)\\&\qquad\qquad x_{T_2}(\omega) \text{ is a r.v. w.r.t. } R((i_1,x_{T_2}^{i_1}(\omega)),\cdot).\end{aligned}$$

and so on.
We denote

$$N_t(\omega) = \sum I_{(t\geq T_k)}$$

Assumption 3 (Non-Zeno executions) *For every starting point $x \in X$, $EN_t < \infty$, for all $t \in \mathbb{R}_+$.*

2.3 Formal Definitions

We can introduce the following definition.

Definition 1. *A General Stochastic Hybrid System (GSHS) is a collection $H = ((Q, d, \mathcal{X}), b, \sigma, Init, \lambda, R)$ where*

- *Q is a countable set of discrete variables;*
- *$d : Q \to \mathbb{N}$ is a map giving the dimensions of the continuous state spaces;*
- *$\mathcal{X} : Q \to \mathbb{R}^{d(\cdot)}$ maps each $q \in Q$ into an open subset X^q of $\mathbb{R}^{d(q)}$;*
- *$b : X(Q, d, \mathcal{X}) \to \mathbb{R}^{d(\cdot)}$ is a vector field;*
- *$\sigma : X(Q, d, \mathcal{X}) \to \mathbb{R}^{d(\cdot)\times m}$ is a $X^{(\cdot)}$-valued matrix, $m \in \mathbb{N}$;*
- *$Init : \mathcal{B}(X) \to [0, 1]$ is an initial probability measure on $(X, \mathcal{B}(S))$;*
- *$\lambda : \overline{X}(Q, d, \mathcal{X}) \to \mathbb{R}^+$ is a transition rate function;*
- *$R : \overline{X} \times \mathcal{B}(\overline{X}) \to [0, 1]$ is a transition measure.*

Following [25], we note that if R_c is a transition measure from $(X \times Q, \mathcal{B}(X \times Q))$ to $(X, \mathcal{B}(X))$ and R_d is a transition measure from $(X, \mathcal{B}(X))$ to $(Q, \mathcal{B}(Q))$ (where Q is equipped with the discrete topology) then one might define a transition measure as follows

$$R(x^i, A) = \sum_{q\in Q} R_d(x^i, q) R_c(x^i, q, A^q)$$

for all $A \in \mathcal{B}(X)$, where $A^q = A \cap (q, X^q)$. Taking in the definition of a GSHS a such kind of reset map, the change of the continuous state at a jump depends on the pre-jump location (continuous and discrete) as well as on the post-jump discrete state.
This construction can be used to prove that the stochastic hybrid processes with jumps, developed in [4], are a particular class of GSHS.

A GSHS execution can be defined as follows.

Definition 2 (GSHS Execution). *A stochastic process $x_t = (q(t), x(t))$ is called a GSHS execution if there exists a sequence of stopping times $T_0 = 0 < T_1 < T_2 \leq \ldots$ such that for each $k \in \mathbb{N}$,*

- *$x_0 = (q_0, x_0^{q_0})$ is a $Q \times X$-valued random variable extracted according to the probability measure $Init$;*
- *For $t \in [T_k, T_{k+1})$, $q_t = q_{T_k}$ is constant and $x(t)$ is a (continuous) solution of the SDE:*

$$dx(t) = b(q_{T_k}, x(t))dt + \sigma(q_{T_k}, x(t))dW_t \tag{4}$$

  where W_t is a the m-dimensional standard Wiener;
- *$T_{k+1} = T_k + S^{i_k}$ where S^{i_k} is chosen according with the survivor function (2).*
- *The probability distribution of $x(T_{k+1})$ is governed by $R\left((q_{T_k}, x(T_{k+1}^-)), \cdot\right)$.*

3 Markov Strings

In this section we formulate a very general class of Markov processes, which will be called *Markov strings*, loosely based on the so-called "melange" operation of Markov processes [23]. A Markov string is a hybrid state 'jump Markov process'. The 'continuous state' component switches back and forth at random moments of times among a countable collections of Markov processes defined on some evolution modes. The 'discrete component' keeps track of the index of which Markov process the continuous component is following. This discrete component plays the role of an 'evolution index'. The continuous state is allowed to jump whenever the evolution index changes. For a Markov string the sojourn time in each mode is given as a stopping time with memoryless property for the process which evolves in that mode. Moreover, the continuous state immediately before a switching between modes is allowed to influence that jump.

3.1 Informal Description

We start with:

1. a countable family of independent Markov processes with some nice properties, for example the strong Markov property, the càdlàg property.

2. a sequence of independent stopping times (for each process is given a stopping time with memoryless property).
3. a renewal kernel is a priory given.

The stopping times play the role of the jump times from one process to another and the renewal kernel gives the distribution of the post-jump state. The probabilistic construction of the Markov string is natural:

1. start with one process, which belongs to the given family;
2. kill the current process at the corresponding stopping time;
3. jump according to the renewal kernel;
4. restart another process (belonging to the given family) from the new state;
5. return to 2. and repeat.

The pieced together process obtained by the above procedure is called Markov string. The main aim of this section is to prove that the Markov string inherits the properties (like the strong Markov property and the càdlàg property) from its component processes.

The Markov string construction is closely related to the mixing operation of Markov processes from [23] and the random evolution process construction from [25].Markov strings differ from the class of processes considered in [23], in that:

1. The jump times are essentially given stopping times, *not necessarily the life times of the component processes*; 2. After a jump, the string is allowed to restart following another process, which might be different from the pre-jump process.
2. The mixing ("melange") operation in [23] is only sketched and the author claims that it can be obtained using the renewal ("renaissance") operation. We consider that the passing from renewal to mixing is not straightforward. It is necessary to emphases the construction of all probabilistic elements associated with the resulted string. Lifting the renewal construction to the mixing construction, remarkable changes should be introduced in the Markov string definitions of the state space, probability space, probabilities on the trajectories.

As well, Markov strings can be obtained by specializing the base process and the 'instantaneous' distribution in the structure of the random evolution processes developed by Siegrist in [25], but the proof of the strong Markov property is not given in [25]. There, the author claims this can be derived from the strong Markov property of revival processes introduced by Ikeda, et. al. in [20]. To our knowledge, this property is completely proved by Meyer, in [23], for revival processes.

3.2 The Ingredients

Suppose that $\mathbb{M}^i = (\Omega^i, \mathcal{F}^i, \mathcal{F}^i_t, x^i_t, \theta^i_t, P^i, P^i_{x^i})$, $i \in Q$ is a countable family of Markov processes. We denote the state space of each $\mathbb{M}^i$ by $(X^i, \mathcal{B}^i)$ and

assume that $\mathcal{B}^i$ is the Borel σ-algebra of X^i if X^i is a topological Hausdorff space. We denote by Δ the cemetery point for all X^i, $i \in Q$. The existence of Δ is assumed for reasons that will be clear below. For each $i \in Q$, the elements $\mathcal{F}^i$, $\mathcal{F}_t^{i,0}$, $\mathcal{F}_t^i$, $\theta_t^i, P^i, P_{x^i}^i$ have the usual meaning as in the Markov process theory.
Let (P_t^i) denote the operator semigroup associated to $\mathbb{M}^i$, which maps $\mathcal{B}^i(X^i)$ into itself, given by

$$P_t^i f^i(x^i) = E_{x^i}^i f^i(x_t^i),$$

where $E_{x^i}^i$ is the expectation w.r.t. $P_{x^i}^i$. Then a function f^i is *p-excessive* $(p > 0)$ w.r.t. $\mathbb{M}^i$ if $f^i \geq 0$ and $e^{-pt}P_t^i f^i \leq f^i$, for all $t \geq 0$ and $e^{-pt}P_t^i f^i \nearrow f^i$ as $t \searrow 0$.

Assumption 4 *For each $i \in Q$, we suppose that:*

1. *$\mathbb{M}^i$ is a strong Markov process.*
2. *P^i is a complete probability.*
3. *The state space X^i is a Borel space.*
4. *$\mathbb{M}^i$ enjoys the càdlàg property, i.e. for each $\omega^i \in \Omega^i$, the sample path $t \mapsto x_t^i(\omega^i)$ is right continuous on $[0, \infty)$ and has left limits on $(0, \infty)$ (inside X_Δ^i).*
5. *The p-excessive functions of $\mathbb{M}^i$ are P^i-a.s. right continuous on trajectories.*

Part 3. implies that the underlying probability space Ω^i can be assumed to be $D_{[0,\infty)}(X^i)$, the space of functions mapping $[0, \infty)$ to X^i which are right continuous functions with left limits. Let us consider ω_Δ^i the cemetery point of Ω^i corresponding to the 'dead' trajectory of $\mathbb{M}^i$ (when the process is trapped to Δ).

In the terminology of [21], parts 1., 3. and 5. of the Assumption 4 imply that each $\mathbb{M}^i$ is a *right process.*

Using this family of Markov processes $\{\mathbb{M}^i\}_{i\in Q}$, we define a new Markov process whose realizations consist of concatenations of realizations for different $\mathbb{M}^i$. To achieve this goal, we need to define the transition mechanism from one process to the others. The jumping mechanism will be driven by:

1. A stopping time (which gives the jump temporal parameter) for each process;
2. A renewal kernel, which gives the post jump state.

Formally, in order to define the desired Markov string, $\mathbb{M}$, we need to give:

1. $(S^i)_{i\in Q}$, where, for each $i \in Q$, S^i is a *stopping time* of $\mathbb{M}^i$,
2. The jumping mechanism between the processes $\mathbb{M}^i$ is governed by a *renewal kernel,* which is a Markovian kernel

$$\Psi : \{\bigcup_{i\in Q} \Omega^i\} \times \mathcal{B}(X) \to [0, 1]$$

Assumption 5 *(i) For each $i \in Q$, S^i is terminal time, i.e. stopping time with the 'memoryless' property:*

$$S^i(\theta_t^i \omega^i) = S^i(\omega^i) - t, \forall t < S^i(\omega^i) \tag{5}$$

(ii) The renewal kernel Ψ satisfies the following conditions: (a) If $S^i(\omega^i) = +\infty$ then $\Psi(\omega^i, \cdot) = \varepsilon_\Delta$ (here, ε_Δ is the Dirac measure corresponding to Δ); (b) If $t < S^i(\omega^i)$ then $\Psi(\theta_t^i \omega^i, \cdot) = \Psi(\omega^i, \cdot)$.

Note that the component processes have the càdlàg property, therefore they may also have jumps, which are not treated separately in the construction of the Markov strings. The sequence of jump times refers to additional jumps, not to the jumps of the trajectories of component processes.

We consider now, for each $i \in Q$, the killed process

$$\widetilde{\mathbb{M}}^i = (\Omega^i, \mathcal{F}^i, \mathcal{F}_t^i, \widetilde{x}_t^i, \widetilde{\theta}_t^i, P^i, P_{x^i}^i)$$

where $\widetilde{x}_t^i(\omega^i) = \begin{cases} x_t^i(\omega^i), & \text{if } t < S^i(\omega^i) \\ \Delta, & \text{if } t \geq S^i(\omega^i) \end{cases}$ and $\widetilde{\theta}_t^i(\omega^i) = \begin{cases} \theta_t^i(\omega^i), & \text{if } t < S^i(\omega^i) \\ \omega_\Delta^i, & \text{if } t \geq S^i(\omega^i) \end{cases}$
In this case, Ω^i should be thought of as a subspace of $\Omega^i \times [0, \infty)$, the above embedding is made through the map $\omega^i \mapsto (\omega^i, S^i(\omega^i))$. The killed process is equivalent with the subprocess of $\mathbb{M}^i$ corresponding to the multiplicative functional $M_t^i = I_{[0,S^i)}(t)$ (see Chapter III, [5]).

3.3 The Construction

Using the elements defined in the Section 3.2 we construct the pieced-together stochastic process $\mathbb{M} = (\Omega, \mathcal{F}, \mathcal{F}_t, x_t, \theta_t, P, P_x)$, which will be called *Markov string*. We have to point out that $\mathbb{M}$ is obtained by the concatenation of the killed processes $\widetilde{\mathbb{M}}^i$.

To completely define the Markov string we need to specify the following elements:

1. $(X, \mathcal{B})$ - the state space;
2. $(\Omega, \mathcal{F}, P)$ - the underlying probability space;
3. $\mathcal{F}_t$ - the natural filtration;
4. θ_t - the translation operator;
5. P_x - Wiener probabilities.

State Space $(X, \mathcal{B})$. The state space will be X defined as follows. X is constructed as the direct sum of spaces X^i, with the same cemetery point Δ, i.e.

$$X = \bigcup_{i \in Q} \{(i, x) | x \in X^i\}. \tag{6}$$

In the same manner as in Section 2, it results that X is a Borel space.

The space X can be endowed with the Borel σ-algebra $\mathcal{B}(X)$ generated by its metric topology. Moreover, we have

$$\mathcal{B}(X) = \sigma\{\bigcup_{i \in Q} \{i\} \times \mathcal{B}^i\}. \tag{7}$$

Then $(X, \mathcal{B}(X))$ is a Borel space, whose Borel σ-algebra $\mathcal{B}(X)$ restricted to each component X^i gives the initial σ-algebra $\mathcal{B}^i$ [10].

We can assume, without loss of generality, that $X^i \cap X^j = \emptyset$ if $i \neq j$. Thus the relations (6) and (7) become

$$X = \bigcup_{i \in Q} X^i; \tag{8}$$

$$\mathcal{B}(X) = \sigma(\bigcup_{i \in Q} \mathcal{B}^i). \tag{9}$$

Therefore, we can assume, as well, that $\Omega^i \cap \Omega^j = \emptyset$ if $i \neq j$.

Probability Space. The space Ω can be thought as the space generated by the concatenation operation defined on the union of the spaces Ω^i (which are pairwise disjoint), i.e. $\Omega = (\bigcup_{i \in Q} \Omega^i)^*$. Note that, for each $i \in Q$, an arbitrary element ω^i of Ω^i must be thought as a trajectory of the killed process $\widetilde{\mathbb{M}}^i$. The cemetery point of Ω is denoted by $\omega_\Delta = (\omega^i_\Delta)_{i \in Q}$. We use to denote by ω (resp. $\widehat{\omega}$ or ω^i) an arbitrary element of Ω (resp. $\bigcup_{i \in Q} \Omega^i$ or Ω^i).

The $\sigma-$algebra $\mathcal{F}$ on Ω will be the smallest $\sigma-$algebra on Ω such that the projection $\pi^i : \Omega \to \Omega^i$ are $\mathcal{F}/\mathcal{F}^i$ measurable, $i \in Q$. The probability P on $\mathcal{F}$ will be defined as a 'product measure'. Let $\widehat{\mathcal{F}}$ be the $\sigma(\bigcup_{i \in Q} \mathcal{F}^i)$ defined on $\bigcup_{i \in Q} \Omega^i$.

Recipe. We give the procedure to construct a sample path of the stochastic process $(x_t)_{t>0}$ with values in X, starting from a fixed initial point $x_0 = x_0^{i_0} \in X^{i_0}$. Let ω^{i_0} be a sample path of the process $(x_t^{i_0})$ starting with x_0. In fact, we give a recipe *to construct a Markov string starting with an initial path* ω^{i_0}. Let $T_1(\omega^{i_0}) = S^{i_0}(\omega^{i_0})$. The event ω and the associated sample path are inductively defined. In the first step

$$\omega = \omega^{i_0}$$

The sample path $x_t(\omega)$ up to the first jump time is now defined as follows:

$$\begin{aligned} &\text{if } T_1(\omega) = \infty : x_t(\omega) = x_t^{i_0}(\omega^{i_0}),\ t \geq 0 \\ &\text{if } T_1(\omega) < \infty : x_t(\omega) = x_t^{i_0}(\omega^{i_0}),\ 0 \leq t < T_1(\omega) \\ &\qquad\qquad\qquad x_{T_1} \text{ is a r.v. according to } \Psi(\omega^{i_0}, \cdot). \end{aligned}$$

The process restarts from $x_{T_1} = x_1^{i_1}$ according to the same recipe, using now the process $(x_t^{i_1})$. Let ω^{i_1} be a sample of the process $(x_t^{i_1})$ starting with $x_1^{i_1}$. Thus, if $T_1(\omega) < \infty$ we define the next jump time

$$T_2(\omega^{i_0}, \omega^{i_1}) = T_1(\omega^{i_0}) + S_{i_2}(\omega^{i_2}).$$

Then, in the second step
$$\omega = \omega^{i_0} * \omega^{i_1}$$

where '$*$' is the concatenation operation of trajectories. The sample path $x_t(\omega)$ between the two jump times is now defined as follows:

$$\begin{aligned} &\text{if } T_2(\omega) = \infty : x_t(\omega) = x^{i_1}_{t-T_1}(\omega^{i_1}),\ t \geq T_1(\omega) \\ &\text{if } T_2(\omega) < \infty : x_t(\omega) = x^{i_1}_t(\omega^{i_1}),\ 0 \leq T_1(\omega) \leq t < T_2(\omega) \\ &\qquad\qquad x_{T_2} \text{ is a r.v. according to } \Psi(\omega^{i_1}, \cdot). \end{aligned}$$

Generally, if $T_k(\omega) = T_k(\omega^{i_0}, \omega^{i_1}, ..., \omega^{i_{k-1}}) <$ with

$$\omega = \omega^{i_0} * \omega^{i_1} * ... * \omega^{i_{k-1}}$$

then the next jump time is

$$T_{k+1}(\omega) = T_{k+1}(\omega^{i_0}, \omega^{i_1}, ..., \omega^{i_k}) = T_k(\omega^{i_0}, \omega^{i_1}, ..., \omega^{i_{k-1}}) + S^{i_k}(\omega^{i_k}) \quad (10)$$

The sample path $x_t(\omega)$ between the two jump times T_k and T_{k+1} is defined as:

$$\begin{aligned} &\text{if } T_{k+1}(\omega) = \infty : x_t(\omega) = x^{i_k}_{t-T_k}(\omega^{i_k}), t \geq T_{k+1}(\omega) \\ &\text{if } T_{k+1}(\omega) < \infty : \begin{array}{l} x_t(\omega) = x^{i_k}_{t-T_k}(\omega^{i_k}), 0 \leq T_k(\omega) \leq t < T_{k+1}(\omega) \\ x_{T_{k+1}} \text{ is a r.v. according to } \Psi(\omega^{i_k}, \cdot). \end{array} \end{aligned} \quad (11)$$

We have constructed a sequence of jump times $0 < T_1 < T_2 < ... < T_n < ...$ Let $T_\infty = \lim_{n\to\infty} T_n$. Then $x_t(\omega) = \Delta$ if $t \geq T_\infty$. A sample path until T_{k_0} (where $k_0 = \min\{k : S^{i_k}(\omega) = \infty\}$) of the process (x_t), starting from a fixed initial point $x_0 = (i_0, x_0^{i_0})$, is obtained as the concatenation:

$$\omega = \omega^{i_0} * \omega^{i_1} * ... * \omega^{i_{k_0-1}}.$$

We denote $N_t(\omega) = \sum I_{(t \geq T_k)}$ the number of jump times in the interval $[0, t]$. To eliminate pathological solutions that take an infinite number of discrete transitions in a finite amount of time (known as Zeno solutions) we impose the following assumption:

Assumption 6 (Non-Zeno dynamics) *For every starting point $x \in X$, $EN_t < \infty$, for all $t \in \mathbb{R}_+$.*

Under Assumption 6, the underlying probability space Ω can be identified with $D_{[0,\infty)}(X)$.

Wiener Probabilities. One might define the expectation $E^x f$, $x \in X$, where f is a $\mathcal{F}$-measurable function on Ω, which depends only on a finite number of variables, by recursion on the number of variables.
Step 1. If $\omega = \omega^{i_0}$and $f(\omega) = f_1(\omega^{i_0})$ with f_1 a $\mathcal{F}^{i_0}$-measurable function on Ω^{i_0}, then

- if $x = x^{i_0} \in X^{i_0}$ then $E_x f = E^{i_0}_{x^{i_0}} f$, where $E^{i_0}_{x^{i_0}}$ is the expectation corresponding to the probability $P^{i_0}_{x^{i_0}}$;
- if $x = x^j \in X^j$, $j \neq i_0$ then $E_x f = 0$.

Step 2. If $\omega = \omega^{i_0} * \omega^{i_1} * ... * \omega^{i_n}$and $f(\omega) = f_n(\omega^{i_0} * \omega^{i_1} * ... * \omega^{i_n})$ with f_n a $\Pi^n_{k=0}\mathcal{F}^{i_k}$-measurable function on $\Pi^n_{k=0}\Omega^{i_k}$ then

$$\begin{aligned} & f_{n-1}(\omega^{i_0} * \omega^{i_1} * ... * \omega^{i_{n-1}}) \\ &= \int_{\Omega^{i_n}} f_n(\omega^{i_0} * \omega^{i_1} * ... * \omega^{i_{n-1}} * \omega^{i_n}) dP^{i_n}_{\Psi(\omega^{i_{n-1}},\cdot)}(\omega^{i_n}); \\ g(\omega) &= f_{n-1}(\omega^{i_0} * \omega^{i_1} * ... * \omega^{i_{n-1}}); \\ E_x f &= E_x g. \end{aligned} \tag{12}$$

Translation Operators. Let us define now the translation operator (θ_t) associated with (x_t). If $t \geq T_\infty(\omega)$, then we take $\theta_t(\omega) = \omega_\Delta$. Otherwise, there exists k such that $T_k(\omega) \leq t < T_{k+1}(\omega)$. In this case we take

$$\theta_t(\omega) = (\theta^{i_k}_{t-T_k(\omega)}(\omega^{i_k}) * \omega^{i_{k+1}} * ...). \tag{13}$$

Lemma 1. *(θ_t) is the translation operator associated with (x_t), i.e.*

$$\theta_s \circ \theta_t = \theta_{s+t}; x_s \circ \theta_t = x_{s+t}.$$

Proof. If $t \geq T_\infty(\omega)$, then $\theta_t(\omega) = \omega_\Delta$ and $x_{s+t}(\omega) = \Delta = x_s(\theta_t(\omega))$.
Suppose that there exist $k, i \geq 0$ such that $T_k(\omega) \leq t < T_{k+1}(\omega)$ and $T_i(\theta_t\omega) \leq s < T_{i+1}(\theta_t\omega)$. Then

$$x_t(\omega) = x^{i_k}_{t-T_k}(\omega^{i_k}); (x_s \circ \theta_t)(\omega) = x^{i_l}_{s-T_l}(\theta^{i_l}_{s-T_l}\omega^{i_l}).$$

Since $\theta_t(\omega)$ is given by (13) and T_{k+1} is given by (10) we obtain

$$\begin{aligned} T_{k+1}(\theta_t\omega) &= S^{i_k}(\theta^{i_k}_{t-T_k(\omega)}(\omega^{i_k})) = S^{i_k}(\omega^{i_k}) - (t - T_k(\omega)) \\ &= T_{k+1}(\omega) - t. \end{aligned}$$

Then

$$T_{i+1}(\theta_t\omega) = T_{k+i+1}(\omega) - t$$

Therefore

$$T_i(\theta_t\omega) \leq s < T_{i+1}(\theta_t\omega) \Leftrightarrow T_{k+i}(\omega) \leq s + t < T_{k+i+1}(\omega).$$

□

Natural Filtrations. Let $(\mathcal{F}_t)$ be the natural filtration with respect to (x_t). The natural filtration $(\mathcal{F}_t)$ on Ω is built such that we have the following definition of $\mathcal{F}_t$-measurability:

Definition 3. *A $\mathcal{F}$-measurable function f on Ω is $\mathcal{F}_t$-measurable if the following property holds:*
For each k, the function $f \cdot I_{\{T_k(\omega)\leq t<T_{k+1}(\omega)\}}$ is equal to $h \circ \eta_k$, where the function $h(\omega^{i_0} * \omega^{i_1} * ... * \omega^{i_k})$ is such that for a fixed $(\widehat{\omega}^{i_0} * \widehat{\omega}^{i_2} * ... * \widehat{\omega}^{i_{k-1}})$ with $T_k(\widehat{\omega}^{i_0} * \widehat{\omega}^{i_2} * ... * \widehat{\omega}^{i_{k-1}}) \leq t$, $\omega^{i_k} \mapsto h(\widehat{\omega}^{i_0} * \widehat{\omega}^{i_2} * ... * \widehat{\omega}^{i_{k-1}} * \omega^{i_k})$ is measurable with respect to $\mathcal{F}^{i_k}_{t-T_k}$.

Because the families of filtrations $(\mathcal{F}^i_t)$ are nondecreasing and right continuous, one can verify that the family $(\mathcal{F}_t)$ has the same properties, as follows.

Proposition 1. *(i) The family $(\mathcal{F}_t)$ is nondecreasing and right continuous.*
(ii) The random variables T_k are stopping times w.r.t. $(\mathcal{F}_t)$.
(iii) Let T a stopping time with respect to $(\mathcal{F}_t)$. For each $k \in \mathbb{N}$, $T \wedge T_k$ is a function on Ω which depends only on $\omega^{i_0} * \omega^{i_1} * ... * \omega^{i_{k-1}}$. On the other hand, if $\omega^{i_0} * \omega^{i_1} * ... * \omega^{i_{k-1}}$ is fixed, the function $(T \wedge T_{k+1} - T_k)^+$ with ω^{i_k} as argument is a stopping time with respect $(\mathcal{F}^{i_k}_t)$.

Proof. The proof can be obtained with small changes from the similar result proofs given in [23] for the case of rebirth processes.□

3.4 Basic Properties

Mainly, in this section we prove that the Markov string (x_t) constructed in Section 3.3 is a right Markov process. The proof engine is based on the Markov property of the discrete time Markov chain (p_n), which will be build in the following.

(p_n) is a discrete time Markov chain associated to (x_t) with the state space $(\bigcup_{i\in Q} \Omega_i, \widehat{\mathcal{F}})$ and the underlying probability space $(\Omega, \mathcal{F})$. The chain (p_n) is essentially 'the $n-th$' step of the process (x_t). If its starting point is ω^{i_0} (a trajectory in Ω^{i_0} starting in $x^{i_0}_0$) then $p_n(\omega) = \omega^{i_n}$.
The transition kernel associated with (p_n) can be defined as follows:

$$H(\widehat{\omega}, A) = P_{\Psi}(\widehat{\omega}, A), A \in \widehat{\mathcal{F}}.$$

The construction of P_x from subsection 3.3 is such that

- H is the transition function of (p_n);
- P_x is the initial probability law of (p_n); i.e. if $\widehat{\omega} \in \bigcup_{i\in Q} \Omega_i$ which starts in $x \in X$

$$P^{\omega}(p_0 \in A) = P_x(A), A \in \mathcal{F}.$$

Let η_k be the projection $(p_0, p_1, ..., p_k)$, i.e. $\eta_k(\omega) = (\omega^{i_0} * \omega^{i_1} * ... * \omega^{i_k})$.

One might construct a jump process (η_t) associated to a Markov string (x_t) following a similar algorithm such that used for Piecewise Deterministic Markov processes, in [10]. We do not have a one-to-one correspondence between the sample paths of (x_t) and (η_t), as in the case of PDMP. Then the jump process will not serve to study the Markov string. Its role is taken by the Markov chain (p_n).

Remark 1. For each k on the set $\{T_k(\omega) \leq t < T_{k+1}(\omega)\}$ we have: $x_t = x^{i_k}_{t-T_k} \circ p_k$.

Proposition 2 (Simple Markov property). *Under Assumptions 4-6, any Markov string is a Markov process.*

Proof. The simple Markov property of (x_t) is equivalent to the following implication [23]:
If f is a positive $\mathcal{F}_t$-measurable function and g is a $\mathcal{F}$-measurable function then

$$E^x[f \cdot g \circ \theta_t] = E^x[f \cdot E^{x_t}[g]]. \tag{14}$$

The identity (14) can be unfolded into two separated equalities

$$E^x[f \cdot g \circ \theta_t \cdot I_{\{t \geq T_\infty\}}] = E^x[f \cdot E^{x_t}[g] \cdot I_{\{t \geq T_\infty\}}] \tag{15}$$

$$E^x[f \cdot g \circ \theta_t \cdot I_{\{T_k(\omega) \leq t < T_{k+1}(\omega)\}}] = E^x[f \cdot E^{x_t}[g] \cdot I_{\{T_k(\omega) \leq t < T_{k+1}(\omega)\}}] \tag{16}$$

The identity (15) is clear because on $\{t \geq T_\infty\}$

$$E^{x_t}[g] = g(\omega_\Delta); \theta_t(\omega) = \omega_\Delta; x_t(\omega) = \Delta.$$

Let us prove now the identity (16). Let $\omega \in \Omega$. By the definition of $\mathcal{F}_t$ we have

$$f(\omega) \cdot I_{\{T_k(\omega) \leq t < T_{k+1}(\omega)\}}(\omega) = h(\omega^{i_0} * \omega^{i_1} * ... * \omega^{i_k}) \tag{17}$$

where h is a measurable function as in the definition 3 and is equal to zero outside of the set $\{T_k(\omega) \leq t < T_{k+1}(\omega)\}$.

In order to prove (16) it is enough to treat the case when the function g depends only on a finite number of variables (because the expectation E^x is defined by the recursion (12)).

We start with the case when the function g depends only on a single variable, ω^{i_0}, i.e. $g(\omega) = a(\omega^{i_0})$, where a is $\mathcal{F}^{i_0}$-measurable on Ω^{i_0}. In this case, the left-hand side of (16) is equal to

$$E^x[f \cdot I_{\{T_k(\omega) \leq t < T_{k+1}(\omega)\}} \cdot a(\theta^{i_k}_{t-T_k(\omega)}(\omega^{i_k}))]. \tag{18}$$

Because the term between [...] depends only on $(\omega^{i_0} * \omega^{i_1} * ... * \omega^{i_k})$, (18) becomes

$$E^x\{\int_{\Omega^{i_k}} h(\omega^{i_0} * \omega^{i_1} * ... * \omega^{i_k}) \cdot a(\theta^{i_k}_{t-T_k(\omega)}(\omega^{i_k}))dP^{i_k}_{\Psi(\omega^{i_{k-1}},\cdot)}(\omega^{i_k})\}. \tag{19}$$

Again, the integrand between {...} depends only on $(\omega^{i_0} * \omega^{i_1} * ... * \omega^{i_{k-1}})$. Since the function $\omega^{i_k} \to h(\omega^{i_0} * \omega^{i_1} * ... * \omega^{i_k})$ is $\mathcal{F}^{i_k}_{t-T_k}$-measurable, we can use the Markov property of the process $\mathbb{M}^{i_k}$ and (19) becomes

$$\int_{\Omega^{i_k}} h(\omega^{i_0} * \omega^{i_1} * ... * \omega^{i_k}) E^{i_k}_{x^{i_k}_{t-T_k}(\omega^{i_k})}[a] dP^{i_k}_{\Psi(\omega^{i_{k-1}},\cdot)}(\omega^{i_k}). \tag{20}$$

Since $x_t(\omega) = x^{i_k}_{t-T_k}(\omega^{i_k})$ on $\{T_k(\omega) \leq t < T_{k+1}(\omega)\}$ the computation of the right-hand side of (16) gives

$$E^x\{h(\omega^{i_0} * \omega^{i_1} * ... * \omega^{i_k}) \cdot E^{i_k}_{x^{i_k}_{t-T_k}(\omega^{i_k})}[a]\} \tag{21}$$

Using the recursive procedure, as before, (21) gives (20).

Suppose now that (16) is established for all functions g which depend only on $(\omega^{i_0} * \omega^{i_1} * ... * \omega^{i_{k-1}})$. We have to prove that (16) is true for

$$g(\omega) = g(\omega^{i_0} * \omega^{i_1} * ... * \omega^{i_k}); k > 0.$$

Let

$$c(\omega) = c(\omega^{i_0} * \omega^{i_1} * ... * \omega^{i_{k-1}}) = \int_{\Omega^{i_k}} b(\omega^{i_0} * \omega^{i_1} * ... * \omega^{i_k}) dP^{i_k}_{\Psi(\omega^{i_{k-1}}, \cdot)}(\omega^{i_k}).$$

Using the recursive procedure, one can check that the functions

$$h(...)g \circ \theta_t \text{ and } h(...)c \circ \theta_t$$

have the same expectations.
On the other hand, the functions

$$h(...)E_{x_t}[g] \text{ and } h(...)E_{x_t}c$$

have the same expectations. Since c depends only on $k-1$ variables, this implies (16) for the general case. □

Proposition 3 (Càdlàg property). *Under Assumptions 4-6, any Markov string* $\mathbb{M} = (\Omega, \mathcal{F}, \mathcal{F}_t, x_t, \theta_t, P, P_x)$ *has the càdlàg property, i.e. for all* $\omega \in \Omega$ *the trajectories* $t \mapsto x_t(\omega)$ *are right continuous on* $[0, \infty)$ *with left limits on* $(0, \infty)$.

Proof. The result is a direct consequence of two facts:

1. The sample paths of (x_t) are obtained by the concatenation of sample paths of component process (i.e. the concatenation is done in such way it preserves the right continuity and the left limits);
2. The component processes enjoy the càdlàg property.

Then the Markov string inherits the càdlàg property.

Proposition 4. *Under Assumptions 4-6, any Markov string is a strong Markov process.*

Proof. Each T_k is a stopping time for (x_t) (see proposition 1 (ii)). For each $k \geq 1$, T_k can be obtained by the following recursion

$$T_{k+1} = T_k + S^{i_k} \circ \theta_{T_k}$$

Let us prove now that the process (x_t) is a strong Markov process. The filtration $(\mathcal{F}_t)$ is nondecreasing and right continuous (see proposition 1 (i)). Then the process (x_t) satisfies the right hypothesis.
Let (P_t) be the semigroup of the whole Markov process (x_t), $P_t g(x) = E_x g(x_t)$, where g is bounded $\mathcal{B}$-measurable function. Let $(U_p)_{p>0}$ the resolvent associated to the semigroup, i.e.

$$U_p g = \int_0^\infty e^{-pt} P_t g dt.$$

It is known that the strong Markov property is equivalent with each from the following assertions [22]:

1. If g is a positive bounded continuous function on X_Δ then $f = U_p g$ $(p > 0)$ is nearly Borel and right continuous on the process trajectories.
2. Each p-excessive function $(p > 0)$ is nearly Borel and right continuous on the process trajectories.

Recall that a real function defined on the state space X_Δ is nearly Borel for the process (x_t) if there exist two Borel function h and h' on X_Δ such that $h' \leq f \leq h$ and

$$P\{\omega | \exists t, h' \circ x_t(\omega) < h \circ x_t(\omega)\} = 0. \tag{22}$$

Let g be a positive bounded continuous function on X. We have $g = \sum_{i \in Q} g^i$, where $g^i = g|_{X^i}$ are bounded continuous functions on X^i. Then $P_t g = \sum_{i \in Q} P_t^i g^i$ and

$$U_p g = \int_0^\infty e^{-pt} P_t g dt = \sum_{i \in Q} \int_0^\infty e^{-pt} P_t^i g^i dt = \sum_{i \in Q} U_p^i g^i.$$

It is known that $f = U_p g$ $(p > 0)$ (the restriction to X) is p-excessive function with respect to (P_t) and for each $i \in Q$ and the function $f^i = U_p^i g^i$ is p-excessive function with respect to (P_t^i). Therefore, f^i is nearly Borel and right continuous on the trajectories of the process (x_t^i). It is clear from the construction that the function f is right continuous on the trajectories of the process (x_t).
Let $h^i, h^{i\prime}$ two Borel functions on X_Δ^i such that $h' \leq f^i \leq h^i$ and

$$h^{i\prime} \circ x_t^i(\omega^i) = h^i \circ x_t^i(\omega^i) P^i - a.s., \forall t \geq 0. \tag{23}$$

Let us consider the function h, h' defined as below:

$$h = \sum_{i \in Q} h^i, h' = \sum_{i \in Q} h^{i\prime}. \tag{24}$$

It is clear that

$$P\{\omega | \exists t \geq T_\infty, h' \circ x_t(\omega) < h \circ x_t(\omega)\} = 0.$$

Let us compute the probability of the following event:

$$A_k = \{\exists t | T_k \leq t < T_{k+1}, h' \circ x_t(\omega) < h \circ x_t(\omega)\}.$$

We have $A_k \in \mathcal{F}$. Let $a_k = I_{A_k}$ which depends only on $\omega^{i_0} * \omega^{i_2} * ... * \omega^{i_k}$. The recursive method to compute the probability of A_k on $\{T_k \leq t < T_{k+1}\}$ gives

$$\int_{\Omega^{i_k}} a_k(\omega^{i_0} * \omega^{i_2} * \cdots * \omega^{i_k}) dP^{i_k}_{\Psi(\omega^{i_{k-1}},\cdot)}(\omega^{i_k}). \tag{25}$$

Since $a_k(\omega^{i_0} * \omega^{i_2} * ... * \omega^{i_k})$ on Ω^{i_k} is exactly the indicator function of

$$B = \{\omega^{i_k} | \exists u < S^{i_k}(\omega^{i_k}), h^{i_k\prime} \circ x^{i_k}_u(\omega) < h^{i_k} \circ x^{i_k}_u(\omega)\}$$

using (23) we obtain that the integral (25) is zero. Therefore the functions h, h' defined by (24) verify the condition (22). Then f will be a nearly Borel function relative to the process (x_t).□

The Propositions 2, 3, 4 can be summarized in the following theorem:

Theorem 1. *Under Assumptions 4-6, any Markov string has the following properties:*
(i) It is a strong Markov process;
(ii) It has the càdlàg property;
(iii) It is a right process.

4 Properties of GSHS

Strong Markov property. GSHS, being constructed as particular Markov strings, they inherit the properties of their diffusion component, namely they are *strong Markov processes* with *càdlàg property.*

Proposition 5 (Strong Markov process). *Under the standard assumptions 1-3, any General Stochastic Hybrid Model H is a strong Markov process.*

Proof. To prove that H is a strong Markov process, it is enough to check that a GSHS is, indeed, a Markov string, i.e. it satisfies the Assumptions 4-6 from the Markov string construction. It is easy to see that

- Assumption 1 implies Assumption 4;
- Assumption 3 implies Assumption 6.

It remains to prove only that Assumption 2 and the construction of a GSHS implies Assumption 5. We can suppose without loss of generality that $\Omega^i \cap \Omega^j = \emptyset$. Then, the kernel Ψ can be defined as follows

$$\Psi : \{\bigcup_{i \in Q} \Omega^i\} \times \mathcal{B}(X) \to [0,1] \text{ such that } \Psi(\omega^i, A) = R(x^i_{S^i(\omega^i)}, A).$$

For any GSHS, we need to check
(a) the memoryless property of kernel, i.e. if $0 < t < S^i(\omega^i)$ then $\Psi(\theta_t^i\omega^i, \cdot) = \Psi(\omega^i, \cdot) \Leftrightarrow R(x^i_{S^i(\theta_t^i\omega^i)}, \cdot) = R(x^i_{S^i(\omega^i)}, \cdot)$.
(b) the memoryless property of the stopping times S^i.
Since the component diffusions are strong Markov processes (b) implies (a). In fact, we have to prove that, if $0 < t < t + s < S^i(\omega^i)$ then stopping times (S^i)

$$P_{x^i}(S^i > t + s|S^i > t) = P_{x_t^i}(S^i > s) \tag{26}$$

We have, for each $i \in Q$,

1. the hitting time of the boundary ∂X^i of the diffusion process (x_t^i) has the memoryless property, i.e. $t^*(\theta_t^i\omega^i) = t_*(\omega^i) - t$.
2. the stopping time $S^{i\prime}$ with the survivor function (3) has the memoryless property because

$$\begin{aligned} P_{x^i}(S^{i\prime} > t + s|S^{i\prime} > t) &= \frac{P_{x^i}\{\omega^i|m^i(\omega^i) > \Lambda^i_{t+s}(\omega^i)\}}{P_{x^i}\{\omega^i|m^i(\omega^i) > \Lambda^i_t(\omega^i)\}} \\ &= \frac{P_{x^i}\{\omega^i|m^i(\omega^i) > \Lambda^i_t(\omega^i) + \Lambda^i_s(\theta^i_t\omega^i)\}}{P_{x^i}\{\omega^i|m^i(\omega^i) > \Lambda^i_t(\omega^i)\}} \\ &= P_{x^i_t}\{\omega^i|m^i(\omega^i) > \Lambda^i_s(\theta^i_t\omega^i)\} \\ &= P_{x^i_t}(S^{i\prime} > s) \end{aligned}$$

(we have used the fact that m^i has the memoryless property, being an exponentially distributed random variable, and the additivity of Λ_t^i w.r.t. t since this is an additive functional).

Since, for each $i \in Q$, the stopping time S^i is the infimum of t^* and $S^{i'}$, the two above facts easily implies the 'memoryless' property of S^i (it is easy to prove that the infimum of two memoryless stopping times is still a memoryless stopping time).
Thus, H is a Markov string obtained by mixing diffusion processes. Therefore, it inherits the strong Markov property from the component diffusions. □

Corollary 1. *Any General Stochastic Hybrid Model H, under the standard assumptions of section 2.2, is a Borel right process .*

Proof. The statement of the corollary is immediate, since the state space is a Lusin space and H is a right process. □

As we discusses in the context of Markov strings, a GSHS might be thought of as a 'restriction' of a random evolution process [25], whose components are diffusion processes defined on different state spaces. We can consider each diffusion component evolving on $\overline{X}$. The first difference is that while a GSHS is defined only on $\bigcup_{i \in Q}\{i\} \times X^i$ a random evolution process should be defined on the entire product space $Q \times \overline{X}$. The second difference is that while for a random evolution process the jump times from one process to another are driven only by transition rates, for a GSHS these might be also boundary hitting times of modes.

However, contrary to [25], GSHS are not always standard processes as the random evolution processes.

The Process Generator. We denote by $\mathcal{B}_b(X)$ the set of all bounded measurable functions $f : X \to \mathbb{R}$. This is a Banach space under the norm $\|f\| = \sup_{x \in X} |f(x)|$. Associated with the semigroup (P_t) is its *strong generator* which is the 'derivative' of P_t at $t = 0$. Let $D(L) \subset \mathcal{B}_b(X)$ be the set of functions f for which the following limit exists $\lim_{t \searrow 0} \frac{1}{t}(P_t f - f)$ and denote this limit Lf. This refers to convergence in the norm $\|\cdot\|$, i.e. for $f \in D(L)$ we have $\lim_{t \searrow 0} ||\frac{1}{t}(P_t f - f) - Lf|| = 0$. Specifying the domain $D(L)$ is an essential part of specifying L.

Proposition 6 (Martingale property). *[10] For $f \in D(L)$ we define the real-valued process $(C_t^f)_{t \geq 0}$ by*

$$C_t^f = f(x_t) - f(x_0) - \int_0^t Lf(x_s) ds. \tag{27}$$

Then for any $x \in X$, the process $(C_t^f)_{t \geq 0}$ is a martingale on $(\Omega, \mathcal{F}, \mathcal{F}_t, P_x)$.

There may be other functions f, not in $D(L)$, for which something akin to (27) is still true. In this way we get the notion of *extended generator* of the process.

Let $D(\widehat{L})$ be the set of measurable functions $f : X \to \mathbb{R}$ with the following property: there exists a measurable function $h : X \to \mathbb{R}$ such that $t \to h(x_t)$ is integrable $P_x - a.s.$ for each $x \in X$ and the process

$$C_t^f = f(x_t) - f(x_0) - \int_0^t h(x_s) ds$$

is a local martingale. Then we write $h = \widehat{L}f$ and call $(\widehat{L}, D(\widehat{L}))$ the extended generator of the process (x_t).

Following [10], for $A \in \mathcal{B}(\overline{X})$ define p, p^* and $\widetilde{p}$ as follows:

$$p(t, A) = \sum_{k=1}^{\infty} I_{(t \geq T_k)} I_{(x_{T_k} \in A)};$$

$$p^*(t) = \sum_{k=1}^{\infty} I_{(t \geq T_k)} I_{(x_{T_k^-} \in \partial X)};$$

$$\widetilde{p}(t, A) = \int_0^t R(x_s, A) \lambda(x_s) ds + \int_0^t R(A, x_{s-}) dp^*(s)$$

$$\widetilde{p}(t, A) = \sum_{T_k \leq t} R(x_{T_k -}, A).$$

Note that p, p^* are counting processes, $p^*(t)$ is counting the number of jumps from the boundary of the process (x_t). $\widetilde{p}(t, A)$ is the compensator of $p(t, A)$ (see [10] for more explanations). The process $q(t, A) = p(t, A) - \widetilde{p}(t, A)$ is a local martingale.

Given a function $f \in \mathcal{C}^1(\mathbb{R}^n, \mathbb{R})$ and a vector field $b : \mathbb{R}^n \to \mathbb{R}^n$, we use $\mathcal{L}_b f$ to denote the Lie derivative of f along b given by $\mathcal{L}_b f(x) = \sum_{i=1}^{n} \frac{\partial f}{\partial x_i}(x) f_i(x)$. Given a function $f \in \mathcal{C}^2(\mathbb{R}^n, \mathbb{R})$, we use $\mathbb{H}^f$ to denote the Hamiltonian operator applied to f, i.e. $\mathbb{H}^f(x) = (h_{ij}(x))_{i,j=1,\cdots,n} \in \mathbb{R}^{n \times n}$, where $h_{ij}(x) = \frac{\partial^2 f}{\partial x_i \partial x_j}(x)$. A^T denotes the transpose matrix of a matrix $A = (a_{ij})_{i,j=1,\cdots,n} \in \mathbb{R}^{n \times m}$ and $Tr(A)$ denotes its trace.

Theorem 2 (GSHS generator). *Let H be an GSHS as in definition 1. Then the domain $D(L)$ of the extended generator L of H, as a Markov process, consists of those measurable functions f on $X \cup \partial X$ satisfying:*

1. *$f : \overline{X} \to \mathbb{R}$, $\mathcal{B}-$measurable such that for each $i \in Q$ the restriction $f^i = f|_{X^i}$ is twice differentiable.*
2. *The boundary condition*

$$f(x) = \int_{\mathbb{X}} f(y) R(x, dy), \ x \in \partial X;$$

3. *$Bf \in L_1^{loc}(p)$ (see [3]) where*

$$Bf(x, s, \omega) := f(x) - f(x_{s-}(\omega)).$$

For $f \in D(L)$, Lf is given by

$$Lf(x) = L_{cont} f(x) + \lambda(x) \int_{\overline{X}} (f(y) - f(x)) R(x, dy) \tag{28}$$

where:

$$L_{cont} f(x) = \mathcal{L}_b f(x) + \frac{1}{2} Tr(\sigma(x)\sigma(x)^T \mathbb{H}^f(x)). \tag{29}$$

Proof. Let $(\widetilde{L}, D(\widetilde{L}))$ be the extended generator of (x_t). We want to show that $(\widetilde{L}, D(\widetilde{L})) = (L, D(L))$. Suppose first that f satisfies 1-3. Then $Bf \in L_1^{loc}(\widetilde{p})$ and $\int_{[0,t] \times \overline{X}} Bf d\widetilde{p} = I_1 + I_2$, where

[3] Following [10], f is in $L_1^{loc}(p)$ if for some sequence of stopping times $\sigma_n \uparrow \infty$

$$E_x \underset{i}{} |f(x_{T_i \wedge \sigma_n}) - f(x_{T_i \wedge \sigma_n -})| < \infty$$

$$I_1 = \int_{[0,t]} \int_{\overline{X}} (f(y) - f(x_s))R(x_s, dy)\lambda(x_s)ds$$
$$I_2 = \int_{[0,t]} \int_{\overline{X}} (f(y) - f(x_{s-}))R(x_{s-}, dy)dp^*(s).$$

Now the support of p^* is contained in the countable set $\{s : x_{s-} \in \partial X\}$ and because of the boundary condition 2. the second integral I_2 vanishes. Thus

$$\int_{[0,t]\times\overline{X}} Bfdq = \sum_{T_k \le t}(f(x_{T_k}) - f(x_{T_k-})) - \int_{[0,t]} \int_{\overline{X}}(f(y) - f(x_s))R(x_s, dy)\lambda(x_s)ds.$$

This is a local martingale because of condition 3. Let T_m denote the last jump time prior or equal to t. Then

$$\sum_{T_k \le t} (f(x_{T_k}) - f(x_{T_k-})) = \{f(x_t) - f(x_{Tm})\} + S_m$$

where

$$S_m = \sum_{k=1}^{m}(f(x_{T_k}) - f(x_{T_{k-1}}))\} - \{f(x_t) - f(x_{Tm}) + \\ + \sum_{k=1}^{m}(f(x_{T_k-}) - f(x_{T_{k-1}}))\}.$$

The first bracketed term on the right is equal to $f(x_t) - f(x)$. Note that $x_{T_k-} = x^{i_{k-1}}_{T_k - T_{k-1}}$, if $x_{T_{k-1}} = (i_{k-1}, x^{i_{k-1}}_{k-1})$. Then Itô-formula gives the second term

$$f(x_{T_k-}) - f(x_{T_{k-1}}) = \int_{T_{k-1}}^{T_k} L_{cont}f(x_s)ds + \int_{T_{k-1}}^{T_k} < \sigma(x_s), \nabla f(x_s) > dW(s).$$

The second term is therefore equal to $\int_0^t L_{cont}f(x_s)ds + \int_0^t < \sigma(x_s), \nabla f(x_s) > dW(s)$ and we obtain

$$C_t^f := f(x_t) - f(x) - \int_0^t Lf(x_s)ds = \int_0^t < \sigma(x_s), \nabla f(x_s) > \\ dW(s) + \int_{[0,t]\times\overline{X}} Bfdq$$

is a local martingale (the sum between a continuous martingale and a discrete martingale), where L is given by (28). Thus $f \in D(\widehat{L})$ and $\widehat{L}f = Lf$.
Conversely, suppose that $f \in D(\widehat{L})$. Then the process $M_t := f(x_t) - f(x) - \int_0^t h(x_s)ds$ is a local martingale, where $h = \widehat{L}f$. Then M_t must be the sum between a continuous martingale M_t^c and a discrete martingale M_t^d. From Theorem (26.12), p.69 [10], we have $M_t^d = M_t^\rho$ for some predictable integrand $\rho \in L_1^{loc}(p)$, where

$$M_t^\rho = \int_{\overline{X}\times\mathbb{R}_+} \rho I_{(s\le t)}dq$$
$$= \sum_{T_k \le t} \rho(x_{T_k}, T_k, \omega)$$
$$- \int_0^t \int_{\overline{X}} \rho(y, s, \omega)\{R(x_s, dy)\lambda(x_s)ds - R(x_{s-}, dy)dp^*(s)\}.$$

Since M_t^d and M_t^ρ agree, their jumps ΔM_t^d and ΔM_t^ρ must agree; these only occur when $t = T_k$ for some k and are given by: $\Delta M_t^d = f(x_t) - f(x_{t-})$; $\Delta M_t^\rho = \rho(x_t, t, \omega) - \int_{\overline{X}} \rho(y, t, \omega) R(x_{t-}, dy) I_{(x_{t-} \in \partial X)}$. Thus $\rho(x_t, t, \omega) = f(x_t) - f(x_{t-})$ on the set $(x_{t-} \notin \partial X)$, which implies that $\rho(x, t, \omega) = f(x) - f(x_{t-})$ for all (x, t) except perhaps a set to which the process 'never jumps', i.e. $G \subset \mathbb{R}_+ \times X$ such that $E_z \int_G p(dt, dx) = 0, \forall z \in X$.
Suppose that $z = x_{t-} \in \partial X$. Then equating ΔM_t^d and ΔM_t^ρ gives $f(x_t) - f(z) = \rho(x_t, t, \omega) - \int_{\overline{X}} \rho(y, t, \omega) R(z, dy)$ and hence $f(x) - f(z) = \rho(x, t, \omega) - \int_{\overline{X}} \rho(y, t, \omega) R(z, dy)$, except on a set $A \in \mathcal{B}(X)$ such that $R(z, A) = 0$. Integrating both sides of the previous equality with respect to $R(z, dx)$, we obtain $\int_{\overline{X}} f(x) R(z, dx) - f(z) = \int_{\overline{X}} \rho(x, t, \omega) R(z, dx) - \int_{\overline{X}} \rho(y, t, \omega) R(z, dy) = 0$.
Thus f satisfies the boundary condition. For fixed z, define $\widetilde{\rho}(x, t, \omega) = \rho(x, t, \omega) - (f(x) - f(z))$.
Using the boundary condition we get

$$\int_{\overline{X}} \widetilde{\rho}(y, t, \omega) R(z, dy) = \int_{\overline{X}} \rho(y, t, \omega) R(z, dy) = \widetilde{\rho}(x, t, \omega).$$

Then $\widetilde{\rho}(x, t, \omega) = \int_{\overline{X}} \widetilde{\rho}(y, t, \omega) R(z, dy)$.
However, the right-hand side does not depend on x, and hence $\widetilde{\rho}(x, t, \omega) = u(t, \omega)$ for some predictable process u. The general expression for ρ is thus

$$\rho(x, t, \omega) = f(x) - f(x_{t-}) + u(t, \omega) I_{(x_{t-} \in \partial X)}.$$

Inserting this in the expression of M_t^ρ we find that M_t^ρ does not depend on u, then we can take $u \equiv 0$, obtaining $\rho = Bf$; hence the part 3 of theorem is satisfied.

Finally, consider the sample paths of M_t, $M_t^{Bf} + M_t^c$, for $t < T_1(\omega)$, starting at $x \in X$. We have

$$M_t = f(x_t(\omega^{i_0})) - f(x) + \int_0^t h(x_s(\omega^{i_0})) ds$$

while, because $p = p^* = 0$ on $[0, T_1)$,
$M_t^{Bf} = - \int_{[0,t]} \int_{\overline{X}} (f(y) - f(x_s(\omega^{i_0}))) R(x_s(\omega^{i_0}), dy) \lambda(x_s(\omega^{i_0})) ds.$
So, since $M_t = M_t^{Bf} + M_t^c$ for all t a.s., it must be the case that $M_t = M_t^c$ for $t \in [0, T_1)$ and the generator coincides with the generator L_{cont} associated to the stochastic equation, the function $f(x_t(\omega^{i_0}))$ should have second order derivatives on $[0, T_1)$. The general case follows by concatenation. Similar calculations show that

$$M_t^{Bf} + M_t^c = f(x_t) - f(x) - \int_0^t Lf(x_s) ds, \ \forall t \geq 0$$

with L given by (28). Hence $f \in D(L)$ and $Lf = \widehat{L}f$.□

5 Conclusions

In this chapter we set up the notion of Markov string, which is roughly speaking, a concatenation of Markov processes. This notion has arisen as a result of our research on stochastic hybrid system modeling [17, 8, 7, 24] and it aims to be a very general formalization of all existing models of stochastic hybrid systems. The Markov string concept has been proved to be a very powerful tool in the studying of the general models of stochastic hybrid processes GSHS introduced at the beginning of the chapter.

One of the main contributions of this work is the proof of the strong Markov property. Since GSHS are a particular class of Markov strings, this property holds also for them.

In the end of this chapter, based on the strong Markov property of GSHS we have developed the extended generator of this model.

Further developments of our model will include two main tracks.

- First it is necessary a study of the reachability problem for GSHS. One possible approach in this direction is the introduction of a bisimulation concept for GSHS. Reachability analysis and model checking are much easier when a concept of bisimulation is available. The state space can be drastically abstracted in some cases.
- Second it is natural to generalize the results on dynamic programming, relaxed controls, control via discrete-time dynamic programming, nonsmooth analysis, from PDMP to GSHS.

References

1. L. Arnold. *Stochastic Differential Equations: Theory and Application.* John Wiley & Sons, 1974.
2. A. Bensoussan and J.L. Menaldi. Stochastic hybrid control. *Journal of Mathematical Analysis and Applications*, 249:261–288, 2000.
3. D.P. Bertsekas and S.E. Shreve. *Stochastic Optimal Control: The Discrete-Time Case.* Athena Scientific, 1996.
4. H.A.P. Blom. Stochastic hybrid processes with hybrid jumps. In *ADHS, Analysis and Design of Hybrid System*, 2003.
5. R.M. Blumenthal and R.K. Getoor. *Markov Processes and Potential Theory.* Academic Press, New York and London, 1968.
6. V.S. Borkar, M.K. Ghosh, and P. Sahay. Optimal control of a stochastic hybrid system with discounted cost. *Journal of Optimization Theory and Applications*, 101(3):557–580, June 1999.
7. M.L. Bujorianu. Extended stochastic hybrid systems. In R. Alur and G. Pappas, editors, *Hybrid Systems: Computation and Control*, number 2993 in LNCS, pages 234–249. Springer Verlag, 2004.
8. M.L. Bujorianu and J. Lygeros. Reachability questions in piecewise deterministic markov processes. In O. Maler and A. Pnueli, editors, *Hybrid Systems: Computation and Control*, number 2623 in LNCS, pages 126–140. Springer Verlag, 2003.

9. M.L. Bujorianu, J. Lygeros, W. Glover, and G. Pola. A stochastic hybrid system modeling framework. Technical Report WP1, Deliverable D1.2, HYBRIDGE, 2002.
10. M.H.A. Davis. *Markov Processes and Optimization.* Chapman & Hall, London, 1993.
11. M.H.A. Davis, V. Dempster, S.P. Sethi, and D. Vermes. Optimal capacity expansion under uncertainty. *Adv. Appl. Prob.*, 19:156–176, 1987.
12. M.H.A. Davis and M.H. Vellekoop. Permanent health insurance: a case study in piecewise-deterministic markov modelling. *Mitteilungen der Schweiz. Vereinigung der Versicherungsmathematiker*, 2:177–212, 1995.
13. M.K. Ghosh, A. Arapostathis, and S.I. Marcus. Optimal control of switching diffusions with application to flexible manufacturing systems. *SIAM Journal on Control Optimization*, 31(5):1183–1204, September 1993.
14. M.K. Ghosh, A. Arapostathis, and S.I. Marcus. Ergodic control of switching diffusions. *SIAM Journal on Control Optimization*, 35(6):1952–1988, November 1997.
15. M.K. Ghosh and A. Bagchi. Modeling stochastic hybrid systems. In *21st IFIP TC7 Conference on System Modelling and Optimization*, 2003.
16. J.P. Hespanha. Stochastic hybrid systems: Application to communication network. In R. Alur and G. Pappas, editors, *Hybrid Systems: Computation and Control*, number 2993 in LNCS, pages 387–401. Springer Verlag, 2004.
17. J. Hu, J. Lygeros, and S. Sastry. Towards a theory of stochastic hybrid systems. In Nancy Lynch and Bruce H. Krogh, editors, *Hybrid Systems: Computation and Control*, number 1790 in LNCS, pages 160–173. Springer Verlag, 2000.
18. I. Hwang, J. Hwang, and C.J. Tomlin. Flight-model-based aircraft conflict detection using a residual-mean interacting multiple model algorithm. In *AIAA Guidance, Navigation, and Control Conference, AIAA-2003-5340*, 2003.
19. HYBRIDGE. Distributed control and stochastic analysis of hybrid system supporting safety critical real-time systems design. http://www.nlr.nl/public/hosted-sites/hybrid.
20. N. Ikeda, M. Nagasawa, and S. Watanabe. Construction of markov processes by piecing out. *Proc. Japan Acad*, 42:370–375, 1966.
21. P.A. Meyer. *Probability and Potentials.* Blaisdell, Waltham Mass, 1966.
22. P.A. Meyer. *Processus de Markov.* Number 26 in LNM. Springer-Verlag, Berlin and Heidelberg and New York, 1967.
23. P.A. Meyer. Renaissance, recollectments, melanges, ralentissement de processus de markov. *Ann. Inst. Fourier*, 25:465–497, 1975.
24. G. Pola, M.L. Bujorianu, J. Lygeros, and M.D. Di Benedetto. Stochastic hybrid models: An overview with applications to air traffic management. In *ADHS, Analysis and Design of Hybrid System*, 2003.
25. K. Siegrist. Random evolution processes with feedback. *Trans. Amer. Math. Soc. Vol.*
26. O. Watkins and J. Lygeros. Safety relevant operational cases in ATM. Technical Report WP1, Deliverable D1.1, HYBRIDGE.

A Background on Markov Processes

Suppose that $\mathbb{M} = (\Omega, \mathcal{F}, \mathcal{F}_t, x_t, \theta_t, P, P_x), \in Q$ is a Markov process. We denote the state space of $\mathbb{M}$ by $(X, \mathcal{B})$ and assume that $\mathcal{B}$ is the Borel σ-algebra of

X if X is a topological Hausdorff space. Let Δ be the cemetery point for X, which is an adjoined point to X, $X_\Delta = X \cup \{\Delta\}$. The existence of Δ is assumed in order to have a probabilistic interpretation of $P_x(x_t \in X) < 1$, i.e. at some 'termination time' $\zeta(\omega)$ when the process $\mathbb{M}$ escapes to and is trapped at Δ. The elements $\mathcal{F}$, $\mathcal{F}_t^0$, $\mathcal{F}_t$, θ_t, P, P_x have the usual meaning, i.e.

- $(\Omega, \mathcal{F}, P)$ denotes the underlying probability space.
- $\mathcal{F}_t^0$ denotes the *natural filtration*, i.e. $\mathcal{F}_t^0 = \sigma\{x_t, s \leq t\}$ and $\mathcal{F}_\infty^0 = \vee_t \mathcal{F}_t^0$.
- $x_t : (\Omega, \mathcal{F}) \to (X, \mathcal{B})$ is a $\mathcal{F}^0/\mathcal{B}$-measurable function for all $t \geq 0$.
- $\theta_t : \Omega \to \Omega$, for all $t \geq 0$, is the *translation operator*, i.e.

$$x_s \circ \theta_t = x_{t+s}, t, s \geq 0$$

- $P_x : (\Omega, \mathcal{F}^0) \to [0,1]$ is a probability measure (so-called *Wiener probability*) such that $P_x(x_t \in E)$ is $\mathcal{B}$-measurable in $x \in X$ for each $t \geq 0$ and $E \in \mathcal{B}$.
- If $\mu \in \mathcal{P}(X_\Delta)$, i.e. μ is a probability measure on $(X, \mathcal{B})$ then we can define

$$P_\mu(\Lambda) = \int_{X_\Delta} P_x(\Lambda)\mu(dx), \Lambda \in \mathcal{F}^0.$$

  We then denote by $\mathcal{F}$ (resp. $\mathcal{F}_t$) the completion of $\mathcal{F}_\infty^0$ (resp. $\mathcal{F}_t^0$) with respect to all P_μ, $\mu \in \mathcal{P}(X_\Delta)$.
- We say that a family $\{\mathcal{M}_t\}$ of sub-σ-algebras of $\mathcal{F}$ is an admissible filtration if $\mathcal{M}_t$ is increasing in t and $x_t \in \mathcal{M}_t/\mathcal{B}$ for each $t \geq 0$. Then $\mathcal{F}_t^0$ is the *minimum admissible filtration.* An admissible filtration $\{\mathcal{M}_t\}$ is *right continuous* if $\mathcal{M}_t = \mathcal{M}_{t+} = \cap\{\mathcal{M}_{t'} | t' > t\}$.
- Given an admissible filtration $\{\mathcal{M}_t\}$, a $[0,\infty]$-valued function τ on Ω is called an $\{\mathcal{M}_t\}$-*stopping time* if $\{\tau \leq t\} \in \mathcal{M}_t$, $\forall t \geq 0$.
- For an admissible filtration $\{\mathcal{M}_t\}$, we say that $\mathbb{M}$ is *strong Markov* with respect to $\{\mathcal{M}_t\}$ if $\{\mathcal{M}_t\}$ is right continuous and

$$P_\mu(x_{\tau+t} \in E | \mathcal{M}_\tau) = P_{x_\tau}(x_t \in E); P_\mu - a.s.$$

  $\mu \in \mathcal{P}(X_\Delta)$, $E \in \mathcal{B}$, $t \geq 0$, for any $\{\mathcal{M}_t\}$-stopping time τ.
- $\mathbb{M}$ has the *càdlàg property* if for each $\omega \in \Omega$, the sample path $t \mapsto x_t(\omega)$ is right continuous on $[0,\infty)$ and has left limits on $(0,\infty)$ (inside X_Δ).
- Let (P_t) denote the operator semigroup associated to $\mathbb{M}$ which maps $\mathcal{B}^b(X)$ (the set of all bounded measurable functions on X) into itself given by

$$P_t f(x) = E_x f(x_t),$$

  where E_x is the expectation with respect to P_x. Then a function f is p-*excessive* if it is non-negative and $e^{-pt} P_t f \leq f$ for all $t \geq 0$ and $e^{-pt} P_t f \nearrow f$ as $t \searrow 0$.

Hybrid Petri Nets with Diffusion That Have Into-Mappings with Generalised Stochastic Hybrid Processes

Mariken H.C. Everdij[1] and Henk A.P. Blom[1]

National Aerospace Laboratory NLR, everdij@nlr.nl, blom@nlr.nl

Summary. Generalised Stochastic Hybrid Processes (GSHPs) are known as the largest class of Markov processes virtually describing all continuous-time processes including diffusion. In general, the state space of a GSHP is of hybrid type, i.e. a Kronecker product of a discrete set and a continuous-valued space. Since Stochastic Petri Nets have proven to be extremely useful in developing continuous-time Markov Chain models for complex practical discrete-valued processes, there is a clear need for a type of hybrid Petri Nets that can play a similar role for developing GSHP models for complex practical problems. To fulfil this need, the report defines a Stochastically and Dynamically Coloured Petri Net (SDCPN), and proves that there exist into-mappings between GSHPs and SDCPNs.

1 Introduction

Davis [6] has introduced Piecewise Deterministic Markov Processes (PDPs) as the most general class of continuous-time Markov processes which include both discrete and continuous processes, except diffusion. A PDP $\{\xi_t\}$ consists of two components: a piecewise constant component $\{\theta_t\}$ and a piecewise continuous valued component $\{x_t\}$, which follows the solution of a θ_t-dependent ordinary differential equation. A jump in $\{\xi_t\}$ occurs when $\{x_t\}$ hits the boundary of a predefined area, or according to a jump rate. If $\{x_t\}$ also makes a jump at a time when $\{\theta_t\}$ switches, this is said to be a hybrid jump.

Bujorianu et al [3] extended this PDP definition to Generalised Stochastic Hybrid Processes (GSHP) by including diffusion by means of Brownian motion. With this extension, between jumps, the process $\{x_t\}$ follows the solution of a θ_t-dependent stochastic (rather than ordinary) differential equation. GSHP forms a powerful and useful class of processes that have strong support in stochastic analysis and control.

A Petri Net is a bipartite graph of places (possible conditions or discrete modes) and transitions (possible mode switches). Tokens, which reside in the places, model which conditions or modes are current. Petri Nets, see e.g. [4], and their many extensions, see e.g. [5] for a good overview, have proven

H.A.P. Blom, J. Lygeros (Eds.): Stochastic Hybrid Systems, LNCIS 337, pp. 31–63, 2006.

to be extremely useful in developing models for various complex practical applications. This usefulness is especially due to their specification power [4], which allows to develop a submodel for each entity of a complex operation, and next to combine the submodels in a constructive way. An example is Stochastic Petri Nets, which have been successfully used in developing continuous-time Markov Chain models for complex practical discrete-valued processes. For this reason, there is a clear need for a type of Petri Nets that can play a similar role for developing PDP or GSHP models for complex practical problems. Several hybrid state Petri Net extensions have been developed in the past. Main classes are:

- Hybrid Petri Net , [1]. Some places have a continuous amount of tokens that may be moved to other places by transitions.
- Fluid Stochastic Petri Net (FSPN), [16]. Some places have a continuous amount of tokens, the flow rate of which is influenced by the discrete part. The discrete part of the FSPN can be mapped to a continuous-time Markov chain.
- Extended Coloured Petri Net (ECPN), [17]. The token colours are real-valued vectors that may follow the solution path of a difference equation.
- High-Level Hybrid Petri Net (HLHPN), [12]. Again, the token colours are real-valued vectors that may follow the solution path of a difference equation, but in addition, a token switch between discrete places may generate a jump in the value of the real-valued vector.
- Differential Petri Nets , [8]. Differential places have a real-valued number of tokens and differential transitions fire with a certain speed that may also be negative.

For none of the above hybrid state Petri Nets it is clear how they relate to PDP. Moreover, none of them include Brownian motion as GSHP does. In order to improve this situation for PDP, Everdij and Blom [10], [11], developed a Petri Net extension named Dynamically Coloured Petri Net (DCPN) and proved that here exist into-mappings between PDPs and DCPNs. In [9], Everdij and Blom showed that this existence of into-mappings extends the power-hierarchy among various model types established by [14], [15]. This is shown in Figure 1, in which the well-known dependability models Reliability Block Diagrams and Fault Trees are at the basis of the hierarchy.

Although PDP form a very general class of continuous-time Markov processes which include both discrete and continuous processes, PDP do not include diffusion. The aim of the current chapter aims to solve this issue by

- including a diffusion term into the PDP definition, following [3], and referred to as GSHP (Generalised Stochastic Hybrid Process);
- introducing an extension of DCPN, referred to as Stochastically and Dynamically Coloured Petri Net (SDCPN), which also covers diffusion;
- and showing that there exist into-mappings between GSHP and SDCPN.

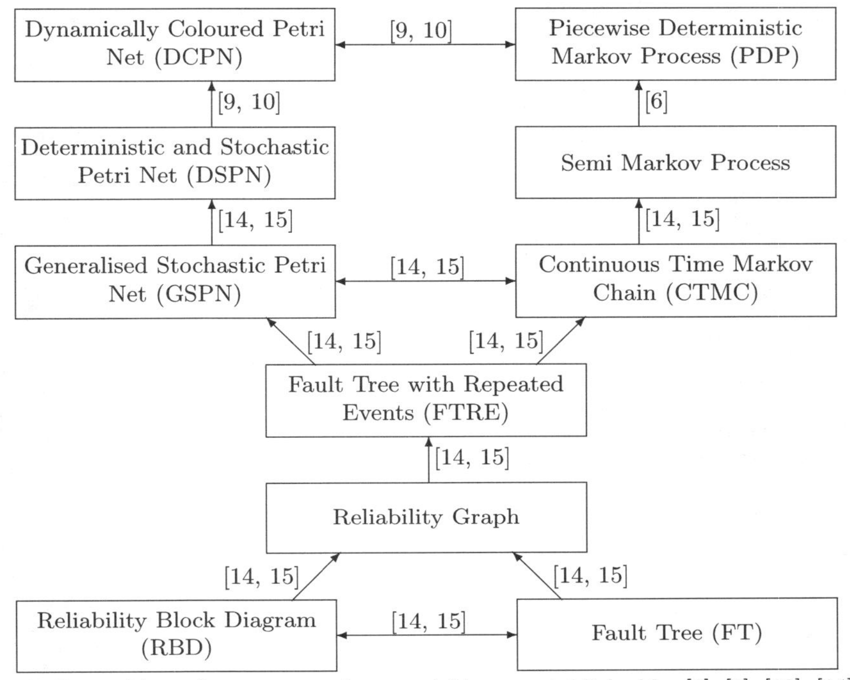

Fig. 1. Power hierarchy among various model types established by [6], [9], [10], [14], and [15]. An arrow from a model to another model indicates that the second model has more modelling power than the first model

The existence of such into-mappings allows combining the specification power of Petri Nets with the stochastic analysis and control power of GSHP. In addition, the into-mappings extend the power hierarchy of Figure 1 with GSHP and with GSHP-related Petri Nets.

The organisation of the paper is as follows. Section 2 briefly describes GSHP. Section 3 defines SDCPN. Section 4 shows that each GSHP can be represented by a SDCPN process. Section 5 shows that each SDCPN process can be represented by a GSHP. Section 6 presents a SDCPN model for a simple aircraft evolution example and its mapping to a GSHP. Section 7 draws conclusions.

2 Generalised Stochastic Hybrid Process

This section presents a definition of Generalised Stochastic Hybrid System (GSHS) and its GSHP solution, see [3]. As much as possible, the notation introduced by Davis [7] for Piecewise Deterministic Markov Process is used.

Definition 1. *A Generalised Stochastic Hybrid System (GSHS) is a nine-tuple* ($\mathbf{K}$, $d(\theta)$, x_0, θ_0, ∂E_θ, g_θ, g_θ^w, λ, Q)*, together with some conditions* C_1 *-* C_4*.*

Below, first the structure of the elements in the tuple and the GSHS conditions are given, next the GSHS execution is explained.

2.1 GSHS Elements

The GSHS elements are defined as follows:

1. $\mathbf{K}$ is a countable set of discrete variables.
2. d is a map from $\mathbf{K}$ into $\mathbb{N}$, giving the dimensions of the continuous state process.
3. For each $\theta \in \mathbf{K}$, E_θ is an open subset of $\mathbb{R}^{d(\theta)}$, and ∂E_θ is its boundary.
4. θ_0 is an initial value in $\mathbf{K}$.
5. x_0 is an initial value in E_{θ_0}.
6. $g_\theta : \mathbb{R}^{d(\theta)} \to \mathbb{R}^{d(\theta)}$ is a vector field.
7. $g_\theta^w : \mathbb{R}^{d(\theta)} \to \mathbb{R}^{d(\theta)} \times \mathbb{R}^b$ is a matrix, with $b \in \mathbb{N}$.
8. $\lambda : E \to \mathbb{R}^+$ is a jump rate function, with $E = \cup_\theta E_\theta$.
9. $Q : E \cup \Gamma^* \to [0,1]$ is a probability measure, with $E = \cup_\theta E_\theta$ and Γ^* the reachable boundary of E.

2.2 GSHS Conditions

Following [3] (Assumptions 1, 2 and 3), the GSHS conditions are:

C_1 g_θ and g_θ^w are such[1] that for each initial state (θ, x) at initial time τ there exists a pathwise unique solution $x_t = \phi_{\theta,x,t-\tau}$ to $dx_t = g_\theta(x_t)dt + g_\theta^w(x_t)dw_t$, where $\{w_t\}$ is b-dimensional standard Brownian motion. If $t_\infty(\theta, x)$ denotes the explosion time of the flow $\phi_{\theta,x,t-\tau}$, i.e. $|\phi_{\theta,x,t-\tau}| \to \infty$ as $t \uparrow t_\infty(\theta, x)$, then it is assumed that $t_\infty(\theta, x) = \infty$ whenever $t_*(\theta, x) = \infty$. In other words, explosions are ruled out.

C_2 With $E = \cup_\theta E_\theta$, $\lambda : E \to \mathbb{R}^+$ is a measurable function such that for all $\xi \in E$, there is $\epsilon(\xi) > 0$ such that $t \to \lambda(\theta, \phi_{\theta,x,t})$ is integrable on $[0, \epsilon(\xi)[$.

C_3 With E as above and Γ^* the reachable boundary of E, Q maps $E \cup \Gamma^*$ into the set of probability measures on $(E, \mathcal{E})$, with $\mathcal{E}$ the Borel-measurable subsets of E, while for each fixed $A \in \mathcal{E}$, the map $\xi \to Q(A; \xi)$ is measurable and $Q(\{\xi\}; \xi) = 0$.

C_4 If $N_t = \sum_k I_{(t \geq \tau_k)}$, then it is assumed that for every starting point ξ and for all $t \in \mathbb{R}^+$, $\mathbb{E}N_t < \infty$. This means, there will be a finite number of jumps in finite time.

---

[1] [3] assumes Lipschitz continuity and boundedness.

2.3 GSHS Execution

The execution of a GSHS generates a Generalised Stochastic Hybrid Process (GSHP) $\{\xi_t\}$, with $\xi_t = (\theta_t, x_t)$, as follows:

For each $\theta \in \mathbf{K}$, consider the stochastic differential equation $dx_t = g_\theta(x_t)dt + g_\theta^w(x_t)dw_t$, where $\{w_t\}$ is b-dimensional standard Brownian motion. Given an initial value $x \in E_\theta$, under GSHS condition C_1, this differential equation has a pathwise unique solution. This means that if at some time instant τ the GSHP state assumes value $\xi_\tau = (\theta_\tau, x_\tau)$, then, as long as no jumps occur, the GSHP state at $t \geq \tau$ is given by $\xi_t = (\theta_t, x_t) = (\theta_\tau, \phi_{\theta_\tau, x_\tau, t-\tau})$, with $\phi_{\theta_\tau, x_\tau, t-\tau} = \int_\tau^t g_{\theta_s}(x_s)dt + \int_\tau^t g_{\theta_s}^w(x_s)dw_s$. At some moment in time, however, the GSHP state value may jump. Such moment is generated by either one of the following events, depending on which event occurs first:

1. A Poisson point process with jump rate $\lambda(\theta_t, x_t)$, $t > \tau$ generates a point.
2. The piecewise continuous process x_t is about to hit the boundary ∂E_{θ_τ} of E_{θ_τ}, $t > \tau$.

At the moment when either of these events occurs, the GSHP state makes a jump. The value of the GSHP state right after the jump is generated by using a transition measure Q, which is the probability measure of the GSHP state after the jump, given the value of the GSHP state immediately before the jump. After this, the GSHP state ξ_t evolves in a similar way from the new value onwards.

The GSHP process is generated by executing a GSHS through time as follows: Suppose at time $\tau_0 \triangleq 0$ the GSHP initial state is $\xi_0 = (\theta_0, x_0)$, then, if no jumps occur, the process state at $t \geq \tau_0$ is given by $\xi_t = (\theta_t, x_t) = (\theta_0, \phi_{\theta_0, x_0, t-\tau_0})$. The complementary distribution function for the time of the first jump (i.e. the probability that the first jump occurs at least $t - \tau_0$ time units after τ_0), also named the survivor function of the first jump, is then given by:

$$G_{\xi_0, t-\tau_0} \triangleq I_{(t-\tau_0 < t_*(\theta_0, x_0))} \cdot \exp\left\{-\int_{\tau_0}^t \lambda(\theta_0, \phi_{\theta_0, x_0, s-\tau_0})ds\right\}, \tag{1}$$

where I is an indicator function and $t_*(\theta_0, x_0)$ denotes the time until the first boundary hit after $t = \tau_0$, which is given by $t_*(\theta_0, x_0) \triangleq \inf\{t - \tau_0 > 0 \mid \phi_{\theta_0, x_0, t-\tau_0} \in \partial E_{\theta_0}\}$. The first factor in Equation (1) is explained by the boundary hitting process: after the process state has hit the boundary, which is when $t - \tau_0 = t_*(\theta_0, x_0)$, this first factor ensures that the survivor function evaluates to zero. The second factor in Equation (1) comes from the Poisson process: this second factor ensures that a jump is generated after an exponentially distributed time with a rate λ that is dependent on the GSHP state.

The time τ_1 until the first jump after τ_0 is generated by drawing a sample from a uniform distribution on $[0, 1]$, and then using a transformation that takes G into account. More formally (see [7], Section 23), the Hilbert cube $\Omega^H = \prod_{i=1}^{\infty} Y_i$, with Y_i a copy of $Y = [0,1]$, provides the canonical space for a countable sequence of independent random variables U_1, U_2, ..., each having uniform $[0,1]$ distribution, defined by $U_i(\omega) = \omega_i$ for elements $\omega = (\omega_1, \omega_2, \ldots) \in \Omega^H$. The complete probability space is $(\Omega, \mathfrak{F}, \mathfrak{P}, \{\mathfrak{F}_t\})$, with $\Omega = \Omega^H \times \Omega^B$, and where Ω^B supports the Brownian motion. Now, define

$$\psi_1(u, \xi_0, \omega) = \begin{cases} \inf\{t : G_{\xi_0, t-\tau_0}(\omega) \leq u\} \\ +\infty \text{ if the above set is empty} \end{cases}$$

and define $\sigma_1(\omega) = \tau_1(\omega) = \psi_1(U_1(\omega), \xi_0, \omega)$, then τ_1 is the time until the first jump.

The value of the hybrid process state to which the jump is made is generated by using the transition measure Q, which is the probability measure of the hybrid state after the jump, given the value of the hybrid state immediately before the jump. The Hilbert cube from above is again used: Let $\psi_2 : [0,1] \times (E \cup \Gamma^*) \to E$, with $E = \cup_\theta E_\theta$ and Γ^* the reachable boundary of E, be a measurable function such that $l\{u : \psi_2(u, \xi) \in B\} = Q(B, \xi)$ for B Borel measurable. Then $\xi_{\tau_1} = \psi_2(U_2(\omega), \xi)$ is a sample from $Q(\cdot, \xi)$.

With this, the algorithm to determine a sample path for the hybrid state process ξ_t, $t \geq 0$, from the initial state $\xi_0 = (\theta_0, x_0)$ on, is in two iterative steps; define $\tau_0 \triangleq 0$ and let for $k = 0$, $\xi_{\tau_k} = (\theta_{\tau_k}, x_{\tau_k})$ be the initial state, then for $k = 1, 2, \ldots$:

Step 1: Draw a sample σ_k from survivor function $G_{\xi_{\tau_{k-1}}, t-\tau_{k-1}}(\omega)$, i.e. $\sigma_k(\omega) = \psi_1(U_{2k-1}(\omega), \xi_{\tau_{k-1}}, \omega)$. Then the time τ_k of the kth jump is $\tau_k = \tau_{k-1} + \sigma_k$. The sample path up to the kth jump is given by

$$\xi_t = (\theta_{\tau_{k-1}}, \phi_{\theta_{\tau_{k-1}}, x_{\tau_{k-1}}, t-\tau_{k-1}}), \quad \tau_{k-1} \leq t < \tau_k \text{ and } \tau_k \leq \infty.$$

Step 2: Draw a multi-dimensional sample ζ_k from transition measure $Q(\cdot; \xi'_{\tau_k})$, where $\xi'_{\tau_k} = (\theta_{\tau_{k-1}}, \phi_{\theta_{\tau_{k-1}}, x_{\tau_{k-1}}, \tau_k - \tau_{k-1}})$, i.e. $\zeta_k = \psi_2(U_{2k}(\omega), \xi'_{\tau_k})$. Then, if $\tau_k < \infty$, the process state at the time τ_k of the kth jump is given by

$$\xi_{\tau_k} = \zeta_k.$$

3 Stochastically and Dynamically Coloured Petri Net (SDCPN)

This section presents a definition of Stochastically and Dynamically Coloured Petri Net (SDCPN). As much as possible, the notation introduced by Jensen [13] for Coloured Petri Net is used.

Definition 2. *A Stochastically and Dynamically Coloured Petri Net (SDCPN) is a 12-tuple* SDCPN $= (\mathcal{P}, \mathcal{T}, \mathcal{A}, \mathcal{N}, \mathcal{S}, \mathcal{C}, \mathcal{V}, \mathcal{W}, \mathcal{G}, \mathcal{D}, \mathcal{F}, \mathcal{I})$, *together with some rules* R_0 - R_4.

Below, first the structure of the elements in the tuple is given, next the SDCPN evolution through time is explained, finally, the SDCPN generated process is outlined.

3.1 SDCPN Elements

The SDCPN elements are defined as follows:

1. $\mathcal{P}$ is a finite set of places. In a graphical notation, places are denoted by circles:

   Place: ◯

2. $\mathcal{T}$ is a finite set of transitions, such that $\mathcal{T} \cap \mathcal{P} = \emptyset$. The set $\mathcal{T}$ consists of 1) a set $\mathcal{T}_G$ of guard transitions, 2) a set $\mathcal{T}_D$ of delay transitions and 3) a set $\mathcal{T}_I$ of immediate transitions, with $\mathcal{T} = \mathcal{T}_G \cup \mathcal{T}_D \cup \mathcal{T}_I$, and $\mathcal{T}_G \cap \mathcal{T}_D = \mathcal{T}_D \cap \mathcal{T}_I = \mathcal{T}_I \cap \mathcal{T}_G = \emptyset$. Notations are:

   Guard transition: ▭

   Delay transition: ▬

   Immediate transition: —

3. $\mathcal{A}$ is a finite set of arcs such that $\mathcal{A} \cap \mathcal{P} = \mathcal{A} \cap \mathcal{T} = \emptyset$. The set $\mathcal{A}$ consists of 1) a set $\mathcal{A}_O$ of ordinary arcs, 2) a set $\mathcal{A}_E$ of enabling arcs and 3) a set $\mathcal{A}_I$ of inhibitor arcs, with $\mathcal{A} = \mathcal{A}_O \cup \mathcal{A}_E \cup \mathcal{A}_I$, and $\mathcal{A}_O \cap \mathcal{A}_E = \mathcal{A}_E \cap \mathcal{A}_I = \mathcal{A}_I \cap \mathcal{A}_O = \emptyset$. Notations are:

   Ordinary arc: ——→

   Enabling arc: ——●

   Inhibitor arc: ——○

4. $\mathcal{N} : \mathcal{A} \to \mathcal{P} \times \mathcal{T} \cup \mathcal{T} \times \mathcal{P}$ is a node function which maps each arc A in $\mathcal{A}$ to a pair of ordered nodes $\mathcal{N}(A)$. The place of $\mathcal{N}(A)$ is denoted by $P(A)$, the transition of $\mathcal{N}(A)$ is denoted by $T(A)$, such that for all $A \in \mathcal{A}_E \cup \mathcal{A}_I$: $\mathcal{N}(A) = (P(A), T(A))$ and for all $A \in \mathcal{A}_O$: either $\mathcal{N}(A) = (P(A), T(A))$ or $\mathcal{N}(A) = (T(A), P(A))$. Further notation:
   - $A(T) = \{A \in \mathcal{A} \mid T(A) = T\}$ denotes the set of arcs connected to transition T, with $A(T) = A_{in}(T) \cup A_{out}(T)$, where
   - $A_{in}(T) = \{A \in A(T) \mid \mathcal{N}(A) = (P(A), T)\}$ is the set of input arcs of T and
   - $A_{out}(T) = \{A \in A(T) \mid \mathcal{N}(A) = (T, P(A))\}$ is the set of output arcs of T. Moreover,
   - $A_{in,O}(T) = A_{in}(T) \cap \mathcal{A}_O$ is the set of ordinary input arcs of T,
   - $A_{in,OE}(T) = A_{in}(T) \cap \{\mathcal{A}_E \cup \mathcal{A}_O\}$ is the set of input arcs of T that are either ordinary or enabling, and
   - $P(A(T))$ is the set of places connected to T by the set of arcs $A(T)$.

Finally, $\{A_i \in \mathcal{A}_I \mid \exists A \in \mathcal{A}, A \neq A_i : \mathcal{N}(A) = \mathcal{N}(A_i)\} = \emptyset$, i.e., if an inhibitor arc points from a place P to a transition T, there is no other arc from P to T.

5. $\mathcal{S}$ is a finite set of colour types. Each colour type is to be written in the form $\mathbb{R}^n$, with n a natural number and with $\mathbb{R}^0 = \emptyset$.
6. $\mathcal{C} : \mathcal{P} \to \mathcal{S}$ is a colour function which maps each place $P \in \mathcal{P}$ to a specific colour type in $\mathcal{S}$.
7. $\mathcal{I} : \mathcal{P} \to \mathcal{C}(\mathcal{P})_{ms}$ is an initialisation function, where $\mathcal{C}(P)_{ms}$ for $P \in \mathcal{P}$ denotes the set of all multisets over $\mathcal{C}(P)$. It defines the initial marking of the net, i.e., for each place it specifies the number of tokens (possibly zero) initially in it, together with the colours they have, and their ordering per place.
8. $\mathcal{V}$ is set of a token colour functions. For each place $P \in \mathcal{P}$ for which $\mathcal{C}(P) \neq \mathbb{R}^0$, it contains a function $\mathcal{V}_P : \mathcal{C}(P) \to \mathcal{C}(P)$ which satisfies conditions that ensure a pathwise unique solution.
9. $\mathcal{W}$ is set of a token colour matrix functions. For each place $P \in \mathcal{P}$ for which $\mathcal{C}(P) \neq \mathbb{R}^0$, it contains a function $\mathcal{W}_P : \mathcal{C}(P) \to \mathcal{C}(P) \times \mathcal{C}'(P)$, which satisfies conditions that ensure a pathwise unique solution, and where $\mathcal{C}'(P)$ collects the Brownian motion terms. Here, $\mathcal{C}'$ maps $\mathcal{P}$ into $\mathbb{R}^b$, with $b \in \mathbb{R}$ a constant.
10. $\mathcal{G}$ is a set of transition guards. For each $T \in \mathcal{T}_G$, it contains a transition guard $\mathcal{G}_T : \mathcal{C}(P(A_{in,OE}(T))) \to \{\text{True, False}\}$. $\mathcal{G}_T(c)$ evaluates to True if c is in the boundary ∂G_T of an open subset G_T in $\mathcal{C}(P(A_{in,OE}(T)))$. Here, if $P(A_{in,OE}(T))$ contains more than one place, e.g., $P(A_{in,OE}(T)) = \{P_i, \dots, P_j\}$, then $\mathcal{C}(P(A_{in,OE}(T)))$ is defined by $\mathcal{C}(P_i) \times \cdots \times \mathcal{C}(P_j)$. If $\mathcal{C}(P(A_{in,OE}(T))) = \mathbb{R}^0$ then $\partial G_T = \emptyset$ and the guard will always evaluate to False.
11. $\mathcal{D}$ is a set of transition enabling rate functions. For each $T \in \mathcal{T}_D$, it contains an integrable transition enabling rate function $\delta_T : \mathcal{C}(P(A_{in,OE}(T))) \to \mathbb{R}_0^+$, which, if T is evaluated from stopping time τ on, specifies a delay time equal to $\mathcal{D}_T(\tau) = \inf\{t \mid e^{-\int_\tau^t \delta_T(c_s)ds} \leq u\}$, where u is a random number drawn from $U[0,1]$ at τ. If $\mathcal{C}(P(A_{in,OE}(T))) = \mathbb{R}^0$ then δ_T is a constant function.
12. $\mathcal{F}$ is a set of firing measures. For each $T \in \mathcal{T}$ it specifies a probability measure $\mathcal{F}_T$ which maps $\mathcal{C}(P(A_{in,OE}(T)))$ into the set of probability measures on $\{0,1\}^{|A_{out}(T)|} \times \mathcal{C}(P(A_{out}(T)))$.

3.2 SDCPN Execution

The execution of a SDCPN provides a series of increasing stopping times, $\tau_0 < \tau_i < \tau_{i+1}$, with for $t \in (\tau_i, \tau_{i+1})$ a fixed number of tokens per place and per token a colour which is the solution of a stochastic differential equation. This number of tokens and the colours of these tokens are generated as follows:

Each token residing in place P has a colour of type $\mathcal{C}(P)$. If a token in place P has colour c at time τ, and if it remains in that place up to time

$t > \tau$, then the colour c_t at time t equals the unique solution of the stochastic differential equation $dc_t = \mathcal{V}_P(c_t)dt + \mathcal{W}_P(c_t)dw_t$ with initial condition $c_\tau = c$.

A transition T is *pre-enabled* if it has at least one token per incoming ordinary and enabling arc in each of its input places and has no token in places to which it is connected by an inhibitor arc; denote $\tau_1^{pre} = \inf\{t \mid T \text{ is pre-enabled at time } t\}$. Consider one token per ordinary and enabling arc in the input places of T and write $c_t \in \mathcal{C}(P(A_{in,OE}(T)))$, $t \geq \tau_1^{pre}$, as the column vector containing the colours of these tokens; c_t may change through time according to its corresponding token colour functions. If this vector is not unique (for example, one input place contains several tokens per arc), all possible such vectors are executed in parallel.

A transition T is *enabled* if it is pre-enabled and a second requirement holds true. For $T \in \mathcal{T}_I$, the second requirement automatically holds true. For $T \in \mathcal{T}_G$, the second requirement holds true when $\mathcal{G}_T(c_t) = \text{True}$. For $T \in \mathcal{T}_D$, the second requirement holds true $\mathcal{D}_T(\tau_1^{pre})$ units after τ_1^{pre}. Guard or delay evaluation of a transition T stops when T is not pre-enabled anymore, and is restarted when it is.

For the evaluation of $\mathcal{D}_T(\tau_1^{pre})$, use is made of a Hilbert cube $\Omega^H = \prod_{i=1}^{\infty} Y_i$, with Y_i a copy of $Y = [0,1]$, which provides the canonical space for a countable sequence of independent random variables U_1, U_2, ..., each having a uniform $[0,1]$ distribution, defined by $U_i(\omega) = \omega_i$ for elements $\omega = (\omega_1, \omega_2, \ldots) \in \Omega^H$. This Hilbert cube applies as follows: Suppose T is a delay transition that is pre-enabled at time τ and has vector of input colours c_t at time $t \geq \tau$. Then transition T is enabled at random time $\inf\{t : \exp\left\{-\int_\tau^t \delta_T(c_s)ds\right\} \leq U_i\}$, with $\inf\{\ \} = +\infty$. The complete probability space is $(\Omega, \mathfrak{F}, \mathfrak{P}, \{\mathfrak{F}_t\})$, with $\Omega = \Omega^H \times \Omega^B$, and where Ω^B supports the Brownian motion.

In case of competing enablings, the following rules apply:

R_0 The firing of an immediate transition has priority over the firing of a guard or a delay transition.

R_1 If one transition becomes enabled by two or more disjoint sets of input tokens at exactly the same time, then it will fire these sets of tokens independently, at the same time.

R_2 If one transition becomes enabled by two or more non-disjoint sets of input tokens at exactly the same time, then the set that is fired is selected randomly.

R_3 If two or more transitions become enabled at exactly the same time by disjoint sets of input tokens, then they will fire at the same time.

R_4 If two or more transitions become enabled at exactly the same time by non-disjoint sets of input tokens, then the transition that will fire is selected randomly.

Here, two sets of input tokens are disjoint if they have no tokens in common that are reserved by ordinary arcs, i.e., they may have tokens in common that are reserved by enabling arcs.

If T is enabled, suppose this occurs at time τ_1, it removes one token per arc in $A_{in,O}(T)$ from each of its input places. At this time τ_1, T produces zero or one token along each output arc: If c_{τ_1} is the vector of colours of tokens that enabled T and (f, a_{τ_1}) is a sample from $\mathcal{F}_T(\cdot; c_{\tau_1})$, then vector f specifies along which of the output arcs of T a token is produced (f holds a one at the corresponding vector components and a zero at the arcs along which no token is produced) and a_{τ_1} specifies the colours of the produced tokens. The colours of the new tokens have sample paths that start at time τ_1.

For drawing the sample from $\mathcal{F}_T(\cdot; c_{\tau_1})$, again use is made of the Hilbert cube Ω^H: Let $\psi_2^T : [0,1] \times \mathcal{C}(P(A_{in,OE}(T))) \to \{0,1\}^{|A_{out}(T)|} \times \mathcal{C}(P(A_{out}(T)))$ be a measurable function such that $l\{u : \psi_2^T(u,c) \in B\} = \mathcal{F}_T(B,c)$ for B in the Borel set of $\{0,1\}^{|A_{out}(T)|} \times \mathcal{C}(P(A_{out}(T)))$. Then a sample from $\mathcal{F}_T(\cdot; c_{\tau_1})$ is given by $\psi_2^T(U_2(\omega), c_{\tau_1})$, if c_{τ_1} is the vector of input colours that enabled T.

In order to keep track of the identity of individual tokens, the tokens in a place are ordered according to the time at which they entered the place, or, if several tokens are produced for one place at the same time, according to the order within the set of arcs $\mathcal{A} = \{A_1, \ldots, A_{|\mathcal{A}|}\}$ along which these tokens were produced (the firing measure produces zero or one token along each output arc).

3.3 SDCPN Stochastic Process

The SDCPN generates a stochastic process which is uniquely defined as follows: The process state at time t is defined by the numbers of tokens in each place, and the colours of these tokens. Provided there is a unique ordering of SDCPN places, and a unique ordering of tokens within a place, this characterisation is unique, except at time instants when one or more transitions fire. To make this characterisation of SDCPN process state unique, it is defined as follows:

- At times t when no transition fires, the number of tokens in each place is uniquely characterised by the vector $(v_{1,t}, \ldots, v_{|\mathcal{P}|,t})$ of length $|\mathcal{P}|$, where $v_{i,t}$ denotes the number of tokens in place P_i at time t and $\{1, \ldots, |\mathcal{P}|\}$ refers to a unique ordering of places adopted for SDCPN. At time instants when one or more transitions fire, uniqueness of $(v_{1,t}, \ldots, v_{|\mathcal{P}|,t})$ is assured as follows: Suppose that τ is such time instant at which one transition or a sequence of transitions fires. Next, assume without loss of generality, that this sequence of transitions is $\{T_1, T_2, \ldots, T_m\}$ and that time is running again after T_m (note that T_1 must be a guard or a delay transition, and T_2 through T_m must be immediate transitions). Then the number of tokens in each place at time t is defined as that vector $(v_{1,t}, \ldots, v_{|\mathcal{P}|,t})$ that occurs after T_m has fired. This construction also ensures that the process $(v_{1,t}, \ldots, v_{|\mathcal{P}|,t})$ has limits from the left and is continuous from the right, i.e., it satisfies the càdlàg property.

- If $(v_{1,t}, \ldots, v_{|\mathcal{P}|,t})$ is the distribution of the tokens among the places of the SDCPN at time t, which is uniquely defined above, then the associated colours of these tokens are uniquely gathered in a vector as follows: This vector first contains all colours of tokens in place P_1, next all colours of tokens in place P_2, etc, until place $P_{|\mathcal{P}|}$, where $\{1, \ldots, |\mathcal{P}|\}$ refers to a unique ordering of places adopted for SDCPN. Within a place the colours of the tokens are ordered according to the unique ordering of tokens within their place defined for SDCPN (see under SDCPN execution above). Since $(v_{1,t}, \ldots, v_{|\mathcal{P}|,t})$ satisfies the càdlàg property, the corresponding vector of token colours does too. An additional case occurs, however, when $(v_{1,t}, \ldots, v_{|\mathcal{P}|,t})$ jumps to the same value again, so that only the process associated with the vector of token colours makes a jump at time τ. In that case, let the process associated with the vector of token colours be defined according to the timing construction as described for $(v_{1,t}, \ldots, v_{|\mathcal{P}|,t})$ above (i.e. at time τ, the process associated with the vector of token colours is defined as that vector of token colours that occurs after the last transition has fired in the sequence of transitions that fire at time τ).

  With this, the SDCPN definition is complete.

4 Generalised Stochastic Hybrid Processes into Stochastically and Dynamically Coloured Petri Nets

This section shows that each Generalised Stochastic Hybrid Process can be represented by a Stochastically and Dynamically Coloured Petri Net, by providing a pathwise equivalent into-mapping from GSHP into the set of SDCPN processes.

Theorem 1. *For any arbitrary Generalised Stochastic Hybrid Process with a finite domain* $\mathbf{K}$ *there exists P-almost surely a pathwise equivalent process generated by a Stochastically and Dynamically Coloured Petri Net* ($\mathcal{P}$, $\mathcal{T}$, $\mathcal{A}$, $\mathcal{N}$, $\mathcal{S}$, $\mathcal{C}$, $\mathcal{I}$, $\mathcal{V}$, $\mathcal{W}$, $\mathcal{G}$, $\mathcal{D}$, $\mathcal{F}$) *satisfying* R_0 *through* R_4.

Proof. Consider an arbitrary GSHP $\{\theta_t, x_t\}$ described by the GSHS elements $\{\mathbf{K}, d(\theta), x_0, \theta_0, \partial E_\theta, g_\theta, \lambda, Q\}$.

First, we construct a SDCPN, the elements $\{\mathcal{P}, \mathcal{T}, \mathcal{A}, \mathcal{N}, \mathcal{S}, \mathcal{C}, \mathcal{I}, \mathcal{V}, \mathcal{W}, \mathcal{G}, \mathcal{D}, \mathcal{F}\}$ and the rules $R_0 - R_4$ of which are characterised in terms of the GSHS elements $\{\mathbf{K}, d(\theta), x_0, \theta_0, \partial E_\theta, g_\theta, \lambda, Q\}$ as follows:

$\mathcal{P} = \{P_\theta; \theta \in \mathbf{K}\}$. Hence, for each $\theta \in \mathbf{K}$ there is one place P_θ.
$\mathcal{T} = \mathcal{T}_G \cup \mathcal{T}_D \cup \mathcal{T}_I$, with $\mathcal{T}_I = \emptyset$, $\mathcal{T}_G = \{T_\theta^G; \theta \in \mathbf{K}\}$, $\mathcal{T}_D = \{T_\theta^D; \theta \in \mathbf{K}\}$. Hence, for each place P_θ there is one guard transition T_θ^G and one delay transition T_θ^D.
$\mathcal{A} = \mathcal{A}_O \cup \mathcal{A}_E \cup \mathcal{A}_I$, with $|\mathcal{A}_I| = 0$, $|\mathcal{A}_E| = 0$, and $|\mathcal{A}_O| = 2|\mathbf{K}| + 2|\mathbf{K}|^2$.

$\mathcal{N}$: The node function maps each arc in $\mathcal{A} = \mathcal{A}_O$ to a pair of nodes. These connected pairs of nodes are: $\{(P_\theta, T_\theta^G); \theta \in \mathbf{K}\} \cup \{(P_\theta, T_\theta^D); \theta \in \mathbf{K}\} \cup \{(T_\theta^G, P_\vartheta); \theta, \vartheta \in \mathbf{K}\} \cup (T_\theta^D, P_\vartheta); \theta, \vartheta \in \mathbf{K}\}$. Hence, each place P_θ has two outgoing arcs: one to guard transition T_θ^G and one to delay transition T_θ^D. Each transition has $|\mathbf{K}|$ outgoing arcs: one arc to each place in $\mathcal{P}$.

$\mathcal{S} = \{\mathbb{R}^{d(\theta)}; \theta \in \mathbf{K}\}$.

$\mathcal{C}$: For all $\theta \in \mathbf{K}$, $\mathcal{C}(P_\theta) = \mathbb{R}^{d(\theta)}$.

$\mathcal{I}$: Place P_{θ_0} contains one token with colour x_0. All other places initially contain zero tokens.

$\mathcal{V}$: For all $\theta \in \mathbf{K}$, $\mathcal{V}_{P_\theta}(\cdot) = g_\theta(\cdot)$.

$\mathcal{W}$: For all $\theta \in \mathbf{K}$, $\mathcal{W}_{P_\theta}(\cdot) = g_\theta^w(\cdot)$.

$\mathcal{G}$: For all $\theta \in \mathbf{K}$, $\partial G_{T_\theta^G} = \partial E_\theta$.

$\mathcal{D}$: For all $\theta \in \mathbf{K}$, $\delta_{T_\theta^D}(\cdot) = \lambda(\theta, \cdot)$. Moreover, for the evaluation of the SDCPN survivor functions, the same Hilbert cube applies as the one applied by the GSHP.

$\mathcal{F}$: If x denotes the colour of the token removed from place P_θ, $(\theta \in \mathbf{K})$, at the transition firing, then for all $\vartheta' \in \mathbf{K}$, $x' \in E_{\vartheta'}$: $\mathcal{F}_{T_\theta^G}(e', x'; x) = Q(\vartheta', x'; \theta, x)$, where e' is the vector of length $|\mathbf{K}|$ containing a one at the component corresponding with arc $(T_\theta^G, P_{\vartheta'})$ and zeros elsewhere. For all $\theta \in \mathbf{K}$, $\mathcal{F}_{T_\theta^D} = \mathcal{F}_{T_\theta^G}$. Moreover, for the evaluation of the SDCPN firing, the same Hilbert cube applies as the one applied by the GSHP.

$R_0 - R_4$: Since there are no immediate transitions in the constructed SDCPN instantiation, rule R_0 holds true. Since there is only one token in the constructed SDCPN instantiation, $R_1 - R_3$ also hold true. Rule R_4 is in effect when for particular θ, transitions T_θ^G and T_θ^D become enabled at exactly the same time. Since λ is integrable, the probability that this occurs is zero, yielding that R_4 holds with probability one. However, if this event should occur, then due to the fact that the firing measures for the guard transition and the delay transition are equal, the application of rule R_4 has no effect on the path of the SDCPN process.

This shows that for any GSHS we are able to construct a SDCPN instantiation. Next, we have to show that the SDCPN execution delivers the 'same' cadlag stochastic process as the GSHS execution does.

In the SDCPN instantiation constructed, initially there is one token in place P_{θ_0}. Because each transition firing removes one token and produces one token, the number of tokens does not change for $t > 0$. Hence, for $t > 0$ there is one token and the possible places for this single token are $\{P_\vartheta; \vartheta \in \mathbf{K}\}$. Figure 2 shows the situation at some time τ_{k-1}, when the GSHP is given by $(\theta_{\tau_{k-1}}, x_{\tau_{k-1}})$. The token resides in place P_{ϑ_i}, which models that $\theta_{\tau_{k-1}} = \vartheta_i$. This token has colour $x_{\tau_{k-1}}$. The colour of the token up to and at the time of the next jump is evaluated according to two steps that are similar to those of GSHP:

Step 1: While the token is residing in place P_{ϑ_i}, its colour x_t changes according to the stochastic flow $\phi_{\vartheta_i, x_{\tau_{k-1}}, t-\tau_{k-1}}$, i.e., $x_t = \phi_{\vartheta_i, x_{\tau_{k-1}}, t-\tau_{k-1}}$ de-

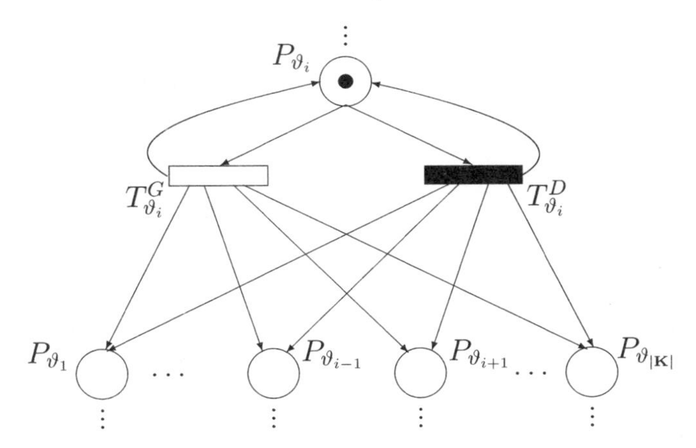

Fig. 2. Part of a Stochastically and Dynamically Coloured Petri Net representing a Generalised Stochastic Hybrid Process

fined on the complete probability space $(\Omega, \mathfrak{F}, \mathfrak{P}, \{\mathfrak{F}_t\})$. Transitions $T^G_{\vartheta_i}$ and $T^D_{\vartheta_i}$ are both pre-enabled and compete for this token which resides in their common input place P_{ϑ_i}. Transition $T^G_{\vartheta_i}$ models the boundary hitting generating a mode switch, while transition $T^D_{\vartheta_i}$ models the Poisson process generating a mode switch. For this, use is made of a random sample from the Hilbert cube. The transition that is enabled first, determines the kind of switch occurring. The time at which this happens is denoted by τ_k.

Step 2: With one, or more (has probability zero), of the transitions enabled at time τ_k, its firing measure is evaluated. For this, use is made of a random sample from the Hilbert cube. The firing measure is such, that if a sample ζ_k from transition measure $Q(\cdot; \vartheta_i, \phi_{\vartheta_i, x_{\tau_{k-1}}, \tau_k - \tau_{k-1}})$, would appear to be $\zeta_k = (\vartheta_j, x)$, then the enabled transition would produce one token with colour $x_{\tau_k} = x$ for place P_{ϑ_j}. The other places get no token.

After this, the above two steps are repeated in the same way from the new state on. The pathwise equivalence of the GSHP and SDCPN processes can be shown from the first stopping time to the next stopping time, and so on. From stopping time to stopping time both processes use the same independent realisations of the random variables U_1, U_2, ..., each having uniform $[0,1]$ distribution, defined by $U_i(\omega) = \omega_i$ for elements $\omega = (\omega_1, \omega_2, \ldots)$ of the Hilbert cube $\Omega^H = \prod_{i=1}^{\infty} Y_i$, with Y_i a copy of $Y = [0,1]$, to generate all random variables in both the GSHP process and the SDCPN process. Hence, from stopping time to stopping time, the GSHP and the associated SDCPN process have equivalent paths and equivalent stopping times. □

5 Stochastically and Dynamically Coloured Petri Nets into Generalised Stochastic Hybrid Processes

Under some conditions, each Stochastically and Dynamically Coloured Petri Net can be represented by a Generalised Stochastic Hybrid Process. In this section this is shown by providing an into-mapping from SDCPN into the set of GSHPs.

Theorem 2. *For each stochastic process generated by a Stochastically and Dynamically Coloured Petri Net* $(\mathcal{P}, \mathcal{T}, \mathcal{A}, \mathcal{N}, \mathcal{S}, \mathcal{C}, \mathcal{I}, \mathcal{V}, \mathcal{W}, \mathcal{G}, \mathcal{D}, \mathcal{F})$ *satisfying* R_0 *through* R_4 *there exists a unique probabilistically equivalent Generalised Stochastic Hybrid Process if the following conditions are satisfied:*

D_1 *There are no explosions, i.e. the time at which a token colour equals* $+\infty$ *or* $-\infty$ *approaches infinity whenever the time until the first guard transition enabling moment approaches infinity.*
D_2 *After a transition firing (or after a sequence of firings that occur at the same time instant) at least one place must contain a different number of tokens, or the colour of at least one token must have jumped*
D_3 *In a finite time interval, each transition is expected to fire a finite number of times, and for* $t \to \infty$ *the number of tokens remains finite.*
D_4 *The initial marking is such, that no immediate transition is initially enabled.*

Proof. For an arbitrary SDCPN that satisfies conditions D_1 – D_4, we first construct a GSHP that is probabilistically equivalent to the SDCPN process. As a preparatory step, the given SDCPN is enlarged as follows: for each guard transition and each place from which that guard transition may be enabled, copy the corresponding places and transitions, including guards and firing measures, and revise the firing measures of the input transitions to these places, such that the new firings ensure that the corresponding guard transitions may be reached from one side only. This step is illustrated with an example:

Example 1. In the picture on the left in Figure 3, transition T_1 (which may be of any type) may fire tokens to place P_1, while transition T_2 is a guard transition that uses these tokens as input. In this example, assume that $\mathcal{C}(P_1) = \mathbb{R}$ and that $\partial G_{T_2} = 3$. This means, transition T_2 is enabled if the colour of the token in place P_1 reaches value 3. This value may be reached from above or from below, depending on whether the initial colour of the token in P_1 is larger or smaller than 3, respectively.

In the picture on the right, place P_1 and transition T_2 have been copied. Transitions T_{2a} and T_{2b} get the same guard as T_2, but transition T_1' gets a new firing measure with respect to T_1: it is similar to the one of T_1, but it delivers a token to place P_{1a} if the colour of this new token is smaller than 3, and it delivers a token to place P_{1b} if its colour is larger than 3. This way, the

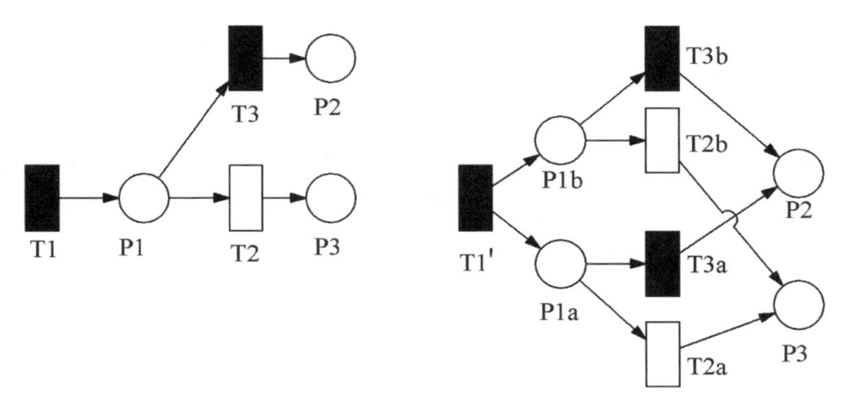

Fig. 3. Example transformation to model SDCPN enlargement

guard of transition T_{2a} is always reached from below, i.e., its input colours are smaller than 3. The guard of transition T_{2b} is always reached from above, i.e., its input colours are larger than 3. The second output transition T_3 of place P_1 also needs to be copied, but the output place of these copies can remain the same as before. (*End of Example*)

(Continuation of proof.) Let this enlarged SDCPN be described by the tuple $(\mathcal{P}, \mathcal{T}, \mathcal{A}, \mathcal{N}, \mathcal{S}, \mathcal{C}, \mathcal{I}, \mathcal{V}, \mathcal{W}, \mathcal{G}, \mathcal{D}, \mathcal{F})$ and satisfy the rules $R_0 - R_4$, and assume that the conditions $D_1 - D_4$ are satisfied. In order to represent this SDCPN by a GSHP, all GSHS elements $\mathbf{K}$, $d(\theta)$, x_0, θ_0, g_θ, g_θ^w, ∂E_θ, λ, Q and the GSHS conditions $C_1 - C_4$ are characterised in terms of this SDCPN:

K: The domain $\mathbf{K}$ for the mode process $\{\theta_t\}$ can be found from the reachability graph (RG) of the SDCPN graph. The nodes in the RG are vectors $V = (v_1, \ldots, v_{|\mathcal{P}|})$, where v_i equals the number of tokens in place P_i, $i = 1, \ldots, |\mathcal{P}|$, where these places are uniquely ordered. The RG is constructed from SDCPN components $\mathcal{P}$, $\mathcal{T}$, $\mathcal{A}$, $\mathcal{N}$ and $\mathcal{I}$. The first node V_0 is found from $\mathcal{I}$, which provides the numbers of tokens initially in each of the places[2]. From then on, the RG is constructed as follows: If it is possible to move in one jump from token distribution V_0 to, say, either one of distributions $V^1, \ldots, V^k$ unequal to V_0, then arrows are drawn from V_0 to (new) nodes $V^1, \ldots, V^k$. Each of $V^1, \ldots, V^k$ is treated in the same way. Each arrow is labelled by the (set of) transition(s) fired at the jump. If a node V^j can be directly reached from V^i by different (sets of) transitions firing, then multiple arrows are drawn from V^i to V^j, each labelled by another (set) of transition(s). Multiple arrows are also drawn if V^j can be directly reached from V^i by firing of one transition, but by different sets of tokens, for example in case this transition has multiple input tokens

[2] Notice that $\mathbf{K}$ has to be constructed for all $\mathcal{I}$ by following the proposed procedure such that is applies for each possible instantiation of the initial token distribution.

per incoming arc in its input places. In this case, the multiple arrows each get this transition as label.
The nodes in the resulting reachability graph, *exclusive* the nodes from which an immediate transition is enabled, form the discrete domain **K** of the GSHP. To emphasise these nodes from which an immediate transition is enabled in the RG picture, they are given in *italics*. Since the number of places in the SDCPN is finite and the number of tokens per place and the number of nodes in the RG are countable, **K** is a countable set, which satisfies the GSHS conditions.

Example 2. As an example, consider the SDCPN graph in Figure 4, which first is enlarged as explained above; the result is Figure 5. The enlarged graph initially has two tokens in place P_{1a} and one in P_3, and the unique ordering of places is $(P_{1a}, P_{1b}, P_2, P_3, P_4)$ such that $V_0 = (2, 0, 0, 1, 0)$. This vector forms the first node of the reachability graph.

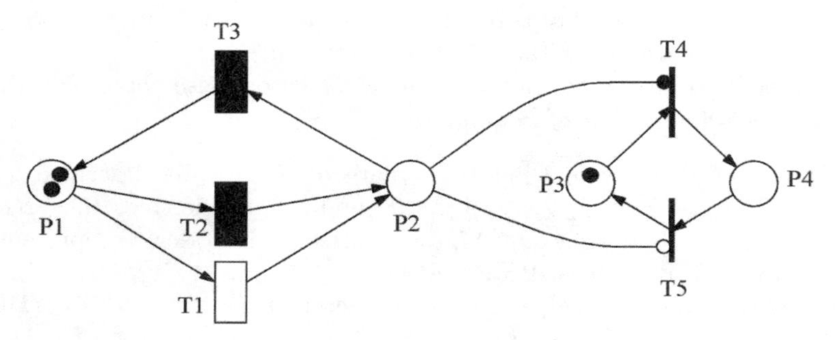

Fig. 4. Example SDCPN to explain reachability graph

Both T_{1a} and T_{2a} are pre-enabled. They both have two tokens per incoming arc in their input place, hence for both transitions, two vectors of input colours are evaluated in parallel. If T_{1a} becomes enabled for one of these input tokens, it removes the corresponding token from P_{1a} and produces a token for P_2 (we assume that all firing measures are such, that each transition will fire a token when enabled, i.e., $\mathcal{F}_T(0, \cdot; \cdot) = 0$), so the new token distribution is $(1, 0, 1, 1, 0)$. Therefore, in the reachability graph two arcs labelled by T_{1a} are drawn from $(2, 0, 0, 1, 0)$ to the new node $(1, 0, 1, 1, 0)$; this duplication of arcs characterises that T_{1a} has evaluated two vectors of input tokens in parallel. The same reasoning holds for transition T_{2a}: two arcs are drawn from $(2, 0, 0, 1, 0)$ to $(1, 0, 1, 1, 0)$. It may also happen that from $(2, 0, 0, 1, 0)$, the guard transition T_{1a} is enabled by its two input tokens at exactly the same time. Due to Rule R_1 it then fires these two tokens at exactly the same time, resulting in node $(0, 0, 2, 1, 0)$. Therefore, an additional arc labelled $T_{1a} + T_{1a}$ is drawn from $(2, 0, 0, 1, 0)$ to $(0, 0, 2, 1, 0)$. Unlike the case for T_{1a}, there is no arc

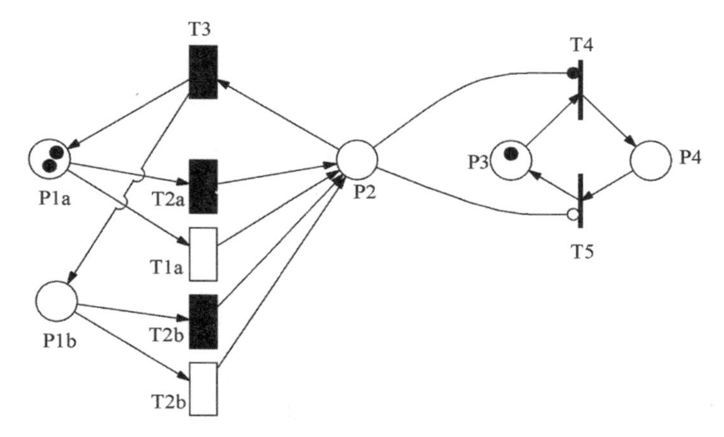

Fig. 5. Example enlarged SDCPN to explain reachability graph

drawn from $(2,0,0,1,0)$ labelled by $T_{2a} + T_{2a}$, since T_{2a} is a delay transition, hence the probability that it is enabled by both its input tokens at the same time is zero. Now consider node $(0,0,2,1,0)$. From this token distribution the immediate transition T_4 is enabled; its firing leads to $(1,0,1,0,1)$. Since node $(1,0,1,1,0)$ enables an immediate transition it is drawn in italics and is excluded from $\mathbf{K}$.
The resulting reachability graph for this example is given in Figure 6. So, for this example, $\mathbf{K} = \{(2,0,0,1,0),\ (0,0,2,0,1),\ (1,0,1,0,1),\ (0,1,1,0,1),\ (1,1,0,1,0),\ (0,2,0,1,0)\}$. (*End of Example*)

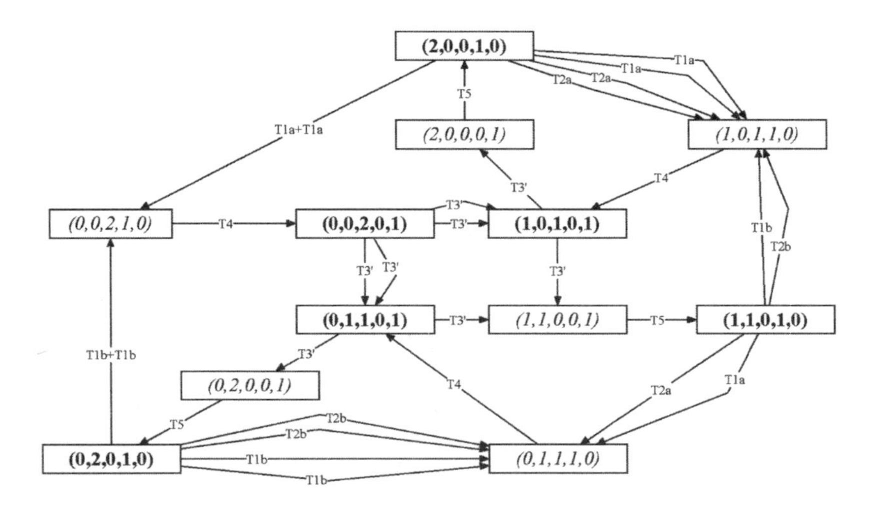

Fig. 6. Example reachability graph

(Continuation of proof.)

$d(\theta)$: The colour of a token in a place P is an element of $\mathcal{C}(P) = \mathbb{R}^{n(P)}$, therefore $d(\theta) = \sum_{i=1}^{|\mathcal{P}|} \theta_i \times n(P_i)$, with $\theta = (\theta_1, \ldots, \theta_{|\mathcal{P}|}) \in \mathbf{K}$, with $\{1, \ldots, |\mathcal{P}|\}$ referring to the unique ordering of places adopted for the SDCPN.

g_θ and g_θ^w: For $x = \text{Col}\{x^1, \ldots, x^{|\mathcal{P}|}\}$, with $x^i \in \mathbb{R}^{\theta_i \times n(P_i)}$, and with $\{1, \ldots,$ $|\mathcal{P}\}$ referring to the unique ordering of places adopted for the SDCPN, g_θ is defined by $g_\theta(x) = \text{Col}\{g_\theta^1(x^1), \ldots, g_\theta^{|\mathcal{P}|}(x^{|\mathcal{P}|})\}$, where for $x^i = \text{Col}\{x^{i1},$ $\ldots, x^{i\theta_i}\}$, with $x^{ij} \in \mathbb{R}^{n(P_i)}$ for all $j \in \{1, \ldots, \theta_i\}$: $g_\theta^i(x^i) = \text{Col}\{\mathcal{V}_{P_i}(x^{i1}),$ $\ldots, \mathcal{V}_{P_i}(x^{i\theta_i})\}$. Here, $j \in \{1, \ldots, \theta_i\}$ refers to the unique ordering of tokens within their place defined for SDCPN (see Section 3). In a similar way, g_θ^w is defined by $g_\theta^w(x) = \text{Diag}\{g_\theta^{w,1}(x^1), \ldots, g_\theta^{w,|\mathcal{P}|}(x^{|\mathcal{P}|})\}$. Since, for all P_i, $\mathcal{V}_{P_i}$ and $\mathcal{W}_{P_i}$ satisfy conditions that ensure existence of a pathwise unique solution without explosion, this also applies to g_θ and g_θ^w.

∂E_θ: For each token distribution θ, the boundary ∂E_θ of subset E_θ is determined from the transition guards corresponding with the set of transitions in $\mathcal{T}_G$ that, under token distribution θ, are pre-enabled (this set is uniquely determined). Without loss of generality, suppose this set of transitions is $T_1, \ldots, T_m$ (note that this set may contain one transition multiple times, if multiple tokens are evaluated in parallel). Suppose $\{P^{i1}, \ldots, P^{ir_i}\}$ are the input places of T_i that are connected to T_i by means of ordinary or enabling arcs. Define $d_i = \sum_{j=1}^{r_i} n(P^{ij})$, then $\partial E_\theta = \partial G'_{T_1} \cup \ldots \cup \partial G'_{T_m}$, where $G'_{T_i} = [G_{T_i} \times \mathbb{R}^{d(\theta)-d_i}] \in \mathbb{R}^{d(\theta)}$. Here $[\cdot]$ denotes a special ordering of all vector elements: Vector elements corresponding with tokens in place P_a are ordered before vector elements corresponding with tokens in place P_b if $b > a$, according to the unique ordering of places adopted for the SDCPN; vector elements corresponding with tokens within one place are ordered according to the unique ordering of tokens within their place defined for SDCPN (see Section 3). If the set of pre-enabled guard transitions is empty, then $\partial E_\theta = \emptyset$.

λ: For each token distribution θ, the jump rate $\lambda(\theta, \cdot)$ is determined from the transition delays corresponding with the set of transitions in $\mathcal{T}_D$ that, under token distribution θ, are pre-enabled (this set is uniquely determined). Without loss of generality, suppose this set of transitions is $T_1, \ldots, T_m$. Then $\lambda(\theta, \cdot) = \sum_{i=1}^{m} \delta_{T_i}(\cdot)$. This equality is due to the fact that the combined arrival process of individual Poisson processes is again Poisson, with an arrival rate equal to the sum of all individual arrival rates. Since δ_T is integrable for all $T \in \mathcal{T}_D$, λ is also integrable. If the set of pre-enabled delay transitions is empty, then $\lambda(\theta, \cdot) = 0$.

Q: For each $\theta \in \mathbf{K}$, $x \in E_\theta$, $\theta' \in \mathbf{K}$ and $x' \in E_{\theta'}$, $Q(\theta', x'; \theta, x)$ is characterised by the reachability graph, the sets $\mathcal{D}$, $\mathcal{G}$ and $\mathcal{F}$ and the rules $R_0 - R_4$. The reachability graph is used to determine which transitions are pre-enabled in token distribution θ; the sets $\mathcal{D}$ and $\mathcal{G}$ and the rules $R_0 - R_4$ are used to determine which pre-enabled transitions will actually fire from state (θ, x); and finally, set $\mathcal{F}$ is used to determine the probability of (θ', x')

being the state after the jump, given state (θ, x) before the jump and the set of transitions that will fire in the jump. Because of its complexity, the characterisation of Q is given in the appendix, but an outline is given next:
Main challenge in the characterisation of Q is the following: In some situations one does not know for certain which transitions will fire in a jump, even if one knows the state (θ, x) before the jump and knows that a jump will occur from (θ, x) to (θ', x'). Hence, in these situations it is not known with certainty which firing measures one should combine in order to construct $Q(\theta', x'; \theta, x)$ from SDCPN elements. However, one does know the following:

- Given θ, one knows which transitions are pre-enabled; this can be read off the reachability graph (i.e. gather the labels of all arrows leaving node θ).
- Given that $\theta \in \mathbf{K}$, no immediate transitions are enabled in θ.
- The probability that a guard transition and a delay transition are enabled at exactly the same time is zero.
- The probability that two delay transitions are enabled at exactly the same time is zero.
- There is a possibility that two or more guard transitions are enabled at exactly the same time. It may even occur (due to rule R_1) that one single guard transition fires twice at the same time.

Hence, the steps to be followed to construct $Q(\theta', x'; \theta, x)$, for any (θ', x', θ, x) are:

1. Determine (using the reachability graph) which transitions are pre-enabled in θ.
2. Consider the guard transitions in this set of pre-enabled transitions and determine which of these are enabled. For a transition T, this is done by considering its vector of input colours (which is part of x) and checking whether this vector has entered the boundary ∂G_T. If the set of enabled guard transitions is not empty, then use rules $R_1 - R_4$ to find out which of these transitions will actually fire with which probability.
   If this set of enabled guard transitions is empty, then one pre-enabled delay transition must be enabled. Use $\mathcal{D}$ to determine for each pre-enabled delay transition the probability with which it will actually fire.
3. Determine which transition firings can actually lead to discrete process state θ' in one jump. This set can be found by identifying in the reachability graph all arrows directly from node θ to θ' and all directed paths from node θ to θ' that pass only nodes that enable immediate transitions (i.e. that pass only nodes in italics).
4. Finally, $Q(\theta', x'; \theta, x)$ is constructed from the firing measures, by conditioning on these arrows and paths from θ to θ'.

θ_0 and x_0: These can be constructed from $\mathcal{I}$, the SDCPN initial marking, which provides the places the tokens are initially in and the colours these tokens have. Hence, $\theta_0 = (v_{1,0}, \dots, v_{|\mathcal{P}|,0})$, where $v_{i,0}$ denotes the initial number of tokens in place P_i, with the places ordered according to the unique ordering adopted for SDCPN, and $x_0 \in I\!\!R^{d(\theta_0)}$ is a vector containing the colours of these tokens. Within a place the colours of the tokens are ordered according to the specification in $\mathcal{I}$. With this, and due to condition D_4 (which prevents different token distributions to be applicable at the initial time), the constructed θ_0 and x_0 are uniquely defined.

C_1: This condition (no explosions) follows from assumption D_1.

C_2: This condition (λ is integrable) follows from the fact that δ_T is integrable for all $T \in \mathcal{T}_D$.

C_3: This condition (Q measurable and $Q(\{\xi\};\xi) = 0$) follows from the assumption that $\mathcal{F}$ is continuous and from assumption D_2.

C_4: This condition ($I\!\!EN_t < \infty$) follows from assumption D_3.

This shows that for any SDCPN satisfying conditions $D_1 - D_4$, we are able to construct unique GSHS elements, and thus a unique GSHS.

Finally, we show that the GSHP process $\{\theta_t, x_t\}$ is probabilistically equivalent to the process generated by the SDCPN:

With the mapping from SDCPN elements into GSHS elements, it is easily shown that the GSHP process $\{\theta_t, x_t\}$ is probabilistically equivalent to the process generated by the SDCPN characterised in Section 3: at each time t the process $\{\theta_t\}$ is probabilistically equivalent to the process $(v_{1,t}, \dots, v_{|\mathcal{P}|,t})$ and the process $\{x_t\}$ is probabilistically equivalent to the process associated with the vector of token colours. This is shown by observing that the initial GSHP state (θ_0, x_0) is probabilistically equivalent to the initial SDCPN state through the mapping constructed above. Moreover, also by the unique mapping of SDCPN elements into GSHS elements, at each time instant after the initial time, the GSHP state is probabilistically equivalent to the SDCPN state: At times t when no jump occurs, the GSHP process evolves according to g_θ and g_θ^w and the SDCPN process evolves according to $\mathcal{V}$ and $\mathcal{W}$. Through the mapping between g_θ and $\mathcal{V}$ and between g_θ^w and $\mathcal{W}$ developed above, these evolutions provide probabilistically equivalent processes. At times when a jump occurs, the GSHP process makes a jump generated by Q, while the SDCPN process makes a jump generated by $\mathcal{F}$. Through the mapping between Q and $\mathcal{F}$ developed above, these jumps provide probabilistically equivalent processes.

6 Example SDCPN and Mapping to GSHP

This section gives a simple example SDCPN model and its mapping to GSHP of the evolution of an aircraft. First, Subsection 6.1 explains how a SDCPN that models a complex operation is generally constructed in three steps. In order to illustrate these steps, Subsection 6.2 presents a simple example of the evolution of one aircraft. Subsection 6.3 gives a SDCPN that models this aircraft evolution and Subsection 6.4 explains the mapping of this SDCPN example in a GSHP.

6.1 SDCPN Construction and Verification Process

A SDCPN modelling a particular operation can be constructed, for example, by first identifying the discrete state space, represented by the places, the transitions and arcs, and next adding the continuous-time-based elements one by one, similar as what one would expect when modelling a GSHP for such operation. However, in case of a very complex operation, with many entities that interact such as occur in air traffic, it is generally more desirable and constructive to do the SDCPN modelling in several iterations, for example in a four-phased approach:

1. In the first phase, each operation entity or agent (for example, a pilot, a navigation system, an aircraft) is modelled separately by one local DCPN (i.e. no Brownian motion components $\mathcal{W}$). Each such entity model is named a Local Petri Net (LPN).
2. In the second phase, the interactions between these entities are modelled, connecting the LPNs, such that these interactions do not change the number of tokens per LPN.
3. In the third phase the Brownian motion components $\mathcal{W}$ are added to the LPNs.
4. In the fourth phase, one verifies whether the conditions D_1 – D_4 under which a mapping to GSHP is guaranteed to exist have been fulfilled. Because of the modularity and fixed number of tokens per LPN, these conditions can easily be verified per LPN, and subsequently per interaction between LPNs.

The additional advantage of this phased approach is that the total SDCPN can be verified simultaneously by multiple domain experts. For example, a Local Petri Net model for a navigation system can be verified by a navigational system expert; a Local Petri Net model for a pilot can be verified by a human factors expert; interactions can be verified by a pilot.

6.2 Aircraft Evolution Example

This subsection presents a simple aircraft evolution example. The next subsections present a SDCPN model and a mapping to GSHP for this example.

Assume the deviation of this aircraft from its intended path depends on the operationality of two of its aircraft systems: the engine system, and the navigation system. Each of these aircraft systems can be in one of two modes: *Working* (functioning properly) or *Not working* (operating in some failure mode). Both systems switch between their modes independently and on exponentially distributed times, with rates δ_3 (engine repaired), δ_4 (engine fails), δ_5 (navigation repaired) and δ_6 (navigation fails), respectively. The operationality of these systems has the following effect on the aircraft path: if both systems are *Working*, the aircraft evolves in *Nominal* mode and the rate of change of the position and velocity of the aircraft is determined by $(\mathcal{V}_1, \mathcal{W}_1)$ (i.e. if z_t is a vector containing this position and velocity then $dz_t = \mathcal{V}_1(z_t)dt + \mathcal{W}_1 dw_t$). If either one, or both, of the systems is *Not working*, the aircraft evolves in *Non-nominal* mode and the position and velocity of the aircraft is determined by $(\mathcal{V}_2, \mathcal{W}_2)$. The factors $\mathcal{W}_1$ and $\mathcal{W}_2$ are determined by wind fluctuations. Initially, the aircraft has a particular position x_0 and velocity v_0, while both its systems are *Working*. The evaluation of this process may be stopped when the aircraft position has *Landed*, i.e. its vertical position and velocity is equal to zero. Once landed, the aircraft is assumed not to depart anymore, hence the rate of change of its position and velocity equals zero.

This simple aircraft evolution example illustrates the kind of difficulty encountered when one wants to model a realistic problem directly as a GSHP. Mathematically one would define three discrete valued processes $\{\kappa_t^1\}$, $\{\kappa_t^2\}$, $\{\kappa_t^3\}$, and an $\mathbb{R}^6$-valued process $\{x_t\}$:

- $\{\kappa_t^1\}$ represents the aircraft evolution mode assuming values in $\{$*Nominal*, *Non-nominal*, *Landed*$\}$;
- $\{\kappa_t^2\}$ represents the navigation mode assuming values in $\{$*Working*, *Not-working*$\}$;
- $\{\kappa_t^3\}$ represents the engine mode assuming values in $\{$*Working*, *Not-working*$\}$;
- $\{x_t\}$ represents the 3D position and 3D velocity of the aircraft

Unfortunately, the process $\{\kappa_t, x_t\}$, with $\kappa_t = \text{Col}\{\kappa_t^1, \kappa_t^2, \kappa_t^3\}$, is not a GSHP, since some κ_t combinations lead to immediate jumps, which is not allowed for GSHP.

6.3 SDCPN Model for the Aircraft Evolution Example

This subsection gives a SDCPN instantiation that models the aircraft evolution example of the previous subsection. In order to illustrate the three-phased approach of subsection 6.1, we first give the Local Petri Net graphs that have been identified in the first phase of the modelling. The entities identified are: Aircraft evolution, Navigation system, and Engine system. This gives us three Local Petri Nets. The resulting graphs are given in Figure 7.

The interactions between the Engine and Navigation Local Petri Net and the Evolution Local Petri Net are modelled by coupling the Local Petri Nets by additional arcs (and, if necessary, additional places or transitions). Here, removal of a token from one Local Petri Net by a transition of another Local Petri Net is prevented by using enabling arcs instead of ordinary arcs for the interactions. The resulting graph is presented in Figure 8. Notice that transition T_1 has to be replaced by two transitions T_{1a} and T_{1b} in order to

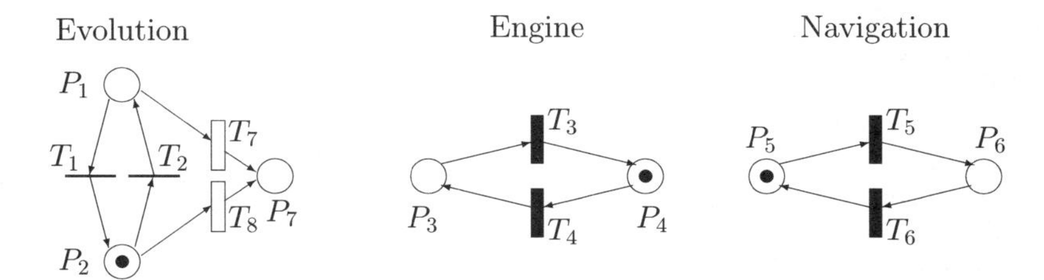

Fig. 7. Local Petri Nets for the aircraft operations example. Place P_1 models Evolution Nominal, P_2 models Evolution Non-nominal, P_3 models Engine system Not working, P_4 models Engine system Working, P_5 models Navigation system Not working, P_6 models Navigation system Working. P_7 models aircraft has landed

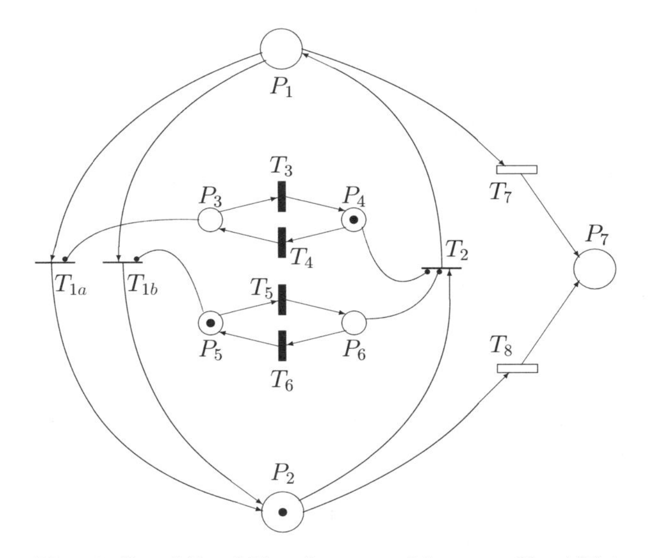

Fig. 8. Local Petri Nets integrated into one Petri Net

allow both the engine and the navigation LPNs to influence transition T_1 separately from each other.

The graph above completely defines SDCPN elements $\mathcal{P}$, $\mathcal{T}$, $\mathcal{A}$ and $\mathcal{N}$, where $\mathcal{T}_G = \{T_7, T_8\}$, $\mathcal{T}_D = \{T_3, T_4, T_5, T_6\}$ and $\mathcal{T}_I = \{T_{1a}, T_{1b}, T_2\}$. The other SDCPN elements are specified below.

- $\mathcal{S}$: Two colour types are defined; $\mathcal{S} = \{\mathbb{R}^0, \mathbb{R}^6\}$.
- $\mathcal{C}$: $\mathcal{C}(P_1) = \mathcal{C}(P_2) = \mathcal{C}(P_7) = \mathbb{R}^6$, hence $n(P_1) = n(P_2) = n(P_7) = 6$. The first three colour components model the longitudinal, lateral and vertical position of the aircraft, the last three components model the corresponding velocities. For places P_3 through P_6, $\mathcal{C}(P_i) = \mathbb{R}^0 = \emptyset$ hence $n(P_i) = 0$.

$\mathcal{I}$: Place P_1 initially has a token with colour $z_0 = (x_0, v_0)'$, with $x_0 \in \mathbb{R}^2 \times (0, \infty)$ and $v_0 \in \mathbb{R}^3 \backslash \text{Col}\{0, 0, 0\}$. Places P_4 and P_6 initially each have a token without colour.

$\mathcal{V}$ and

$\mathcal{W}$: The token colour functions for places P_1, P_2 and P_7 are determined by $(\mathcal{V}_1, \mathcal{W}_1)$, $(\mathcal{V}_2, \mathcal{W}_2)$, and $(\mathcal{V}_7, \mathcal{W}_7)$, respectively, where $(\mathcal{V}_7, \mathcal{W}_7) = (0, 0)$. For places P_3 – P_6 there is no token colour function.

$\mathcal{G}$: Transitions T_7 and T_8 have a guard that is defined by $\partial G_{T_7} = \partial G_{T_8} = \mathbb{R}^2 \times \{0\} \times \mathbb{R}^2 \times \{0\}$.

$\mathcal{D}$: The enabling rates for transitions T_3, T_4, T_5 and T_6 are $\delta_{T_3}(\cdot) = \delta_3$, $\delta_{T_4}(\cdot) = \delta_4$, $\delta_{T_5}(\cdot) = \delta_5$ and $\delta_{T_6}(\cdot) = \delta_6$, respectively.

$\mathcal{F}$: Each transition has a unique output place, to which it fires to their output place a token with a colour (if applicable) equal to the colour of the token removed, i.e. for all T, $\mathcal{F}_T(1, \cdot; \cdot) = 1$.

6.4 Mapping to GSHP

In this subsection, the SDCPN aircraft evolution example is mapped to a GSHP, following the construction in the proof of Theorem 2. Because the boundaries of the guard transitions T_7 and T_8 (i.e. $\partial G_{T_7} = \partial G_{T_8} = \mathbb{R}^2 \times \{0\} \times \mathbb{R}^2 \times \{0\}$) are always reached from one side only, there is no need to first enlarge the SDCPN for these guard transitions (see Section 5).

The SDCPN of Figure 8 has seven places hence the reachability graph has elements that are vectors of length 7. Since there is always one token in the set of places $\{P_1, P_2, P_7\}$, one token in $\{P_3, P_4\}$ and one token in $\{P_5, P_6\}$, the reachability graph has $3 \times 2 \times 2 = 12$ nodes, see Figure 9. However, four nodes are excluded from $\mathbf{K}$: nodes $(1, 0, 1, 0, 0, 1, 0)$, $(0, 1, 0, 1, 0, 1, 0)$ and $(1, 0, 0, 1, 1, 0, 0)$ enable immediate transitions, and node $(1, 0, 1, 0, 1, 0, 0)$ cannot be reached since it requires the enabling of a delay transition that is competing with an immediate transition, while due to SDCPN rule R_0, an immediate transition always gets priority. Therefore, $\mathbf{K}$ consists of the remaining 8 nodes $\{m_1, m_2, m_3, m_4, m_5, m_6, m_7, m_8\}$, which are specified in Table 1.

Table 1. Discrete modes in $\mathbf{K}$

Node	Engine	Navigation	Evolution
$m_1 = (1, 0, 0, 1, 0, 1, 0)$	*Working*	*Working*	*Nominal*
$m_2 = (0, 1, 1, 0, 0, 1, 0)$	*Not working*	*Working*	*Non-nominal*
$m_3 = (0, 1, 1, 0, 1, 0, 0)$	*Not working*	*Not working*	*Non-nominal*
$m_4 = (0, 1, 0, 1, 1, 0, 0)$	*Working*	*Not working*	*Non-nominal*
$m_5 = (0, 0, 0, 1, 0, 1, 1)$	*Working*	*Working*	*Landed*
$m_6 = (0, 0, 1, 0, 0, 1, 1)$	*Not working*	*Working*	*Landed*
$m_7 = (0, 0, 1, 0, 1, 0, 1)$	*Not working*	*Not working*	*Landed*
$m_8 = (0, 0, 0, 1, 1, 0, 1)$	*Working*	*Not working*	*Landed*

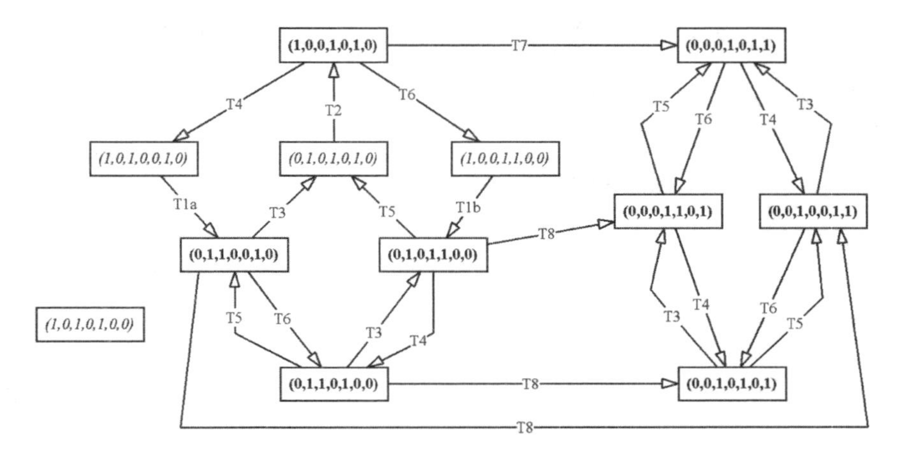

Fig. 9. Reachability graph for the SDCPN of Figure 8

Following Section 5, for each $\theta = (\theta_1, \ldots, \theta_7) \in \mathbf{K}$, the value of $d(\theta)$ equals $d(\theta) = \sum_{i=1}^{|\mathcal{P}|} \theta_i \times n(P_i)$. Since there is always one token in the set of places $\{P_1, P_2, P_7\}$, hence $\theta_1 + \theta_2 + \theta_7 = 1$, and since $n(P_1) = n(P_2) = n(P_7) = 6$ and $n(P_3) = n(P_4) = n(P_5) = n(P_6) = 0$, we find for all θ that $d(\theta) = 6$.

Since initially there is a token in places P_1, P_4 and P_6, the initial mode θ_0 equals $\theta_0 = m_1 = (1, 0, 0, 1, 0, 1, 0)$. The GSHP initial continuous state value equals the vector containing the initial colours of all initial tokens. Since the initial colour of the token in Place P_1 equals z_0, and the tokens in places P_4 and P_6 have no colour, the GSHP initial continuous state value equals z_0.

Following Section 5, with $\theta = (\theta_1, \ldots, \theta_7) \in \mathbf{K}$, for $x = \text{Col}\{x^1, \ldots, x^7\}$, with $x^i \in \mathbb{R}^{\theta_i \times n(P_i)}$, the function g_θ is defined by $g_\theta(x) = \text{Col}\{g_\theta^1(x^1), \ldots, g_\theta^7(x^7)\}$, where for $x^i = \text{Col}\{x^{i1}, \ldots, x^{i\theta_i}\}$, with $x^{ij} \in \mathbb{R}^{n(P_i)}$ for all $j \in \{1, \ldots, \theta_i\}$: $g_\theta^i(x^i)$ satisfies $g_\theta^i(x^i) = \text{Col}\{\mathcal{V}_{P_i}(x^{i1}), \ldots, \mathcal{V}_{P_i}(x^{i\theta_i})\}$. Since there is at most one token in each place, θ_i is either zero or one, hence either $x^i = \emptyset$ or $x^i = x^{i1}$. Since there is no token colour function for places $\{P_3, P_4, P_5, P_6\}$ and there is only one token in $\{P_1, P_2, P_7\}$, $g_\theta(x) = \mathcal{V}_1$ for $\theta = m_1$, $g_\theta(x) = \mathcal{V}_2$ for $\theta \in \{m_2, m_3, m_4\}$, and $g_\theta(x) = 0$ otherwise. In a similar way, $g_\theta^w(x) = \mathcal{W}_1$ for $\theta = m_1$, $g_\theta^w(x) = \mathcal{W}_2$ for $\theta \in \{m_2, m_3, m_4\}$, and $g_\theta^w(x) = 0$ otherwise, see Table 2.

The boundary ∂E_θ is determined from the transitions guards that, under token distribution θ, are enabled. This yields: for $\theta = m_1$, $\partial E_\theta = \partial G_{T_7} = \mathbb{R}^2 \times \{0\} \times \mathbb{R}^2 \times \{0\}$; for $\theta \in \{m_2, m_3, m_4\}$, $E_\theta = \partial G_{T_8} = \mathbb{R}^2 \times \{0\} \times \mathbb{R}^2 \times \{0\}$; for $\theta \in \{m_5, m_6, m_7, m_8\}$, $\partial E_\theta = \emptyset$.

The jump rate $\lambda(\theta, \cdot)$ is determined from the enabling rates corresponding with the set of delay transitions in $\mathcal{T}_D$ that, under token distribution θ, are pre-enabled. At each time, always two delay transitions are pre-enabled: either

T_3 or T_4 and either T_5 or T_6. Hence $\lambda(\theta, \cdot) = \sum_{i=j,k} \delta_{T_i}(\cdot)$ if T_j and T_k are pre-enabled. See Table 2 for the resulting λ's.

The probability measure Q is determined by the reachability graph, the sets $\mathcal{D}$, $\mathcal{G}$ and $\mathcal{F}$ and the rules $R_0 - R_4$. In Table 3, $Q(\zeta; \xi) = p$ denotes that if ξ is the value of the GSHP before the hybrid jump, then, with probability p, ζ is the value of the GSHP immediately after the jump.

Table 2. Example GSHS components $g_\theta(\cdot)$, $g_\theta^w(\cdot)$ and λ as a function of θ

θ	$g_\theta(\cdot)$	$g_\theta^w(\cdot)$	λ
m_1	$\mathcal{V}_1(\cdot)$	$\mathcal{W}_1(\cdot)$	$\delta_4 + \delta_6$
m_2	$\mathcal{V}_2(\cdot)$	$\mathcal{W}_2(\cdot)$	$\delta_3 + \delta_6$
m_3	$\mathcal{V}_2(\cdot)$	$\mathcal{W}_2(\cdot)$	$\delta_3 + \delta_5$
m_4	$\mathcal{V}_2(\cdot)$	$\mathcal{W}_2(\cdot)$	$\delta_4 + \delta_5$
m_5	0	0	$\delta_4 + \delta_6$
m_6	0	0	$\delta_3 + \delta_6$
m_7	0	0	$\delta_3 + \delta_5$
m_8	0	0	$\delta_4 + \delta_5$

Table 3. Example GSHS component Q

For $z \notin \partial E_{m_1}$:	$Q(m_2, z; m_1, z) = \frac{\delta_4}{\delta_4+\delta_6}$, $Q(m_4, z; m_1, z) = \frac{\delta_6}{\delta_4+\delta_6}$
For $z \in \partial E_{m_1}$:	$Q(m_5, z; m_1, z) = 1$
For $z \notin \partial E_{m_2}$:	$Q(m_3, z; m_2, z) = \frac{\delta_6}{\delta_3+\delta_6}$, $Q(m_1, z; m_2, z) = \frac{\delta_3}{\delta_3+\delta_6}$
For $z \in \partial E_{m_2}$:	$Q(m_6, z; m_2, z) = 1$
For $z \notin \partial E_{m_3}$:	$Q(m_4, z; m_3, z) = \frac{\delta_3}{\delta_3+\delta_5}$, $Q(m_2, z; m_3, z) = \frac{\delta_5}{\delta_3+\delta_5}$
For $z \in \partial E_{m_3}$:	$Q(m_7, z; m_3, z) = 1$
For $z \notin \partial E_{m_4}$:	$Q(m_3, z; m_4, z) = \frac{\delta_4}{\delta_4+\delta_5}$, $Q(m_1, z; m_4, z) = \frac{\delta_5}{\delta_4+\delta_5}$
For $z \in \partial E_{m_4}$:	$Q(m_8, z; m_4, z) = 1$
For all z:	$Q(m_6, z; m_5, z) = \frac{\delta_4}{\delta_4+\delta_6}$, $Q(m_8, z; m_5, z) = \frac{\delta_6}{\delta_4+\delta_6}$
For all z:	$Q(m_7, z; m_6, z) = \frac{\delta_6}{\delta_3+\delta_6}$, $Q(m_5, z; m_6, z) = \frac{\delta_3}{\delta_3+\delta_6}$
For all z:	$Q(m_8, z; m_7, z) = \frac{\delta_3}{\delta_3+\delta_5}$, $Q(m_6, z; m_7, z) = \frac{\delta_5}{\delta_3+\delta_5}$
For all z:	$Q(m_7, z; m_8, z) = \frac{\delta_4}{\delta_4+\delta_5}$, $Q(m_5, z; m_8, z) = \frac{\delta_5}{\delta_4+\delta_5}$

From a mathematical perspective, the GSHP model has clear advantages. However, the GSHP model does not show the structure of the SDCPN. Because of this, the SDCPN model of Subsection 6.3 is simpler to comprehend and to verify against the aircraft evolution example description of Subsection 6.2. These complementary advantages from both perspectives tend to increase with the complexity of the operation considered.

7 Conclusions

Generalised Stochastic Hybrid Processes (GSHPs) can be used to describe virtually all complex continuous-time stochastic processes. However, for complex practical problems it is often difficult to develop a GSHP model, and have it verified both by mathematical and by multiple operational domain experts. This paper has introduced a novel Petri Net, which is named Stochastically and Dynamically Coloured Petri Net (SDCPN) and has shown that under some mild conditions, any SDCPN generated process can be mapped into a probabilistically equivalent GSHP. Moreover, it is shown that any GSHP with a finite discrete state domain can be mapped into a pathwise equivalent process which is generated by a executing a GSHS. A consequence of both results is that there exist into-mappings between GSHPs and SDCPN processes. The development of a SDCPN model for complex practical problems has similar specification advantages as basic Petri Nets have over automata [4].

The key result of this paper is that this is the first time that proof of the existence of into-mappings between GSHPs and Petri Nets has been established. This significantly extends the modelling power hierarchy of [14],[15] in terms of Petri Nets and Markov processes, see Figure 10.

To the authors' best knowledge, SDCPN is the only hybrid Petri Net that incorporates Brownian motion. Moreover, SDCPN and DCPN are the only hybrid Petri Nets for which into-mappings with hybrid state Markov processes are known. Due to the existence of these into-mappings, GSHP theoretical results like stochastic analysis, stability and control theory, also apply to SDCPN stochastic processes. The mapping of SDCPN into GSHP implies that any specific SDCPN stochastic process can be analysed as if it is a GSHP, often without the need to first apply the transformation into a GSHP as we did for the aircraft evolution example in Section 6. Because of this, for accident risk modelling in air traffic management, in [2] SDCPNs are adopted for their specification power and for their GSHP inherited stochastic analysis power.

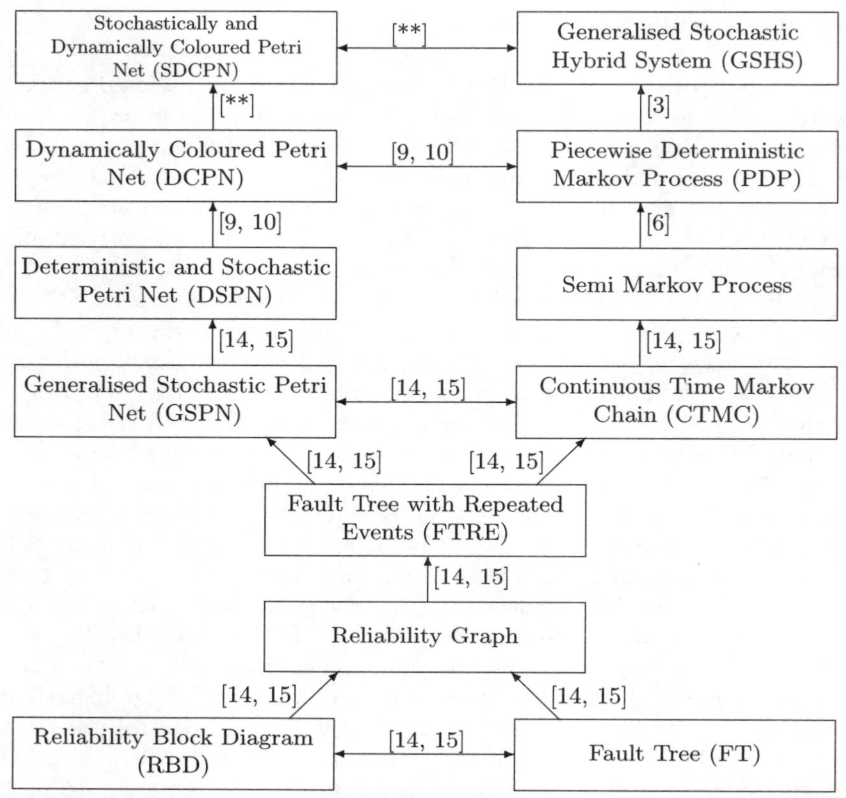

Fig. 10. Power hierarchy among various model types established by [6], [9], [10], [14], [15], [3] and the current paper (denoted by [**]). An arrow from a model to another model indicates that the second model has more modelling power than the first model

References

1. J. Le Bail, H. Alla, and R. David. Hybrid Petri nets. *Eropean Control Conference, Grenoble, France*, pages 1472–1477, 1991.
2. H.A.P. Blom, G.J. Bakker, P.J.G. Blanker, J. Daams, M.H.C. Everdij, and M.B. Klompstra. Accident risk assessment for advanced ATM. 2nd *USA/Europe Air Traffic Management R&D Seminar, Orlando*, 1998. Also in: *Air Transportation Systems Engineering*, AIAA, Eds. G.L. Donohue, A.G. Zellweger, AIAA, pp. 463-480 (2001).
3. M.L. Bujorianu, J. Lygeros, W. Glover, and G. Pola. A stochastic hybrid system modelling framework. Technical report, University of Cambridge and University of L'Aquila, May 2003. Hybridge report D1.2, Also a chapter in this book.
4. C.G. Cassandras and S. Lafortune. *Introduction to Discrete Event Systems.* Kluwer Academic Publishers, 1999.

5. R. David and H. Alla. Petri nets for the modeling of dynamic systems - a survey. *Automatica*, 30(2):175–202, 1994.
6. M.H.A. Davis. Piecewise Deterministic Markov Processes: a general class of non-diffusion stochastic models. *Journal Royal Statistical Soc. (B)*, 46:353–388, 1984.
7. M.H.A. Davis. *Markov models and optimization.* Chapman and Hall, 1993.
8. I. Demongodin and N.T. Koussoulas. Differential Petri nets: Representing continuous systems in a discrete-event world. *IEEE Transactions on Automatic Control*, 43(4), 1998.
9. M.H.C. Everdij and H.A.P. Blom. Petri nets and hybrid state Markov processes in a power-hierarchy of dependability models. *Proc. IFAC Conference on Analysis and Design of Hybrid System (ADHS03), Saint-Malo, Brittany, France*, pages 355–360, June 2003.
10. M.H.C. Everdij and H.A.P. Blom. Piecewise Deterministic Markov Processes represented by Dynamically Coloured Petri Nets. *Stochastics*, 77(1):1–29, February 2005.
11. M.H.C. Everdij, H.A.P. Blom, and M.B. Klompstra. Dynamically Coloured Petri Nets for air traffic management safety purposes. *Proc. 8th IFAC Symposium on Transportation Systems, Chania, Greece*, pages 184–189, 1997.
12. A. Giua and E. Usai. High-level hybrid etri nets: a definition. *Proceedings 35th Conference on Decision and Control, Kobe, Japan*, pages 148–150, 1996.
13. K. Jensen. *Coloured Petri Nets: Basic concepts, analysis methods and practical use*, volume 1. Springer-Verlag, 1992.
14. M. Malhotra and K.S. Trivedi. Power-hierarchy of dependability-model types. *IEEE Transactions on Reliability*, R-43(3):493–502, 1994.
15. J.K. Muppala, R.M. Fricks, and K.S. Trivedi. Techniques for system dependability evaluation. In W. Grasman, editor, *Computational probability*, pages 445–480. Kluwer Academix Publishers, The Netherlands, 2000.
16. K.S. Trivedi and V.G. Kulkarni. FSPNs: Fluid stochastic Petri nets. In M. Ajmone Marsan, editor, *Proceedings 14th International Conference on Applications and theory of Petri Nets*, volume 691 of *Lecture notes in Computer Science*, pages 24–31. Springer Verlag, Heidelberg, 1993.
17. Y.Y. Yang, D.A. Linkens, and S.P. Banks. Modelling of hybrid systems based on extended coloured Petri nets. In P. Antsaklis et al, editor, *Hybrid Systems II*, pages 509–528. Springer, 1995.

A Characterisation of Q in Terms of SDCPN Elements

In this appendix, Q is characterised in terms of SDCPN, as part of the characterisation in Appendix C of GSHP in terms of SDCPN.

For each $\theta \in \mathbf{K}$, $x \in E_\theta$, $\theta' \in \mathbf{K}$ and $A \subset E_{\theta'}$, the value of $Q(\theta', A; \theta, x)$ is a measure for the probability that if a jump occurs, and if the value of the GSHP just prior to the jump is (θ, x), then the value of the GSHP just after the jump is in (θ', A). Measure $Q(\theta', A; \theta, x)$ is characterised in terms of the SDCPN by the reachability graph (RG) (see Appendix C), elements $\mathcal{D}$, $\mathcal{G}$ and Rules $R_0 - R_4$ and the set $\mathcal{F}$, as below. This is done in four steps:

1. Determine which transitions are pre-enabled in (θ, x).

2. Determine for each pre-enabled transition the probability with which it is enabled in (θ, x).
3. Determine for each pre-enabled transition whether its firing can possibly lead to discrete state θ'.
4. Use the results of the previous two steps and the set of firing functions to characterise Q.

Step 1: Determine which transitions are pre-enabled in (θ, x).

Consider all arrows in the RG leaving node θ. These arrows are labelled by names of transitions which are pre-enabled in θ, for example T_1 (if T_1 is pre-enabled in θ), T_1+T_2 (if T_1 and T_2 are both pre-enabled and there is a non-zero probability that they fire at exactly the same time), etc. Therefore the arrows leaving θ may be characterised by these labels. Denote the multi-set of arrows, characterised by these labels, by $\mathcal{B}_\theta$. This set is a multi-set since there may exist several arrows with the same label (e.g. if one transition is pre-enabled by different sets of input tokens). We use notation $B \in \mathcal{B}_\theta$ for an element B of $\mathcal{B}_\theta$ (e.g. $B = T_1$ represents an arrow with T_1 as label), and notation $T \in B$ for a transition T in label B (e.g. as in $B = T + T_1$).

Step 2: Determine for each pre-enabled transition the probability with which it is enabled in (θ, x).

Given that a jump occurs in (θ, x), the set of transitions that will actually fire in (θ, x) is not empty, and is given by one of the labels in $\mathcal{B}_\theta$. In the following, we determine, for all $B \in \mathcal{B}_\theta$, the probability $p_B(\theta, x)$ that all transitions in label B will fire.

- Denote the vector of input colours of transition T in a particular label by c_T^x. For a transition in a label this vector is unique since we consider transitions with multiple vectors of input colours separately in the multi-set $\mathcal{B}_\theta$.
- Consider the multi-set $\mathcal{B}_\theta^G = \{B \in \mathcal{B}_\theta | \forall T \in B : T \in \mathcal{T}_G \text{ and } c_T^x \in \partial G_T\}$.
- If $\mathcal{B}_\theta^G \neq \emptyset$ then this set contains all transitions that are enabled in (θ, x). Rules $R_1 - R_4$ are used (R_0 is not applicable) to determine for each $B \in \mathcal{B}_\theta^G$ the probability with which the transitions in label B will actually fire:
  - Rules R_1 and R_3 are used as follows: if B is such that there exists $B' \in \mathcal{B}_\theta^G$ such that the transitions in B form a real subset of the set of transitions in B', then $p_B(\theta, x) = 0$. The set of thus eliminated labels B is denoted by $\mathcal{B}_\theta^{R_{1,3}}$.
  - Rules R_2 and R_4 are used as follows: If the multi-set $\mathcal{B}_\theta^G - \mathcal{B}_\theta^{R_{1,3}}$ contains m elements, then each of these labels gets a probability $p_B(\theta, x) = 1/m$.

- If $\mathcal{B}_\theta^G = \emptyset$ then only Delay transitions can be enabled in (θ, x). Consider the multi-set $\mathcal{B}_\theta^D = \{B \in \mathcal{B}_\theta | \forall T \in B : T \in \mathcal{T}_D\}$. Each $B \in \mathcal{B}_\theta^D$ consists of one delay transition, with $p_B(\theta, x) = \frac{\delta_B(c_B^x)}{\sum\limits_{T \in \mathcal{B}_\theta^D} \delta_T(c_T^x)}$.

Step 3: Determine for each pre-enabled transition whether its firing can possibly lead to discrete state θ'.

In the RG, consider nodes θ and θ' and delete all other nodes that are elements of $\mathbf{K}$, including the arrows attached to them. Also, delete all nodes and arrows that are not part of a directed path from θ to θ'. The residue is named $\mathrm{RG}_{\theta\theta'}$. Then, if θ and θ' are not connected in $\mathrm{RG}_{\theta\theta'}$ by at least one path, a jump from (θ, x) to a state in (θ', A) is not possible.

Step 4: Use the results of the previous two steps and the set of firing functions to characterise Q.

From the previous step we have

- $Q(\theta', A; \theta, x) = 0$ if θ and θ' are not connected in $\mathrm{RG}_{\theta\theta'}$ by at least one path.

If θ and θ' are connected then in $\mathrm{RG}_{\theta\theta'}$ one or more paths from θ to θ' can be identified. Each such path may consist of only one arrow, or of sequences of directed arrows that pass nodes that enable immediate transitions. All arrows are labelled by names of transitions, therefore the paths between θ and θ' may be characterised by the labels on these arrows, i.e. by the transitions that consecutively fire in the jump from θ to θ'. Denote the multi-set of paths, characterised by these labels, by $\mathcal{L}_{\theta\theta'}$. Examples of elements of $\mathcal{L}_{\theta\theta'}$ are T_1 (if T_1 is pre-enabled in θ and its firing leads to θ'), $T_1 + T_2$ (if there is a non-zero probability that T_1 and T_2 will fire at exactly the same time, and their combined firing leads to θ'), $T_4 \circ T_3$ (if T_3 is pre-enabled in θ, its firing leads to the immediate transition T_4 being enabled, and the firing of T_4 leads to θ'), etc.

Next, we factorise Q by conditioning on the path $L \in \mathcal{L}_{\theta\theta'}$ along which the jump is made. Under the condition that a jump occurs:

$$Q(\theta', A; \theta, x) = \sum_{L \in \mathcal{L}_{\theta\theta'}} p_{\theta', x' | \theta, x, L}(\theta', A \mid \theta, x, L) \times p_{L|\theta,x}(L \mid \theta, x),$$

where $p_{\theta', x' | \theta, x, L}(\theta', A \mid \theta, x, L)$ denotes the conditional probability that the SDCPN state immediately after the jump is in (θ', A), given that the SDCPN state just prior to the jump equals (θ, x), given that the set of transitions L fires to establish the jump. Moreover, $p_{L|\theta,x}(L \mid \theta, x)$ denotes the conditional probability that the set of transitions L fires, given that the SDCPN state immediately prior to the jump equals (θ, x).

In the remainder of this appendix, first $p_{L|\theta,x}(L \mid \theta, x)$ is characterised for each $L \in \mathcal{L}_{\theta\theta'}$. Next, $p_{\theta',x'|\theta,x,L}(\theta', A \mid \theta, x, L)$ is characterised for each $L \in \mathcal{L}_{\theta\theta'}$.

Characterisation of $p_{L|\theta,x}(L \mid \theta, x)$ for each $L \in \mathcal{L}_{\theta\theta'}$

First, assume that $\mathcal{L}_{\theta\theta'}$ does not contain immediate transitions. This yields: each $L \in \mathcal{L}_{\theta\theta'}$ either contains one or more guard transitions, or one delay transition (other combinations occur with zero probability). In particular, $\mathcal{L}_{\theta\theta'}$ is a subset of $\mathcal{B}_\theta$ defined earlier. Then $p_{L|\theta,x}(L \mid \theta, x)$ is determined by $p_{L|\theta,x}(L \mid \theta, x) = \frac{p_L(\theta,x)}{\sum_{B \in \mathcal{L}_{\theta\theta'}} p_B(\theta,x)}$, with $p_B(\theta, x)$ defined earlier.

Next, consider the situations where $\mathrm{RG}_{\theta\theta'}$ may also contain nodes that enable immediate transitions. If L is of the form $L = T_j \circ T_k$, with T_j an immediate transition, then $p_{L|\theta,x}(L \mid \theta, x) = p_{T_k|\theta,x}(T_k \mid \theta, x)$, with the right-hand-side constructed as above for the case without immediate transitions. The same value $p_{T_k|\theta,x}(T_k \mid \theta, x)$ follows for cases like $L = T_m \circ T_j \circ T_k$, with T_j and T_m immediate transitions. However, if the firing of T_k enables more than one immediate transition, then the value of $p_{T_k|\theta,x}(T_k \mid \theta, x)$ is equally divided among the corresponding paths. This means, for example, that if there are $L_1 = T_j \circ T_k$ and $L_2 = T_m \circ T_k$ then $p_{L_1|\theta,x}(L_1 \mid \theta, x) = p_{L_2|\theta,x}(L_2 \mid \theta, x) = \frac{1}{2} p_{T_k|\theta,x}(T_k \mid \theta, x)$.

With this, $p_{L|\theta,x}(L \mid \theta, x)$ is uniquely characterised.

Characterisation of $p_{\theta',x'|\theta,x,L}(\theta', A \mid \theta, x, L)$ for each $L \in \mathcal{L}_{\theta\theta'}$

For probability $p_{\theta',x'|\theta,x,L}(\theta', A \mid \theta, x, L)$, first notice that both (θ, x) and (θ', x') represent states of the complete SDCPN, while the firing of L changes the SDCPN only locally. This yields that in general, several tokens stay where they are when the SDCPN jumps from θ to θ' while the set L of transitions fires.

- $p_{\theta',x'|\theta,x,L}(\theta', A \mid \theta, x, L) = 0$ if for all $x' \in A$, the components of x and x' that correspond with tokens not moving to another place when transitions L fire, are unequal.

In all other cases:

- Assume L consists of one transition T that, given θ and x, is enabled and will fire. Define again c_T^x as the vector containing the colours of the input tokens of T; c_T^x may not be unique. For each c_T^x that can be identified, a sample from $\mathcal{F}_T(\cdot, \cdot; c_T^x)$ provides a vector e' that holds a one for each output arc along which a token is produced and a zero for each output arc along which no token is produced, and it provides a vector c' containing the colours of the tokens produced. These elements together define the size of the jump of the SDCPN state. This gives:

$$p_{\theta',x'|\theta,x,L}(\theta', A \mid \theta, x, L) = \sum_{c_T^x} \int_{(e',c')} \mathfrak{F}_T(e', c'; c_T^x) \times \mathbf{I}_{(\theta',A;e',c',c_T^x)},$$

where $\mathbf{I}_{(\theta',A;e',c',c_T^x)}$ is the indicator function for the event that if tokens corresponding with c_T^x are removed by T and tokens corresponding with (e', c') are produced, then the resulting SDCPN state is in (θ', A).

- If L consists of several transitions $T_1, \ldots, T_m$ that, given θ and x, will all fire at the same time, then the firing measure $\mathfrak{F}_T$ in the equation above is replaced by a product of firing measures for transitions $T_1, \ldots, T_m$:

$$p_{\theta',x'|\theta,x,L}(\theta', A \mid \theta, x, L) = \sum_{c_{T_1}^x,\ldots,c_{T_k}^x} \int_{(e_1',c_1'),\ldots,(e_k',c_k')} \mathfrak{F}_{T_1}(e_1', c_1'; c_{T_1}^x) \times \cdots \times$$

$$\times \mathfrak{F}_{T_k}(e_k', c_k'; c_{T_k}^x) \times \mathbf{I}_{(\theta',A;e_1',c_1',c_{T_1}^x,\ldots,e_k',c_k',c_{T_k}^x)},$$

where $\mathbf{I}_{(\theta',A;e_1',c_1',c_{T_1}^x,\ldots,e_k',c_k',c_{T_k}^x)}$ denotes indicator function for the event that the combined removal of $c_{T_1}^x$ through $c_{T_k}^x$ by transitions T_1 through T_k, respectively, and the combined production of (e_1', c_1') through (e_k', c_k') by transitions T_1 through T_k, respectively, leads to a SDCPN state in (θ', A).

- If L is of the form $L = T_j \circ T_k$, with T_j an immediate transition, then the result is:

$$p_{\theta',x'|\theta,x,L}(\theta', A \mid \theta, x, L) = \sum_{c_{T_k}^x} \int_{(e_j',c_j',c_j,e_k',c_k')} \mathfrak{F}_{T_j}(e_j', c_j'; c_j) \times \mathfrak{F}_{T_k}(e_k', c_k'; c_{T_k}^x) \times$$

$$\times \mathbf{I}_{(\theta',A;e_j',c_j',e_k',c_k',c_T^x)},$$

where $\mathbf{I}_{(\theta',A;e_j',c_j',e_k',c_k',c_T^x)}$ denotes indicator function for the event that the removal of $c_{T_k}^x$ and the production of (e_k', c_k') by transition T_k leads to T_j having a vector of colours of input tokens c_j and the subsequent removal of c_j and the production of (e_j', c_j') by transition T_j leads to a SDCPN state in (θ', A).

- In cases like $L = T_m \circ T_j \circ T_k$, with T_j and T_m immediate transitions, the firing functions of this sequence of transitions are multiplied in a similar way as above.

With this, probability measure Q of the constructed GSHP is uniquely characterised in terms of SDCPN elements.

Communicating Piecewise Deterministic Markov Processes

Stefan Strubbe[1] and Arjan van der Schaft[1]

Department of Applied Mathematics, University of Twente
P.O. Box 217, 7500 AE Enschede, The Netherlands,
s.n.strubbe@math.utwente.nl, a.j.vanderschaft@math.utwente.nl

Summary. In this chapter we introduce the automata framework CPDP, which stands for Communicating Piecewise Deterministic Markov Processes. CPDP is developed for compositional modelling and analysis for a class of stochastic hybrid systems. We define a parallel composition operator, denoted as $|^P_A|$, for CPDPs, which can be used to interconnect component-CPDPs, to form the composite system (which consists of all components, interacting with each other). We show that the result of composing CPDPs with $|^P_A|$ is again a CPDP (i.e., the class of CPDPs is closed under $|^P_A|$). Under certain conditions, the evolution of the state of a CPDP can be modelled as a stochastic process. We show that for these CPDPs, this stochastic process can always be modelled as a PDP (Piecewise Deterministic Markov Process) and we present an algorithm that finds the corresponding PDP of a CPDP. After that, we present an extended CPDP framework called value-passing CPDP. This framework provides richer interaction possibilities, where components can communicate information about their continuous states to each other. We give an Air Traffic Management example, modelled as a value-passing CPDP and we show that according to the algorithm, this CPDP behavior can be modelled as a PDP. Finally, we define bisimulation relations for CPDPs. We prove that bisimilar CPDPs exhibit equal stochastic behavior. Bisimulation can be used as a state reduction technique by substituting a CPDP (or a CPDP component) by a bisimulation-equivalent CPDP (or CPDP component) with a smaller state space. This can be done because we know that such a substitution will not change the stochastic behavior.

1 Introduction

Many real-life systems nowadays are complex hybrid systems. They consist of multiple components 'running' simultaneously, having both continuous and discrete dynamics and interacting with each other. Also, many of these systems have a stochastic nature. An interesting class of stochastic hybrid systems is formed by the Piecewise Deterministic Markov Processes (PDPs), which were introduced in 1984 by Davis (see [3, 4]). Motivation for considering PDP systems is two-fold. First, almost all stochastic hybrid processes

H.A.P. Blom, J. Lygeros (Eds.): Stochastic Hybrid Systems, LNCIS 337, pp. 65–104, 2006.

that do not include diffusions can be modelled as a PDP, and second, PDP processes have nice properties (such as the strong Markov property) when it comes to stochastic analysis. (In [4] powerful analysis techniques for PDPs have been developed). However, PDPs cannot communicate or interact with other PDPs. In order to let PDPs communicate and interact with other PDP's the aim of this paper is to develop a way of opening the structure of PDPs accordingly to this purpose.

In this chapter we present a theory of the automata framework Communicating Piecewise Deterministic Markov Processes (CPDPs, introduced in [12]). A CPDP automaton can be seen as a PDP type process enhanced with interaction/communication possibilities (see [14] for the relation between PDPs and CPDPs). Also, CPDPs can be seen as a generalization of Interactive Markov Chains (IMCs, see [8]). To show the relation of CPDP with IMC, we describe in Section 2 how the CPDP model originated from the IMC model. This section ends with a formal definition of the CPDP model.

CPDPs are designed for communication/interaction with other CPDPs. In Section 3 we describe how CPDPs can be interconnected by using so called parallel composition operators. The use of these parallel composition operators is very common in the field of process algebra (see for example [11] and [9]). We make use of the active/passive composition operators from [13]. We show how composition of CPDPs originates from composition of IMCs. We state the result that the result of composing two CPDPs is again a member of the class of CPDPs. This means that the behavior of two (or more) simultaneously evolving CPDPs, which communicate with each other, can be expressed as a single CPDP. In this way, a complex CPDP can be modelled in a compositional way by modelling its components (as CPDPs) and by selecting the right composition operators to interconnect the component-CPDPs.

Section 4 concerns the relation between CPDPs and PDPs. A PDP is a stochastic process. The behavior of a CPDP can in general not be described by a stochastic process because 1. a CPDP can have multiple hybrid jumps (i.e. the hybrid state discontinuously jumps to another hybrid state) at the same time instant and 2. a CPDP can have nondeterminism, which means that certain choices that influence the state evolution are unmodelled instead of probabilistic as in PDPs. In order to guarantee that the state evolution of a CPDP can be modelled by a stochastic process (and can then be stochastically analyzed), we introduce the concept of scheduler. A scheduler can be seen as a supervisor, which makes probabilistic choices to resolve non-determinism of the CPDP). Then we give an algorithm to check whether a CPDP with scheduler can be converted into a CPDP (with scheduler) that has only one hybrid jump per time instant (i.e. hybrid jumps of multiplicity greater than one are converted to hybrid jumps of multiplicity one). Finally we show that the evolution of the state of a CPDP with scheduler, whose hybrid jumps all have multiplicity one, can be modelled as a PDP. The contents of this section are based on [5]).

In Section 5, we enrich the communication mechanism of CPDPs with so called value passing. With this notion of value passing, a CPDP can receive information about the output variables of other CPDPs. The enriched framework is called value-passing CPDPs. Value-passing is a concept that is successfully used for several process algebra models (see for example [1] and [9] for application of value-passing to the specification language LOTOS). In Section 6 we give an ATM (Air Traffic Management) example of a value passing CPDP. We also apply the algorithm of Section 4 to show that this value-passing CPDP can be converted to a PDP. The ATM-example was first modelled as a Dynamically Coloured Petri Net (DCPN) (see the chapter at pp. 325–350 of this book). DCPN is a Petri net formalism, which has also been designed for compositional specification of PDP-type systems (see [6] and [7] for the DCPN model).

Section 7 is about compositional state reduction by bisimulation. Bisimulation, which we define for CPDP in this section, is a notion of external equivalence. This means that two bisimilar CPDPs cannot be discriminated by an external agent that observes the values of the output variables of the CPDP and interacts with the CPDP. The bisimulation notion that we use is a probabilistic bisimulation (see [10] and [2] for probabilistic bisimulation in the contexts of probabilistic transition systems and probabilistic timed automata). The main result in this section is the bisimulation-substitution-theorem which states that replacing a component of a complex CPDP by another bisimilar component does not change the complex system (up to bisimilarity). In this way we can perform compositional state reduction by reducing the state space of the individual components (via bisimulation). The contents of this section are based on [15]).

The chapter ends in Section 8 with conclusions and a small discussion on compositional modelling and analysis in the context of stochastic hybrid systems.

2 The CPDP Model

In this section we describe how the CPDP model originates from the IMC model. We start with describing the IMC model.

2.1 Interactive Markov Chains

An IMC (Interactive Markov Chain) is a quadruple $(L, \Sigma, \mathcal{A}, \mathcal{S})$, where L is the set of locations (or discrete states), Σ is the set of actions (or events), $\mathcal{A}$ is the set of interactive transitions and consists of triples (l, a, l') with $l, l' \in L$ and $a \in \Sigma$, and $\mathcal{S}$ is the set of Markovian (or spontaneous) transitions and consists of triples (l, λ, l') with $l, l' \in L$ and $\lambda \in \mathbb{R}^+$.

In Figure 1 we see an IMC with two locations, l_1 and l_2, with two interactive transitions (pictured as solid arrows) labelled with event a and with

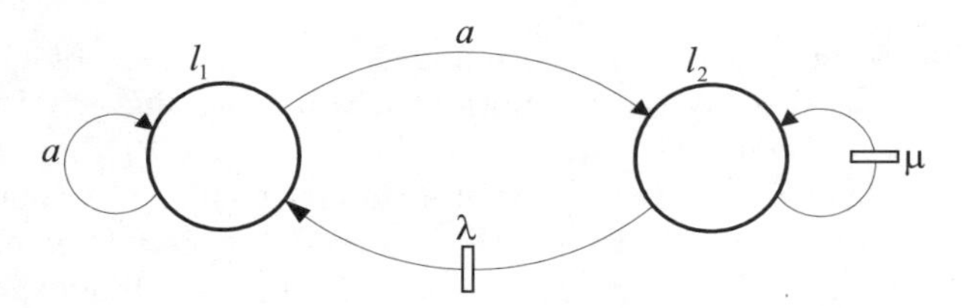

Fig. 1. Interactive Markov Chain

two Markovian transitions (pictured as solid arrows with a little box) labelled with rates λ and μ.

The semantics of the IMC of Figure 1 is as follows: suppose that l_1 in Figure 1 is the initial location (at time $t = 0$). Two things can happen: either the interactive transition labelled a from l_1 to l_2 is taken, or the interactive transition labelled a from l_1 to itself is taken. Note that the choice between these two transitions is not modelled in the IMC, is not determined by the IMC, therefore non-determinism is present at this point (later we will call this form internal non-determinism). Also the time when one of the a-transitions is taken is not modelled (and is therefore left non-deterministic). Suppose that at some time t_1 the a-transition to l_2 is taken. Then at the same time t_1 the process arrives in l_2 (i.e. transitions do not consume time). In l_2 there are two possibilities: either the Markovian transition from l_2 to l_1 with rate λ is taken or the Markovian transition from l_2 to itself with rate μ is taken. In this case neither the choice between these two transitions nor the time of the transition is non-deterministic. The choice and the time are determined probabilistically by a race of Poisson processes: as soon as the process arrives in l_2, two Poisson processes are started with constant rates λ and μ. The process that generates the first point then determines the time and the transition to be taken. Recall that the probability density function of the time of the first point generated by a Poisson process with constant rate λ is equal to $\lambda e^{-\lambda t}$. Suppose that the Poisson process of the λ-transition generates a point after one second and that the Poisson process of the μ-transition generates a point after two seconds, then at time $t = t_1 + 1$ the λ transition is taken which brings the process back to l_1.

2.2 From IMC to CPDP

The first step we could take for transforming the IMC model into the CPDP model is assigning continuous dynamics to the locations. If, in Figure 1, we assign the input/output system $\dot{x} = f_1(x), y = g_1(x)$, with x and y taking value in $\mathbb{R}$ and f_1 and g_1 continuous mappings from $\mathbb{R}$ to $\mathbb{R}$, to l_1 and we assign $\dot{x} = f_2(x), y = g_2(x)$ (with x and y of the same dimensions as x and y of l_1) to l_2, then the resulting process can be pictured as in Figure 2

Suppose that the input/output systems of l_1 and l_2 have given initial states x_1 and x_2 respectively. Then the semantics of the process of Figure 2 would

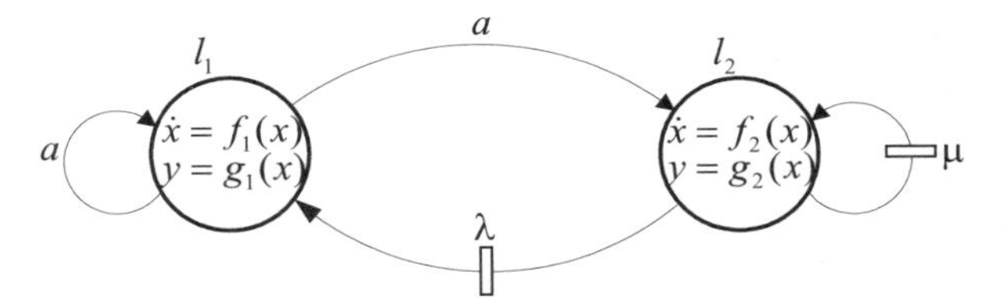

Fig. 2. Interactive Markov Chain enriched with continuous dynamics

be the same as the process of Figure 1, except that when the process is in l_1, then there are continuous variables x and y evolving according to f_1 and g_1 and when the process jumps to l_2, variable x is reset to x_2 (the initial continuous state of l_2) and x and y will then evolve according to f_2 and g_2.

So far, there is little interaction between the discrete dynamics (i.e. the transitions) and the continuous dynamics (i.e. the input/output systems). The transitions are executed independently of the (values of the) continuous variables. The evolution of the continuous variables depends on the transitions as far as it concerns the reset: after every transition, the state variable x is reset to a given value.

In the field of Hybrid Systems, the systems that are studied typically do have (much) interaction between the discrete and the continuous dynamics. In the next step towards the CPDP model, we add some of these interaction possibilities to the model of Figure 2: we add guards, we add reset maps and we allow that the (Poisson) rate of Markovian transitions depends on the value of the continuous variables (and might therefore be non-constant in time).

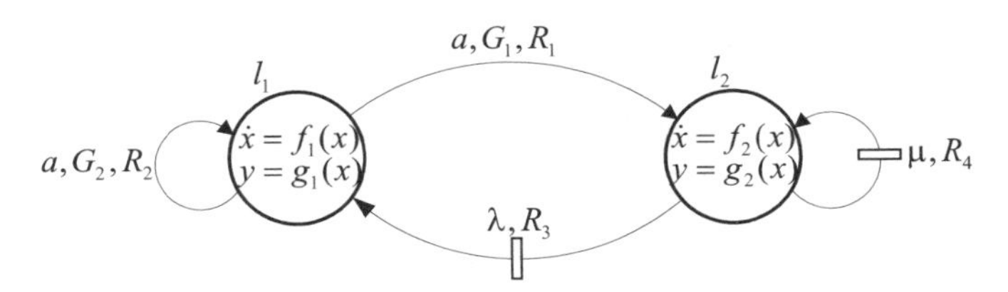

Fig. 3. Interactive Markov Chain enriched with continuous dynamics and discrete/continuous interaction

Guards

We add a guard to each interactive transition. In Figure 3, G_1 and G_2 are the guards. We define a guard of a transition α as a subset of the continuous state space of the origin location of α. In Figure 3 the origin location of the a-transition from l_1 to l_2, is l_1 and therefore G_1 is a subset of $\mathbb{R}$, which is the state space of x at location l_1. The meaning of guard G_1 is that the a-transition to l_2 may not be executed when the value of x (at location l_1) does not lie in G_1 and it may be executed when $x \in G_1$. Via the guards, interactive transitions depend on the continuous variables.

Reset maps

We add reset maps to each interactive and each Markovian transition. A reset map of a transition α probabilistically resets the value of the state of the target location of α, at the moment that α is executed. Therefore, a reset map is a probability measure on the state space of the target location. We also allow to have different (reset) probability measures for different values of the state variables just before the transition is taken. Suppose that the a-transition to l_2 is taken at the moment that the variable x (at l_1) equals $\hat{x}$. Then $R_1(\hat{x})$ is a probability measure that chooses the new value of x at l_2.

Poisson jump rates

We let Poisson jump rates of a Markovian transition depend (continuously) on the state value of the origin location. In Figure 3, λ, whose transition has origin location l_2, is thus a function from $\mathbb{R}$ (the state space of l_2) to $\mathbb{R}$. If $\lambda(\hat{x}_1) > \lambda(\hat{x}_2)$, then this can be interpreted as: the probability that the Poisson process (corresponding to λ) generates a point within a small time interval when $x = \hat{x}_1$ is bigger than the probability of the generation of a point within the same small time interval when $x = \hat{x}_2$. Suppose that (for example after the a-transition from l_1) x in l_2 is at time t_1 reset to $\hat{x}$. Let $x(t)$ (with $x(t_1) := \hat{x}$) be the value of variable x at time t when x evolves along the vectorfield f_2. Then, the probability density function of the time of the first point generated by the Poisson process with rate $\lambda(x(t))$ is equal to $\lambda(x(t))e^{-\int_0^t \lambda(x(s))ds}$.

2.3 Interaction Between Concurrent Processes

The generality of the model of Figure 3 is in fact the generality that we want as far as it concerns the modelling of non-composite systems (i.e. systems that consist of only one component). However, the main aim of the modelling framework that we develop, is compositional modelling. A framework is suitable for compositional modelling if it is possible to model each component of the (composite) system separately and interconnect these separate component-models such that the result describes the behavior of the composite system. With components of a system we mean parts of the system that are running/working simultaneously. For example an Air Traffic Management system that includes multiple (flying) aircraft, where each aircraft forms one subsystem, consists (partly) of subsystems (or components) that 'run' simultaneously. In many composite systems, the components are not independent of each another, but are able to interact with each other and consequently to influence each other. In an ATM system, one aircraft might send a message (via radio) to another aircraft, which might change the course of the aircraft that receives the message. This is a broadcasting kind of interaction/communication, where there is a clear distinction between the active partner (the one that sends the message) and the passive partner (the one that receives the message). We want to add the possibility of broadcasting communication to the model of Figure 3. In order to do so, we add another type of transition to the model called *passive transitions*. This addition brings us to the class of CPDPs (Communicating Piecewise Deterministic Markov Processes), which will be formally defined after the next paragraph.

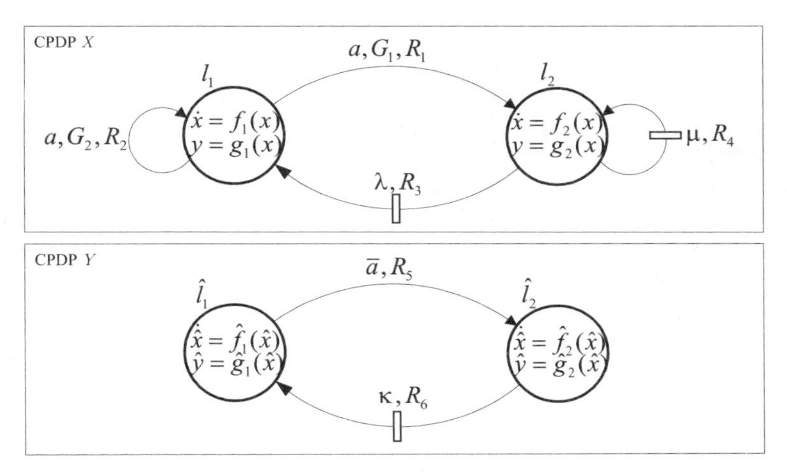

Fig. 4. Two CPDP automata. CPDP Y has a passive transition with label $\bar{a}$.

In Figure 4 we see two CPDPs. CPDP X is the one from Figure 3 and does not have passive transitions. CPDP Y has a passive transition from $\hat{l}_1$ to $\hat{l}_2$ and has a spontaneous transition from $\hat{l}_2$ to $\hat{l}_1$. The passive transition is pictured as a solid arrow, the bar on top of the event label ($\bar{a}$ in Figure 4) denotes that the event is a passive event and that the transition is therefore a passive transition. The passive transition with event $\bar{a}$ reflects that the message a is received. A message a can only be received if some other CPDP has broadcast a message a. Now we can interpret the label a above an interactive transition as: if this transition is executed, the message a is broadcast. We assume that broadcasting and receiving of a message happens instantly (i.e. does not consume time).

For CPDPs, we use the term *active transition* instead of the IMC term *interactive transition* to stress the distinction between activeness and passiveness of transitions. The CPDP terminology for *Markovian transition* is *spontaneous transition*.

2.4 Definition of CPDP

We now give the formal definition of CPDP as an automaton.

Definition 1. *A CPDP is a tuple $(L, V, \nu, W, \omega, F, G, \Sigma, \mathcal{A}, \mathcal{P}, \mathcal{S})$, where*

- *L is a set of locations*
- *V is a set of state variables. With $d(v)$ for $v \in V$ we denote the dimension of variable v. $v \in V$ takes its values in $\mathbb{R}^{d(v)}$.*
- *W is a set of output variables. With $d(w)$ for $w \in W$ we denote the dimension of variable w. $w \in W$ takes its values in $\mathbb{R}^{d(w)}$.*
- *$\nu : L \to 2^V$ maps each location to a subset of V, which is the set of state variables of the corresponding location.*
- *$\omega : L \to 2^W$ maps each location to a subset of W, which is the set of output variables of the corresponding location.*
- *F assigns to each location l and each $v \in \nu(l)$ a mapping from $\mathbb{R}^{d(v)}$ to $\mathbb{R}^{d(v)}$, i.e. $F(l, v) : \mathbb{R}^{d(v)} \to \mathbb{R}^{d(v)}$. $F(l, v)$ is the vector field that determines the evolution of v for location l (i.e. $\dot{v} = F(l, v)$ for location l).*
- *G assigns to each location l and each $w \in \omega(l)$ a mapping from $\mathbb{R}^{d(v_1)+\cdots+d(v_m)}$ to $\mathbb{R}^{d(w)}$, where v_1 till v_m are the state variables of location l. $G(l, w)$ determines the output equation of w for location l (i.e. $w = G(l, w)$).*
- *Σ is the set of communication labels. $\bar{\Sigma}$ denotes the 'passive' mirror of Σ and is defined as $\bar{\Sigma} = \{\bar{a} | a \in \Sigma\}$.*
- *$\mathcal{A}$ is a finite set of active transitions and consists of five-tuples (l, a, l', G, R), denoting a transition from location $l \in L$ to location $l' \in L$ with communication label $a \in \Sigma$, guard G and reset map R. G is a closed subset of the state space of l. The reset map R assigns to each point in G for each variable $v \in \nu(l')$ a probability measure on the state space (and its Borel sets) of v for location l'.*
- *$\mathcal{P}$ is a finite set of passive transitions of the form $(l, \bar{a}, l', R)$. R is defined on the state space of l (as the R of an active transition is defined on the guard space).*
- *$\mathcal{S}$ is a finite set of spontaneous transitions and consists of four-tuples (l, λ, l', R), denoting a transition from location $l \in L$ to location $l' \in L$ with jump-rate λ and reset map R. The jump rate λ (i.e. the Poisson rate of the Poisson process of the spontaneous transition) is a mapping from the state space of l to $\mathbb{R}_+$. R is defined on the state space of l as it is done for passive transitions.*

Example 1. CPDP X of Figure 4 is defined as:

$(L_X, V_X, \nu_X, W_X, \omega_X, F_X, G_X, \Sigma, \mathcal{A}_X, \mathcal{P}_X, \mathcal{S}_X)$ with $L_X = \{l_1, l_2\}$, $V_X = \{x\}$, $\nu_X(l_1) = \nu_X(l_2) = \{x\}$, $W_X = \{y\}$, $\omega_X(l_1) = \omega_X(l_2) = \{y\}$, $F_X(l_1, x) = f_1(x)$ and $F_X(l_2, x) = f_2(x)$, $G_X(l_1, x) = g_1(x)$ and $G_X(l_2, x) = g_2(x)$, $\Sigma = \{a\}$, $\mathcal{A}_X = \{(l_1, a, l_2, G_1, R_1), (l_1, a, l_1, G_2, R_2)\}$,$\mathcal{P}_X = \emptyset$, $\mathcal{S}_X = \{(l_2, \lambda, l_1, R_3), (l_2, \mu, l_2, R_4)\}$. CPDP Y of Figure 4 is defined as:

$(L_Y, V_Y, \nu_Y, W_Y, \omega_Y, F_Y, G_Y, \Sigma, \mathcal{A}_Y, \mathcal{P}_Y, \mathcal{S}_Y)$ with $L_Y = \{\hat{l}_1, \hat{l}_2\}$, $V_Y = \{\hat{x}\}$, $\nu_Y(\hat{l}_1) = \nu_Y(\hat{l}_2) = \{\hat{x}\}$, $W_Y = \{\hat{y}\}$, $\omega_Y(\hat{l}_1) = \omega_Y(\hat{l}_2) = \{\hat{y}\}$, $F_Y(\hat{l}_1, \hat{x}) = \hat{f}_1(\hat{x})$ and $F_Y(\hat{l}_2, \hat{x}) = \hat{f}_2(\hat{x})$, $G_Y(\hat{l}_1, \hat{x}) = \hat{g}_1(\hat{x})$ and $G_Y(\hat{l}_2, \hat{x}) = \hat{g}_2(\hat{x})$, $\Sigma = \{a\}$, $\mathcal{A}_Y = \emptyset, \mathcal{P}_Y = \{(\hat{l}_1, \bar{a}, \hat{l}_2, R_5)\}$, $\mathcal{S}_Y = \{(\hat{l}_2, \kappa, \hat{l}_1, R_6)\}$.

For a CPDP X with $v \in V_X$, where V_X is the set of state variables of X, we call $\mathbb{R}^{d(v)}$ the state space of state variable v. We call $\{(v = r) | r \in \mathbb{R}^{d(v)}\}$ the valuation space of v and each $(v = r)$ for $r \in \mathbb{R}^{d(v)}$ is called a valuation. We call $\{(v_1 = r_1, v_2 = r_2, \cdots, v_m = r_m) | r_i \in \mathbb{R}^{d(v_i)}\}$, where v_1 till v_m are the variables from $\nu(l)$, the valuation space or state space of location l and each $(v_1 = r_1, \cdots, v_m = r_m)$ is called a valuation or state of l. A valuation (state) is an unordered tuple (e.g. $(v_1 = 0, v_2 = 1)$ is the same valuation as $(v_2 = 1, v_1 = 0)$). We denote the valuation space of l by $val(l)$. We call $\{(l, x) | l \in L, x \in val(l)\}$ the state space of a CPDP with location set L and valuation spaces $val(l)$. Each state of a CPDP consists of a location (which comes from a discrete set) and a valuation (which comes from a continuum), therefore we call the state (state space) of a CPDP also hybrid state (hybrid state space). The state space of a location l with $\nu(l) = \{v_1, \cdots, v_m\}$ can be seen as $\mathbb{R}^{d(v_1)+\cdots+d(v_m)}$, because the state space is (topologically) homeomorphic to $\mathbb{R}^{d(v_1)+\cdots+d(v_m)}$ with homeomorphism $\pi_l : val(l) \to \mathbb{R}^{d(v_1)+\cdots+d(v_m)}$ with $\pi_l((v_1 = r_1, \cdots, v_m = r_m)) = (r_1, \cdots, r_m)$. We use unordered tuples for the valuations (states) because this will turn out to be helpful for the composition operation and for some other definitions and proofs.

3 Composition of CPDPs

In the process algebra and concurrent processes literature it is common to define a *parallel composition operator*, normally denoted by $||$. $||$ has as its arguments two processes, say X and Y, of a certain class of processes. The result of the composition operation, denoted by $X||Y$, is again a process that falls within the same class of processes (i.e. the specific class of processes is closed under $||$). The main idea of using this kind of composition operator is that the process $X||Y$ describes the behavior of the composite system that consists of components X and Y (which might interact with each other).

3.1 Composition for IMCs

The interaction-mechanism used for IMCs (see [8]) is not broadcasting interaction but is *interaction via shared events.* This means that if X and Y are two interacting IMCs and a is (by definition) a shared event, then an interactive a-transition of X can only be executed when at the same time an a-transition of Y is executed (and vice versa). In other words, an a-transition of X has to synchronize with an a transition of Y (and vice versa). Markovian transitions, and interactive transitions with labels that are (by definition) not shared events, can be executed independently of the other component. This notion of interaction for IMC is formalized by a parallel composition operator. If we define A as the set of shared events and we denote the corresponding IMC composition operator by $||_A$, then $||_A$ is defined as follows:

Definition 2. *Let $X = (L_X, \Sigma, \mathcal{A}_X, \mathcal{S}_X)$ and $Y = (L_Y, \Sigma, \mathcal{A}_Y, \mathcal{S}_Y)$ be two IMCs, having the same set of events. Let $A \subset \Sigma$ be the set of shared events. Then $X||_A Y$ is the IMC $(L, \Sigma, \mathcal{A}, \mathcal{S})$, where $L := \{l_1||_A l_2 \mid l_1 \in L_X, l_2 \in L_Y\}$ and where $\mathcal{A}$ and $\mathcal{S}$ are the smallest sets that satisfy the following (structural operational) composition rules:*

$$1. \frac{l_1 \xrightarrow{a} l_1', l_2 \xrightarrow{a} l_2'}{l_1||_A l_2 \xrightarrow{a} l_1'||_A l_2'} (a \in A), \tag{1}$$

$$2a. \frac{l_1 \xrightarrow{a} l_1'}{l_1||_A l_2 \xrightarrow{a} l_1'||_A l_2} (a \notin A), 2b. \frac{l_2 \xrightarrow{a} l_2'}{l_1||_A l_2 \xrightarrow{a} l_1||_A l_2'} (a \notin A), \tag{2}$$

$$3a. \frac{l_1 \xrightarrow{\lambda} l_1'}{l_1||_A l_2 \xrightarrow{\lambda} l_1'||_A l_2}, 3b. \frac{l_2 \xrightarrow{\lambda} l_2'}{l_1||_A l_2 \xrightarrow{\lambda} l_1||_A l_2'}. \tag{3}$$

Here, $l_1 \xrightarrow{a} l_1'$ means $(l_1, a, l_1') \in \mathcal{A}_X$, $l_2 \xrightarrow{a} l_2'$ means $(l_2, a, l_2') \in \mathcal{A}_Y$, $l_1 \xrightarrow{\lambda} l_1'$ means $(l_1, \lambda, l_1') \in \mathcal{S}_X$, $l_2 \xrightarrow{\lambda} l_2'$ means $(l_2, \lambda, l_2') \in \mathcal{S}_Y$, $l_1||l_2 \xrightarrow{a} l_1'||l_2'$ means $(l_1||l_2, a, l_1'||l_2') \in \mathcal{A}$, $l_1||l_2 \xrightarrow{\lambda} l_1'||l_2$ means $(l_1||l_2, \lambda, l_1'||l_2) \in \mathcal{S}$, etc. Furthermore, $\frac{B}{C}(A)$ should be read as "If A and B, then C", and $\frac{B_1,B_2}{C}(A)$ should be read as: if A and B_1 and B_2, then C.

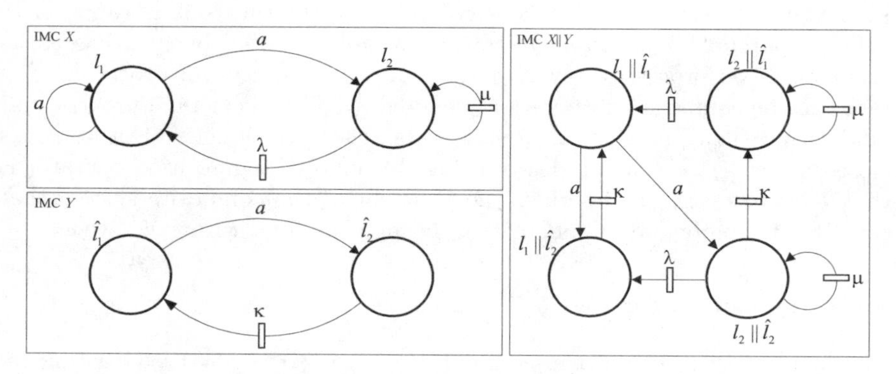

Fig. 5. Composition of two IMCs

In Figure 5, we see on the left two IMCs, X and Y, and we see on the right the IMC $X||Y$, where $||$ is used as shorthand notation for $||_{\{a\}}$. We now check that indeed $X||Y$ expresses the combined behavior of IMCs X and Y

interacting on shared event a: suppose that X and Y initially start in locations l_1 and $\hat{l}_1$ respectively. In $X||Y$, this joint initial location is represented by the location named $l_1||\hat{l}_1$. For a transition to be executed, there are two possibilities: 1. X takes the a transition to l_1 while Y at the same time takes the a-transition to $\hat{l}_2$, 2. X takes the a transition to l_2 while Y at the same time takes the a-transition to $\hat{l}_2$. Note that, since a is a shared event, it is not possible that X takes an a-transition, while Y idles (i.e. stays in location $\hat{l}_1$). Case 1 and 2 are in $X||Y$ represented by the a-transitions to locations $l_1||\hat{l}_2$ and $l_2||\hat{l}_2$ respectively. Note that in cases 1 and 2 one a-transition in $X||Y$ reflect two combined (or synchronized) transitions, one in X and one in Y. If case 2 is executed, then right after the synchronized a-transitions (of X and Y) three Poisson processes are started. Two from X (with parameters λ and μ) and one from Y (with parameter κ). In $X||Y$ this is reflected by the three Markovian transitions at location $l_2||\hat{l}_2$. Suppose that the λ-process generates the first jump. Then X jumps to location l_1 and Y stays in location $\hat{l}_2$, waiting for the κ-process to generate a jump to location $\hat{l}_1$. In $X||Y$ this is reflected by taking the λ-transition to location $l_1||\hat{l}_2$. Then in location $l_1||\hat{l}_2$ again a Poisson process with parameter κ is started. One could question whether this correctly reflects the behavior of the composite system, because when X jumps to l_1, Y stays in $\hat{l}_2$ and the κ-Poisson process keeps running and is not started again as happens in location $l_1||\hat{l}_2$. That indeed starting the κ-process again reflects correctly the composite behavior is due to the fact that the exponential probability distribution (of the Poisson process) is memoryless, which means that, if R_κ denotes a random variable with exponential distribution function $-e^{\kappa t}$, then

$$\Pr(R_\kappa > \hat{t} + t | R_\kappa > \hat{t}) = \Pr(R_\kappa > t),$$

where $\Pr(A|B)$ denotes the conditional probability of A given B. We know that when X takes the λ-transition after having spent $\hat{t}$ time units in location l_2, then the κ-process did not generate a jump before $\hat{t}$ time units, i.e. $R_\kappa > \hat{t}$. Therefore it is correct to start the κ process again in location $l_1||\hat{l}_2$. (We will see that the situation for composition of CPDPs will be similar when it comes to restarting Poisson processes after an executed transition). The reader can check that the part of $X||Y$ we did not explain here also correctly reflects the composite behavior of X and Y.

3.2 Composition of CPDPs

We have distinguished two kinds of communication: communication via shared events and communication via active/passive events. For CPDP we want to allow both types of interaction. Some interactions of communicating systems can better be modelled through shared events and some interactions can better be modelled through active/passive events. We refer to [13] for a discussion on this issue. This means that also for two interacting CPDPs, we use a set

A (which is a subset of the set of active events Σ) which contains the events that are used as shared events. Then the active events not in A together with the passive events (i.e. the ones in $\bar{\Sigma}$) can be used for active/passive communication. In Figure 6 we see the CPDP $X||Y$, with $||$ shorthand for $||_\emptyset$ (i.e. we choose to have no shared events for this composition), which reflects the composite behavior of X and Y of Figure 4.

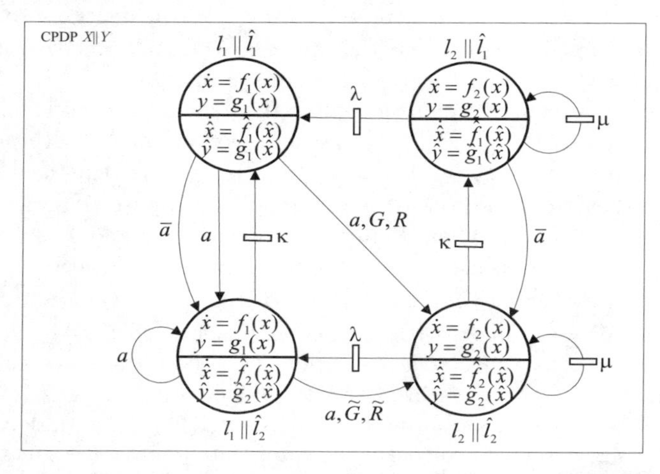

Fig. 6. Composition of two CPDPs (Most guards and reset maps are not drawn)

The communication, reflected by CPDP $X||Y$ of Figure 6, is only through active/passive events (and not through shared events). We will now argue that $X||Y$ of Figure 6 indeed reflects the composite behavior of X and Y interacting via active a and passive $\bar{a}$ events and should therefore be the result of composing X with Y for $A = \emptyset$: suppose X and Y initially start in l_1 and $\hat{l}_1$ respectively, which is reflected by location $l_1||\hat{l}_1$ of $X||Y$. Note that $l_1||\hat{l}_1$ contains the continuous dynamics of both l_1 and $\hat{l}_1$. One possibility is that X executes the a-transition to l_2. Since a is an active event and is not a shared event, X can execute this transition independently of Y. By executing this transition, the message a is send by X. Y has a $\bar{a}$-transition at location $\hat{l}_1$, which means that at $\hat{l}_1$, Y is able to receive the message a. This means that when x executes the a-transition to l_2, Y receives the signal a and synchronizes its $\bar{a}$ transition on the a-transition of X. In Figure 6 this synchronized transition is reflected by the a-transition from $l_1||\hat{l}_1$ to $l_2||\hat{l}_2$. This transition broadcasts signal a which reflects the broadcasting of a by X. $l_1||\hat{l}_1 \xrightarrow{a,G,R} l_2||\hat{l}_2$ (i.e. the a-transition from $l_1||\hat{l}_1$ to $l_2||\hat{l}_2$) can be executed when $x \in G_1$, with G_1 from Figure 4. There is no condition for $\hat{x}$ (i.e. the passive transition can always be taken as soon as an active a-message is broadcast). Therefore G should be equal to $G_1 \times \mathbb{R}^{d(\hat{x})}$. The reset map R should reset x

via R_1 (of Figure 4) and should reset $\hat{x}$ via R_6 (of Figure 4). The probability measures of R_1 and R_6 are independent therefore we can use the product probability measure for $R(x, \hat{x}) = R_1(x) \times R_6(\hat{x})$, where x and $\hat{x}$ are elements from the state spaces of l_1 and $\hat{l}_1$ respectively.

We discuss a few more transitions of $X||Y$:

- $l_1||\hat{l}_2 \xrightarrow{a,\tilde{G},\tilde{R}} l_2||\hat{l}_2$: this transition reflects that X executes the active a-transition to l_2 while Y does not receive the a-message because Y has no $\bar{a}$-transition at location $\hat{l}_2$. Again $\tilde{G}$ should be equal to $G_1 \times \mathbb{R}^{d(\hat{x})}$. $\tilde{R}$ should reset x according to R_1 and should leave $\hat{x}$ unaltered. Therefore $\tilde{R}(x, \hat{x}) = R_1(x) \times Id_{\hat{x}}$, where $Id_{\hat{x}}$ is the identity probability measure for which the set $\{\hat{x}\}$ has probability one (i.e. the probability that $\hat{x}$ stays unaltered after the reset is one).
- $l_1||\hat{l}_2 \xrightarrow{a} l_1||\hat{l}_2$: this transition reflects that X executes $l_1 \xrightarrow{a,G_2,R_2} l_1$ while Y receives no message a. (We do not specify guard and reset map of this transition here).
- $l_2||\hat{l}_2 \xrightarrow{\lambda,\tilde{R}'} l_1||\hat{l}_2$ (reset map $\tilde{R}'$ is not drawn in Figure 6): this transition reflects that X executes the spontaneous λ-transition from l_2 to l_1, while Y stays unaltered. $\tilde{R}'(x, \hat{x})$ should be equal to $R_3(x) \times Id_{\hat{x}}$, with R_3 from Figure 4. Here we have a similar situation as with IMC: after this λ-transition, the κ-process of Y is restarted. As for the IMC case, this is correct because the Poisson process is memoryless. Note that the random variable that belongs to this CPDP κ-process depends on the state where the κ-process is started: if at t_0 the κ-process is activated at state $x(t_0)$ (i.e. a hybrid jump to state $x(t_0)$ took place at time t_0), then the random variable $R_\kappa(x(t_0))$, which denotes the amount of time t after t_0 until κ generates a jump, given that κ is activated at $x(t_0)$, has probability density function $\kappa(x(t_0+t))\mathrm{e}^{-\int_0^t \lambda(x(t_0+s))\mathrm{d}s}$, which is different for different values of t_0. For this situation we get

$$\Pr(R_\kappa(x(t_0)) > \hat{t} + t | R_\kappa(x(t_0)) > \hat{t}) = \Pr(R_\kappa(x(t_0 + \hat{t})) > t),$$

  from which we see that it is correct to (re)activate the κ-process after the transition at state $x(t_0 + \hat{t})$ when it is given that the κ-process that was activated at state $x(t_0)$ did not generate a jump within $\hat{t}$ time units.
- $l_1||\hat{l}_1 \xrightarrow{\bar{a}} l_1||\hat{l}_2$: this transition reflects that Y can also receive a-messages that are not broadcast by X but by some other component Z that we might want to add to the composition $X||Y$. (Then we get the composite model $(X||Y)||Z$).

Because from Figures 4 and 6 we now have an understanding how a CPDP composition operator $||$ should map two CPDPs (X and Y) to a new CPDP ($X||Y$), we are ready to formalize the composition operation. We give a definition of the operator denoted by $|_A^P|$, where A is the set of shared active events and P is the set of shared passive events. So far we did not see the distinction between shared and non-shared passive events. This distinction is only useful when there are more than two components involved. Suppose we have a composite system with three components. Component one has an active transition with label a and can therefore potentially send the message a. Components two and three both have passive transitions with label $\bar{a}$, therefore they both can potentially receive the message a. Now, if $\bar{a}$ is a shared event of components two and three, then it is possible that both can at the same time receive the signal a of component one (which results into three synchronizing transitions, one active and two passive transitions). If $\bar{a}$ is not a shared event of components two and three, then this means that only one of the components two and three may receive the signal a of component one (i.e. it is not allowed that the three transitions synchronize, only synchronization of one active with one passive transition is allowed). For a discussion on the use of this distinction between shared and non-shared passive events, we refer to [13]. Before we give the definition of composition of CPDPs, we first look at the composition rules (i.e. the operational semantics) of the operator $|_A^P|$.

Suppose we have two CPDPs, X and Y, which interact under the set of shared active events A and the set of shared passive events P. If $a \in A$, then an a-transition in X can be executed only when at the same time an a-transition in Y can be executed. This is expressed by the following composition rule, which is the analogy of the IMC composition rule 1 in (1).

$$r1. \frac{l_1 \overset{a,G_1,R_1}{\longrightarrow} l_1', l_2 \overset{a,G_2,R_2}{\longrightarrow} l_2'}{l_1 |_A^P| l_2 \overset{a,G_1 \times G_2, R_1 \times R_2}{\longrightarrow} l_1' |_A^P| l_2'} (a \in A).$$

The synchronized transition, in the CPDP $X|_A^P|Y$, has guard $G_1 \times G_2$, which expresses that if one of the two guards G_1 and G_2 is not satisfied, then the synchronized transition can not be executed. The reset map is constructed via the product probability measures $R_1 \times R_2$, which expresses that R_1 independently resets the state variables of l_1' of X and R_2 independently resets the state variables of l_2' of Y.

If $a \notin A$, then active a-transitions can be executed independently and passive $\bar{a}$-transitions can synchronize on a-transitions of other components. This is expressed by the following composition rule.

$$r2. \frac{l_1 \overset{a,G_1,R_1}{\longrightarrow} l_1', l_2 \overset{\bar{a},R_2}{\longrightarrow} l_2'}{l_1 |_A^P| l_2 \overset{a,G_1 \times val(l_2), R_1 \times R_2}{\longrightarrow} l_1' |_A^P| l_2'} (a \notin A).$$

The guard of the synchronized transition equals $G_1 \times val(l_2)$, where $val(l_2)$ denotes the state space of location l_2. This expresses that there is no guard condition on the passive transition (i.e. it may always synchronize when an active a-partner is available). We also need the mirror rule $r2'$:

$$r2'. \frac{l_1 \overset{\bar{a},R_1}{\longrightarrow} l_1', l_2 \overset{a,G_2,R_2}{\longrightarrow} l_2'}{l_1 |_A^P| l_2 \overset{a,val(l_1) \times G_2, R_1 \times R_2}{\longrightarrow} l_1' |_A^P| l_2'} (a \notin A).$$

If $a \notin A$, then an a-transition can be executed also when there is no passive $\bar{a}$-transition available in the other component (A signal can be broadcast also when there is no receiver to receive the message). This is expressed by the following rule r3 and its mirror $r3'$ which we will not explicitly state. The IMC analogy are rules 2a and 2b in (2).

$$r3.\frac{l_1 \stackrel{a,G_1,R_1}{\longrightarrow} l_1', l_2 \stackrel{\bar{a}}{\not\longrightarrow}}{l_1|_A^P|l_2 \stackrel{a,G_1\times val(l_2),R_1\times Id}{\longrightarrow} l_1'|_A^P|l_2}(a \notin A).$$

Here Id is the identity probability measure, which does not change the state value of l_2 with probability one.

The following three rules $r4$,$r5$ and $r6$ concern the passive transitions of $X|_A^P|Y$. A passive $\bar{a}$-transition of $X|_A^P|Y$ reflects that either X or Y can receive an a-message from a component Z that we might want to add to the composition. If $\bar{a} \in P$ and X can execute a $\bar{a}$-transition from location l_1 and Y can execute a $\bar{a}$-transition from location l_2. Then if X is in l_1 and Y is in l_2 and an a-message is broadcast (by the other component Z), then the two passive transitions will be executed at the same time (of the a-message) and will therefore synchronize. This is expressed by the following rule.

$$r5.\frac{l_1 \stackrel{\bar{a},R_1}{\longrightarrow} l_1', l_2 \stackrel{\bar{a},R_2}{\longrightarrow} l_2'}{l_1|_A^P|l_2 \stackrel{\bar{a},R_1\times R_2}{\longrightarrow} l_1'|_A^P|l_2'}(\bar{a} \in P).$$

If $\bar{a} \in P$, but only one component has a $\bar{a}$-transition to receive the message a from Z, then this component will receive the message while the other component stays unchanged. This is expressed by the following rule $r6$ (and its mirror $r6'$ which we do not explicitly state here).

$$r6.\frac{l_1 \stackrel{\bar{a},R_1}{\longrightarrow} l_1', l_2 \stackrel{\bar{a}}{\not\longrightarrow}}{l_1|_A^P|l_2 \stackrel{\bar{a},R_1\times Id}{\longrightarrow} l_1'|_A^P|l_2}(\bar{a} \in P)$$

If $\bar{a} \notin P$, then two passive $\bar{a}$-transitions cannot synchronize because only one is allowed to receive the message a from Z. Therefore these passive $\bar{a}$-transitions of X and Y remain in the composition (to potentially receive an a-message from Z) but will not synchronize. This is expressed by the following rules $r4$ and $r4'$.

$$r4.\frac{l_1 \stackrel{\bar{a},R_1}{\longrightarrow} l_1'}{l_1|_A^P|l_2 \stackrel{\bar{a},R_1\times Id}{\longrightarrow} l_1'|_A^P|l_2}(\bar{a} \notin P), \quad r4'.\frac{l_2 \stackrel{\bar{a},R_2}{\longrightarrow} l_2'}{l_1|_A^P|l_2 \stackrel{\bar{a},Id\times R_2}{\longrightarrow} l_1|_A^P|l_2'}(\bar{a} \notin P)$$

Finally we need one more composition rule $r7$ (and its mirror $r7'$) to express that spontaneous transitions of X and Y remain in the composition $X|_A^P|Y$ (as we have seen in the discussion on Figure 6). The IMC analogy of these rules are rules 3a and 3b in (3).

$$r7.\frac{l_1 \stackrel{\lambda_1,R_1}{\longrightarrow} l_1'}{l_1|_A^P|l_2 \stackrel{\hat{\lambda}_1,R_1\times Id}{\longrightarrow} l_1'|_A^P|l_2}, \quad r7'.\frac{l_2 \stackrel{\lambda_2,R_2}{\longrightarrow} l_2'}{l_1|_A^P|l_2 \stackrel{\hat{\lambda}_2,Id\times R_2}{\longrightarrow} l_1|_A^P|l_2'}.$$

Here $\hat{\lambda}_1$ and $\hat{\lambda}_2$ are defined on the combined state space of locations l_1 and l_2 and equal $\hat{\lambda}_1(x_1,x_2) = \lambda_1(x_1)$ and $\hat{\lambda}_2(x_1,x_2) = \lambda_2(x_2)$, where x_1 and x_2 are states of l_1 and l_2 respectively.

Definition 3. *If* $X = (L_X, V_X, \nu_X, W_X, \omega_X, F_X, G_X, \Sigma, \mathcal{A}_X, \mathcal{P}_X, \mathcal{S}_X)$ *and* $Y = (L_Y, V_Y, \nu_Y, W_Y, \omega_Y, F_Y, G_Y, \Sigma, \mathcal{A}_Y, \mathcal{P}_Y, \mathcal{S}_Y)$ *are two CPDPs that have the same set of events* Σ *and if we have* $V_X \cap V_Y = W_X \cap W_Y = \emptyset$, *then* $X|_A^P|Y$ *is defined as the CPDP* $(L, V, \nu, W, \omega, F, G, \Sigma, \mathcal{A}, \mathcal{P}, \mathcal{S})$, *where*

- $L = \{l_1|_A^P|l_2 \mid l_1 \in L_X, l_2 \in L_Y\}$,
- $V = V_X \cup V_Y$, $W = W_X \cup W_Y$,
- $\nu(l_1|_A^P|l_2) = \nu(l_1) \cup \nu(l_2)$, $\omega(l_1|_A^P|l_2) = \omega(l_1) \cup \omega(l_2)$,
- $F(l_1|_A^P|l_2, v)$ *equals* $F_X(l_1, v)$ *if* $v \in \nu_X(l_1)$ *and equals* $F_Y(l_2, v)$ *if* $v \in \nu_Y(l_2)$.
- $G(l_1|_A^P|l_2, w)$ *equals* $G_X(l_1, w)$ *if* $w \in \omega_X(l_1)$ *and equals* $G_Y(l_2, w)$ *if* $w \in \omega_Y(l_2)$.
- $\mathcal{A}$, $\mathcal{P}$ *and* $\mathcal{S}$ *contain and only contain the transitions that are the result of applying one of the rules r1,r2,r2',r3,r3',r4,r4',r5,r6,r6',r7 and r7', defined above.*

Example 2. It can be checked that, according to Definition 3, CPDP $X||Y$ from Figure 6 is indeed the resulting CPDP of composing X and Y from Figure 4 with composition operator $|_A^P|$, where $A = \emptyset$ and $P = \bar{\Sigma}$. Note that any other $P \subset \bar{\Sigma}$ would give the same result because X has no passive transitions and therefore it is not relevant for the composition of X and Y whether passive transitions synchronize or not (which is determined by P).

In order to prove that, for certain A and P, the composition operator $|_A^P|$ is commutative and associative, we need to introduce an equivalence notion, that equates CPDPs that are exactly the same except that the locations may have different names. We call this equivalence notion, in the line of [2], *isomorphism* and we define it as follows.

Definition 4. *Two CPDPs* $X = (L_X, V, \nu_X, W, \omega_X, F_X, G_X, \Sigma, \mathcal{A}_X, \mathcal{P}_X, \mathcal{S}_X)$ *and* $Y = (L_Y, V, \nu_Y, W, \omega_Y, F_Y, G_Y, \Sigma, \mathcal{A}_Y, \mathcal{P}_Y, \mathcal{S}_Y)$, *with shared* V,W *and* Σ, *are* isomorphic *if there exists a bijection* $\pi : L_X \to L_Y$ *such that, for all* $l \in L_X$, $\nu_X(l) = \nu_Y(\pi(l))$, $\omega_X(l) = \omega_Y(\pi(l))$, $F_X(l, v) = F_Y(\pi(l), v)$ *for all* $v \in \nu(l)$, $G_X(l, w) = G_Y(\pi(l), w)$ *for all* $w \in \omega(l)$, *for any* a,$\bar{a}$,λ,l', G *and* R *we have that:* $(l, a, l', G, R) \in \mathcal{A}_X$ *if and only if* $(\pi(l), a, \pi(l'), G, R) \in \mathcal{A}_Y$, $(l, \bar{a}, l', R) \in \mathcal{P}_X$ *if and only if* $(\pi(l), \bar{a}, \pi(l'), R) \in \mathcal{A}_Y$, $(l, \lambda, l', R) \in \mathcal{S}_X$ *if and only if* $(\pi(l), \lambda, \pi(l'), R) \in \mathcal{S}_Y$.

We now state a result on the commutativity and associativity of the composition operators $|^P_A|$. The operator $|^P_A|$ is called commutative if for all CPDPs X and Y we have that $X|^P_A|Y$ is isomorphic to $Y|^P_A|X$. The operator $|^P_A|$ is called associative if for all CPDPs X,Y and Z we have that $(X|^P_A|Y)|^P_A|Z$ is isomorphic to $X|^P_A|(Y|^P_A|Z)$.

Theorem 1. *The composition operator $|^P_A|$ is commutative for all A and P. $|^P_A|$ is associative if and only if for all $a \in \Sigma$ we have: if $\bar{a} \notin P$ then $a \in A$.*

Proof. The proof of this theorem in the context of active/passive labelled transition systems can be found on www.cs.utwente.nl/˜strubbesn. The proof can easily be generalized to the context of CPDPs.

If we have n CPDPs X_i $(i = 1 \cdots n)$ with events-set Σ that are composed via an associative operator $|^P_A|$, then the order of composition does not influence the resulting CPDP and therefore we can write $X_1|^P_A|X_2|^P_A| \cdots X_{n-1}|^P_A|X_n$ to unambiguously (up to isomorphism) denote the resulting composite CPDP.

4 PDP-Semantics of CPDPs

Under certain conditions, the state evolution of a CPDP can be modelled as a stochastic process. In this section we give the exact conditions under which this is true. We also prove that the stochastic process may always be chosen of the PDP-type. In order to achieve this result, we first need to make a distinction between guarded CPDP states and unguarded CPDP states.

Definition 5. *A state (l, x) of a CPDP X is called* guarded, *if there exists an active transition with origin location l such that x is an element of the guard of this transition. A CPDP state is* unguarded *if it is not guarded.*

If we execute a CPDP X from some initial hybrid state (l_0, x_0) then the first part of the state trajectory (i.e. the evolution of the state variables in time) and of the output trajectory (i.e. the evolution of the output variables in time) is determined by F_X and G_X respectively. This is the case until the first transition is executed, which might cause a jump (i.e. discontinuity) in the state/output trajectories. We choose that at these points of discontinuity, the state/output trajectories have the *cadlag* property, which means that at these points the trajectories are continuous from the right and have limits from the left. If then at $t = t_1$, X executes a transition which resets the state to a unguarded state x_1, then the value of the state trajectory at $t = t_1$ equals x_1 (and the value of the output trajectory equals the output value of x_1). If the state after reset x_1 is guarded, then it is possible that at the same time t_1 from state x_1 another active transition is executed. If this transition resets the state to a unguarded state x'_1, then the value of the state trajectory at t_1 equals x'_1. If this transition resets the state to an guarded state x'_1, then another active transition can be executed, etc. We see that the CPDP model allows multiple transitions at the same time instant.

Formally, let $E := \{(l,x)|l \in L_X, x \in val(l)\}$ be the state space of CPDP X, where $val(l)$ denotes the space of all valuations for the state variables of location l. The trajectories of X are elements of the space $D_E[0,\infty[$ which is the space of right-continuous E-valued functions on $\mathbb{R}_+$ with left-hand limits. According to [4], a metric can be defined on E such that $(E, \mathcal{B}(E))$, with $\mathcal{B}(E)$ the set of Borel sets of E under this metric, is a Borel space (i.e. a subset of a complete separable metric space) and each Borel set B is such that for each $l \in L_X$, $\{x|(l,x) \in B\}$ (i.e. the restriction of B to l) is a Borel set of the Euclidean state space $val(l)$ of location l. Therefore, the concept of continuity within a location (i.e. for sets $\{(l,x)|x \in val(l)\}$) coincides with the standard (Euclidean) concept of continuity.

The CPDP model exhibits non-determinism. This means that at certain time instants of the execution of a CPDP (from some initial state) choices have to be made which are neither deterministic (like a differential equation deterministically determines (a part of) the state trajectory) nor stochastic (i.e. a probability measure can be used to make a probabilistic choice). These non-deterministic choices are simply unmodelled. We distinguish two sources of non-determinism for the CPDP: 1. The choice when an active transition is taken. 2. The choice which active transition is taken. To resolve non-determinism of type 1, we use, in the line of [8], the *maximal progress* strategy, which means that as soon as the state enters a guard area (i.e. at the first time instant that the state is guarded), an active transition has to be executed. To resolve non-determinism of type 2, we use a socalled *scheduler* S which

1. assigns to each guarded state x a probability measure on the set of all active transitions that have x as an element of their guard (i.e. the set of all active transitions that are allowed to be executed from state x) and
2. assigns to each pair $(x, \bar{a})$, with x any state and $\bar{a} \in \bar{\Sigma}$ such that there is a $\bar{a}$-transition at the location of x, a probability measure on the set of all $\bar{a}$-transitions at the location of x.

In other words, if an active transition has to be executed from state x, S probabilistically chooses which active transition is executed and if an active a triggers a $\bar{a}$-transition, then S probabilistically chooses which $\bar{a}$-transition is executed.

For identifying the stochastic process of a CPDP, we only look at *closed* CPDPs, which are CPDPs that have no passive transitions. Closed CPDPs are called closed because we assume that they represent the whole system (i.e. no more other component-CPDPs will be added). Therefore closed CPDPs should have no passive transitions because passive transitions can only be executed when another component triggers it (via an active transition). The order of finding the stochastic behavior of the composite system is therefore: first compose the different components. Then remove all passive transitions of the resulting CPDP. This results in a closed CPDP where, under maximal progress and scheduler S, all choices for the execution of the CPDP are made probabilistically. One could question whether the evolution of the state can, for closed CPDPs, be modelled as a stochastic process. We can state a condition on the CPDP under which this is not possible: if with non-zero probability we can reach an guarded state x where with non-zero probability an infinite sequence of active transitions can be chosen such that each transition resets the state within the guard of the next transition, then the trajectory of this execution deadlocks (i.e. time does not progress anymore after reaching x at some time $\hat{t}$ and therefore the trajectory is not defined for time instants after time $\hat{t}$). Trajectories of stochastic processes do not deadlock like this, therefore this state evolution cannot be modelled by a stochastic process.

In order to find the stochastic process of a closed CPDP, we would first like to state decidable conditions on a CPDP, which guarantee that the probability that an execution deadlocks (i.e. comes at a point where time does not progress anymore) is zero.

4.1 The Stochastic Process of a Closed CPDP

Suppose we have a closed CPDP X with location set L_X and active transition set $\mathcal{A}_X$. The CPDP operates under maximal progress and under scheduler S. We write $S_x(\alpha)$ for the probability that active transition α is taken when an active transition is executed at state x. We assume that the CPDP has no spontaneous transitions. The case 'with spontaneous transitions' is treated at the end of this section.

We call the jump of a CPDP from the current state to another unguarded state via a sequence of active transitions a hybrid jump. We call the number of active transitions involved in a hybrid jump the multiplicity of the hybrid jump. For example, if at state x_1 a transition α is taken to x_1', which lies in the guard of transition β, and immediately transition β is taken to a unguarded state x_1'', then this hybrid jump from x_1 to x_1'' has multiplicity two.

We need to introduce the concept of total reset map. $R_{tot}(B, x)$ denotes the probability of jumping into $B \in \mathcal{B}(E)$ when an active jump takes place at state x. We have that

$$R_{tot}(B, x) = \sum_{\alpha \in \mathcal{A}_{l_x \rightarrow}} [S_x(\alpha) R_\alpha(B \cap val(l'_\alpha), x)],$$

where $\mathcal{A}_{l_x \rightarrow}$ is the set of all active transitions that leave the location of x. We define the total guard $G_{tot,l}$ of location l as the union of the guards of all active transitions with origin location l. It can be seen now that for the stochastic executions (i.e. generating trajectories during simulation) of X it is enough to know R_{tot} and $G_{tot,l}$ (for all $l \in L_X$) instead of $\mathcal{A}_X$: a trajectory that starts in (l_0, x_0) evolves until it hits G_{tot,l_0} at some state (l_0, x_1). From x_1 we determine the target state (l_1, x_1') of the (first step of the) hybrid jump by drawing a sample from $R_{tot}(\cdot, x_1)$. If x_1' is unguarded, the next piecewise deterministic part of the trajectory is determined by the differential equations of the state variables of location l_1 until G_{tot,l_1} is hit. If x_1' is guarded, we directly draw a new target state (l_1', x_1'') from $R_{tot}(\cdot, x_1')$, etc. Therefore, if two closed CPDPs that are isomorphic except for the active transition set, and they have the same total reset map and the same total guards, then the stochastic behaviors (concerning the state trajectories) of the two CPDPs are the same and consequently if some stochastic process models the state evolution of one CPDP, then it also models the state evolution of the other CPDP.

Finding the stable and unstable parts of an active transition

Take any $\alpha \in \mathcal{A}_X$. We now show how to split up α in a stable part α_s and an unstable part α_u such that the stochastic behavior of X does not change.

We define G_{α_s} as the set of all $x \in G_\alpha$ (i.e. all x in the guard of α) such that $R_\alpha(val_s(l_\alpha'), x) \neq 0$, where $val_s(l_\alpha')$ is the unguarded part of the state space of the target location of α. Then for all $x \in G_{\alpha_s}$ we define

$$R_{\alpha_s}(B, x) := \frac{R_\alpha(B \cap val_s(l_\alpha'), x)}{R_\alpha(val_s(l_\alpha'), x)},$$

$$S_x(\alpha_s) := S_x(\alpha) R_\alpha(val_s(l_\alpha'), x).$$

The scheduler works on α_s as $S_x(\alpha_s)$ (as defined above).

We define G_{α_u} as the set of all $x \in G_\alpha$ such that $R_\alpha(val_u(l_\alpha'), x) \neq 0$. For all $x \in G_{\alpha_u}$ we define

$$R_{\alpha_u}(B, x) := \frac{R_\alpha(B \cap val_u(l_\alpha'), x)}{R_\alpha(val_u(l_\alpha'), x)},$$

$$S_x(\alpha_s) := S_x(\alpha) R_\alpha(val_u(l_\alpha'), x).$$

The scheduler works on α_u as $S_x(\alpha_u)$ (as defined above).

It can be seen that replacing α by α_s and α_u does not change the total reset map.

Resolving hybrid jumps of multiplicity greater than one

For any $n \in \mathbb{N}$ we will now define T_s^n and T_u^n. T_s^n is a set of stable transitions representing hybrid jumps of multiplicity n and T_u^n is a set of unstable transitions representing hybrid jumps of multiplicity n. A stable transition is a transition that always jumps to the unguarded state space of the target location. An unstable transition always jumps to the guarded state space. A stable transition is stable in the sense that after the hybrid jump caused by the transition, no other hybrid jump will happen immediately and therefore we are sure that a stable transition will not cause an explosion of hybrid jumps (i.e. a hybrid jump of multiplicity infinity). An unstable transition does not need to induce such a blow up of hybrid jumps, but potentially it can.

We define T_s^1 as the set of all active transitions α_s (with $\alpha \in \mathcal{A}_X$) such that $G_{\alpha_s} \neq \emptyset$ and we define T_u^1 as the set of all active transitions α_u (with $\alpha \in \mathcal{A}_X$) such that $G_{\alpha_u} \neq \emptyset$.

We introduce the following notations. $P_x(B \circ \beta \circ \alpha)$ denotes the probability that, given that an active jump takes place at state x, transition α is executed followed directly by transition β jumping into the set $B \in \mathcal{B}(val(l'_\beta))$. It can be seen that

$$P_x(B \circ \beta \circ \alpha) = S_x(\alpha) \int_{\hat{x} \in G_\beta} S_{\hat{x}}(\beta) R_\beta(B, \hat{x}) \mathrm{d}R_\alpha(\hat{x}, x).$$

We will now inductively determine the sets T_s^n and T_u^n. Suppose the sets T_s^{n-1} and T_u^{n-1} and T_s^1 and T_u^1 are given. Now, for any $\alpha \in T_u^{n-1}$, $\beta \in T_s^1 \cup T_u^1$ such that $l'_\alpha = l_\beta$, we define $G_{\beta \circ \alpha}$ as all $x \in G_\alpha$ such that $R_\alpha(G_\beta, x) \neq 0$. Then, for all $x \in G_{\beta \circ \alpha}$ we define

$$S_x(\beta \circ \alpha) := P_x(val(l'_\beta) \circ \beta \circ \alpha),$$

$$R_{\beta \circ \alpha}(B, x) := \frac{P_x(B \circ \beta \circ \alpha)}{S_x(\beta \circ \alpha)}.$$

If $G_{\beta \circ \alpha} \neq \emptyset$ and $\beta \in T_s^1$ then we add transition $\beta \circ \alpha$, with guard, reset map and scheduler as above, to T_s^n. If $G_{\beta \circ \alpha} \neq \emptyset$ and $\beta \in T_u^1$ then we add transition $\beta \circ \alpha$, with guard, reset map and scheduler as above, to T_u^n.

Finding the PDP that models the state evolution of the CPDP

If we define, for $z \in \{s, u\}$ and $B \in \mathcal{B}(E)$,

$$R_{tot,z}^n(B, x) := \sum_{\{\alpha \in T_z^n | l_\alpha = l_x\}} [S_x(\alpha) R_\alpha(B \cap val(l'_\alpha), x)],$$

with $B \cap val(l'_\alpha)$ sloppy notation for $\{x | x \in val(l'_\alpha), (l'_\alpha, x) \in B\}$, then it can be seen that for any $n \in \mathbb{N}$ we have

$$R_{tot}(B, x) = \sum_{i=1}^{n} [R_{tot,s}^i(B, x)] + R_u^n(B, x),$$

with other words, if X^n is isomorphic to CPDP X, except that the active transition set of X^n equals $T_s^1 \cup T_s^2 \cup \cdots \cup T_s^n \cup T_u^n$ (which need not be isomorphic to $\mathcal{A}_X$), then the total reset maps of X and X^n are the same for all n.

We are now ready to state the theorem which gives necessary and sufficient conditions on the CPDP such that the state evolution can be modelled by a stochastic process. Also, the theorem says that if the state evolution can be modelled by a stochastic process, then it can be modelled by a stochastic process from the class of PDPs. The proof of the theorem makes use of the results from [14].

Theorem 2. *Let X^n be derived from X as above. Let $R^n_{tot,s}$ denote the total stable reset map of X^n. The state evolution of X can be modelled by a stochastic process if and only if $R(E,x) := \lim_{n\to\infty} R^n_{tot,s}(E,x) = 1$ for all $x \in E_u$, with E_u the guarded part of E. If this condition is satisfied, then the PDP with the same state space as X, with invariants $E^0_l = val(l)\backslash G_{tot,l}$ and with transition measure $Q(B,x) = R(B,x)$, models the state evolution of X.*

Proof. From the text above and from the results of [14], it is clear that if $R(E,x) = 1$ for all x, then the PDP suggested by the theorem models the state evolution of X. If for some $x \in E$, $R(E,x) < 1$, then it can be seen that this must mean that there exists a hybrid jump with multiplicity infinity such that the probability of this hybrid jump at x is greater than zero. This means that (from x) there is a deadlock probability (i.e. time does not progress anymore) greater than zero, which means that the state evolution of X cannot be modelled by a stochastic process (as we saw before).

Corollary 1. *If for some $n \in N$ we have that $T^n_u = \emptyset$, then the multiplicity of the hybrid jumps of X is bounded by n and the state of X exhibits a PDP behavior, with the same PDP as the corresponding PDP of X^n (which can be constructed according to [14] because all hybrid jumps of X^n have multiplicity one).*

The case including spontaneous transitions

Now we treat the case where there are also spontaneous transitions present. Let X be a CPDP without passive and spontaneous transitions and let $\hat{X}$ be an isomorphic copy of X together with a set of spontaneous transitions $\mathcal{S}_{\hat{X}}$. Suppose that the multiplicity of the hybrid jumps of X is bounded by n. Let $\hat{X}^n$ be an isomorphic copy of X^n together with the following spontaneous transitions: for any spontaneous transition $(l,\lambda,l',R) \in \mathcal{S}_{\hat{X}}$ we add to $\hat{\mathcal{S}}$, which denotes the set of spontaneous transitions of $\hat{X}^n$, the transition $(l,\lambda,L,\hat{R})$, where, for $B \in \mathcal{B}(E)$, $\hat{R}(B,x) :=$

$$R(B \cap Inv_s(l'),x) + \sum_{\{\alpha \in \mathcal{A}_{X^n} | l_\alpha = l\}} \int_{\hat{x} \in G_\alpha} S_{\hat{x}}(\alpha) R_\alpha(B \cap val(l'_\alpha)) \mathrm{d}R(\hat{x},x).$$

Note that all transitions from $\mathcal{A}_{X^n}$ are stable. Also note that $(l,\lambda,L,\hat{R})$ is not a standard CPDP transition, but a transition that represents a Poisson process in location l with jump-rate λ and with reset map $\hat{R}$, which can jump to multiple locations. Therefore we write L instead of l' in the tuple of the transition.

It is known that the superposition of two (or more) Poisson processes is again a Poisson process (see, in the context of CPDP, [14] for a proof of this result). This means that if we combine all spontaneous transitions of $\hat{X}^n$ with origin location l to one spontaneous transition $(l,\lambda_l,L,\hat{R}_{tot,l})$, with

$$\lambda_l(x) = \sum_{\alpha \in \hat{\mathcal{S}}_{l\rightarrow}} \lambda_\alpha(x),$$

and

$$\hat{R}_{tot,l}(B,x) = \sum_{\alpha \in \hat{\mathcal{S}}_{l\rightarrow}} [\frac{\lambda_\alpha(x)}{\lambda_l(x)} R_\alpha(B,x)],$$

and if we replace all spontaneous transitions by these combined spontaneous transitions, then the stochastic behavior (concerning the evolution of the state) will not change. Now it can be easily seen that if we add jump rate $\lambda(l,x) = \lambda_l(x)$ to the PDP that models the state evolution of X and we let, for unguarded states (l,x), the transition measure $Q(B,(l,x)) = \hat{R}_{tot,l}(B,x)$, then this PDP will model the state evolution of $\hat{X}$.

5 Value-Passing CPDPs

In the CPDP-model as it is defined so far, it is not possible that one component can inform another component about the value of its state or output variables. In Dynamically Colored Petri Nets (see [6]), this is possible. In this section we introduce an addition to the CPDP model, which adds this feature of communicating state data. We chose to follow a standard method of data communication, called *value-passing*. Value-passing has been defined for different models like LOTOS ([9]). Value-passing can be seen as a natural extension to (the standard) communication through shared events because it is also expressed through "shared events"/"synchronization of active transitions".

5.1 Definition of Value-Passing CPDP

We introduce a new definition for CPDP, which makes communication of state data possible.

Definition 6. *A value-passing CPDP is a tuple* $(L, V, W, \nu, \omega, F, G, \Sigma, \mathcal{A}, \mathcal{P}, \mathcal{S})$, *where all elements except* $\mathcal{A}$ *are defined as in Definition 1 and where* $\mathcal{A}$ *is a finite set of active transitions that consists six-tuples* (l, a, l', G, R, vp), *denoting a transition from location* $l \in L$ *to location* $l' \in L$ *with communication label* $a \in \Sigma$, *guard* G, *reset map* R *and value-passing element* vp. G *is a subset of the state space of* l. vp *can be equal to either* $!Y$, $?U$ *or* $\emptyset$. *For the case* $!Y$, Y *is an ordered tuple* $(w_1, w_2, \cdots, w_m)$ *where* $w_i \in w(l)$ *for* $i = 1 \cdots m$, *meaning that this transition can pass the values of the variables from* Y *(in this specific order) to other transitions in other components. For the case* $?U$, *we have* $U \subset \mathbb{R}^n$ *for some* $n \in \mathbb{N}$, *meaning that this transition asks for input a tuple of the form of* Y *with total dimension* n *(i.e.* $\sum_{i=1..m} d(w_i) = n$*) such that the valuation of* Y *lies in* U. *The reset map* R *assigns to each point in*

$G \times U$ (for the case $vp = ?U$) or to each point in G (for the cases $vp = !Y$ and $vp = \emptyset$) for each state variable $v \in \nu(l')$ a probability measure on the state space of v at location l'.

We formalize the notion of state data communication by adding three composition rules to $|^P_A|$ called $r1data$,$r2data$ and $r2data'$:

$$r1data. \frac{l_1 \overset{a,G_1,R_1,v_1}{\longrightarrow} l_1', l_2 \overset{a,G_2,R_2,v_2}{\longrightarrow} l_2'}{l_1|^P_A|l_2 \overset{a,G_1|G_2,R_1 \times R_2, v_1|v_2}{\longrightarrow} l_1'|^P_A|l_2'} (a \in A, v_1|v_2 \neq \perp).$$

Here, $l_1 \overset{a,G_1,R_1,v_1}{\longrightarrow} l_1'$ means $(l_1, a, l_1', G_1, R_1, v_1) \in \mathcal{A}_X$ with $v_1 \neq \emptyset$. Active transitions with value passing identifier equal to $\emptyset$ will be denoted as before (like $l_1 \overset{a,G_1,R_1}{\longrightarrow} l_1'$ for example). Furthermore, $v_1|v_2$ is defined as: $v_1|v_2 := !Y$ if $v_1 = !Y$ and $v_2 := ?U$ and dim(U)=dim(Y) or if $v_2 = !Y$ and $v_1 := ?U$ and dim(U)=dim(Y); $v_1|v_2 := ?(U_1 \cap U_2)$ if $v_1 = ?U_1$ and $v_2 = ?U_2$ and dim(U_1)=dim(U_2); $v_1|v_2 := \perp$ otherwise. Here $\perp$ means that v_1 and v_2 are not compatible.

$G_1|G_2$ is, only when $v_1|v_2 \neq \perp$, defined as follows: $G_1|G_2 := (G_1 \cap U) \times G_2$ if $v_1 = !Y$ and $v_2 = ?U$; $G_1|G_2 := G_1 \times (G_2 \cap U)$ if $v_1 = ?U$ and $v_2 = !Y$; $G_1|G_2 := G_1 \times G_2$ if $v_1 = ?U_1$ and $v_2 = ?U_2$. Here, $G \cap U$, which is abuse of notation, contains all state valuations x such that $x \in G$ and $Y(x) \in U$, where $Y(x)$ is the value of the ordered tuple Y according to valuation x.

In these definitions of $v_1|v_2$ and $G_1|G_2$ we see an interplay between the state guards G_1,G_2 and the input guards U_1,U_2: in the synchronization of an $(l_1, a, l_1', G_1, R_1, !Y)$ transition with a $(l_2, a, l_2', G_2, R_2, ?U)$ transition, U restricts the guard G_1 such that the Y-part of G_1 lies in U. This restriction can not be coded in $v_1|v_2$ (as it is done in the $?U_1$-$?U_2$-case), therefore we need to code it in the state guards.

Composition rules $r2data$ and $r2data'$ are defined as follows.

$$r2data. \frac{l_1 \overset{a,G_1,R_1,v_1}{\longrightarrow} l_1'}{l_1|^P_A|l_2 \overset{a,G_1 \times val(l_2), R_1 \times Id, v_1}{\longrightarrow} l_1'|^P_A|l_2} (a \notin A).$$

The mirror of $r2data$ is then defined as:

$$r2data'. \frac{l_2 \overset{a,G_2,R_2,v_2}{\longrightarrow} l_2'}{l_1|^P_A|l_2 \overset{a,val(l_1) \times G_2, Id \times R_2, v_2}{\longrightarrow} l_1|^P_A|l_2'} (a \notin A).$$

Definition 7. *If $X = (L_X, V_X, \nu_X, W_X, \omega_X, F_X, G_X, \Sigma, \mathcal{A}_X, \mathcal{P}_X, \mathcal{S}_X)$ and $Y = (L_Y, V_Y, \nu_Y, W_Y, \omega_Y, F_Y, G_Y, \Sigma, \mathcal{A}_Y, \mathcal{P}_Y, \mathcal{S}_Y)$ are two value passing CPDPs that have the same set of events Σ and if we have $V_X \cap V_Y = W_X \cap W_Y = \emptyset$, then $X|^P_A|Y$ is defined as in Definition 3 except that besides the rules r1,r2,r2',r3,r3',r4,r4',r5,r6,r6',r7 and r7' for the operator $|^P_A|$ we also have the rules $r1data$,$r2data$ and $r2data'$.*

6 Value Passing CPDP and CPDP-to-PDP Conversion: An ATM Example

6.1 ATM Example of Value Passing CPDP

In Figure 7 we see five value-passing CPDPs: *CurrentGoal*, *AudioAlert*, $Memory$, $HMI{-}PF$ and $TaskPerformance$. Together, these five components form a part of a system that models the behavior of a pilot which is controlling a flying aircraft. This pilot is called the pilot-flying. (Normally, there is also another pilot in the cockpit called the pilot-not-flying who is not directly controlling the aircraft). This example comes from Chapter 16 of this book, where it is modelled as a Dynamically Coloured Petri Net (DCPN). In this section we model an abstract version of this system as a value-passing CPDP. We first give a global description of the system. After that we give a more detailed description of each CPDP component.

There are seven distinct goals defined for the pilot-flying, C1 till C7. Which goal should be achieved by the pilot at which time depends on the situation. If at some time t_1, the pilot is working on goal C1 (which is: collision avoidance) then CPDP *CurrentGoal* is in location l_1 with $k = 1$ (the value of k equals the number of the goal) and CPDP $TaskPerformance$ is in the top location (meaning that the pilot is performing tasks for some goal while the bottom loction means that the pilot is not working an a goal). If the pilot is working on goal C2 (which is: emergency actions), then $k = 2$ and then the value q denotes which specific emergency action is executed (if $k \neq 2$ then q, which is not relevant then, equals zero). The pilot can switch to another goal in two ways:

1. He achieved a goal and is ready for a new goal. He 'looks' at the memory-unit whether there is another goal that needs to be achieved. In that case the pilot starts working on the goal in the memory-unit with the highest priority (C_1 has priority over C_2, C_2 over C_3 etc.), unless he sees on the display of $HMI{-}PF$, which is a failure indicator device, that certain aircraft-systems are not working properly. In the latter case the pilot should switch to goal C2 (emergency action).
2. The pilot is working on a goal, while CPDP *AudioAlert*, which is a communication device that can communicate alert messages, sends an *alert*-message. This message contains a value (communicated via value-passing communication) which denotes the interrupt-goal. CPDP *CurrentGoal* receives this message and if the interrupt goal has higher priority than the goal that is worked on, the pilot switches to the interrupt-goal. If the interrupt-goal has lower priority, the goal is stored into the memory-unit.

We now briefly say how the interactions between the five components are modelled: CPDP *CurrentGoal* reads the memory and the failure-indicators via value-passing-synchronization on events *getmem* and $getHMI$ respectively (see Figure 7). *CurrentGoal* receives alert-messages via value-passing-synchronization on event *alert*. $TaskPerformance$ sends the active signal

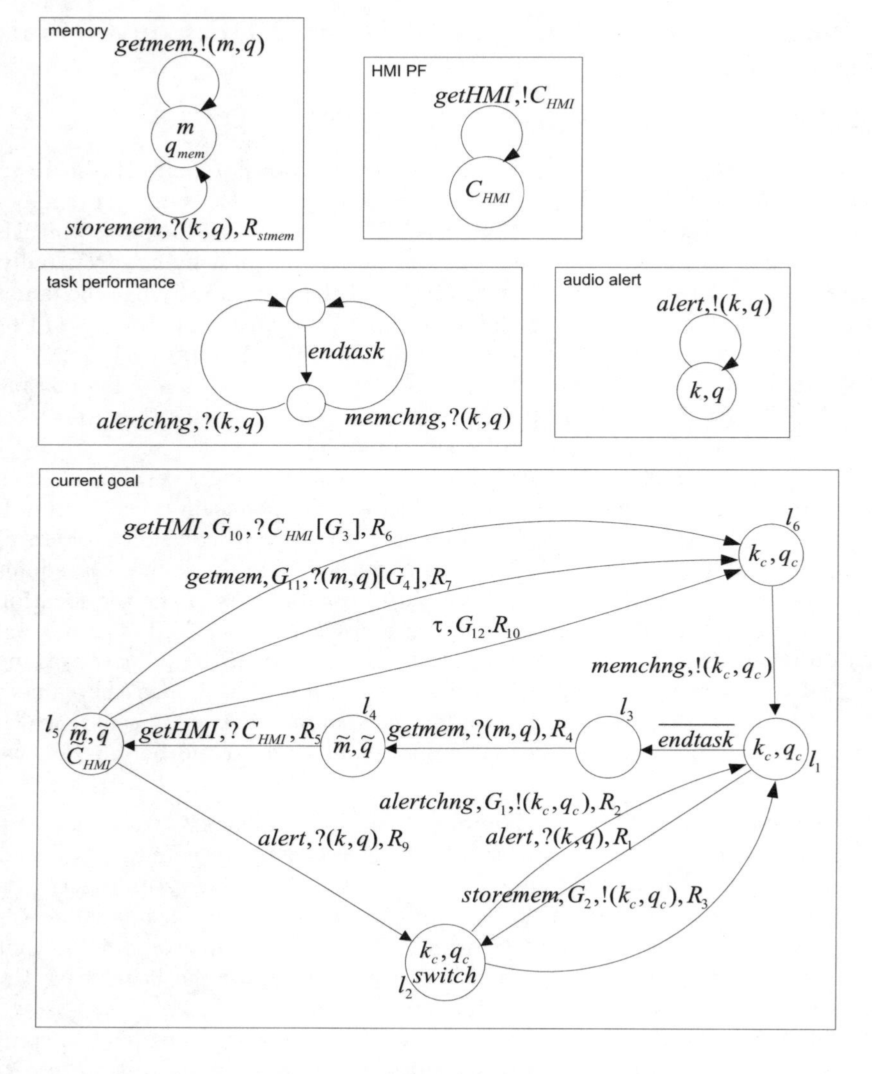

Fig. 7. CPDP pilot flying model

endtask as soon as the pilot finished the last task of the goal he was working on, this signal is received by *CurrentGoal* via a passive $\overline{endtask}$-transition. *CurrentGoal* stores a value in the memory-unit *Memory* via a value-passing-synchronization on event *storemem*. Finally, *CurrentGoal* communicates to *TaskPerformance* that a new goal is started because of an alert-message or because a new goal was retrieved from the memory, via value-passing-synchronization on events *alertchng* and *memchng* respectively.

The five CPDPs are interconnected via composition operators of the $|^P_A|$ type as

$$(((CurrentGoal|_{A_1}|AudioAlert)|_{A_2}|Memory)$$
$$|_{A_3}|TaskPerformance)|_{A_4}|HMI{-}PF, \tag{4}$$

with $A_1 := \{alert\}$, $A_2 := \{getmem, storemem\}$, $A_3 := \{alertchng, memchng\}$ and $A_4 = \{getHMI\}$. We now describe each of the five CPDPs in more detail.

CPDP *HMI-PF* has one location with one variable named C_{HMI}. The value of this variable indicates whether there is a failure in one of the five systems (indicated by *HMI-PF*). C_{HMI} consists of five components C^i_{HMI} $(i = 1, 2, 3, 4, 5)$ which all have either value $true$ or $false$ (with $true$ indicating a failure for the corresponding system). There is only one transition, which is an unguarded active transition from the only location to itself with label $getHMI$ and with output C_{HMI}. This transition is used only to send the state information to the component *CurrentGoal*, therefore the reset map of this transition does not change the state C_{HMI}. Note that for the CPDPs in this ATM-example, we do not define output variables. We assume that for every state variable used in active transitions we have an output variable copy defined.

CPDP $AudioAlert$ has one location with two variables named k and q. $k \in \{1, 2, 3, 4, 5, 6, 7\}$ and $q \in \{1, 2, 3, 4, 5, 6\}$. These values represent the interrupt goal (and failure in case $k = 2$). There is one active transition with label $alert$ and with outputs k and q. This transition should normally be guarded (where the guard is satisfied as soon as an alert signal should be sent), but at the abstraction level of our model we do not model this. Also the reset map of this transition is not specified here.

CPDP $Memory$ has one location with two variables named m and q_{mem}. m is a variable with seven components (m_1 till m_7 for the goals C1 till C7) which can have value ON and OFF. (In the DCPN model of this system there is also the value $LATER$ for m_4 and m_5 which we do not consider in the CPDP). q_{mem} is a variable with six components (for the six failures) taking values in $\{0, 1\}$. There are two active transitions. The unguarded transition with label $getmem$ and output m and q_{mem} is used to send information to *CurrentGoal*, therefore the reset map leaves the state unaltered. The unguarded transition with label $storemem$ and input k and q is used by *CurrentGoal* to change the memory state. (Note that we write $?(k, q)$ to denote inputs of the combined state-space of k and q which is $?\mathbb{R}^2$ because $k, q \in \mathbb{R}$). The reset map R_{stmem} of this transition changes m_k (with k the received input) to ON and changes q^q_{mem} (with q the received input) to 1.

CPDP *TaskPerformance* has two locations, $Idle$ and $Busy$, both without variables. When the system switches from $Busy$ to $Idle$, the active transition with label *endtask* is executed. The system can switch from $Idle$ to $Busy$ via two transitions: 1. Via the active input transition with label $alertchng$ and inputs k and q. This happens when *CurrentGoal* executes an active output transition with label *alertchng* due to having received a signal from *AudioAlert*. (Normally *TaskPerformance* should use the information from the inputs k and q via the reset map of the transition, but we do not model that at our level of abstraction). 2. Via the active input transition with label *memchng* and inputs k and q. This happens when *CurrentGoal* executes an active output transition with label *memchng* due to the situation where the pilot is idling and a new goal is retrieved by *CurrentGoal* from the memory.

CPDP *CurrentGoal* is the only CPDP that we have modelled in detail. *CurrentGoal* has six locations, named l_1 till l_6. We will now describe each location:

- Location l_1 has two variables named k_c and q_c. The process is in this location when one of the goals is being achieved (i.e. *TaskPerformance* is in location *Busy*) and the values of k_c and q_c represent the current goal and (in case $k_c = 2$) current failure. There are two outgoing transitions: 1. An unguarded active input transition to l_2 labelled *alert* with inputs k and q, synchronizing on an *alert* signal from *AudioAlert*, with reset map

$$R_1 := \begin{cases} k_c := k, q_c := q, switch := true & \text{if k < k}_\text{c} \\ k_c := k_c, q_c := q_c.switch := false & \text{else.} \end{cases}$$

  2. A passive transition to l_3 labelled $\overline{endtask}$, synchronizing on an *endtask* signal from *TaskPerformance*.
- The process is in location l_2 when (1) after having received the *alert* signal the current goal needs to be changed (according to the *alert* signal) or when (2) the interrupt goal (from the *alert* signal) needs to be stored in memory. (1) is the case when $switch = true$, (2) is the case when $switch = false$. Therefore, $G_1 := \{(k_c, q_c, switch)|switch = true\}$, $G_2 := \{(k_c, q_c, switch)|switch = false\}$, with G_1 the guard of the active output transition labelled *alertchng* with outputs k_c and q_c and reset map R_2 and with G_2 the guard of the active output transition labelled *storemem* with outputs k_c and q_c and reset map R_3. R_2 and R_3 are the same and do the following reset: $k_c := k_c$, $q_c := q_c$. Note that, under maximal progress, the process jumps immediately to location l_1 as soon as it arrives in location l_2, causing also a synchronizing transition in either *TaskPerformance* (with label *alertchng*) or *Memory* (with label *storemem*).
- The process arrives in location l_3 after the *endtask* signal. Then the pilot should check the memory whether there are other goals that need to be achieved. With the unguarded active input transition with label *getmem* and inputs m and q and reset map R_4, the process jumps to location l_4 while retrieving the memory state (m, q). The reset map R_4 stores this (m, q) in $(\tilde{m}, \tilde{q})$.
- Before executing a goal from the memory, the pilot should first check *HMI-PF* to see whether there are indications for failing devices. This happens in the transition to l_5 on the label *getHMI* while retrieving the *HMI-PF*

state C_{HMI}. The reset R_5 stores C_{HMI} together with $\tilde{m}$ and $\tilde{q}$ in the state of l_5.

- From location l_5 there is an active transition to l_6 with label τ and guard $G_{12} := \{(\tilde{m}, \tilde{q}, \tilde{C}_{HMI}) | \tilde{C}^i_{HMI} = true$ for some $i = 1, 2, 3, 4, 5$ or $\tilde{m}^i = ON$ for some $i < 7\}$. Under maximal progress, this τ-transition is taken immediately after arriving in l_5 when the *Memory* and *HMI-PF* states give reason to work on a new goal. The reset map R_{10} resets $k_c := 2, q_c := r$ if $S := \{i | i \leq 5, \tilde{C}^i_{HMI} = true\} \neq \emptyset$, where r is randomly chosen from the set S, otherwise R_{10} resets $k_c := \min\{i | m_i = ON\}, q_c := 0$. If the guard G_{12} is not satisfied in l_5, then this means that the pilot should wait until an *alert* signal is received or until either the *Memory* state or the *HMI-PF* state changes such that the pilot should work on a new goal. On an *alert* signal from *AudioAlert* the transition to l_2 is taken where R_9 is equal to R_1. The active input transition to l_6 labelled *getmem* waits till the *Memory* state has changed such that the input-guard G_4 is satisfied, where $G_4 := \{(m, q) | m^i = ON$ for some $2 \neq i < 7\}$. The reset map R_7 resets $k_c := \min\{i | m_i = ON\}, q_c := 0$. The active input transition to l_6 labelled *getHMI* waits till the *HMI-PF* state has changed such that the input-guard G_3 is satisfied, where $G_3 := \{C_{HMI} | C^i_{HMI} = true$ for some $i = 1, 2, 3, 4, 5\}$. The reset map R_6 resets $k_c := 2, q_c := r$ with r randomly chosen from $S := \{i | i \leq 5, \tilde{C}^i_{HMI} = true\} \neq \emptyset$.
- If the process arrives in location l_6, then this means that the state of l_6 represents the goal that should immediately be worked on by the pilot. Therefore, the unguarded active transition to l_1 labelled *memchng* is taken immediately (under maximal progress). The outputs k_c and q_c are accepted by the *memchng* transition in *TaskPerformance*. The reset map of the output *memchng* transition copies the state of l_6 to the state of l_1.

6.2 Examples of Value-Passing-CPDP to PDP Conversion

We follow the algorithm from Section 4.1 to check whether the CPDP ATM-example of Section 6, which has no spontaneous transitions, can be converted to a PDP.

Example 3 (ATM). We assume that the system modelled by (4) is closed (i.e. no more components will be connected). This means that we remove the passive transitions in the composite CPDP (which are some $\overline{endtask}$ transitions). It can be seen that the composite CPDP does not have active input-transitions. We assume that time will elapse in the locations of *AudioAlert* and *TaskPerformance*. Both may have (different) extra dynamics of the form $\dot{x} = f(x)$, then the guards of transitions *alert* and *endtask* depend on x. We assume that the transitions *alert*, *alertchng* and *memchng* are stable. Note that location l_1 is unguarded, that locations l_2,l_3,l_4 and l_6 are guarded and that location l_5 has both an unguarded and a guarded state space.

First we look at T^1_s: the stable parts of the transitions that represent hybrid jumps of multiplicity one. For this example we have

$$T_s^1 = \{storemem, alertchng, memchng, getHMI_{s,45}\},$$

where these names correspond to the transitions with the same label in Figure 7: *storemem* represents the transition from l_2 to l_1 synchronized with the transition with the same label in component *memory*. $getHMI_{s,45}$ corresponds to the stable part, which is the part that does not jump into guard G_{12}, of the transition between l_4 and l_5 synchronizing with the transition in *HMI-PF*, etc. Because R_5 makes a copy of C_{HMI},m and q, we get that the guard of $getHMI_{s,45}$ equals $val(l_4)\backslash G_{12}$ and the guard of $getHMI_{u,45}$, the unstable part, equals G_{12}. Furthermore, we have for this example

$$T_u^1 = \{alert_{12}, alert_{52}, getmem_{34}, getmem_{56}, getHMI_{u,45}, getHMI_{56},$$

$$endtask\},\ \ T_s^2 = \{alertchng \circ alert_{12}, alertchng \circ alert_{52}, storemem \circ alert_{12},$$

$$storemem \circ alert_{52}, memchng \circ \tau, memchng \circ getHMI, memchng \circ getmem,$$

$$getHMI_s \circ getmem\},$$

where $getHMI_s \circ getmem$ denotes the transition that represents the hybrid jump of multiplicity two that consists of *getmem* from l_3 to l_4 followed directly by the stable part of $getHMI$ from l_4 to l_5, etc. Then,

$$T_u^2 = \{getmem \circ endtask, getHMI_u \circ getmem, \tau \circ getHMI\},$$

$$T_s^3 = \{memchng \circ \tau \circ getHMI_u, getHMI_s \circ getmem \circ endtask\},$$

$$T_u^3 = \{getHMI_u \circ getmem \circ endtask, \tau \circ getHMI_u \circ getmem\},$$

$$T_s^4 = \{memchng \circ \tau \circ getHMI_u \circ getmem\},$$

$$T_u^4 = \{\tau \circ getHMI_u \circ getmem \circ endtask\}.$$

$$T_s^5 = \{memchng \circ \tau \circ getHMI_u \circ getmem \circ endtask\},$$

$$T_u^5 = \emptyset.$$

We see, when X denotes the composite CPDP, that X^5 (i.e. the CPDP that has active transitions $(\cup_{i=1}^5 T_s^i) \cup T_u^5$) has no unstable transitions. This means that X^5 can directly be converted to a PDP, which then is the corresponding PDP of X.

To prove that the composite CPDP of this ATM example can be converted to a PDP, it would also have been enough to show that the CPDP does not have cycles such that the locations of the cycle all have guarded parts. It is clear that a cycle in component *Current goal* should include location l_1, which is an unguarded location. It can easily be seen that in the composite CPDP the two (product)locations that contain l_1 are both unguarded and that any cycle in the composite CPDP should contain one of these two locations. Therefore this composite CPDP does not have transitions with multiplicity infinity and should therefore be convertable to a PDP. (However, if we want to specify this PDP, we still have to do the algorithm or something similar).

Because the algorithm terminates on the ATM-example above, we know that the ATM-example has a PDP behavior. However, it is possible that the algorithm does not terminate, while the CPDP does exhibit a PDP behavior. We now give an example of this.

Example 4. Let CPDP X have one location, l_1. The state-space of l_1 is $[0,1]$, the continuous dynamics of l_1 is the clock dynamics $\dot{x} = 1$. From l_1 to l_1 there is one active transition with guard G and reset map R. $G = [\frac{1}{2}, 1]$. For $x \in G$, $R(\{0\}, x) = \frac{1}{2}$ and $R(A, x) = |A \cap [\frac{1}{2}, 1]|$ for $A \in \mathcal{B}([0,1]\backslash\{0\})$. This means that from an x in G, the reset map jumps to 0 with probability $\frac{1}{2}$ and jumps uniformly into $[\frac{1}{2}, 1]$ with probability $\frac{1}{2}$. It can easily be seen that for X we have that $T_u^n \neq \emptyset$ for all $n \in \mathbb{N}$. This means that the algorithm explained above does not terminate for this example. Still, according to Theorem 2, X expresses a PDP behavior, because for $x \in G$, $R([0,1], x) = \lim_{n\to\infty} R^n_{tot,s}([0,1], x) = \frac{1}{2} + \frac{1}{2} \cdot \frac{1}{2} + \frac{1}{2} \cdot \frac{1}{2} \cdot \frac{1}{2} + \cdots = 1$.

7 Bisimulation for CPDPs

In this section we define bisimulation relations for CPDPs. Bisimulation is an equivalence relation. The idea of bisimulation is that two CPDPs are bisimulation-equivalent if for an external agent the CPDPs cannot be distinguished from each other. We assume here that an external agent cannot see the state-value of a CPDP but it does see the output-value of a CPDP and it does also see the events (including possible value passing information) of active transitions. We assume that the behavior of the external agent can be modelled as another CPDP. Thus, if CPDPs X_1 and X_2 are bisimilar (i.e. bisimulation-equivalent), then $X_1|^P_A|Y$ and $X_2|^P_A|Y$ behave externally equivalently for each external-agent-CPDP Y and each operator of the form $|^P_A|$. *External equivalent behavior* will be defined later in this section, but for the intuitive understanding, we will already give two examples here.

1. Suppose the initial states of CPDPs X_1, X_2 are given. If then, for some CPDP Y (with some initial state) and some $|^P_A|$, the probability that the output-value of $X_1|^P_A|Y$ equals $\hat{w}$ at time $\hat{t}$, is different from the probability that the output-value of $X_2|^P_A|Y$ equals $\hat{w}$ at time $\hat{t}$, then X_1 and X_2 are not bisimilar.

2. As an example of two bisimilar CPDPs, we compare CPDP X from Figure 4 to CPDP $\tilde{X}$ from Figure 8. We let $\tilde{\lambda}$, $\tilde{\mu}$, all $\tilde{G}_i$ and all $\tilde{R}_i$ be copies of λ,μ,G_i and R_i from Figure 4, i.e. $\tilde{\lambda}$, $\tilde{\mu}$, $\tilde{G}_i$ and the $\tilde{x}$-resets of $\tilde{R}_i$ do not depend on $\bar{x}$. The $\bar{x}$ resets of $\tilde{R}_i$ are not relevant here and may therefore be chosen arbitrarily (like $\bar{x} := 0$ for each $\tilde{R}_i$). Thus, we get $\tilde{\lambda}(\tilde{x}, \bar{x}) = \lambda(\tilde{x})$, $\tilde{G}_i = \{(\tilde{x}, \bar{x}) | \tilde{x} \in G_i\}$, etc. Then, the only difference between X and $\tilde{X}$, if we regard $\tilde{x}$ as a copy of x, is that the locations of $\tilde{X}$ have another state variable $\bar{x}$ (evolving along vectorfields $\bar{f}_1$ and $\bar{f}_2$). But this extra variable $\bar{x}$ does not influence the output y, which only depends on x (or $\tilde{x}$), and it also

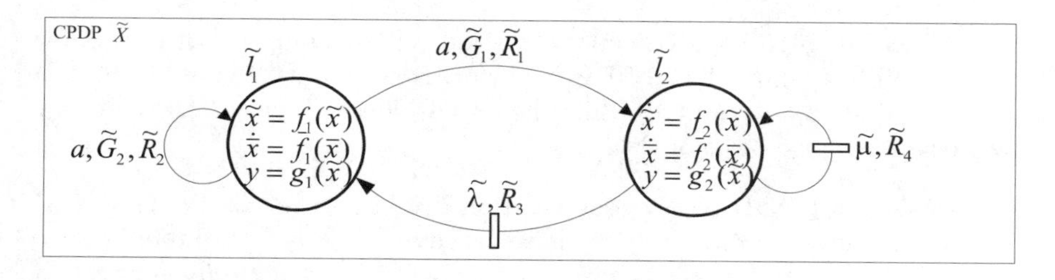

Fig. 8. CPDP $\tilde{X}$ (bisimulation equivalent to CPDP X of Figure 4)

does not influence hybrid jumps because it does not influence the guards of the transitions, the Poisson processes and the resets of x (or $\tilde{x}$). It is intuitively clear then that CPDPs X and $\tilde{X}$ cannot be distinguished by an external agent. After the formal definition of bisimulation for CPDPs, we will show that X and $\tilde{X}$ are indeed bisimilar.

X can be seen as a state reduced equivalent of $\tilde{X}$ because the state space of X is smaller (i.e. the variable $\bar{x}$ is not present in X). More formally, we could say that we have state reduction because each state x of X represents a whole set of states $\{(\tilde{x}, \bar{x})|\tilde{x} = x\}$ of $\tilde{X}$ (i.e. the state valuation $(x = 1)$ of X for example, represents the set of state valuations $\{(\tilde{x} = 1, \bar{x} = r)|r \in \mathbb{R}\}$ of $\tilde{X}$). State valuation $(\tilde{x} = 1, \bar{x} = 0)$ is for example equivalent to state valuation $(\tilde{x} = 1, \bar{x} = 1)$ because the external behavior of $\tilde{X}$ that starts/continues from $(\tilde{x} = 1, \bar{x} = 0)$ is the same as the external behavior of $\tilde{X}$ that starts/continues from $(\tilde{x} = 1, \bar{x} = 1)$. We could say therefore that $\{(\tilde{x} = 0, \bar{x} = r)|r \in \mathbb{R}\}$ forms an equivalence class of states. In the formal definition of bisimulation for CPDPs, we will see that we can indeed use this concept of equivalence classes of states. Before we do that, we need to introduce the technical concepts of *induced equivalence relation*, *measurable relation* and *equivalent (probability) measure*.

We define the equivalence relation on X that is induced by a relation $\mathcal{R} \subset X \times Y$ with the property that $\pi_1(\mathcal{R}) = X$ and $\pi_2(\mathcal{R}) = Y$, where $\pi_i(\mathcal{R})$ denotes the projection of $\mathcal{R}$ on the i-th component, as the transitive closure of $\{(x, x')|\exists y \text{ s.t. } (x, y) \in \mathcal{R} \text{ and } (x', y) \in \mathcal{R}\}$. We write $X/_{\mathcal{R}}$ and $Y/_{\mathcal{R}}$ for the sets of equivalence classes of X and Y induced by $\mathcal{R}$. We denote the equivalence class of $x \in X$ by $[x]$. We will now define the notions of *measurable relation* and of *equivalent measure*.

Definition 8. *Let $(X, \mathcal{X})$ and $(Y, \mathcal{Y})$ be Borel spaces and let $\mathcal{R} \subset X \times Y$ be a relation such that $\pi_1(\mathcal{R}) = X$ and $\pi_2(\mathcal{R}) = Y$. Let $\mathcal{X}^*$ be the collection of all $\mathcal{R}$-saturated Borel sets of X, i.e. all $B \in \mathcal{X}$ such that any equivalence class of X is either totally contained or totally not contained in B. It can be checked that $\mathcal{X}^*$ is a σ-algebra. Let*

$$\mathcal{X}^*/_{\mathcal{R}} = \{[A]|A \in \mathcal{X}^*\},$$

where $[A] := \{[a] | a \in A\}$. *Then* $(X/_{\mathcal{R}}, \mathcal{X}^*/_{\mathcal{R}})$, *which is a measurable space, is called the* quotient space *of* X *with respect to* $\mathcal{R}$. *A unique bijective mapping* $f : X/_{\mathcal{R}} \to Y/_{\mathcal{R}}$ *exists, such that* $f([x]) = [y]$ *if* $(x, y) \in \mathcal{R}$. *We say that the relation* $\mathcal{R}$ *is* measurable *if for all* $A \in \mathcal{X}^*/_{\mathcal{R}}$ *we have* $f(A) \in \mathcal{Y}^*/_{\mathcal{R}}$ *and vice versa.*

If a relation on $X \times Y$ is measurable, then the quotient spaces of X and Y are homeomorphic (under bijection f from Definition 8). We could say therefore that under a measurable relation X and Y have a shared quotient space. In the field of descriptive set theory, a relation $\mathcal{R} \subset X \times Y$ is called measurable if $\mathcal{R} \in \mathcal{B}(X \times Y)$ (i.e. $\mathcal{R}$ is a Borel set of the space $X \times Y$). This definition does not coincide with our definition of measurable relation. In fact, many interesting measurable relations are not Borel sets of the product space $X \times Y$.

Definition 9. *Suppose we have measures* P_X *and* P_Y *on Borel spaces* $(X, \mathcal{X})$ *and* $(Y, \mathcal{Y})$ *respectively. Suppose that we have a measurable relation* $\mathcal{R} \subset X \times Y$. *The measures* P_X *and* P_Y *are called equivalent with respect to* $\mathcal{R}$ *if we have* $P_X(f_X^{-1}(A)) = P_Y(f_Y^{-1}(f(A)))$ *for all* $A \in \mathcal{X}^*/_{\mathcal{R}}$ *(with* f *as in Definition 8 and with* f_X *and* f_Y *the mappings that map* X *and* Y *to* $X/_{\mathcal{R}}$ *and* $Y/_{\mathcal{R}}$ *respectively).*

As an example, we show that relation $\mathcal{R} = \{(x, (\tilde{x}, \bar{x})) | x = \tilde{x}\}$ on $val(X) \times val(\tilde{X})$, where $val(X)$ and $val(\tilde{X})$ denote the state spaces of CPDPs X and $\tilde{X}$ of Figures 4 and 8, is a measurable relation and that the reset maps $R_i(x)$ and $\tilde{R}_i(\tilde{x}, \bar{x})$ are equivalent measures under this relation if $f([x]) = ([\tilde{x}, \bar{x}])$: the induced equivalence relation of $\mathcal{R}$ on X equals $\{\{x\} | x \in val(X)\}$, i.e. each single valuation forms an equivalence class of X. The induced equivalence relation of $\mathcal{R}$ on $\tilde{X}$ equals $\{\{(\tilde{x} = q, \bar{x} = r) | r \in \mathbb{R}\} | q \in \mathbb{R}\}$. The saturated Borel sets of X are all Borel sets of X, the saturated Borel sets $\tilde{X}$ are all sets of the form $B \times \mathbb{R}$ with B a Borel set for the state $\tilde{x}$ (i.e. a Borel set of $\mathbb{R}$). The bijective mapping f from Definition 8 maps each saturated Borel set B of X to the saturated Borel set $B \times \mathbb{R}$ of Y, from which follows, according to Definition 8, that $\mathcal{R}$ is measurable.

If states x and $(\tilde{x}, \bar{x})$ are equivalent (i.e. $f([x]) = [(\tilde{x}, \bar{x})]$), then the measures $R_i(\cdot, x)$ and $\tilde{R}_i(\cdot, (\tilde{x}, \bar{x}))$ are equivalent because R_i and $\tilde{R}_i$ are defined such that for each (saturated borel set of X) $B \in \mathcal{B}(\mathbb{R})$ we have $R_i(B, x) = \tilde{R}_i(B \times \mathbb{R}, (\tilde{x}, \bar{x}))$.

In order to define bisimulation for CPDPs we also need to introduce the notions of *combined reset map* and *combined jump rate function*: we consider CPDP (without value passing) $X = (L, V, W, v, w, F, G, \Sigma, \mathcal{A}, \mathcal{P}, \mathcal{S})$, with hybrid state space $E = E_s \cup E_u$, together with scheduler S. We define R, which we call the combined reset map, as follows. R assigns to each triplet (l, x, a) with $(l, x) \in E_u$ and with $a \in \Sigma$ such that $l \xrightarrow{a}$ (i.e. there exists an active transition labelled a leaving l), a measure on E. This measure $R(l, x, a)$ is for any l' and any Borel set $A \subset val(l')$ defined as:

$$R(l,x,a)(l',A) = \sum_{\alpha \in \mathcal{A}_{l,a,l'}} S(l,x)(\alpha)R_\alpha(A,x),$$

where $\mathcal{A}_{l,a,l'}$ denotes the set of active transitions from l to l' with label a and (l',A) denotes the set $\{(l',x)|x \in A\}$. (This measure is uniquely extended to all Borel sets of E). Now, for $A \in \mathcal{B}(E)$, $R(l,x,a)(A)$ equals the probability of jumping into A via an active transition with label a given that the jump takes place at (l,x).

Furthermore, R assigns to each triplet $(l,x,\bar{a})$ with $(l,x) \in E$ and with $\bar{a} \in \bar{\Sigma}$ such that $l \xrightarrow{\bar{a}}$, a measure on E, which for any l' and any Borel set $A \subset val(l')$ is defined as:

$$R(l,x,\bar{a})(l'_1,A) = \sum_{\alpha \in \mathcal{P}_{l,\bar{a},l'}} S(l,x)(\alpha)R_\alpha(A,x).$$

(This measure is uniquely extended to all Borel sets of E). Now, $R(l,x,\bar{a})(A)$, with $A \in \mathcal{B}(E)$, equals the probability of jumping into A if a passive transition with label $\bar{a}$ takes place at (l,x).

We define the combined jump rate function λ for CPDP X as

$$\lambda(l,x) = \sum_{\alpha \in \mathcal{S}_{l\rightarrow}} \lambda_\alpha(l,x),$$

with $(l,x) \in E$.

Finally, for spontaneous jumps, R assigns to each $(l,x) \in E$ such that $\lambda(l,x) \neq 0$, a probability measure on E, which for any l' and any Borel set $A \subset val(l')$ is defined as:

$$R(l,x)(l'_1,A) = \sum_{\alpha \in \mathcal{S}_{l\rightarrow l'}} \frac{\lambda_\alpha(l,x)}{\lambda(l,x)} R_\alpha(A,x).$$

(This measure is uniquely extended to all Borel sets of E). Now we are ready to give the definition of bisimulation for CPDPs.

Definition 10. *Suppose we have CPDPs $X = (L_X, V_X, W, v_X, w_X, F_X, G_X, \Sigma, A_X, P_X, S_X)$ and $Y = (L_Y, V_Y, W, v_Y, w_Y, F_Y, G_Y, \Sigma, A_Y, P_Y, S_Y)$ with shared W and Σ and with schedulers S_X and S_Y. A measurable relation $\mathcal{R} \subset val(X) \times val(Y)$ is a bisimulation if $((l_1,x),(l_2,y)) \in \mathcal{R}$ implies that*

1. *$\omega_X(l_1) = \omega_Y(l_2)$, for all $w \in \omega_X(l_1)$ we have $G_X(l_1,x,w) = G_Y(l_2,y,w)$, $\lambda(l_1,x) = \lambda(l_2,y)$ (with λ the combined jump rate function defined on both $val(X)$ and $val(Y)$).*
2. *$(\phi_{l_1}(t,x), \phi_{l_2}(t,y)) \in \mathcal{R}$ (with $\phi_l(t,z)$ the state at time t when the state equals z at time zero).*
3. *If $\lambda(l_1,x) = \lambda(l_2,y) \neq 0$, then $R(l_1,x)$ and $R(l_2,y)$ are equivalent probability measures with respect to $\mathcal{R}$.*

4. *For any $\bar{a} \in \bar{\Sigma}$ we have that either both $l_1 \not\xrightarrow{\bar{a}}$ and $l_2 \not\xrightarrow{\bar{a}}$ or else $R(l_1, x, \bar{a})$ and $R(l_2, y, \bar{a})$ are equivalent probability measures.*
5. *For any $a \in \Sigma$ we have that either both $l_1 \not\xrightarrow{a}$ and $l_2 \not\xrightarrow{a}$ or else $R(l_1, x, a)$ and $R(l_2, y, a)$ are equivalent measures.*

X with initial state (l_1, x) and Y with initial state (l_2, y) are bisimilar if $((l_1, x), (l_2, y))$ is contained in some bisimulation.

Definition 10 formalizes what we mean by equivalent external behavior. It can now be seen that, according to Definition 10, CPDP X (from Figure 4) with initial state (l_x, x) (for some l_x and some $x \in val(l_x)$) together with some scheduler S_X, and CPDP $\tilde{X}$ (from Figure 8) with initial state $(l_{\tilde{x}}, (\tilde{x}, \bar{x}))$ (with $l_{\tilde{x}} = l_x$ and $\tilde{x} = x$ and $\bar{x} \in \mathbb{R}$) together with scheduler $S_{\tilde{X}}(\tilde{l}, (\tilde{x} = q, \bar{x} = r))(\tilde{\alpha}) := S_X(l, x = q)(\alpha)$ (where $\tilde{\alpha}$ is the transition of $\tilde{X}$ that corresponds according to Figures 4 and 8 to transition α of X) are bisimilar under the relation $\mathcal{R} = \{(x, (\tilde{x}, \bar{x})) | x = \tilde{x}\}$ on $val(X) \times val(\tilde{X})$ (which was already shown to be a measurable relation).

We now state a theorem which justifies our notion of bisimulation when it concerns the stochastic behavior. It says that if two closed CPDPs are bisimilar, then the stochastic processes that model the output evolution of the CPDPs are equivalent (in the sense of indistinguishability).

Theorem 3. *The stochastic processes of the outputs of two bisimilar closed CPDPs (with their schedulers), whose quotient spaces are Borel spaces, can be realized such that they are indistinguishable.*

Proof. The proof can be found in [15]. There, *invariants* are used instead of guards. It can be seen that the proof is still valid if the invariant of a location is defined as the unguarded state space of that location.

It can easily be seen that if two non-closed CPDPs are bisimilar, then if we close both CPDPs (i.e. if we remove all passive transitions), then the closed CPDPs are still bisimilar and, by Theorem 3, the stochastic processes that model the output evolution of the CPDPs are equivalent.

We now state a theorem which justifies our notion of bisimulation when it concerns the interaction behavior. It says that two bisimilar CPDPs interact in an equivalent way (with any other CPDP) by stating that substituting a CPDP-component (in a composition context with multiple components) by another, but bisimilar, component, results in a composite CPDP that is bisimilar to the original composite CPDP. Checking bisimilarity between two composite CPDPs can only be done if both composite CPDPs have their own schedulers. Therefore we first have to investigate how a scheduler of a composite CPDP can be composed from the schedulers of the components.

It appears that the schedulers of the components do not contain enough information to define the scheduler of the composite CPDP. We illustrate this with Figure 9, where we see two CPDPs, X and Y, with schedulers S_X and

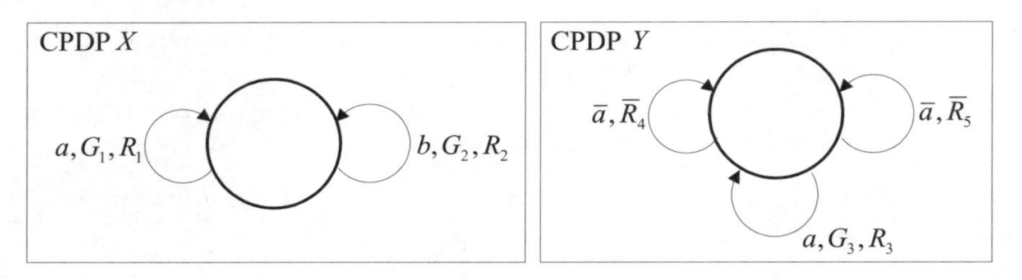

Fig. 9. Example concerning internal/external scheduling

S_Y. Suppose we connect X and Y via composition operator $|_{\emptyset}^{\bar{\Sigma}}|$. If $x \in G_1$ and $x \notin G_2$ and $y \notin G_3$, then the scheduler S of $X|_{\emptyset}^{\bar{\Sigma}}|Y$ is at (x, y) determined because (a, G_1, R_1) is the only transition that is enabled at (x, y), therefore the scheduler has to choose this transition. However, this a-transition will trigger one of the two $\bar{a}$-transitions of Y. Thus, the scheduler still has to choose between the transitions $(a, G_1 \times val(Y), R_1 \times \bar{R}_4)$ (i.e. the synchronization of (a, G_1, R_1) with $(\bar{a}, \bar{R}_4)$) and $(a, G_1 \times val(Y), R_1 \times \bar{R}_5)$. Here we should respect S_Y which is defined to make a choice between the two passive transitions. Thus we get,

$$S(x, y)(a, G_1 \times val(Y), R_1 \times \bar{R}_i) = S_Y(y, \bar{a})(\bar{a}, \bar{R}_i), \quad i \in \{4, 5\}.$$

If $x \notin G_1$ and $x \in G_2$ and $y \in G_3$, then at state (x, y), two active transitions of $X|_{\emptyset}^{\bar{\Sigma}}|Y$ are enabled: $(b, G_2 \times val(Y), R_2 \times Id)$ and $(a, val(X) \times G_3, Id \times R_3)$. S_X and S_Y give no information how to choose between the b-transition and the a-transition. We call this case a case of *external scheduling* (i.e. the choice cannot be made by the internal schedulers, the schedulers of the individual components). Thus, besides the internal schedulers S_X and S_Y, we need a strategy for external scheduling. We define this as follows.

Definition 11. *ESS is an external scheduling strategy for $X|_A^P|Y$ with internal schedulers S_X and S_Y if ESS assigns to each state (x, y) a mapping from the set of event pairs EP to $[0, 1]$, where*

$$EP := \{[\alpha, \beta] | \alpha = \beta \in \Sigma, \alpha \in \Sigma \wedge \beta = *, \alpha = * \wedge \beta \in \Sigma,$$

$$\alpha \in \Sigma \wedge \beta = \bar{\alpha}, \alpha = \bar{\beta} \wedge \beta \in \Sigma, \alpha = \beta \in \bar{\Sigma}, \alpha \in \bar{\Sigma} \wedge \beta = *, \alpha = * \wedge \beta \in \bar{\Sigma}\},$$

which respects the transition structure of $X|_A^P|Y$.

We explain the meaning of external scheduling strategy by using the example of Figure 9: if ESS is an external scheduling strategy for $X|_{\emptyset}^{\bar{\Sigma}}|Y$ and $ESS(x, y)([a, \bar{a}]) = 1$, then the set of transitions of the form $(a, G_x \times val(Y), R_x \times \bar{R}_y)$ (with (a, G_x, R_x) an a-transition of X and $(\bar{a}, \bar{R}_y)$ a $\bar{a}$-transition of Y) at state (x, y) get probability one. The probabilities of the individual transitions of this form are determined by the internal schedulers.

If we have $ESS(x,y)([a,\bar{a}]) > 0$ with $x \notin G_1$, then ESS does not respect the transition structure, because for $x \notin G_1$ no a-transition of X can be executed, and is therefore not a valid external scheduling strategy, etc. In general, an external scheduling strategy does not have to respect the internal schedulers where it concerns the choice between active transitions (within one component) labelled with different events, but it has to respect the internal schedulers where it concerns the passive transitions and the choice between active transitions (in one component) with the same event-label. The choice to allow to ignore internal schedulers where it concerns active transitions with different event-labels, has been made because first, in some cases it is not clear what it means to respect the internal schedulers and second, this freedom does not influence the result of the bisimulation-substitution-theorem that we state after the following example about a scheduler that does respect the internal schedulers as much as possible.

Example 5. Suppose we have two CPDPs X and Y with schedulers S_X and S_Y, which we interconnect with composition operator $|^{\Sigma}_{\emptyset}|$. A valid external scheduling strategy would be:

- For states (x,y) with $x \in val_u(X)$ (i.e. the guarded states of X) and $y \in val_s(Y)$ the choice for the active transition of X is made by S_X. (Which passive transitions synchronize depends on Y and S_Y)
- For states (x,y) with $x \in val_s(X)$ and $y \in val_u(Y)$ the choice for the active transition of Y is made by S_Y. (Which passive transitions synchronize depends on X and S_X)
- For states (x,y) with $x \in val_u(X)$ and $y \in val_u(Y)$, the choice for the active transition (of X or Y) is determined with probability half by S_X and with probability half by S_Y. (Which passive transitions synchronize depends on X,Y, S_X and S_Y).

Note that the strategy of Example 5 will not work in case $A \neq \emptyset$. Also, in general, the composition of two schedulers under an external scheduling strategy, which results in a internal schedular for the composite system (as in Example 5), is not commutative and not associative.

Theorem 4. *Suppose we have three CPDPs, X_1,X_2 and Y, with schedulers S_{X_1}, S_{X_2} and S_Y. Suppose $\mathcal{R} \subset val(X_1) \times val(X_2)$ is a bisimulation and $val(X_1)/_{\mathcal{R}}$ and $val(X_2)/_{\mathcal{R}}$ (i.e. the quotient spaces of X_1 and X_2 under $\mathcal{R}$) are Borel spaces. Then,*

$$\mathcal{R}' := \{((x_1,y),(x_2,y))|(x_1,x_2) \in \mathcal{R}, y \in val(Y)\}$$

is a bisimulation on $(val(X_1) \times val(Y)) \times (val(X_2) \times val(Y))$ for the CPDPs $X_1|^P_A|Y$ and $X_2|^P_A|Y$ with external scheduling strategies ESS_1 and ESS_2 such that $ESS_1(x_1,y) = ESS_1(x_2,y)$ if $(x_1,x_2) \in \mathcal{R}$. Furthermore, $(val(X_1) \times val(Y))/_{\mathcal{R}'}$ and $(val(X_2) \times val(Y))/_{\mathcal{R}'}$ are Borel spaces.

Proof. The proof can be found, mutatis mutandis, in [15].

With Theorem 4, we can use bisimulation as a compositional reduction technique: suppose we want to perform stochastic analysis on a (closed) composite CPDP that consists of multiple components. To reduce the state space of this complex system, we can reduce (by bisimulation) each component individually and put the reduced state component back in the composition. In this way the state of the composite CPDP will be reduced as soon as one or more of the components are state reduced. We know that the stochastic behavior of the output evolution is not changed by bisimulation, therefore we can perform the stochastic analysis on the (closed) state reduced composite CPDP.

Bisimulation for value-passing CPDPs

The definition of bisimulation can also be defined for value-passing CPDPs. We will not do that here, but we are convinced that it can be shown that with small extensions to the operation of schedulers (such that they can handle value-passing), and to the definitions of combined reset map and external scheduling strategies, the Theorems 3 and 4 also apply to the case of value-passing CPDPs. However, this result still has to be achieved.

8 Conclusions and Discussion

In this chapter we introduced the CPDP automata framework. CPDPs are automata with labelled transitions and spontaneous (stochastic) transitions. The locations of a CPDP are enriched with state and output variables. Each state variable (of a specific location) evolves according to a specified differential equation. State variables are probabilistically reset after a transition has been executed. CPDPs can interact/communicate with each other via the event-labels of the labelled transitions. For the extended framework value-passing-CPDP, event labels may even hold information about the output variables. We defined a bisimulation notion for CPDP. We proved that bisimilar CPDPs exhibit equivalent stochastic and interaction behavior. Therefore, bisimulation can be used as a compositional state reduction technique.

This means that we can take a component from a complex CPDP, find a state reduced bisimilar component and put the state reduced component back in the composition. The problem however is: how to find a state reduced bisimilar component? For certain classes of systems, like for IMC (see [8]) and for linear input/output systems (see [16]), (decidable) algorithms have been developed to find maximal (i.e. maximally state reduced) bisimulations. Since CPDPs are very general in the stochastics and the continuous dynamics, we can not expect that similar algorithms can be developed for CPDPs also. However, we can try to find subclasses of CPDPs that do allow automatic generation of maximal bisimulations. Any complex CPDP can then in principle be state reduced by finding the components that allow automatic generation of bisimulations and replace these components with their maximal bisimilar equivalents.

Bisimulation can be seen as a compositional analysis technique, i.e. it uses the composition structure in order to make analysis easier. Other compositional analysis techniques should benefit from the composition structure in their specific ways. In our CPDP model there is a clear distinction between the different components of a complex system and it is formalized how the composite behavior is constituted from the components and from the interaction mechanisms (i.e. the composition operators) that interconnect the components. Since we have this clear and formal composition structure (including a clear operational semantics for the composition operation), we think our model might be suitable for developing compositional analysis techniques.

References

1. T. Bolognesi and E. Brinksma. Introduction to the iso specification language lotos. *Comp. Networks and ISDN Systems*, 14:25–59, 1987.
2. P. R. D'Argenio. *Algebras and Automata for Timed and Stochastic Systems.* PhD thesis, University of Twente, 1997.
3. M. H. A. Davis. Piecewise Deterministic Markov Processes: a general class of non-diffusion stochastic models. *Journal Royal Statistical Soc. (B)*, 46:353–388, 1984.
4. M. H. A. Davis. *Markov Models and Optimization.* Chapman & Hall, London, 1993.
5. S.N. Strubbe et al. On control of complex stochastic hybrid systems. Technical report, Twente University, 2004. http://www.nlr.nl/public/hosted-sites/hybridge/.
6. M. H. C. Everdij and H. A. P. Blom. Petri-nets and hybrid-state markov processes in a power-hierarchy of dependability models. In *Proceedings IFAC Conference on Analysis and Design of Hybrid Systems ADHS 03*, 2003.
7. M.H.C. Everdij and H.A.P. Blom. Piecewise deterministic Markov processes represented by dynamically coloured Petri nets. *Stochastics: An International Journal of Probability and Stochastic Processes*, 77(1):1–29, February 2005.
8. H. Hermanns. *Interactive Markov Chains*, volume 2428 of *Lecture Notes in Computer Science.* Springer, 2002.
9. M. Haj-Hussein L. Logrippo, M. Faci. An introduction to lotos: Learning by examples. *Comp. Networks and ISDN Systems*, 23(5):325–342, 1992.
10. K. G. Larsen and A. Skou. Bisimulation through probabilistic testing. *Information and Computation*, 94:1–28, 1991.
11. R. Milner. *Communication and Concurrency.* Prentice Hall, 1989.
12. S. N. Strubbe, A. A. Julius, and A. J. van der Schaft. Communicating Piecewise Deterministic Markov Processes. In *Proceedings IFAC Conference on Analysis and Design of Hybrid Systems ADHS 03*, 2003.
13. S. N. Strubbe and R. Langerak. A composition operator for complex control systems. Submitted to Formal Methods conference 2005, 2005.

14. S. N. Strubbe and A. J. van der Schaft. Stochastic equivalence of CPDP-automata and Piecewise Deterministic Markov Processes. Accepted for the IFAC world congress, 2005.
15. S.N. Strubbe and A.J. van der Schaft. Bisimulation for communicating piecewise deterministic markov processes (cpdps). In *HSCC 2005*, volume 3414 of *Lecture Notes in Computer Science*, pages 623–639. Springer, 2005.
16. A.J. van der Schaft. Bisimulation of dynamical systems. In *HSCC 2004*, volume 2993 of *Lecture Notes in Computer Science*, pages 555–569. Springer, 2004.

Part II

Analytical Approaches

A Stochastic Approximation Method for Reachability Computations

Maria Prandini[1] and Jianghai Hu[2]

[1] Dipartimento di Elettronica e Informazione, Politecnico di Milano, Piazza Leonardo da Vinci 32, 20133 Milano, Italy, prandini@elet.polimi.it

[2] School of Electrical and Computer Engineering, Purdue University, West Lafayette, IN 47906, USA, jianghai@purdue.edu

Summary. We develop a grid-based method for estimating the probability that the trajectories of a given stochastic system will eventually enter a certain target set during a –possibly infinite– look-ahead time horizon. The distinguishing feature of the proposed methodology is that it rests on the approximation of the solution to stochastic differential equations by using Markov chains. From an algorithmic point of view, the probability of entering the target set is computed by appropriately propagating the transition probabilities of the Markov chain backwards in time starting from the target set during the time horizon of interest. We consider air traffic management as an application example. Specifically, we address the problem of estimating the probability that two aircraft flying in the same region of the airspace get closer than a certain safety distance and that an aircraft enters a forbidden airspace area. In this context, the target set is the set of unsafe configurations for the system, and we are estimating the probability that an unsafe situation occurs.

1 Introduction

In general terms, a reachability problem consists of determining if the trajectories of a given system starting from some set of initial states will eventually enter a pre-specified set.

An important application of reachability analysis is the verification of the correctness of the behavior of a system, which makes reachability analysis relevant in a variety of control applications. In particular, in many safety-critical applications a certain region of the state space is "unsafe", and one has to verify that the system state keeps outside this unsafe set. If the outcome of safety verification is negative, then some action has to be taken to appropriately modify the system.

Given the unsafe set and the set of initial states, a safety verification problem can be reformulated as either a forward reachability problem or a backward reachability problem. Forward reachability consists in determining the set of states that a given system can reach starting from some set of

H.A.P. Blom, J. Lygeros (Eds.): Stochastic Hybrid Systems, LNCIS 337, pp. 107–139, 2006.

initial states. Conversely, backward reachability consists in determining the set of initial states starting from which the system will eventually enter a given target set of states. One can perform safety verification by checking either that the forward reachable set is disjoint from the unsafe set or that the backward reachable set leading to the unsafe set is disjoint from the set of initial states.

One method for safety verification is the model checking approach, which verifies safety by constructing forward/backward reachable sets based on a model of the system. The main issue of this approach is the ability to "compute" with sets, i.e., to represent sets and propagate them through the system dynamics. This process can be made fully automatic. Model checkers have in fact been developed for different classes of deterministic systems.

In the case of deterministic finite automata, sets can be represented by enumeration, and forward (backward) reachable sets can be computed starting from the given initial (target) set and adding one-step successor (predecessor) till convergence is achieved. Termination of the algorithm is guaranteed since the state space is finite. Safety verification is then "decidable" for this class of systems, that is, there does exist a computational procedure that decides in a finite number of steps whether safety is verified or not for an arbitrary deterministic finite automata. The technical challenge for the verification of deterministic finite automata is to devise algorithms and data structure to handle large state spaces.

In the case of hybrid systems, two key issues arise due to the uncountable number of states in the continuous state space: i) set representation and propagation by continuous flow is generally difficult; and ii) the state space is not finite, hence termination of the algorithm for reachable set computation is not guaranteed ([32]). Decidability results have been proven for certain classes of hybrid systems by using discrete abstraction consisting in building a finite automaton that is "equivalent" to the original hybrid system for the purpose of safety verification ([2]).

Exact methods for reachability computations exist only for a restricted class of hybrid systems with simple dynamics. In the case of more complex dynamics, approximation methods have been developed, which can be classified as "over-approximation" and "asymptotic approximation" methods.

The over-approximation methods aim at obtaining efficient over-approximations of reachable sets. The main idea is to start from sets that are easy to represent in a compact form and approximating the system dynamics so that the sets obtained through the direct or inverse evolution of the approximated system admit the same representation of the starting sets, while ensuring over-approximation of the reachable sets of the original system. Polyhedral and ellipsoidal methods ([4, 19]) belong to this category of approximation approaches.

The asymptotic approximation methods aim at obtaining an approximation of the reachable sets that converges to the true reachable sets as some accuracy parameter tends to zero. Level set methods and gridding techniques belong to this category. In level set methods, sets are represented as the zero sublevel set of an appropriate function. The evolution of the boundary of this set through the system dynamics can be described through a Hamilton-Jacobi-Isaacs partial differential equation. An approximation to the reachable set is then obtained by a suitable numerical approximation of this equation ([26, 25]). In [30] a Markov chain approximation of a deterministic system is introduced to perform reachability analysis. The Markov chain is obtained by gridding the state space of the original system and defining the transition probabilities over the so-obtained discrete set of states so as to guarantee that admissible trajectories of the original system correspond to trajectories with non zero probability of the Markov chain. If the probability that the Markov chain enters the unsafe set is zero, then, one can conclude that the original system is safe. However, if such probability is not zero, the original system may still be safe.

In all approaches, reachability computations become more intensive as the dimension of the continuous state space grows. This is particularly critical in asymptotic approximation methods. On the other hand, the over-approximation methods have to be designed based on the characteristics of the specific system under study, and generally provide solutions to the safety verification problem that are too conservative when the system dynamics is complex and the reachable sets have arbitrary shapes. In comparison, the asymptotic approximation methods can be applied to general classes of systems and they do not require a specific shape for the reachable sets.

In many control applications, the dynamics of the system under study is subjected to the perturbation of random noises that are either inherent or present in the environment. These systems are naturally described by stochastic models, whose trajectories occur with different probabilities. For this class of systems, one can adopt either a worst-case approach or a probabilistic approach to safety verification. In the worst-case approach to safety verification, one requires all the admissible trajectories of the system to be outside the unsafe set, regardless of their probability, thus ignoring the stochastic nature of the system. In [20], for example, the system is stochastic because of some random noise signal affecting the system dynamics. However, the noise process is assumed to be bounded and is treated as if it were a deterministic signal taking values in a known compact set for the purpose of reachability computations. In the probabilistic approach to safety verification, one allows some trajectories of the system to enter the unsafe set if this event has low probability, thus avoiding the conservativeness of the worst-case approach.

A probabilistic approach to safety verification can be useful within a structured alerting system where alarms of different severity are issued depending on the level of criticality of the situation. For systems operating in a highly dynamic uncertain environment, safety has to be repeatedly verified on-line based on the updated information on the system behavior. In these applications it is then very important to have some measure of criticality for evaluating whether the selected control input is appropriate or a corrective action should be taken to timely steer the system out of the unsafe set. A natural choice for the measure of criticality is the probability of intrusion into the unsafe set within a finite/infinite time horizon: the higher the probability of intrusion, the more critical the situation.

In this chapter, we describe a methodology for probabilistic reachability analysis of a certain class of stochastic hybrid systems governed by stochastic differential equations with time-driven jumps. The distinguishing feature of the proposed methodology is that it rests on the approximation of the solution to stochastic differential equations by using Markov chains. The basic idea is to construct a Markov chain whose state space is obtained by discretizing the original space into grids. For properly chosen transition probabilities, the Markov chain converges weakly to the solution to the stochastic differential equation as the discretization step approaches zero. Therefore, an approximation of the probability of interest can be obtained by computing the corresponding quantity for the Markov chain.

From an algorithmic point of view, we propose a backward reachability algorithm which computes for each state an estimate of the probability that the system will enter the unsafe set starting from that state by appropriately propagating the transition probabilities of the Markov chain backwards in time starting from the unsafe set during the time horizon of interest.

According to the classification of safety verification approaches mentioned above, our approach can be described as an asymptotic approximation probabilistic model checking method based on backward reachability computations.

We shall consider the problem of conflict prediction in Air Traffic Management (ATM) as an application example.

2 Stochastic Approximation Method

2.1 Formulation of the Reachability Problem

Consider an n-dimensional system whose dynamics is governed by the stochastic differential equation

$$dS(t) = a(S,t)dt + b(S)\Gamma\, dW(t), \qquad (1)$$

during the time interval $T = [0, t_f]$, where 0 is the current time instant, and t_f is a positive real number (possibly infinity) representing the look-ahead time horizon. Function $a : \mathbb{R}^n \times T \to \mathbb{R}^n$ is the drift term, function $b : \mathbb{R}^n \to \mathbb{R}^{n \times n}$ is the diffusion term, and Γ is a diagonal matrix with positive entries, which modulates the variance of the standard n-dimensional Brownian motion $W(\cdot)$.

We suppose that $b : \mathbb{R}^n \to \mathbb{R}^{n \times n}$ is a continuous function, whereas $a : \mathbb{R}^n \times T \to \mathbb{R}^n$ is continuous in its first argument and only piecewise continuous in its second argument.

Let $\mathcal{D} \subset \mathbb{R}^n$ be a set representing the unsafe region for the system.

Our objective is to evaluate the probability that $S(t)$ enters $\mathcal{D}$ starting from some initial state $S(0)$ during the time interval $T = [0, t_f]$.

Since $\mathcal{D}$ represents an unsafe region, which, in the ATM application introduced later, corresponds to a region where a conflict takes place, in the sequel we shall refer to the probability of interest:

$$P\{S(t) \in \mathcal{D} \text{ for some } t \in T\}, \tag{2}$$

as the *probability of conflict.*

To evaluate the probability of conflict (2) numerically, we consider an open domain $\mathcal{U} \subset \mathbb{R}^n$ that contains $\mathcal{D}$ and has compact support. $\mathcal{U}$ should be large enough so that the situation can be declared safe once S ends up outside $\mathcal{U}$. With reference to the domain $\mathcal{U}$, the probability of entering the unsafe set $\mathcal{D}$ can be expressed as

$$P_c := P\{S \text{ hits } \mathcal{D} \text{ before hitting } \mathcal{U}^c \text{ within the time interval } T\}, \tag{3}$$

where $\mathcal{U}^c$ denotes the complement of $\mathcal{U}$ in $\mathbb{R}^n$. Implicit in the above definition is that if S hits neither $\mathcal{D}$ nor $\mathcal{U}^c$ during T, no conflict occurs.

For the purpose of computing (3), we can assume that in equation (1), S is defined on the open domain $\mathcal{U} \setminus \mathcal{D}$ with initial condition $S(0)$, and that it is stopped as soon as it hits the boundary $\partial\mathcal{U} \cup \partial\mathcal{D}$.

2.2 Markov Chain Approximation: Weak Convergence Result

We now describe an approach to approximate the solution $S(\cdot)$ to equation (1) defined on $\mathcal{U} \setminus \mathcal{D}$ with absorption on the boundary $\partial\mathcal{U} \cup \partial\mathcal{D}$. The idea is to discretize $\mathcal{U} \setminus \mathcal{D}$ into grid points that constitute the state space of a Markov chain. By carefully choosing the transition probabilities, the solution to the Markov chain will converge weakly to that of the stochastic differential equation (1) as the grid size approaches zero. Therefore, at a small grid size, a good estimate of the probability P_c in (3) is provided by the corresponding quantity associated with the Markov chain, which is much easier to compute.

To define the Markov chain, we first need to introduce some notations.

Let $\Gamma = \text{diag}(\sigma_1, \sigma_2, \dots, \sigma_n)$, with $\sigma_1, \sigma_2, \dots, \sigma_n > 0$.

Fix a grid size $\delta > 0$. Denote by $\delta\mathbb{Z}^n$ the integer grids of $\mathbb{R}^n$ scaled properly, more precisely,

$$\delta\mathbb{Z}^n = \{(m_1\eta_1\delta, m_2\eta_2\delta, \dots, m_n\eta_n\delta) |\, (m_1, m_2, \dots, m_n) \in \mathbb{Z}^n\},$$

where η_i, $i = 1, \dots, n$, are defined as $\eta_i := \frac{\sigma_i}{\bar{\sigma}}$, $i = 1, \dots, n$, with $\bar{\sigma} = \max_i \sigma_i$. For each grid point $q \in \delta\mathbb{Z}^n$, define the immediate neighbors set

$$\mathcal{N}_q = \{q + (i_1\eta_1\delta, i_2\eta_2\delta, \dots, i_n\eta_n\delta) |\, (i_1, i_2, \dots, i_n) \in \mathcal{I}\}, \tag{4}$$

where $\mathcal{I} \subseteq \{0, 1, -1\}^n \setminus \{(0, 0, \dots, 0)\}$. The immediate neighbors set $\mathcal{N}_q$ is a subset of all the points in $\delta\mathbb{Z}^n$ whose distance from q along the coordinate axis x_i is at most $\eta_i\delta$, $i = 1,\dots,n$. The larger the cardinality of $\mathcal{N}_q$, the more intensive the computations. For the convergence result to hold, different choices for $\mathcal{N}_q$ are possible, which depend, in particular, on the diffusion term b in (1). For the time being, consider the immediate neighbors set as given. We shall then see possible choices for it in some specific cases.

Define $\mathcal{Q} = (\mathcal{U} \setminus \mathcal{D}) \cap \delta\mathbb{Z}^n$, which consists of all those grid points in $\delta\mathbb{Z}^n$ that lie inside $\mathcal{U}$ but outside $\mathcal{D}$. The interior of $\mathcal{Q}$, denoted by $\mathcal{Q}^0$, consists of all those points in $\mathcal{Q}$ which have all their neighbors in $\mathcal{Q}$. The boundary of $\mathcal{Q}$ is defined to be $\partial\mathcal{Q} = \mathcal{Q} \setminus \mathcal{Q}^0$, and is the union of two disjoint sets: $\partial\mathcal{Q} = \partial\mathcal{Q}_{\mathcal{U}} \cup \partial\mathcal{Q}_{\mathcal{D}}$, where points in $\partial\mathcal{Q}_{\mathcal{U}}$ have at least one neighbor outside $\mathcal{U}$, and points in $\partial\mathcal{Q}_{\mathcal{D}}$ have at least one neighbor inside $\mathcal{D}$. If a point satisfies both the conditions, then we assign it only to $\partial\mathcal{Q}_{\mathcal{D}}$. This will eventually lead to an overestimation of the probability of conflict. However, if $\mathcal{U}$ is chosen to be large enough, the overestimation error is negligible.

We now define a Markov chain $\{Q_k,\ k \geq 0\}$ on the state space $\mathcal{Q}$. Denote by $\Delta t > 0$ the amount of time elapsing between any two successive discrete time steps k and $k+1$, $k \geq 0$. $\{Q_k,\ k \geq 0\}$ is a time-inhomogeneous Markov chain such that:

1. each state in $\partial\mathcal{Q}$ is an absorbing state, i.e., the state of the chain remains unchanged after it hits any of the states $q \in \partial\mathcal{Q}$:
$$P\{Q_{k+1} = q' \,|\, Q_k = q\} = \begin{cases} 1, & q' = q \\ 0, & \text{otherwise} \end{cases}$$
2. starting from a state q in $\mathcal{Q}^0$, the chain jumps to one of its neighbors in $\mathcal{N}_q$ or stays at the same state according to transition probabilities determined by its current location q and the current time step k:
$$P\{Q_{k+1} = q' \,|\, Q_k = q\} = \begin{cases} p^k_{q'}(q), & q' \in \mathcal{N}_q \cup \{q\} \\ 0, & \text{otherwise}, \end{cases} \tag{5}$$
where $p^k_{q'}(q)$ are functions of the drift and diffusion terms evaluated at q and time $k\Delta t$.

Set $\Delta t = \lambda\delta^2$, where λ is some positive constant.

Let the Markov chain be at state $q \in \mathcal{Q}^0$ at some time step k. Define
$$m^k_q = \tfrac{1}{\Delta t} E\{Q_{k+1} - Q_k \,|\, Q_k = q\},$$
$$V^k_q = \tfrac{1}{\Delta t} E\{(Q_{k+1} - Q_k)(Q_{k+1} - Q_k)^T \,|\, Q_k = q\}.$$
Suppose that as $\delta \to 0$,
$$\begin{aligned} m^k_q &\to a(s, k\Delta t), \\ V^k_q &\to b(s)\Gamma^2 b(s)^T, \end{aligned} \tag{6}$$
$\forall s \in \mathcal{U} \setminus \mathcal{D}$, where for each $\delta > 0$ q is a point in $\mathcal{Q}^0$ closest to s.

If the chain $\{Q_k, k \geq 0\}$ starts from a point $\bar{q} \in \mathcal{Q}^0$ closest to $S(0)$, then by Theorem 8.7.1 in [6] (see also [31]), we conclude that

Proposition 1. *Fix $\delta > 0$ and consider the corresponding Markov chain $\{Q_k, k \geq 0\}$. Denote by $\{Q(t), t \geq 0\}$ the stochastic process that is equal to Q_k on the time interval $[k\Delta t, (k+1)\Delta t)$ for all k, where $\Delta t = \lambda\delta^2$. Suppose that as $\delta \to 0$, the equations (6) are satisfied. Then as $\delta \to 0$, $\{Q(t), t \geq 0\}$ converges weakly to the solution $\{S(t), t \geq 0\}$ to equation (1) defined on $\mathcal{U} \setminus \mathcal{D}$ with absorption on the boundary $\partial\mathcal{U} \cup \partial\mathcal{D}$.* □

Remark 1. As the grid size δ decreases, the time interval between consecutive discrete time steps has to decrease for the stochastic process $S(\cdot)$ to be approximated by a Markov chain with one-step successors limited to the immediate neighbors set. It is then not surprising that the time interval Δt is a decreasing function of the grid size δ for the convergence result to hold. □

Let $k_f := \lfloor \frac{t_f}{\Delta t} \rfloor$ be the largest integer not exceeding $t_f/\Delta t$ ($k_f = \infty$ if $t_f = \infty$). As a result of Proposition 1, a good approximation to the probability of conflict P_c in (3) is given by

$$\begin{aligned} P_{c,\delta} &:= P\{Q_{k_f} \in \partial\mathcal{Q}_{\mathcal{D}}\} \\ &= P\{Q_k \text{ hits } \partial\mathcal{Q}_{\mathcal{D}} \text{ before hitting } \partial\mathcal{Q}_{\mathcal{U}} \text{ within } 0 \leq k \leq k_f\}, \end{aligned}$$

with the chain $\{Q_k, k \geq 0\}$ starting from a point $\bar{q} \in \mathcal{Q}$ closest to $S(0)$, for a small δ.

2.3 Examples of Transition Probability Functions

In this section, we describe a possible choice for the immediate neighbors set and the transition probabilities that is effective in guaranteeing that equations (6) (and, hence, the converge result) hold. We distinguish between two different structures of the diffusion term b that will fit the ATM application example.

Decoupled noise components

Suppose that the matrix b in equation (1) has the following form: $b(s) = \beta(s)I$, where $\beta : \mathbb{R}^n \to \mathbb{R}$ and I is the identity matrix of size n.

Equation (1) then takes the form:

$$dS(t) = a(S,t)dt + \beta(S)\Gamma dW(t).$$

Since each component of the n-dimensional Brownian motion $W(\cdot)$ directly affects a single component of $S(\cdot)$, the immediate neighbors set $\mathcal{N}_q$, $q \in \delta\mathbb{Z}^n$, can be taken as the set of points along each one of the x_i, $i = 1, \ldots, n$, directions whose distance from q is $\eta_i\delta$, $i = 1, \ldots, n$, respectively. For each $q \in \delta\mathbb{Z}^n$, $\mathcal{N}_q$ is then composed of the following $2n$ elements:

$$\begin{array}{ll} q_{1_+} = q + (+\eta_1\delta, 0, \ldots, 0), & q_{1_-} = q + (-\eta_1\delta, 0, \ldots, 0), \\ q_{2_+} = q + (0, +\eta_2\delta, \ldots, 0), & q_{2_-} = q + (0, -\eta_2\delta, \ldots, 0), \\ \vdots & \vdots \\ q_{n_+} = q + (0, 0, \ldots, +\eta_n\delta), & q_{n_-} = q + (0, 0, \ldots, -\eta_n\delta), \end{array}$$

Figure 1 plots the case when $n = 3$. Each grid point has six immediate neighbors (q_{1_-}, q_{1_+}, q_{2_-}, q_{2_+}, q_{3_-}, and q_{3_+}): two (q_{1_-} and q_{1_+}) at a distance $\eta_1\delta$ along direction x_1, two (q_{2_-} and q_{2_+}) at a distance $\eta_2\delta$ along direction x_2, and two (q_{3_-} and q_{3_+}) at a distance $\eta_3\delta$ along direction x_3.

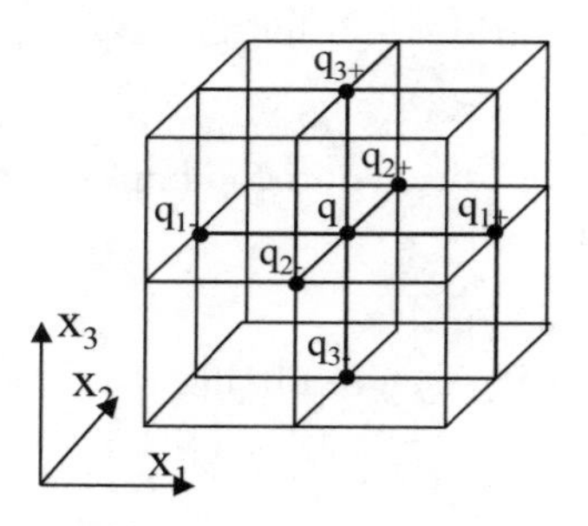

Fig. 1. Neighboring grid points in the three dimensional case.

We now define the transition probabilities in (5):
If $q \in \mathcal{Q}^0$, then

$$P\{Q_{k+1} = q' | \, Q_k = q\} =$$

$$\begin{cases} p_q^k(q) = \dfrac{\xi_0^k(q)}{C_q^k}, & q' = q \\ p_{q_{i_+}}^k(q) = \dfrac{\exp(\delta\xi_i^k(q))}{C_q^k}, & q' = q_{i_+}, \ i = 1, \ldots, n \\ p_{q_{i_-}}^k(q) = \dfrac{\exp(-\delta\xi_i^k(q))}{C_q^k}, & q' = q_{i_-}, \ i = 1, \ldots, n \\ 0, & \text{otherwise}, \end{cases} \quad (7)$$

where

$$\xi_0^k(q) = \tfrac{2}{\lambda\bar{\sigma}^2\beta(q)^2} - 2n \quad \xi_i^k(q) = \tfrac{[a(q,k\Delta t)]_i}{\eta_i\bar{\sigma}^2\beta(q)^2}, \quad i = 1, \ldots, n$$

$$C_q^k = 2\textstyle\sum_{i=1}^n \operatorname{csh}(\delta\xi_i^k(q)) + \xi_0^k(q).$$

λ is a positive constant that has to be chosen small enough such that $\xi_0^k(q)$ defined above is positive for all $q \in \mathcal{Q}$ and all $k \geq 0$. In particular, this is guaranteed if

$$0 < \lambda \leq (n\sigma_1^2 \max_{s \in \mathcal{U} \setminus \mathcal{D}} \beta(s)^2)^{-1}. \tag{8}$$

As for Δt, we set $\Delta t = \lambda \delta^2$.

A direct computation shows that, with this choice for the neighboring set, the transition probabilities, and Δt, for each $q \in \mathcal{Q}^0$ and $k \geq 0$

$$m_q^k = \frac{2}{\lambda \delta C_q^k} \begin{bmatrix} \eta_1 \mathrm{sh}(\delta \xi_1^k(q)) \\ \eta_2 \,\mathrm{sh}(\delta \xi_2^k(q)) \\ \vdots \\ \eta_n \,\mathrm{sh}(\delta \xi_n^k(q)) \end{bmatrix},$$

$$V_q^k = \frac{2}{\lambda C_q^k} \mathrm{diag}(\eta_1^2 \mathrm{csh}\,(\delta \xi_1^k(q)), \eta_2^2 \mathrm{csh}(\delta \xi_2^k(q)), \dots, \eta_n^2 \mathrm{csh}(\delta \xi_n^k(q)))\,.$$

It is then easily verified that the equations in (6) are satisfied, which in turn leads to the weak convergence result in Proposition 1.

Coupled noise components

We consider the case when the dimension n of S is even and matrix $\Gamma = \mathrm{diag}(\sigma_1, \sigma_2, \dots, \sigma_n)$ satisfies $\sigma_h = \sigma_{h+n/2} > 0$, $h = 1, \dots, n/2$. Moreover, we assume that the diffusion term b in equation (1) takes the following form

$$b(s) = \begin{bmatrix} I & \alpha(s)\, I \\ \alpha(s)\, I & I \end{bmatrix}^{1/2}$$

with $\alpha : \mathbb{R}^n \to [0, 1]$. The components h and $h + n/2$ of $S(\cdot)$ are then both directly affected only by the components h and $h + n/2$ of $W(\cdot)$, for every $h = 1, 2, \dots, n/2$. Based on this observation, the immediate neighbors set $\mathcal{N}_q$, $q \in \delta\mathbb{Z}^n$, can be chosen as follows:

$$\mathcal{N}_q = \{q + (i_1\eta_1\delta, i_2\eta_2\delta, \dots, i_n\eta_n\delta) |\; (i_1, i_2, \dots, i_n) \in \mathcal{I}\},$$

where $\mathcal{I} = \{(i_1, i_2, \dots, i_n) |\; \exists h$ such that $i_h = \pm 1, i_{h+n/2} = \pm 1, i_j = 0, \forall j \neq h, h + n/2\}$. The $2n$ elements of $\mathcal{N}_q$ have the following expression

$$\begin{aligned}
q_{1_{++}} &= q + (+\eta_1\delta, 0, \dots, 0, +\eta_1\delta, 0, \dots, 0) \\
q_{1_{--}} &= q + (-\eta_1\delta, 0, \dots, 0, -\eta_1\delta, 0, \dots, 0) \\
q_{1_{+-}} &= q + (+\eta_1\delta, 0, \dots, 0, -\eta_1\delta, 0, \dots, 0) \\
q_{1_{-+}} &= q + (-\eta_1\delta, 0, \dots, 0, +\eta_1\delta, 0, \dots, 0) \\
q_{2_{++}} &= q + (0, +\eta_2\delta, \dots, 0, 0, +\eta_2\delta, \dots, 0) \\
q_{2_{--}} &= q + (0, -\eta_2\delta, \dots, 0, 0, -\eta_2\delta, \dots, 0) \\
q_{2_{+-}} &= q + (0, +\eta_2\delta, \dots, 0, 0, -\eta_2\delta, \dots, 0) \\
q_{2_{-+}} &= q + (0, -\eta_2\delta, \dots, 0, 0, +\eta_2\delta, \dots, 0) \\
&\vdots \\
q_{(n/2)_{++}} &= q + (0, 0, \dots, 0, +\eta_{n/2}\delta, \dots, 0, 0, \dots, 0, +\eta_{n/2}\delta) \\
q_{(n/2)_{--}} &= q + (0, 0, \dots, 0, -\eta_{n/2}\delta, \dots, 0, 0, \dots, 0, -\eta_{n/2}\delta) \\
q_{(n/2)_{+-}} &= q + (0, 0, \dots, 0, +\eta_{n/2}\delta, \dots, 0, 0, \dots, 0, -\eta_{n/2}\delta) \\
q_{(n/2)_{-+}} &= q + (0, 0, \dots, 0, -\eta_{n/2}\delta, \dots, 0, 0, \dots, 0, +\eta_{n/2}\delta),
\end{aligned}$$

where we used the fact that $\eta_i = \frac{\sigma_i}{\bar{\sigma}} = \frac{\sigma_{i+n/2}}{\bar{\sigma}} = \eta_{i+n/2}$, $i = 1, \ldots, n/2$.

We now define the transition probabilities in (5):

If $q \in \mathcal{Q}^0$, then

$$P\{Q_{k+1} = q' | Q_k = q\} =$$

$$\begin{cases} p_q^k(q) = \dfrac{\xi_0^k(q)}{C}, & q' = q \\ p_{q_{i++}}^k(q) = \dfrac{(1+\alpha(q)) \exp(\delta \xi_{i++}^k(q))}{C \mathrm{csh}(\delta \xi_{i++}^k(q))}, & q' = q_{i++}, \; i = 1, \ldots, n/2 \\ p_{q_{i--}}^k(q) = \dfrac{(1+\alpha(q)) \exp(-\delta \xi_{i++}^k(q))}{C \mathrm{csh}(\delta \xi_{i++}^k(q))}, & q' = q_{i--}, \; i = 1, \ldots, n/2 \\ p_{q_{i+-}}^k(q) = \dfrac{(1-\alpha(q)) \exp(\delta \xi_{i+-}^k(q))}{C \mathrm{csh}(\delta \xi_{i+-}^k(q))}, & q' = q_{i+-}, \; i = 1, \ldots, n/2 \\ p_{q_{i-+}}^k(q) = \dfrac{(1-\alpha(q)) \exp(-\delta \xi_{i+-}^k(q))}{C \mathrm{csh}(\delta \xi_{i+-}^k(q))}, & q' = q_{i-+}, \; i = 1, \ldots, n/2 \\ 0, & \text{otherwise,} \end{cases} \tag{9}$$

where

$$\begin{aligned} \xi_0^k(q) &= \tfrac{4}{\lambda \bar{\sigma}^2} - 2n, \\ \xi_{i++}^k(q) &= \tfrac{[a(q,k\Delta t)]_i + [a(q,k\Delta t)]_{i+n/2}}{\eta_i \bar{\sigma}^2 (1+\alpha(q))}, \quad i = 1, \ldots, n/2 \\ \xi_{i+-}^k(q) &= \tfrac{[a(q,k\Delta t)]_i - [a(q,k\Delta t)]_{i+n/2}}{\eta_i \bar{\sigma}^2 (1-\alpha(q))}, \quad i = 1, \ldots, n/2 \\ C &= \tfrac{4}{\lambda \bar{\sigma}^2}, \end{aligned}$$

λ is a positive constant that has to be chosen small enough such that $\xi_0^k(q)$ defined above is positive for all $q \in \mathcal{Q}$ and all $k \geq 0$. In particular, this is guaranteed if

$$0 < \lambda \leq (\bar{\sigma}^2 n/2)^{-1}. \tag{10}$$

The time elapsed between successive jumps is set equal to $\Delta t = \lambda \delta^2$.

It can be verified that, with this choice for the neighboring set, the transition probabilities, and Δt, for each $q \in \mathcal{Q}^0$ and each $k \geq 0$,

$$m_q^k = \frac{2}{\lambda\delta C}\begin{bmatrix} \eta_1(1+\alpha(q))\frac{\mathrm{sh}(\delta\xi^k_{1_{++}}(q))}{\mathrm{csh}(\delta\xi^k_{1_{++}}(q))} + \eta_1(1-\alpha(q))\frac{\mathrm{sh}(\delta\xi^k_{1_{+-}}(q))}{\mathrm{csh}(\delta\xi^k_{1_{+-}}(q))} \\ \vdots \\ \eta_{n/2}(1+\alpha(q))\frac{\mathrm{sh}(\delta\xi^k_{(n/2)_{++}}(q))}{\mathrm{csh}(\delta\xi^k_{(n/2)_{++}}(q))} + \eta_{n/2}(1-\alpha(q))\frac{\mathrm{sh}(\delta\xi^k_{(n/2)_{+-}}(q))}{\mathrm{csh}(\delta\xi^k_{(n/2)_{+-}}(q))} \\ \eta_1(1+\alpha(q))\frac{\mathrm{sh}(\delta\xi^k_{1_{++}}(q))}{\mathrm{csh}(\delta\xi^k_{1_{++}}(q))} - \eta_1(1-\alpha(q))\frac{\mathrm{sh}(\delta\xi^k_{1_{+-}}(q))}{\mathrm{csh}(\delta\xi^k_{1_{+-}}(q))} \\ \vdots \\ \eta_{n/2}(1+\alpha(q))\frac{\mathrm{sh}(\delta\xi^k_{(n/2)_{++}}(q))}{\mathrm{csh}(\delta\xi^k_{(n/2)_{++}}(q))} - \eta_{n/2}(1-\alpha(q))\frac{\mathrm{sh}(\delta\xi^k_{(n/2)_{+-}}(q))}{\mathrm{csh}(\delta\xi^k_{(n/2)_{+-}}(q))} \end{bmatrix},$$

$$V_q^k = \begin{bmatrix} I & \alpha(q)I \\ \alpha(q)I & I \end{bmatrix} \Gamma^2$$

So if $\delta \to 0$ and we always choose q to be a point in $\mathcal{Q}^0$ closest to a fixed $s \in \mathcal{U} \setminus \mathcal{D}$, then

$$\begin{aligned} m_q^k &\to a(s, k\Delta t) \\ V_q^k &\to \begin{bmatrix} I & \alpha(q)I \\ \alpha(q)I & I \end{bmatrix} \Gamma^2 = b(s)\Gamma^2 b(s)^T. \end{aligned}$$

Therefore, we conclude that Proposition 1 holds in this case as well.

2.4 An Iterative Algorithm for Reachability Computations

We next describe an iterative procedure to compute the probability $P_{c,\delta}$ that approximates the probability of conflict P_c in (3):

$$\begin{aligned} P_{c,\delta} &:= P\{Q_{k_f} \in \partial\mathcal{Q}_{\mathcal{D}}\} \\ &= P\{Q_k \text{ hits } \partial\mathcal{Q}_{\mathcal{D}} \text{ before hitting } \partial\mathcal{Q}_{\mathcal{U}} \text{ within } 0 \le k \le k_f\}, \end{aligned}$$

with the chain $\{Q_k, k \ge 0\}$ starting from a point $\bar{q} \in \mathcal{Q}$ closest to $S(0)$.

We address both the finite and infinite horizon cases ($k_f < \infty$ and $k_f = \infty$).

Let

$$P_{c,\delta}^{(k)}(q) := P\{Q_{k_f} \in \partial\mathcal{Q}_D | \, Q_k = q\}, \tag{11}$$

be a set of functions defined on $\mathcal{Q}$ and indexed by $k = 0, 1, \dots, k_f$. Since the chain $\{Q_k, k \ge 0\}$ starts at $\bar{q}$ at $k = 0$, the desired quantity $P_{c,\delta}$ can be expressed in terms of the introduced functions as $P_{c,\delta}^{(0)}(\bar{q})$. The procedures described below determine the whole set of functions $P_{c,\delta}^{(k)} : \mathcal{Q} \to \mathbb{R}$ for $k =$

$0, 1, \ldots, k_f$. This has the advantage that at any future time $t \in [0, t_f]$ an estimate of the probability of conflict over the new time horizon $[t, t_f]$ is readily available, eliminating the need for re-computation. As a matter of fact, for each $t \in [0, t_f]$, $P_{c,\delta}^{(\lfloor t/\Delta t \rfloor)} : \mathcal{Q} \to \mathbb{R}$ represents an estimate of the probability of conflict over the time horizon $[t, t_f]$ as a function of the value taken by the state at time t.

To compute $P_{c,\delta}^{(0)}$, fix a k such that $0 \leq k < k_f$. It is easily seen then that $P_{c,\delta}^{(k)} : \mathcal{Q} \to \mathbb{R}$ satisfies the following recursive equation

$$P_{c,\delta}^{(k)}(q) = \begin{cases} p_q^k(q) P_{c,\delta}^{(k+1)}(q) + \sum_{q' \in \mathcal{N}_q} p_{q'}^k(q) P_{c,\delta}^{(k+1)}(q'), & q \in \mathcal{Q}^0 \\ 1, & q \in \partial \mathcal{Q}_{\mathcal{D}} \\ 0, & q \in \partial \mathcal{Q}_{\mathcal{U}}. \end{cases} \tag{12}$$

This is the key equation to compute $P_{c,\delta}^{(0)}$.

Finite horizon

In the finite horizon case ($k_f < \infty$), the probability $P_{c,\delta} = P_{c,\delta}^{(0)}(\bar{q})$ can be computed by iterating equation (12) backward k_f times starting from $k = k_f - 1$ and using the initialization $P_{c,\delta}^{(k_f)}(q) = \bar{P}(q)$, $q \in S$, where

$$\bar{P}(q) = \begin{cases} 1, & \text{if } q \in \partial S_D \\ 0, & \text{otherwise.} \end{cases} \tag{13}$$

The reason for the above initialization is obvious considering the definition (11) of $P_{c,\delta}^{(k)}$.

The procedure to compute an approximation of P_c in the finite horizon case is summarized in the following algorithm.

Algorithm 1 *Given $S(0)$, $a : \mathbb{R}^n \times T \to \mathbb{R}^n$, $b : \mathbb{R}^n \to \mathbb{R}^{n \times n}$, Γ, and $\mathcal{D}$, then*

1. *Select the open set $\mathcal{U} \subset \mathbb{R}^n$ containing $\mathcal{D}$, and fix $\delta > 0$.*
2. *Define the Markov chain $\{Q_k, k \geq 0\}$ with state space $\mathcal{Q} = (\mathcal{U} \setminus \mathcal{D}) \cap \delta\mathbb{Z}^n$ and appropriate transition probabilities.*
3. *Set $\bar{k} = k_f$ and initialize $P_{c,\delta}^{(\bar{k})}$ with $\bar{P}$ defined in equation (13).*
4. *For $k = \bar{k}-1, \ldots, 0$, compute $P_{c,\delta}^{(k)}$ from $P_{c,\delta}^{(k+1)}$ according to equation (12).*
5. *Choose a point $\bar{q}$ in S closest to $S(0)$ and set $P_{c,\delta} = P_{c,\delta}^{(0)}(\bar{q})$.*

As for the choice of the grid size δ, one has to take into consideration different aspects:

i) In a time interval of length Δt, the maximal distance that the Markov chain can travel is $\eta_i \delta$ along the direction x_i, $i = 1, \ldots, n$. Thus given $\mathcal{U}$, for the diffusion process $S(t)$ to be approximated by the Markov chain, the component along the x_i axis $|[a(\cdot,\cdot)]_i|$ of $a(\cdot,\cdot)$ has to be upper bounded roughly by $\frac{\eta_i \delta}{\Delta t}$ over $\mathcal{U} \setminus \mathcal{D} \times T$, for any $i = 1, \ldots, n$. In view of Remark 1, this condition translates into upper bounds on the admissible values for δ. In particular, in the aircraft safety analysis case $\Delta t = \lambda \delta^2$, hence $\delta \leq \min_i \frac{\eta_i}{\lambda |[a(\cdot,\cdot)]_i|}$. Thus, fast diffusion processes cannot be simulated by Markov chains corresponding to large δ's.

ii) For a fixed grid size δ, the size of the state space $\mathcal{Q}$ is of the order of $1/\delta^n$, so each iteration in Algorithm 1 takes a time proportional to $1/\delta^n$. The number of iterations is given by $k_f \simeq t_f/\Delta t$. If Δt is proportional to δ^2 as in the safety analysis case, the running time of Algorithm 1 is proportional to $1/\delta^{n+2}$.

Therefore, for small δ's the running time may be too long, but large δ's may not allow for the simulation of fast moving processes. A suitable δ is a compromise between these two conflicting requirements.

Infinite horizon

In the infinite horizon case $k_f = \infty$, hence Algorithm 1 cannot be applied directly since it would take infinitely many iterations. In this section we consider a special case in which this difficulty can be easily overcome.

We start by rewriting the iteration law (12) in matrix form. Arrange the sequence $\{P_{c,\delta}^{(k)}(q),\, q \in \mathcal{Q}^0\}$ into a long column vector according to some fixed ordering of the points in $\mathcal{Q}^0$, and denote it by $\mathbf{P}_{c,\delta}^{(k)} \in \mathbb{R}^{|\mathcal{Q}^0|}$. Here $|\mathcal{Q}^0|$ is the cardinality of $\mathcal{Q}^0$. Then equation (12) can be written as

$$\mathbf{P}_{c,\delta}^{(k)} = \mathbf{A}^{(k)} \mathbf{P}_{c,\delta}^{(k+1)} + \mathbf{b}^{(k)} \tag{14}$$

for suitably chosen matrix $\mathbf{A}^{(k)} \in \mathbb{R}^{|\mathcal{Q}^0| \times |\mathcal{Q}^0|}$ and vector $\mathbf{b}^{(k)} \in \mathbb{R}^{|\mathcal{Q}^0|}$. Note that $\mathbf{A}^{(k)}$ is a sparse positive matrix with the property that the sum of its elements on each row is smaller than or equal to 1, where equality holds if and only if that row corresponds to a point in $(\mathcal{Q}^0)^0$, the interior of $\mathcal{Q}^0$ consisting of all those points in $\mathcal{Q}^0$ whose immediate neighbors all belong to $\mathcal{Q}^0$. On the other hand, $\mathbf{b}^{(k)}$ is a positive vector with nonzero elements on exactly those rows corresponding to points on the boundary $\partial(\mathcal{Q}^0) = \mathcal{Q}^0 \setminus (\mathcal{Q}^0)^0$ of $\mathcal{Q}^0$. Both $\mathbf{A}^{(k)}$ and $\mathbf{b}^{(k)}$ depend on the grid size δ. We do not write it explicitly to simplify the notation.

Suppose that from some time instant t_c on, $a(s,t)$, $s \in \mathbb{R}^n$, $t \in T$, remain constant in time. Under this assumption, we have that $\mathbf{A}^{(k)} \equiv \mathbf{A}$ and $\mathbf{b}^{(k)} \equiv \mathbf{b}$ for $k > k_c := \lfloor \frac{t_c}{\Delta t} \rfloor$. Hence, for $k > k_c$ equation (14) becomes

$$\mathbf{P}_{c,\delta}^{(k)} = \mathbf{A} \mathbf{P}_{c,\delta}^{(k+1)} + \mathbf{b}. \tag{15}$$

We next address the problem of computing $\mathbf{P}_{c,\delta}^{(k_c+1)}$. Once we have determined $\mathbf{P}_{c,\delta}^{(k_c+1)}$, we can execute Algorithm 1 with step 2 replaced by

2'. Set $\bar{k} = k_c + 1$ and initialize $P_{c,\delta}^{(\bar{k})}$ with $P_{c,\delta}^{(k_c+1)}$.

to determine the approximation $P_{c,\delta}^{(0)}(\bar{q})$ of P_c.

The procedure to compute $\mathbf{P}_{c,\delta}^{(k_c+1)}$ rests on the following lemma.

Lemma 1. *The eigenvalues of* $\mathbf{A}$ *are all in the interior of the unit disk of the complex plane.* □

Proof. Suppose that $\mathbf{A}$ has an eigenvalue γ with $|\gamma| \geq 1$, and let $\mathbf{v}$ be an eigenvector such that $\mathbf{A}\mathbf{v} = \gamma\mathbf{v}$. Assume that $|\mathbf{v}_i| = \max(|\mathbf{v}_1|, \ldots, |\mathbf{v}_{|\Omega^0|}|)$ for some i. Then

$$|\mathbf{v}_i| \leq |\gamma\mathbf{v}_i| = |\,[\mathbf{A}\mathbf{v}]_i| \leq \sum_{j=1}^{|\Omega^0|} \mathbf{A}_{ij}|\mathbf{v}_j| \leq \sum_{j=1}^{|\Omega^0|} \mathbf{A}_{ij}|\mathbf{v}_i| \leq |\mathbf{v}_i|,$$

which is possible only if $|\mathbf{v}_1| = \cdots = |\mathbf{v}_{|\Omega^0|}|$. However, this leads to a contradiction since by changing i in the above equation to one such that $\sum_{j=1}^{|\Omega^0|} \mathbf{A}_{ij} < 1$, one gets $|\mathbf{v}_i| < |\mathbf{v}_i|$. □

Based on Lemma 1, we draw the following facts regarding equation (15):

Lemma 2. *Consider equation*

$$\mathbf{P}^{(k)} = \mathbf{A}\mathbf{P}^{(k+1)} + \mathbf{b}. \tag{16}$$

i) There is a unique $\mathbf{P} \in \mathbb{R}^{|\Omega^0|}$ *satisfying*

$$\mathbf{P} = \mathbf{A}\mathbf{P} + \mathbf{b}. \tag{17}$$

ii) Starting from any initial value $\mathbf{P}^{(k_0)}$ *at some* k_0 *and iterating equation (16) backward in time,* $\mathbf{P}^{(k)}$ *converges to the fix point* $\mathbf{P}$ *as* $k \to -\infty$. *Moreover, if* $\mathbf{P}^{(k_0)} \geq \mathbf{P}$, *then* $\mathbf{P}^{(k)} \geq \mathbf{P}$ *for all* $k \leq k_0$. *Conversely, if* $\mathbf{P}^{(k_0)} \leq \mathbf{P}$, *then* $\mathbf{P}^{(k)} \leq \mathbf{P}$ *for all* $k \leq k_0$. *Note that here the symbols* $\geq$ *and* $\leq$ *denote component-wise comparison between vectors.* □

Proof. $\mathbf{P} = (I - \mathbf{A})^{-1}\mathbf{b}$ since $I - \mathbf{A}$ is invertible by Lemma 1. Define $\mathbf{e}^{(k)} = \mathbf{P}^{(k)} - \mathbf{P}$. Then $\mathbf{e}^{(k)} = \mathbf{A}\mathbf{e}^{(k+1)}$. So by Lemma 1, $\mathbf{e}^{(k)}$ converges to 0 as $k \to -\infty$. The last conclusion is a direct consequence of the fact that all components of the matrix $\mathbf{A}$ and vector $\mathbf{b}$ are nonnegative. □

Lemma 2 shows that equation (15) admits a fixed point $\mathbf{P}$ to which $\mathbf{P}^{(k)}$ obtained by iterating from any initial condition converges as $k \to -\infty$. Such a fixed point is in fact the desired quantity $\mathbf{P}_{c,\delta}^{(k_c+1)}$. Thus one way of computing $\mathbf{P}_{c,\delta}^{(k_c+1)}$ is to solve the linear equation $(I - \mathbf{A})\mathbf{P}_{c,\delta}^{(k_c+1)} = \mathbf{b}$ directly, using sparse matrix computation tools if possible. In our simulations, we determined $\mathbf{P}_{c,\delta}^{(k_c+1)}$ by iterating equation (16) starting at some k_0 from two initial conditions $\mathbf{P}_l^{(k_0)}$ and $\mathbf{P}_u^{(k_0)}$ that are respectively a lower bound and an upper bound of $\mathbf{P}$ (for example, one can choose $\mathbf{P}_l^{(k_0)}$ to be identically 0 on $\mathcal{Q}^0$ and $\mathbf{P}_u^{(k_0)}$ to be identically 1 on $\mathcal{Q}^0$). By Lemma 2, the iterated results at every $k \leq k_0$ for the two initial conditions will provide a lower bound and an upper bound of $\mathbf{P}_{c,\delta}^{(k_c+1)}$, respectively, which also converge toward each other (hence to $\mathbf{P}_{c,\delta}^{(k_c+1)}$ as well) as $k \to -\infty$. By running the iterations for the upper and lower bounds in parallel we can determine an approximation of $\mathbf{P}_{c,\delta}^{(k_c+1)}$ within any accuracy.

Remark 2. As $\delta \to 0$, the size of the matrix $\mathbf{A}$ becomes larger. Moreover, the ratio $|(\mathcal{Q}^0)^0|/|\mathcal{Q}^0| \to 1$. Hence $\mathbf{A}$ will have an eigenvalue close to 1 whose corresponding eigenvector is close to $(1, \ldots, 1)$. This causes slower convergence for the iteration (16) and numerical problems for the solution to the fixed point equation (17). □

2.5 Extension to the Case When the Initial State Is Uncertain

The procedure for estimating P_c can be easily extended to the case when the initial state $S(0)$ is not known precisely.

Suppose that $S(0)$ is described as a random variable with distribution $\mu_S(s)$, $s \in \mathcal{U} \setminus \mathcal{D}$. Then, the probability of entering the unsafe set $\mathcal{D}$ can be expressed as

$$P_c = \int_{\mathcal{U} \setminus \mathcal{D}} p_c(s) d\mu_S(s), \tag{18}$$

where $p_c : \mathcal{U} \setminus \mathcal{D} \to [0, 1]$ is defined by

$$p_c(s) := P\{S \text{ hits } \mathcal{D} \text{ before hitting } \mathcal{U}^c \text{ within the time interval } T | \, S(0) = s\}.$$

For each $s \in \mathcal{U} \setminus \mathcal{D}$, $p_c(s)$ is the probability of entering the unsafe set $\mathcal{D}$ over the time horizon T when $S(0) = s$ and is exactly the quantity estimated with $P_{c,\delta}^{(0)}$ in the iterative procedure proposed in Section 2.4. The integral (18) then reduces to a finite summation when approximating the map p_c with $P_{c,\delta}^{(0)}$.

3 Application to Aircraft Conflict Prediction

In the current centralized ATM system, aircraft are prescribed to follow certain flight plans, and Air Traffic Controllers (ATCs) on the ground are responsible for ensuring aircraft safety by issuing trajectory specifications to the pilots. The flight plan assigned to an aircraft is "safe" if by following it the aircraft will not get into any conflict situation.

Conflict situations arise, for example, when an aircraft gets closer than a certain distance to another aircraft or it enters some forbidden region of the airspace. In the sequel, these conflicts are shortly referred to as "aircraft-to-aircraft conflict" and "aircraft-to-airspace conflict", respectively.

The procedure used to prevent the occurrence of a conflict in ATM typically consists of two phases, namely, aircraft conflict detection and aircraft conflict resolution. Automated tools are currently being studied to support ATCs in performing these tasks. A comprehensive overview of the methods proposed in the literature for aircraft-to-aircraft conflict detection can be found in [18] .

In automated conflict detection, models for predicting the aircraft future position are introduced and the possibility that a conflict would happen within a certain time horizon is evaluated based on these models ([34, 27, 28, 7]). If a conflict is predicted, then the aircraft flight plans are modified in the conflict resolution phase so as to avoid the actual occurrence of the predicted conflict. The cost of the resolution action in terms of, for example, delay, fuel consumption, deviation from originally planned itinerary, is usually taken into account when selecting a new flight plan ([10, 33, 23, 13, 17, 24, 35, 16]).

The conflict detection issue can be formulated as a probabilistic safety verification problem, where the objective is to evaluate if the flight plan assigned to an aircraft is "safe". Safety can be assessed by estimating the probability that a conflict will occur over some look-ahead time horizon. In practice, once a prescribed threshold value of the probability of conflict is surpassed, an alarm of corresponding severity should be issued to the air traffic controllers/pilots to warn them on the level of criticality of the situation [34].

There are several factors that combined make this conflict analysis problem highly complicated, and as such impossible to solve analytically. Aircraft flight plans can be, in principle, arbitrary motions in the three dimensional airspace, and they are generally more complex than the simple planar linear motions assumed in [28, 8] when determining analytic expressions for the probability of an aircraft-to-aircraft conflict. Also, forbidden airspace areas may have an arbitrary shape, which can also change in time, as, for example, in the case of a storm that covers an area of irregular shape that evolves dynamically. Finally, and probably most importantly, the random perturbation to the aircraft motion is spatially correlated. Wind is a main source of uncertainty on the aircraft position, and if we consider two aircraft, the closer the aircraft, the larger the correlation between the wind perturbations. Although this last factor is known to be critical, it is largely ignored in the current literature on aircraft safety studies, probably because it is difficult to model and analyze. The methods proposed in the literature to compute the probability of conflict are generally based on the description of the aircraft future positions first proposed in [27]. In [27], each aircraft motion is described as a Gaussian random process whose variance grows in time, and the processes modeling the motions of different aircraft are assumed to be uncorrelated. However, this assumption may be unrealistic in practice, and can cause erroneous evaluations of the probability of conflict, since the correlation between the wind perturbations affecting the aircraft positions is stronger when two aircraft are closer to each other. To our knowledge, the first attempt to model the wind perturbation to the aircraft motion for ATM applications was done in [22], which inspired this work.

The model introduced for predicting the aircraft future position incorporates the information on the aircraft flight plan, and takes into account the presence of wind as the main source of uncertainty on the aircraft actual motion. We address the general case when the aircraft might change altitude during its flight. Modeling altitude changes is important not only because the aircraft changes altitude when it is inside a Terminal Radar Approach Control (TRACON) area, but also because altitude changes can be used as resolution maneuvers to avoid, e.g., severe perturbation areas or conflict situations with other aircraft ([29],[21],[17]).

It is important to note that we do not address issues related to a possible discrepancy between the flight plan at the ATC level and that set by the pilot on board of the aircraft. Modeling this aspect would require a more complex stochastic hybrid model than the one introduced here, where the hybrid component of the system is mainly due to changes in the aircraft dynamics at the way-points prescribed by their flight plan. Detecting situation awareness errors in fact requires modeling ATC and pilots by hybrid systems, and building an observer for the overall hybrid system obtained by composing the hybrid models of the agents and the aircraft.

The results illustrated here have appeared in [9], [11], [12], and [14].

3.1 Model of the Aircraft Motion

In this section we introduce a kinematic model of the aircraft motion to predict the aircraft future position during the time interval $T = [0, t_f]$.

The airspace and the aircraft position at time $t \in T$ are $\mathbb{R}^3$ and $X(t) \in \mathbb{R}^3$, respectively. We assume that the flight plan assigned to the aircraft is specified in terms of a velocity profile $u : T \to \mathbb{R}^3$, meaning that at time $t \in T$ the aircraft plans to fly at a velocity $u(t)$. Since, according to the common practice in ATM systems, aircraft are advised to travel at constant speed piecewise linear motions specified by a series of way-points, the velocity profile u is taken to be a piecewise constant function.

We suppose that the main source of uncertainty in the aircraft future position during the time interval T is the wind which affects the aircraft motion by acting on the aircraft velocity. The wind contribution to the velocity of the aircraft is due to the *wind speed.* Note that here we adopt the ATM terminology and use the word 'speed' for the velocity vector.

The wind speed can be further decomposed into two components: i) a deterministic term representing the nominal wind speed, which may depend on the aircraft location and time t, and is assumed to be known to the ATC through measurements or forecast; and ii) a stochastic term representing the effect of air turbulence and errors in the wind speed measurements and forecast.

As a result of the above discussion, the position X of the aircraft during the time horizon T is governed by the following stochastic differential equation:

$$dX(t) = u(t)dt + f(X,t)dt + \Sigma(X,t)dB(X,t), \tag{19}$$

initialized with the aircraft current position $X(0)$.

We next explain the different terms appearing in equation (19).

First of all, $f : \mathbb{R}^3 \times T \to \mathbb{R}^3$ is a time-varying vector field on $\mathbb{R}^3$: for a fixed $(x,t) \in \mathbb{R}^3 \times T$, $f(x,t)$ represents the nominal wind speed at position x and at time t. We call f the *wind field.*

$B(\cdot,\cdot)$ is a time-varying random field on $\mathbb{R}^3 \times T$ modeling (the integral of) air turbulence perturbations to aircraft velocity as well as wind speed forecast errors. It can be thought of as the time integral of a Gaussian random field correlated in space and uncorrelated in time. Formally, $B(\cdot,\cdot)$ has the following properties:

i) for each fixed $x \in \mathbb{R}^3$, $B(x,\cdot)$ is a standard 3-dimensional Brownian motion. Hence $dB(x,t)/dt$ can be thought of as a 3-dimensional white noise process;
ii) $B(\cdot,\cdot)$ is time increment independent. This implies, in particular, that the collections of random variables $\{B(x,t_2) - B(x,t_1)\}_{x \in \mathbb{R}^3}$ and $\{B(x,t_4) - B(x,t_3)\}_{x \in \mathbb{R}^3}$ are independent for any $t_1, t_2, t_3, t_4 \in T$, with $t_1 \leq t_2 \leq t_3 \leq t_4$;
iii) for any $t_1, t_2 \in T$ with $t_1 \leq t_2$, $\{B(x,t_2)-B(x,t_1)\}_{x \in \mathbb{R}^3}$ is an (uncountable) collection of Gaussian random variables with zero mean and covariance

$$E\{[B(x,t_2)-B(x,t_1)][B(y,t_2)-B(y,t_1)]^T\} = \rho(x-y)(t_2-t_1)I_3, \ \forall x,y \in \mathbb{R}^3,$$

where I_3 is the 3-by-3 identity matrix, and $\rho : \mathbb{R}^3 \to \mathbb{R}$ is a continuous function with $\rho(0) = 1$ and $\rho(x)$ decreases to zero as $x \to \infty$. In addition, ρ has to be non-negative definite in the sense that the k-by-k matrix $[\rho(x_i - x_j)]_{i,j=1}^k$ is non-negative definite for arbitrary $x_1, \ldots, x_k \in \mathbb{R}^3$ and positive integer k. See [1] for other equivalent conditions of this non-negative definite requirement.

Remark 3. Typically the wind field f is supposed to satisfy some continuity property. This condition, together with the monotonicity assumption on the spatial correlation function ρ, is introduced to model the fact that the closer two points in space, the more similar the wind speeds at those points, and, as the two points move farther away from each other, their wind speeds become more and more independent.

The spatial correlation function $\rho : \mathbb{R}^3 \to \mathbb{R}$ can be taken to be $\rho(x) = \exp(-c_h\|x\|_h - c_v\|x\|_v)$ for some $c_v \geq c_h > 0$, where the subscripts h and v stand for "horizontal" and "vertical", and $\|(x_1,x_2,x_3)\|_h := \sqrt{x_1^2 + x_2^2}$ and $\|(x_1,x_2,x_3)\|_v := |x_3|$ for any $(x_1,x_2,x_3) \in \mathbb{R}^3$. This is to model the fact that the wind correlation in space is weaker in the vertical direction.

Exponentially decaying spatial correlation functions are a popular choice for random field models in geostatistics [15]. This choice is actually suitable for ATM applications. In [5], the wind field prediction made by the Rapid Update Cycle (RUC [3]) developed at the National Oceanic and Atmospheric Administration (NOAA) Forecast System Laboratory (FOL) is compared with the empirical data collected by the Meteorological Data Collection Reporting System (MDCRS) near Denver International Airport. The result of this comparison is that the spatial correlation statistics of the wind field prediction errors is adequately described by an exponentially decaying function of the horizontal separation.

As a random field, $B(\cdot,\cdot)$ is Gaussian, stationary in space (its finite dimensional distributions remain unchanged when the origin of $\mathbb{R}^3$ is shifted), and isotropic in the horizontal directions (its finite dimensional distributions are invariant with respect to changes of orthonormal coordinates in the horizontal directions).

Finally, $\Sigma : \mathbb{R}^3 \times T \to \mathbb{R}^{3\times 3}$ modulates the variance of the random perturbation to the aircraft velocity. We assume that $\Sigma(\cdot,\cdot)$ is a constant diagonal matrix Σ given by $\Sigma := \mathrm{diag}(\sigma_h, \sigma_h, \sigma_v)$, for some constant $\sigma_h, \sigma_v > 0$. Note that after the modulation of Σ the random contribution of the wind to the aircraft velocity remains isotropic horizontally. However, its variance in the vertical direction can be different from that in the horizontal ones.

Equation (19) can then be rewritten as

$$dX(t) = u(t)dt + f(X,t)dt + \Sigma\, dB(X,t) \tag{20}$$

with initial condition $X(0)$.

Based on model (20) of the aircraft motion, we shall derive the equations to study the aircraft-to-aircraft and aircraft-to-airspace problems.

Note that this simplified model of the aircraft motion does not take into account the feedback control action of the flight management system (FMS), which tries to reduce the tracking error with respect to the planned trajectory. However, the algorithm described based on this model can be extended to address also the case when a model of the FMS is included.

3.2 Aircraft-to-Aircraft Conflict Problem

Consider two aircraft, say "aircraft 1" and "aircraft 2", flying in the same region of the airspace during the time interval $T = [0, t_f]$.

According to the ATM definition, a two-aircraft encounter is conflict-free if the two aircraft are either at a horizontal distance greater than r or at a vertical distance greater than H during the whole duration of the encounter, where r and H are prescribed quantities [29] . Currently, $r = 5$ nautical miles (nmi) for en-route airspace and $r = 3$ nmi inside the TRACON area, whereas $H = 1000$ feet (ft). If the two aircraft get closer than r horizontally and H vertically at some $t \in T$, then, an aircraft-to-aircraft conflict occurs.

Denote the position of aircraft 1 and aircraft 2 by X_1 and X_2, respectively. Based on (20), the evolutions of $X_1(\cdot)$ and $X_2(\cdot)$ over the time interval T are governed by

$$dX_1(t) = u_1(t)dt + f(X_1,t)dt + \Sigma\, dB(X_1,t), \tag{21}$$
$$dX_2(t) = u_2(t)dt + f(X_2,t)dt + \Sigma\, dB(X_2,t), \tag{22}$$

starting from the initial positions $X_1(0)$ and $X_2(0)$.

The probability of conflict can be expressed in terms of the relative position $Y := X_2 - X_1$ of the two aircraft as

$$P\{Y(t) \in \mathcal{D} \text{ for some } t \in T\}, \tag{23}$$

where $\mathcal{D} \in \mathbb{R}^3$ is the closed cylinder of radius r and height $2H$ centered at the origin.

Affine case

Let the wind field $f(x,t)$ be affine in x, i.e.,

$$f(x,t) = R(t)x + d(t), \quad \forall x \in \mathbb{R}^3,\ t \in T,$$

where $R : T \to \mathbb{R}^{3\times 3}$ and $d : T \to \mathbb{R}^3$ are continuous functions. We shall show that in this case we can refer to a simplified model for the two-aircraft system to compute the probability of conflict.

Since the positions of the two aircraft, X_1 and X_2, are governed by equations (21) and (22), by subtracting (21) from (22), we have that the relative position $Y = X_2 - X_1$ of aircraft 1 and aircraft 2 is governed by

$$dY(t) = v(t)dt + R(t)Y(t)dt + \Sigma d[B(X_2,t) - B(X_1,t)], \tag{24}$$

where $v := u_2 - u_1$ is the nominal relative velocity. $B(\cdot,\cdot)$ can be rewritten in the Karthunen-Loeve expansion as

$$B(x,t) = \sum_{n=0}^{\infty} \sqrt{\lambda_n}\phi_n(x)B_n(t),$$

where $\{B_n(t)\}_{n\geq 0}$ is a series of independent three-dimensional standard Brownian motions, and $\{(\lambda_n, \phi_n(x))\}_{n\geq 0}$ is a complete set of eigenvalue and eigenfunction pairs for the integral operator $\phi(x) \mapsto \int_{\mathbb{R}^3} \rho(s-x)\phi(s)\, ds$, i.e.,

$$\begin{cases} \lambda_n\phi_n(x) = \int_{\mathbb{R}^3} \rho(s-x)\phi_n(s)\, ds, \\ \rho(x-y) = \sum_{n=0}^{\infty} \lambda_n\phi_n(x)\phi_n(y), \end{cases} \quad \forall x, y \in \mathbb{R}^3. \tag{25}$$

Fix $x_1, x_2 \in \mathbb{R}^3$ and let $y = x_2 - x_1$. Define

$$Z(t) := B(x_2, t) - B(x_1, t) = \sum_{n=0}^{\infty} \sqrt{\lambda_n}[\phi_n(x_2) - \phi_n(x_1)]B_n(t). \tag{26}$$

$Z(t)$ is a Gaussian process with zero mean and covariance

$$E\{[Z(t_2) - Z(t_1)][Z(t_2) - Z(t_1)]^T\} = 2[1 - \rho(y)](t_2 - t_1)I_3, \quad \forall t_1 \leq t_2,$$

where the last equation follows from (25) and the fact that $\rho(0) = 1$. Note also that $Z(0) = 0$. Therefore, in terms of distribution we have

$$Z(t) \stackrel{d}{=} \sqrt{2[1 - \rho(y)]}\, W(t), \tag{27}$$

where $W(t)$ is a standard 3-dimensional Brownian motion.

As a result, (24) can then be approximated weakly by

$$dY(t) = v(t)dt + R(t)Y(t)dt + \sqrt{2[1 - \rho(Y)]}\Sigma\, dW(t). \tag{28}$$

By this we mean that the stochastic process $Y(t) = X_2(t) - X_1(t)$ obtained by subtracting the solution to (21) from the solution to (22) initialized respectively with $X_1(0)$ and $X_2(0)$ has the same distribution as the solution to (28) initialized with $Y(0) = X_2(0) - X_1(0)$.

Equation (28) is a particular case of (1) with $S = Y$, $\Gamma = \Sigma$, $a(y,t) = v(t) + R(t)y$, and $b(y) = \sqrt{2[1 - \rho(y)]}I$, with the discontinuity in a caused by the discontinuity in the aircraft flight plan at the prescribed timed way-points. Given that $b(y) = \beta(y)I$ with $\beta(y) := \sqrt{2[1 - \rho(y)]}$, we can apply Algorithm 1 to estimate the probability of conflict (23) with the transition probabilities of the approximating Markov chain given by (7).

Examples of 2-D aircraft-to-aircraft conflict prediction

We consider two aircraft flying in the same region of the airspace at a fixed altitude. The two-aircraft system is described by equations (21) and (22), with X_1 and X_2 denoting the two aircraft positions and taking values in $\mathbb{R}^2$. Note that the model described in Section 3.1 refers to the 3D flight case, where the aircraft positions take value in $\mathbb{R}^3$. However, it can be easily reformulated for the 2D case by minor modifications. In the 2D case, a conflict occurs when $Y = X_2 - X_1$ enters the unsafe set $\mathcal{D} = \{y \in \mathbb{R}^2 : \|y\| \leq r\}$.

In the following examples the safe distance r is set equal to 3, whereas the spatial correlation function ρ and matrix Σ are given by $\rho(y) = \exp(-c\|y\|)$, $y \in \mathbb{R}^2$ and $\Sigma = \sigma I$, where c and σ are positive constants. In all the plots of the estimated probability of conflict, the reported level curves refer to values $0.1, 0.2, \ldots, 0.9$.

Unless otherwise stated, in all of the examples in this subsection we use the following parameters:

The time interval of interest is $T = [0, 40]$. The relative velocity of the two aircraft during the time horizon T is given by

$$v(t) = \begin{cases} (2,0), & 0 \leq t < 10; \\ (0,1), & 10 \leq t < 20; \\ (2,0), & 20 \leq t \leq 40. \end{cases}$$

The parameter σ is equal to 1.

Based on the values of T and $v(t)$, $t \in T$, the domain $\mathcal{U}$ is chosen to be the open rectangle $(-80, 10) \times (-40, 10)$. The grid size is $\delta = 1$, hence the sampling time interval is $\Delta t = \lambda\delta^2 = (4\sigma^2)^{-1}\delta^2 = 0.25$.

λ appearing in (7) is set equal to $\lambda = (4\sigma^2)^{-1}$.

Example 1. We consider the case when the wind field is identically zero: $f(x,t) = 0$, for all $t \in T$, $x \in \mathbb{R}^2$. We set $c = 0.2$ in the spatial correlation function ρ. In Figure 2 we plot the level curves of the estimated probability of conflict over the time horizon $[t, t_f]$ as a function of the aircraft relative position at time t. As one can expect, the probability of conflict over $[t, t_f]$ takes higher values along the nominal path, which is the path traced by a point that starts from the origin at time $t_f = 40$ and moves backward in time according to the nominal relative velocity $v(\cdot)$ until time t. Furthermore, as the relative positions between the aircraft at time t move farther away from that path, the probability of conflict decreases. Experiments (not reported here) show that the smaller the variance parameter σ, the faster this decrease.

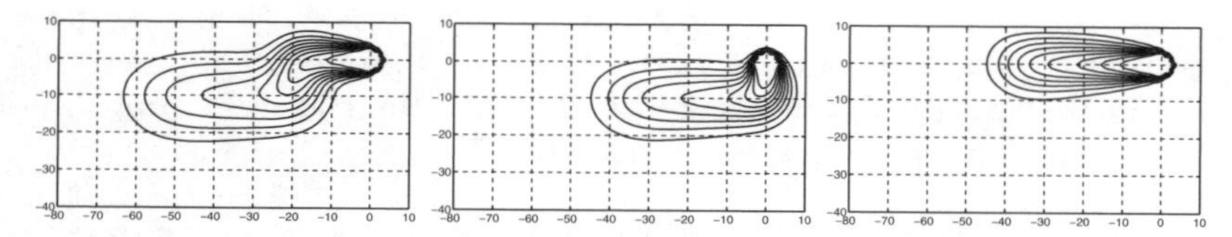

Fig. 2. Example 1. Level curves of the estimated probability of conflict over the time horizon $[t, 40]$ ($c = 0.2$). Left: $t = 0$. Center: $t = 10$. Right: $t = 20$.

Example 2. This example differs from the previous one only in the value of c, which is now set equal to $c = 0.05$. Then $\rho(y) = \exp(-0.05\|y\|)$ for $y \in \mathbb{R}^2$, which decreases much more slowly than in the previous case as $\|y\|$ increases. Since ρ characterizes the strength of spatial correlation in the random field $B(\cdot,\cdot)$, this means that the random components of the wind contributions to the two aircraft velocities tend to be more correlated to each other than in Example 1. In Figure 3, we plot the level curves of the estimated probability of conflict over $[t, t_f]$ in the cases $t = 0$, $t = 10$, and $t = 20$. One can see that, compared to the plots in Figure 2, the regions with higher probability of conflict in Figure 3 are more concentrated along the nominal path, which is especially evident near the origin. In a sense, this implies that the current approaches to estimating the probability of conflict, based on the assumption of independent wind perturbations to the aircraft velocities, could be pessimistic. The intuitive explanation of this phenomenon is that random wind perturbations to the aircraft velocities with larger correlations are more likely to cancel each other, resulting in more predictable behaviors and hence smaller probability of conflict.

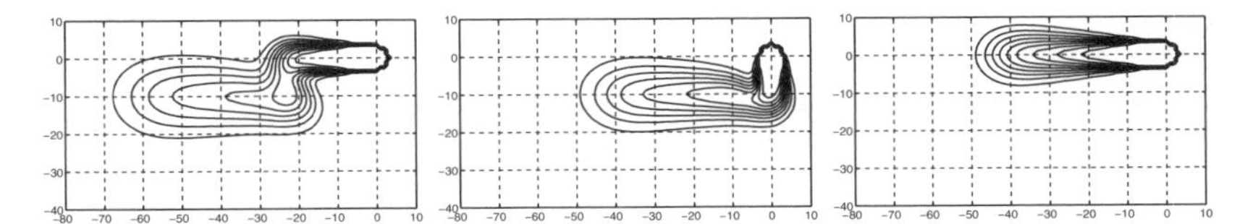

Fig. 3. Example 2. Level curves of the estimated probability of conflict over the time horizon $[t, 40]$ ($c = 0.05$). Left: $t = 0$. Center: $t = 10$. Right: $t = 20$.

Example 3. In this example, we choose $c = 0.05$ as in Example 2. However, we assume that there is a nontrivial affine wind field f defined by

$$f(x,t) = R(t)[x - z(t)], \quad x \in \mathbb{R}^2,\ t \in [0,40],$$

where

$$R(t) \equiv \frac{1}{50}\begin{bmatrix} 0 & 1 \\ -1 & 0 \end{bmatrix}, \quad z(t) = \begin{bmatrix} 3t \\ t^2/5 \end{bmatrix}.$$

The wind field f can be viewed as a windstorm swirling clockwise, whose center $z(t)$ accelerates along a curve during T. In fact, the choice of $z(t)$ will have no effect on the probability of conflict since it does not affect the aircraft relative position. In the first row of Figure 4, we plot the wind field f in the region $[-100, 200] \times [-100, 200]$ at the time instant $t = 0$ and the level curves of the estimated probability of conflict over $[t, t_f]$, at $t = 0$. In the second and third rows we represent similar plots for $t = 10$ and $t = 20$, respectively. One can see that, compared to the results in Figure 3, the regions with high probability of conflict are "bent" counterclockwise, and the farther away from the origin, the more the bending. This is because the net effect of the wind field f on the relative velocity v of the two aircraft is RY, which points clockwise when the relative position Y is in the third quarter of the Cartesian plane.

Example 4. Suppose now that in Example 3 we change the ending epoch t_f from 40 to infinity, and assume that the relative velocity v remains constant and equal to $(2,0)^T$ from time 20 on. For this infinite horizon problem, we can obtain an estimate of the probability of conflict at time $t = 0, 10, 20$ as drawn from top to bottom in Figure 5. Note that, unlike in the previous examples, the regions with high probability of conflict extend outside the domain $\mathcal{U}$ and are truncated. This is the price we pay to evaluate numerically the probability of conflict.

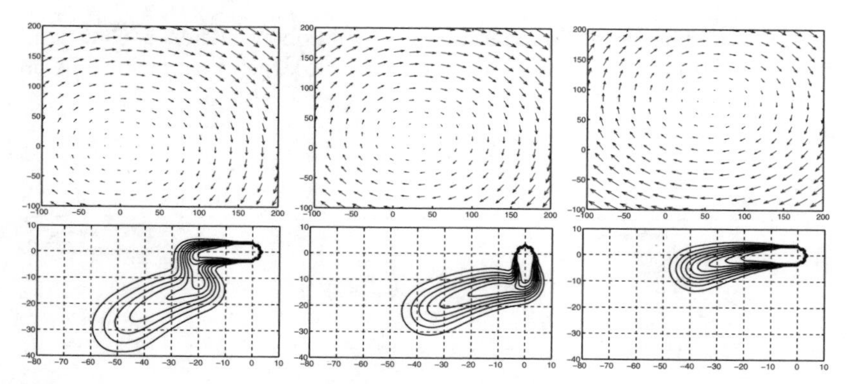

Fig. 4. Example 3. Wind field at time t, and level curves of the estimated probability of conflict over the time horizon $[t, 40]$ ($c = 0.05$). Left: $t = 0$. Center: $t = 10$. Right: $t = 20$.

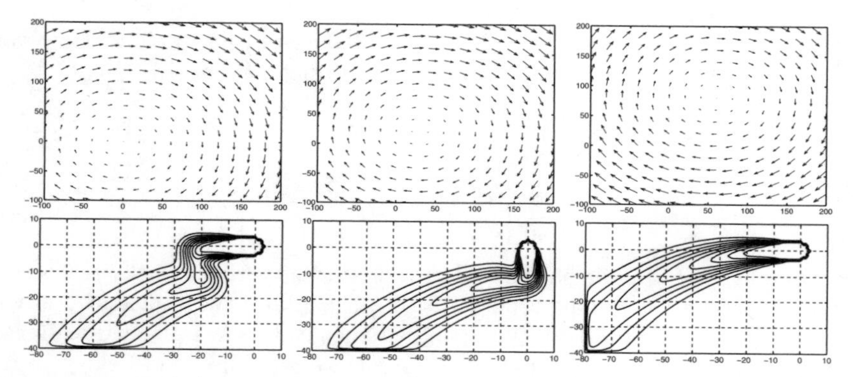

Fig. 5. Example 4. Wind field at time t, and level curves of the estimated probability of conflict over the time horizon $[t, \infty]$ ($c = 0.05$). Left: $t = 0$. Center: $t = 10$. Right: $t = 20$.

Examples of 3-D aircraft-to-aircraft conflict prediction

We consider a two-aircraft encounter where the aircraft positions X_1 and X_2 take values in $\mathbb{R}^3$ and are governed by equations (21) and (22).

The wind field f is assumed to be identically zero. A conflict occurs when $Y = X_2 - X_1$ enters the unsafe set $\mathcal{D} = \{y \in \mathbb{R}^2 : \|y\|_h \leq r, \|y\|_v \leq H\}$. Here we set $r = 3$ and $H = 1$.

We consider the case when $\rho(y) = \exp(-c_h\|y\|_h - c_v\|y\|_v)$, $y \in \mathbb{R}^3$, with c_h and c_v positive constants, and the matrix Σ is given by $\Sigma = \mathrm{diag}(\sigma_h, \sigma_h, \sigma_v)$, where $\sigma_h = 1$ and $\sigma_v = 0.5$.

We evaluate the probability that a conflict situation occurs within the time horizon $T = [0, 40]$, when the relative velocity of the two aircraft during T is given by

$$v(t) = \begin{cases} (2,0,0), & 0 \leq t < 5; \\ (0,1,1), & 5 \leq t \leq 10. \end{cases}$$

Based on the values taken by T, $v(\cdot)$, r and H, we choose the domain $\mathcal{U}$ to be $\mathcal{U} = (-30,15) \times (-15,10) \times (-15,10)$. We set the discretization step size $\delta = 1$, and $\lambda = (6\sigma_h^2)^{-1} = 1/6$. Thus $\Delta t = \lambda\delta^2 = 1/6$.

Figure 6 represents the estimated probability of conflict over the time horizon $[0,10]$ as a function of the relative position of the two aircraft at time t. The plots refer to the cases when $c_h = 0.2$, $c_v = 0.5$ and $c_h = 0.05$, $c_v = 0.05$ shown column-wise from left to right. In each column, we have the three dimensional isosurface at value 0.2 of the estimated probability of conflict viewed from different angles. The relevance of isosurfaces is that, in practice, once the relative position of the two aircraft is within the isosurface at a prescribed threshold value, an alarm of corresponding severity should be issued to the pilots to warn them on the level of criticality of the situation ([34]).

Note that when the parameters c_h and c_v of the spatial correlation function ρ are set equal to $c_h = c_v = 0.05$, the wind spatial correlation is increased. As a consequence of this fact, the isosurface at 0.2 concentrates more tightly along the deterministic path that leads to a conflict, and it extends longer as well.

General case

If no assumption is made on the wind field $f(x,t)$, to compute the probability of conflict (23), it no longer suffices to consider only the relative position of the two aircraft as in the affine case. Instead, we have to keep track of the two aircraft positions.

Define

$$\hat{X} = \begin{bmatrix} X_1 \\ X_2 \end{bmatrix} \in \mathbb{R}^6.$$

Then equations (21) and (22) can be written in terms of $\hat{X}$ as a single equation:

$$d\hat{X}(t) = \hat{u}(t)dt + \hat{f}(\hat{X},t)dt + \hat{\Sigma}d\hat{B}(\hat{X},t), \tag{29}$$

where we set

$$\hat{\Sigma} := \begin{bmatrix} \Sigma & 0 \\ 0 & \Sigma \end{bmatrix}, \ \hat{u}(t) := \begin{bmatrix} u_1(t) \\ u_2(t) \end{bmatrix}, \ \hat{f}(\hat{X},t) = \begin{bmatrix} f(X_1,t) \\ f(X_2,t) \end{bmatrix}, \ \hat{B}(\hat{X},t) := \begin{bmatrix} B(X_1,t) \\ B(X_2,t) \end{bmatrix}.$$

Fix $\hat{x} \in \mathbb{R}^6$. Let $\hat{Z}(t) := \hat{\Sigma}\,\hat{B}(\hat{x},t)$. $\{\hat{Z}(t), t \geq 0\}$ is a Gaussian process with zero mean and covariance

$$E[\hat{Z}(t)\hat{Z}(t)^T] = \begin{bmatrix} t\,I_3 & \hat{\rho}(\hat{x})\,t\,I_3 \\ \hat{\rho}(\hat{x})\,t\,I_3 & t\,I_3 \end{bmatrix} \hat{\Sigma}^2,$$

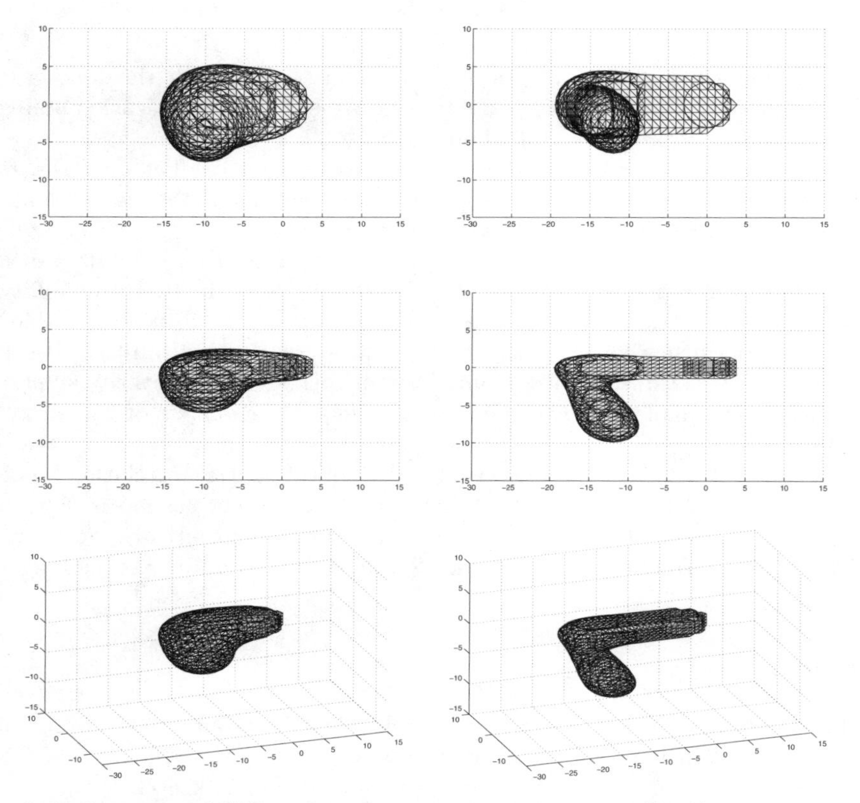

Fig. 6. Estimated probability of conflict over the time horizon $[0, 10]$: isosurface at value 0.2. Left: $c_h = 0.2$ and $c_v = 0.5$. Right: $c_h = 0.05$ and $c_v = 0.05$. First row: top view. Second row: side view. Third row: three dimensional plot.

with $\hat{\rho}(\hat{x}) := \rho(x_1 - x_2)$, with $\hat{x} := (x_1, x_2)$. Analogously to the previous section, in terms of distribution, $\hat{Z}(t) \overset{d}{\simeq} \sigma(\hat{x})\hat{\Sigma}\,\hat{W}(t)$, where $\hat{W}(t)$ is a standard Brownian motion in $\mathbb{R}^6$, and

$$\sigma(\hat{x}) := \begin{bmatrix} I_3 & \hat{\rho}(\hat{x})\, I_3 \\ \hat{\rho}(\hat{x})\, I_3 & I_3 \end{bmatrix}^{1/2} \in \mathbb{R}^{6\times 6}.$$

As a result, (29) becomes

$$d\hat{X}(t) = \hat{u}(t)dt + \hat{f}(\hat{X}, t)dt + \sigma(\hat{X})\hat{\Sigma}\, d\hat{W}(t). \tag{30}$$

Equation (29) is a particular case of (1) with $S = \hat{X}$, $\Gamma = \hat{\Sigma}$, $a(\hat{x}, t) = \hat{u}(t) + \hat{f}(\hat{x}, t)$, and $b(\hat{x}) = \sigma(\hat{x})$. In this case, we can apply Algorithm 1 to estimate the probability of conflict (23) with the transition probabilities of the approximating Markov chain given by (9).

Example 5. In this example, we consider two aircraft flying in the same region of the airspace at a fixed altitude. The safe distance r is set equal to 3, whereas the spatial correlation function ρ and matrix Σ are given by $\rho(y) = \exp(-c\|y\|)$, $y \in \mathbb{R}^2$ and $\Sigma = \sigma I$, where $c = 1$ and $\sigma = 2$.

The time interval of interest is $T = [0, 20]$. The velocities of the two aircraft during the time horizon T are supposed to be constant and given by

$$u_1(t) = \begin{bmatrix} 4 \\ 0 \end{bmatrix}, \quad u_2(t) = \begin{bmatrix} 2 \\ 0 \end{bmatrix}, \quad 0 \leq t \leq 20.$$

The wind field is assumed to depend only on the spatial coordinate $x \in \mathbb{R}^2$ as follows

$$f(x,t) = \begin{bmatrix} \frac{\exp[([x]_1+20)/2]-1}{\exp[([x]_1+20)/2]+1} \\ 0 \end{bmatrix}.$$

where $[x]_1$ is the first component of x. Under this wind field model, the wind direction is along the $[x]_1$ axis from right to left on the half-plane with $[x]_1 < -20$, and from left to right on the half-plane with $[x]_1 > -20$. The maximal strength $\|f(x,t)\|$ of the wind is 1, which is achieved when $[x]_1 \to \pm\infty$.

Based on the values taken by T, and $u_1(t), u_2(t)$, $t \in T$, we set $\mathcal{U} := \mathcal{U}_1 \times \mathcal{U}_2$, with $\mathcal{U}_1$ and $\mathcal{U}_2$ open rectangles $\mathcal{U}_1 = (-100, 30) \times (-24, 24)$ and $\mathcal{U}_2 = (-60, 80) \times (-16, 16)$. Finally, we set $\lambda = (2\sigma^2)^{-1} = 0.125$ and $\delta = 1.5$, so that $\Delta t = \lambda\delta^2 = 9/32$.

In Figure 7, we plot the level curves of the estimated probability of conflict as a function of the initial position of aircraft 1, for five different initial positions of aircraft 2: $(-40, 0)$, $(-30, 0)$, $(-20, 0)$, $(0, 0)$, and $(20, 0)$, moving from top to bottom in the figure. On each row, the figure on the left side corresponds to the probability of conflict as computed by Algorithm 1. Since we use a relative coarse grid $\delta = 1.5$, the level curves are not smooth. For better visualization, we plot on the right side the level curves of a smoothed version of the probability of conflict maps, whose value at each grid point $w \in \mathcal{U}_1 \cap \delta\mathbb{Z}^2$ is the average value of the probability of conflict at w and its four immediate neighboring points w_{1-}, w_{1+}, w_{2-}, w_{2+}. In effect, this is equivalent to passing the original probability of conflict map through a low pass filter. This also corresponds to assuming that there is uncertainty in the initial position of aircraft 1, such that it is equally probable that aircraft 1 occupies its nominal position and the four immediate neighboring grid points.

In the reported example, we see that, unlike the affine wind field case, the probability of conflict in general depends on the initial positions of both aircraft, not just on their initial relative position. If the probability of conflict would depend only on the aircraft initial relative position, then the level curves in the plots of Figure 7 will be all identically shaped and one could be obtained from another by translation of an amount given by the difference between the

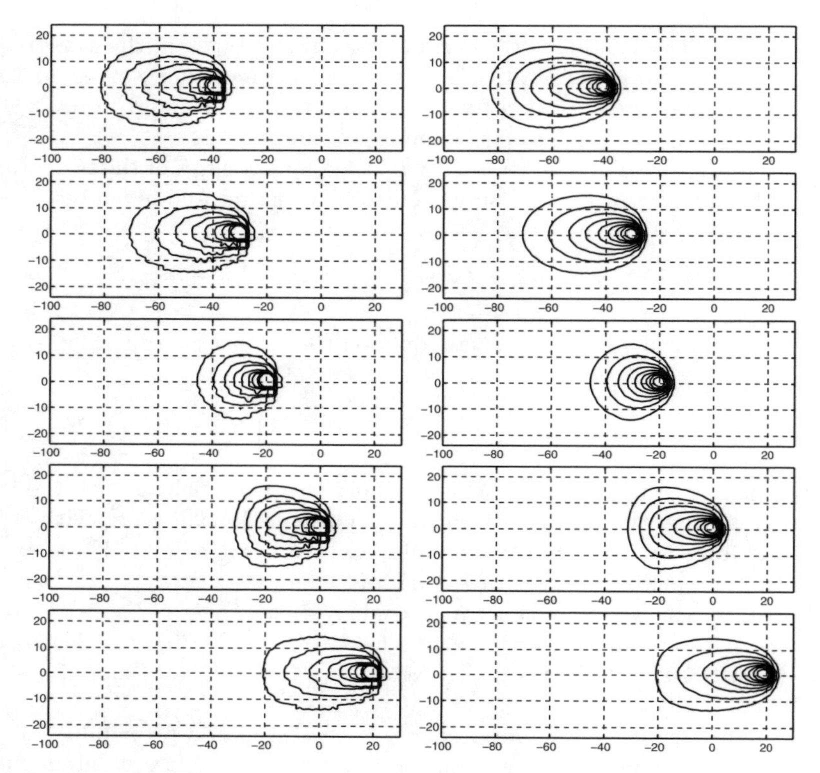

Fig. 7. Example 5. Left: Level curves of the estimated probability of conflict over the time horizon $[0, 20]$ as a function of the initial position of aircraft 1 for fixed initial position of aircraft 2 (from top to bottom: $(-40, 0)$, $(-30, 0)$, $(-20, 0)$, $(0, 0)$, and $(20, 0)$). Right: Level curve of a smooth version of the corresponding quantity on the left. (Non-affine wind field)

corresponding initial positions of aircraft 2, which is obviously not the case in Figure 7.

The dependence of the probability of conflict on the initial positions of both aircraft rather than simply their relative position is more eminent at those places where there is a large acceleration (or deceleration) in wind components, i.e., at those places with higher degree of nonlinearity in the wind field. If the nonlinearity of the wind field is relatively small, the two-aircraft system could be described in terms of the their relative position, significantly reducing the computation time.

3.3 Aircraft-to-Airspace Conflict Problem

An aircraft-to-airspace conflict occurs when the aircraft enters a forbidden area of the airspace. For a variety of reasons, an aircraft trajectory is constrained to limited spaces during a flight. Large sectors of airspace over Europe are "no-go" because of, for example, Special Use Airspace (SUA) areas in the military airspace or separation buffers around strategically important objects. Airspace restrictions can also originate dynamically due to severe weather conditions or high traffic congestion causing some airspace area to exceed its maximal capacity. The management of air traffic as density increases around the restricted areas is then crucial to avoid aircraft-to-airspace conflicts.

Consider an aircraft flying in some region of the airspace. An aircraft-to-airspace conflict occurs if the aircraft enters the prohibited area within the look-ahead time horizon T. If this area can be described by a set $\mathcal{D} \subset \mathbb{R}^3$, then this problem can be formulated as the estimation of the probability

$$P\{X(t) \in \mathcal{D} \text{ for some } t \in T\} \tag{31}$$

where $X(t)$ is the aircraft position at time $t \in T$ and is obtained by (20) initialized with $X(0)$.

Note that we are considering a single aircraft, and, for each fixed $x \in \mathbb{R}^n$, $B(x, \cdot)$ is a standard 3-dimensional Brownian motion, and $B(\cdot, \cdot)$ is time increment independent and stationary. We can then replace $B(\cdot, \cdot)$ with a standard Brownian motion $W(\cdot)$, and refer to

$$dX(t) = u(t)dt + f(X, t)dt + \Sigma\, dW(t), \tag{32}$$

initialized with $X(0)$, for the purpose of computing the probability in (31).

Equation (32) is a particular case of (1) with $S = X$, $\Gamma = \Sigma$, $a(x,t) = u(t) + f(x,t)$, and $b(x) = I$. In this case, we can apply Algorithm 1 to estimate the probability of conflict (23) with the transition probabilities of the approximating Markov chain given by (7).

Example 6. Suppose that an aircraft is flying along the x_1-axis while climbing up at an accelerated rate according to the flight plan $u(t) = (3/2, 0, 2t/75)$, $t \in T = [0, 15]$. The wind field f is assumed to be identically zero. The matrix Σ is given by $\Sigma = \text{diag}(\sigma_h, \sigma_h, \sigma_v)$, where $\sigma_h = 1$ and $\sigma_v = 0.5$.

Consider a prohibited airspace area $\mathcal{D}$ given by the union of two ellipsoids specified by $\{(x_1, x_2, x_3) \in \mathbb{R}^3 : 2(x_1 + 4)^2 + (x_2 - 4)^2 + 10x_3^2 \le 9\}$ and $\{(x_1, x_2, x_3) \in \mathbb{R}^3 : x_1^2 + 2(x_2 + 5)^2 + 10x_3^2 \le 16\}$, in the (x_1, x_2, x_3) Cartesian coordinate system with x_3 representing the flight level.

Figure 8 shows the plots of the isosurface at value 0.2 of the probability of conflict as a function of the aircraft initial position, at time $t = 0$, $t = 5$, and $t = 10$, viewed from three different angles. The probability of conflict is estimated through Algorithm 1 with $\mathcal{U} = (-38, 6) \times (-15, 11) \times (-6, 3)$ and $\delta = 1$.

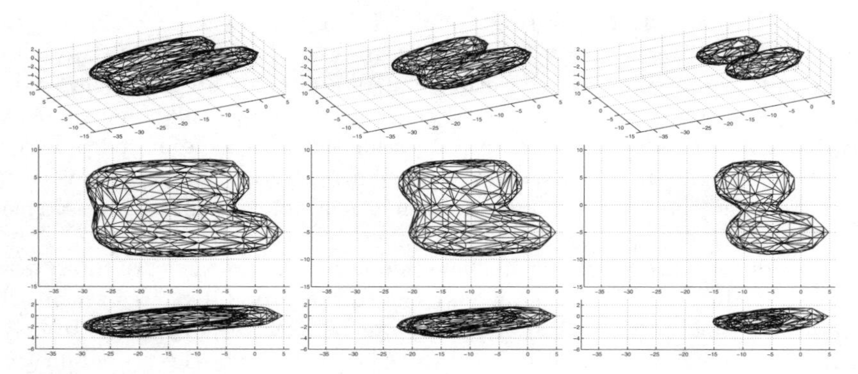

Fig. 8. Estimated probability of conflict over the time horizon $[t, 15]$: isosurface at value 0.2. Left: $t = 0$. Center: $t = 5$. Right: $t = 10$. First row: 3D plot. Second row: top view. Third row: side view.

4 Conclusions

In this work, we describe a novel grid-based method for estimating the probability that the trajectories of a system governed by a stochastic differential equation with time-driven jumps will enter some target set during some possibly infinite look-ahead time horizon. The distinguishing feature of the proposed method is that it is based on a Markov chain approximation scheme, integrating a backward reachability computation procedure.

This method is applied to estimate the probability that two aircraft flying in the same region of the airspace get closer than a certain safety distance and the probability that an aircraft enters a forbidden airspace area. The intended application is aircraft conflict detection, with the final objective of supporting air traffic controllers in detecting potential conflict situations so as to improve the efficiency of the air traffic management system in terms of airspace usage.

It is worth noticing that, though we provide as an application example air traffic control, our results may have potentials in other safety-critical contexts, where the safety verification problem can be reformulated as that of verifying if a given stochastic system trajectories will eventually enter some unsafe set.

Grid-based methods are generally computationally intensive. On the other hand, the outcome of the proposed grid-based algorithm is a map that associates to each admissible initial condition of the system the corresponding estimate of the probability of entering the unsafe set, which could be used not only for detecting an unsafe situation, but also for designing an appropriate action to timely steer the system outside the unsafe set. One could, for example, force the system to slide along a certain isosurface depending on the trust level.

References

1. R.J. Adler. *The Geometry of Random Fields.* John Wiley & Sons, 1981.
2. R. Alur, T. Henzinger, G. Lafferriere, and G.J. Pappas. Discrete abstractions of hybrid systems. *Proceedings of the IEEE*, 88(2):971–984, 2000.
3. S.G. Benjamin, K. J. Brundage, P. A. Miller, T. L. Smith, G. A. Grell, D. Kim, J. M. Brown, T. W. Schlatter, and L. L. Morone. The Rapid Update Cycle at NMC. In *Proc. Tenth Conference on Numerical Weather Prediction*, pages 566–568, Portland, OR, Jul. 1994.
4. A. Chutinan and B.H. Krogh. Verification of infinite-state dynamic systems using approximate quotient transition systems. *IEEE Transactions on Automatic Control*, 46(9):1401–1410, 2001.
5. R.E. Cole, C. Richard, S. Kim, and D. Bailey. An assessment of the 60 km rapid update cycle (RUC) with near real-time aircraft reports. Technical Report NASA/A-1, MIT Lincoln Laboratory, Jul. 1998.
6. R. Durrett. *Stochastic calculus: A practical introduction.* CRC Press, 1996.
7. H. Erzberger, R.A. Paielli, D.R. Isaacson, and M.M. Eshow. Conflict detection and resolution in the presence of prediction error. In *Proc. of the 1st USA/Europe Air Traffic Management R & D Seminar*, Saclay, France, June 1997.
8. J. Hu, J. Lygeros, M. Prandini, and S. Sastry. Aircraft conflict prediction and resolution using Brownian Motion. In *Proc. of the* 38th *Conf. on Decision and Control*, Phoenix, AZ, December 1999.
9. J. Hu and M. Prandini. Aircraft conflict detection: a method for computing the probability of conflict based on Markov chain approximation. In *European Control Conf.*, Cambridge, UK, September 2003.
10. J. Hu, M. Prandini, and S. Sastry. Optimal coordinated maneuvers for three dimensional aircraft conflict resolution. *Journal of Guidance, Control and Dynamics*, 25(5):888–900, 2002.
11. J. Hu, M. Prandini, and S. Sastry. Aircraft conflict detection in presence of spatially correlated wind perturbations. In *AIAA Guidance, Navigation, and Control Conference and Exhibit*, Austin, USA, August 2003.
12. J. Hu, M. Prandini, and S. Sastry. Probabilistic safety analysis in three dimensional aircraft flight. In *Proc. of the* 42nd *Conf. on Decision and Control*, Maui, USA, December 2003.
13. J. Hu, M. Prandini, and S. Sastry. Optimal coordinated motions for multiple agents moving on a plane. *SIAM Journal on Control and Optimization*, 42(2):637–668, 2003.
14. J. Hu, M. Prandini, and S. Sastry. Aircraft conflict prediction in presence of a spatially correlated wind field. *IEEE Transactions on Intelligent Transportation Systems*, 6(3):326–340, 2005.
15. E. H. Isaaks and R.M. Srivastava. *An Introduction to Applied Geostatistics.* Oxford University Press, 1989.
16. J. Kosecka, C. Tomlin, G.J. Pappas, and S. Sastry. Generation of Conflict Resolution Maneuvers For Air Traffic Management. In *Proc. of the IEEE Conference on Intelligent Robotics and System '97*, volume 3, pages 1598–1603, Grenoble, France, September 1997.
17. J. Krozel and M. Peters. Strategic conflict detection and resolution for free flight. In *Proc. of the* 36th *Conf. on Decision and Control*, volume 2, pages 1822–1828, San Diego, CA, December 1997.

18. J.K. Kuchar and L.C. Yang. A review of conflict detection and resolution modeling methods. *IEEE Transactions on Intelligent Transportation Systems, Special Issue on Air Traffic Control - Part I*, 1(4):179–189, 2000.
19. A.B. Kurzhanski and P. Varaiya. Ellipsoidal techniques for reachability analysis. In B. Krogh and N. Lynch, editors, *Hybrid Systems: Computation and Control*, Lecture Notes in Computer Science, pages 202–214. Springer Verlag, 2000.
20. A.B. Kurzhanski and P. Varaiya. On reachability under uncertainty. *SIAM J. Control Optim.*, 41(1):181–216, 2002.
21. J. Lygeros and N. Lynch. On the formal verification of the TCAS conflict resolution algorithms. In *Proc. of the* 36th *Conf. on Decision and Control*, pages 1829–1834, San Diego, CA, December 1997.
22. J. Lygeros and M. Prandini. Aircraft and weather models for probabilistic conflict detection. In *Proc. of the* 41st *Conf. on Decision and Control*, Las Vegas, NV, December 2002.
23. F. Medioni, N. Durand, and J.M. Alliot. Air traffic conflict resolution by genetic algorithms. In *Proc. of the Artificial Evolution, European Conference (AE 95)*, pages 370–383, Brest, France, September 1995.
24. P.K. Menon, G.D. Sweriduk, and B. Sridhar. Optimal strategies for free-flight air traffic conflict resolution. *Journal of Guidance, Control, and Dynamics*, 22(2):202–211, 1999.
25. I. Mitchell, A. Bayen, and C. Tomlin. Validating a Hamilton-Jacobi approximation to hybrid system reachable sets. In A. Sangiovanni-Vincentelli and M. Di Benedetto, editors, *Hybrid Systems: Computation and Control*, Lecture Notes in Computer Science, pages 418–432. Springer Verlag, 2001.
26. I. Mitchell and C. Tomlin. Level set methods for computation in hybrid systems. In B. Krogh and N. Lynch, editors, *Hybrid Systems: Computation and Control*, Lecture Notes in Computer Science, pages 310–323. Springer Verlag, 2000.
27. R.A. Paielli and H. Erzberger. Conflict probability estimation for free flight. *Journal of Guidance, Control, and Dynamics*, 20(3):588–596, 1997.
28. M. Prandini, J. Hu, J. Lygeros, and S. Sastry. A probabilistic approach to aircraft conflict detection. *IEEE Transactions on Intelligent Transportation Systems, Special Issue on Air Traffic Control - Part I*, 1(4):199–220, 2000.
29. Radio Technical Commission for Aeronautics. Minimum operational performance standards for traffic alert and collision avoidance system (TCAS) airborn equipment. Technical Report RTCA/DO-185, RTCA, September 1990. Consolidated Edition.
30. J. Schrder and J. Lunze. Representation of quantised systems by the Frobenius-Perron operator. In A. Sangiovanni-Vincentelli and M. Di Benedetto, editors, *Hybrid Systems: Computation and Control*, Lecture Notes in Computer Science, pages 473–486. Springer Verlag, 2001.
31. D.W. Stroock and S.R.S. Varadhan. *Multidimensional Diffusion Processes.* Springer-Verlag, 1979.
32. C. Tomlin, I. Mitchell, A. Bayen, and M. Oishi. Computational techniques for the verification and control of hybrid systems. *Proceedings of the IEEE*, 91(7):986–1001, 2003.
33. C. Tomlin, G.J. Pappas, and S. Sastry. Conflict resolution for air traffic management: A study in multi-agent hybrid systems. *IEEE Transactions on Automatic Control*, 43(4):509–521, 1998.

34. L.C. Yang and J. Kuchar. Prototype conflict alerting system for free fligh. In *Proc. of the AIAA 35th Aerospace Sciences Meeting, AIAA-97-0220*, Reno, NV, January 1997.
35. Y. Zhao and R. Schultz. Deterministic resolution of two aircraft conflict in free flight. In *Proc. of the AIAA Guidance, Navigation, and Control Conference, AIAA-97-3547*, New Orleans, LA, August 1997.

Critical Observability of a Class of Hybrid Systems and Application to Air Traffic Management

Elena De Santis, Maria D. Di Benedetto, Stefano Di Gennaro, Alessandro D'Innocenzo, and Giordano Pola

Department of Electrical Engineering and Computer Science, Center of Excellence DEWS
University of L'Aquila, Poggio di Roio, 67040 – L'Aquila, Italy
desantis,dibenede,digennar,adinnoce,pola@ing.univaq.it

Summary. We present a novel observability notion for switching systems that model safety–critical systems, where a set of states – called critical states – must be detected immediately since they correspond to hazards that may lead to catastrophic events. Some sufficient and some necessary conditions for critical observability are derived. An observer is proposed for reconstructing the hybrid state evolution of the switching system whenever a critical state is reached. We apply our results to the runway crossing control problem, i.e., the control of aircraft that cross landing or take–off runways. In the hybrid model of the system, five agents are present; four are humans, each modeled as hybrid systems, subject to situation awareness errors.

1 Introduction

The class of hybrid control problems is extremely broad (it contains continuous control problems as well as discrete event control problems as special cases). Hence, it is very difficult to devise a general yet effective strategy to solve them. Research in the area of hybrid systems addresses significant application domains to develop further understanding of the implications of the hybrid model on control algorithms and to evaluate whether using this formalism can be of substantial help in solving complex, real–life, control problems (see e.g. [12] and the references therein).

An application that has benefited greatly from this modelling paradigm is the design of embedded controllers for transportation systems. In particular, power–train control is one of the most interesting and challenging problem in embedded system design. In [2], we presented a general framework for power–train control based on hybrid models and demonstrate that it is possible to find effective control laws with guaranteed properties without resorting to average–value models. By using hybrid systems modelling and synthesis,

H.A.P. Blom, J. Lygeros (Eds.): Stochastic Hybrid Systems, LNCIS 337, pp. 141–170, 2006.

solutions to several challenging control problems were proposed (see e.g. the Fast Force Transient problem [3], the cut–off problem [1], the digital idle speed control problem [10]). These problems were solved by means of a power–train full state feedback. Since, in most cases, state measurements are not available, the synthesis of a state observer is of fundamental importance to make the hybrid control algorithms really applicable.

Another application of hybrid modelling in transportation systems that can potentially improve the quality of present solutions is the design of Air–Traffic Management systems. The objective of Air–Traffic Management is to ensure the safe and efficient operation of aircraft. The stress placed on the present systems by the ever increasing air traffic has forced the authorities to plan for an overhaul of ATM to make them more reliable, safer and more efficient. A move in this direction requires more automation and a more sophisticated monitoring and control system. Automation and control require in turn a precise formulation of the problem. In this context, variables that can be measured or estimated have to be identified together with safety indices and objective functions. To make things more complex, the behavior of ATM depends critically on the actions of humans who control the operations that are very difficult to observe, measure, model, and predict.

Error detection and control must rely upon robust state estimation techniques, thus providing a strong motivation for a rigorous approach to observability and detectability based on tests of affordable computational complexity. Other motivations are the necessity of developing controllers for assisting human operators in detecting critical situations and avoiding propagation of errors that could lead to catastrophic events. In fact, in an ATM closed–loop system with mixed computer–controlled and human–controlled subsystems, recovery from non–nominal situations implies the existence of an outer control loop that has to identify critical situations and act accordingly to prevent them to evolve into accidents. Estimation methods and observer design techniques are essential in this regard for the design of a control strategy for error propagation avoidance and/or error recovery.

Observability has been extensively studied both in the continuous ([22], [25]) and in the discrete domains (see e.g. [29], [30], [36]). In particular, Sontag in [32] defined different observability concepts and analyzed their relations for polynomial systems. More recently, various researchers have approached the study of observability for hybrid systems, but the definitions and the testing criteria for it varied depending on the class of systems under consideration and on the knowledge that is assumed at the output. Vidal et al. [35] considered autonomous switching systems and proposed a definition of observability based on the concept of indistinguishability of continuous initial states and discrete state evolutions from the outputs in free evolution. Incremental observability was introduced in [6] for the class of piecewise affine (PWA) systems. Incremental observability means that different initial states always give different outputs independently of the applied input. In [5], the notion of generic final–state determinability proposed by Sontag [32] was extended to hybrid systems and sufficient conditions were given for linear hybrid systems. In [8], we introduced a notion of observability and detectability for the class of switching systems, based on the reconstructability of the hybrid state evolution, knowing the hybrid outputs, for some suitable continuous inputs. In [4], a methodology was presented for the design of dynamic observers of hybrid systems, which reconstructs the discrete state and the continuous state from the knowledge of the continuous and discrete outputs. In [17],[18], extensions of [4] were derived. In [21] the definitions of observability of [34] and the results of [4] on the design of an observer for deterministic hybrid systems were extended to discrete–time stochastic linear autonomous hybrid systems.

In some safety-critical applications, such as Air Traffic Management (ATM), we need to determine the actual state of the system immediately, as a delay in determining the state may lead to unsafe or even catastrophic behavior of the system. For this reason, some authors [28] extended the definition of observability to capture this urgency. In particular, in [14], [15] a notion of critical observability referred to the discrete dynamics was introduced, considering a subset of critical (discrete) states of the hybrid system. An observer based on this definition of observability was designed for fault and error detection in prescribed time horizon.

In this paper, we extend the work presented above to a class of hybrid systems, linear switching systems with minimum and maximum dwell time. The choice of this particular subclass of hybrid systems is motivated by the following considerations:

i) switching systems are an appropriate abstraction for modelling important complex systems such as ATM systems (e.g. [14], [15]) or automotive engines (e.g. [1], [2], [10]);
ii) the semantics of switching systems allows the derivation of necessary and sufficient computable observability conditions that become sufficient for the general class of hybrid systems where the transitions may depend on the continuous component of the hybrid state.

The paper is organized as follows. In Section 2, we review a set of formal definitions for switching systems. In Section 3, we propose a general definition of observability, based on the possibility of reconstructing the hybrid system state. We then give some necessary and sufficient testable conditions for observability. As a special case, we introduce the notion of *critical observability* and in Section 4, we offer conditions for checking observability properties and for the existence of observers. Furthermore, we consider in Section 5 as a non trivial case–study, the so–called active runway crossing control problem. In particular, we concentrate on the design of an observer for generating an alarm when critical situations occur, e.g., an aircraft crossing the runway when another aircraft is taking off. In Section 6, we offer some concluding remarks.

2 Linear Switching Systems

In this paper, we consider the class of linear switching systems that are a special case of hybrid systems, as defined in [26].

In a general hybrid system, an invariance condition may be associated with each discrete state. Given a discrete location, when the continuous state does not satisfy the corresponding invariance condition, a transition has to take place. A guard condition may be associated with each transition and has to be satisfied for that transition to be enabled. Switching systems may be seen as abstractions of hybrid systems, where we assume that the transitions do not depend on the value of the continuous state (that is, for any transition, the 'guard condition' is the continuous state space) and, for any discrete state, the 'invariance condition' is the continuous state space associated to that discrete state. The continuous state space associated with each discrete state is characterized by its own dimension that is not necessarily the same for all the discrete states.

Definition 1. *A linear switching system* $\mathcal{S}$ *is a tuple* $(\Xi, \Xi_0, \Theta, S, E, R, \Upsilon)$ *where:*

- $\Xi = \bigcup_{q_i \in Q} \{q_i\} \times \mathbb{R}^{n_i}$ *is the hybrid state space, where*
  ◦ $Q = \{q_i, i \in J\}$ *is the discrete state space and* $J = \{1, 2, \cdots, N\}$*;*
  ◦ $\mathbb{R}^{n_i}$ *is the continuous state space associated with* $q_i \in Q$*;*
- $\Xi_0 = \bigcup_{q_i \in Q_0} \{q_i\} \times X_i^0 \subset \Xi$ *is the set of all initial hybrid states;*
- $\Theta = \Sigma \times \mathbb{R}^m$ *is the hybrid input space, where*
  ◦ $\Sigma = \{\sigma_1, \cdots, \sigma_r\}$ *is the finite set of discrete uncontrolled inputs;*
  ◦ $\mathbb{R}^m$ *is the continuous input space;*
- S *is a mapping that associates to any discrete state* $q_i \in Q$*, the following continuous–time linear system*

$$\dot{x}(t) = A_i x(t) + B_i u(t), \qquad y(t) = C_i x(t), \qquad i \in J \tag{1}$$

  with $A_i \in \mathbb{R}^{n_i \times n_i}$*,* $B_i \in \mathbb{R}^{n_i \times m}$*,* $C_i \in \mathbb{R}^{p \times n_i}$*,* $x \in \mathbb{R}^{n_i}$ *the continuous state,* $u \in \mathbb{R}^m$ *the continuous input and* $y \in \mathbb{R}^p$ *the continuous output;*
- $E \subset Q \times \Sigma \times Q$ *is a collection of transitions;*
- $R \colon E \times \Xi \to \Xi$ *is the reset function;*
- $\Upsilon = \Psi_E \times \Psi_Q \times \mathbb{R}^p$ *is the output space, where:*
  ◦ $\Psi_E = \{\epsilon, \psi_E^1, \cdots, \psi_E^{N_1}\}$ *is the output space associated with the transitions by means of the function* $\eta \colon E \to \Psi_E$*;* ϵ *is the unobservable output;*
  ◦ $\Psi_Q = \{\psi_Q^1, \cdots, \psi_Q^{N_2}\}$ *is the output space associated with the discrete states by means of the function* $h \colon Q \to \Psi_Q$*;*
  ◦ $\mathbb{R}^p$ *is the continuous output space.*

We now formally define the semantics of linear switching systems. First of all we assume that *the discrete disturbance is not available for measurements*, thus yielding a non–deterministic system, and that the class of admissible continuous inputs is the set $\mathcal{U}$ of piecewise continuous control functions $u : \mathbb{R} \to \mathbb{R}^m$. Following [26], we recall that a hybrid time basis τ is an infinite or finite sequence of sets $I_j = \{t \in \mathbb{R} : t_j \leq t \leq t'_j\}$, with $t'_j = t_{j+1}$; let be $\mathrm{card}\,(\tau) = L + 1$. If $L < \infty$, then t'_L can be finite or infinite. Time t'_j is said to be a *switching time* and the symbol $\mathcal{T}$ denotes the set of all hybrid time bases. The switching system temporal evolution is then defined as follows.

Definition 2. *An execution of* $\mathcal{S}$ *is a collection* $\chi = (\xi_0, \tau, \sigma, u, \xi)$ *with* $\xi_0 = (q_0, x_0) \in \Xi_0$, $\tau \in \mathcal{T}$, $\sigma \colon \mathbb{N} \to \Sigma$, $u \in \mathcal{U}$, $\xi \colon \mathbb{R} \times \mathbb{N} \to \Xi$, *where the hybrid state evolution* ξ *is defined as follows:*

$$\begin{aligned}
&\xi(t_0, 0) = \xi_0,\\
&\xi(t_{j+1}, j+1) = R(e_j, \xi(t'_j, j)),\\
&e_j = (q(j), \sigma(j), q(j+1)) \in E,\\
&x(t, j) = x(t),
\end{aligned}$$

where $q : \mathbb{N} \to Q$, $e_j = (q\,(j)\,, \sigma\,(j)\,, q\,(j+1)) \in E$ *and* $x\,(t)$ *is the (unique) solution at time* t *of the dynamical system* $S\,(q\,(j))$, *with initial time* t_j, *initial condition* $x\,(t_j, j)$ *and control law* u. *The observed output evolution of* $\mathcal{S}$ *is defined by the function* $y^o \colon \mathbb{R} \to \Upsilon$, *such that*

$$y^o\,(t) = \begin{cases} (\eta\,(e_{j-1})\,, h(q\,(j)), C_i x(t,j))\,, & \textit{if } t = t_j,\\ (\epsilon, h(q(j)), C_i x(t,j))\,, & \textit{if } t \in (t_j, t'_j), \end{cases}$$

where $\eta\,(e_{-1}) = \epsilon$.

We denote by $\mathcal{Y}^o$ the class of functions $y^o \colon \mathbb{R} \to \Upsilon$. Given a control $u \in \mathcal{U}$ and the initial hybrid state $\xi_0 = (q_0, x_0)$, the resulting executions are called executions of $\mathcal{S}$ with initial hybrid state ξ_0.

We assume the existence of a *minimum dwell time* [27] before which no discrete input causes a transition, and of a *maximum dwell time* [8] before which a transition certainly occurs.

Assumption 7 *(Minimum and maximum dwell time) Given the linear switching system* $\mathcal{S}$, *there exist* $\Delta_m > 0$ *and* $\Delta_M > 0$, *called respectively minimum and maximum dwell time, so that any execution* $\chi = (\xi_0, \tau, \sigma, u, \xi)$ *has to satisfy the condition*

$$\Delta_m \leq t'_j - t_j \leq \Delta_M, \quad \forall j = 0, 1, \cdots, L-1. \tag{2}$$

The existence of a minimum dwell time is a widely used assumption in the analysis of switching systems (e.g. [27], [24] and the references therein), and models the inertia of the system to react to an external (discrete) input. The existence of a maximum dwell time is related to the so–called liveness property of the system and is widely used in the context of Discrete Event Systems (DES) (e.g. [29]). Moreover, as shown in [10], minimum and maximum dwell times offer a method for approximating hybrid systems by means of switching systems.

An execution is *infinite* if card $(\tau) = \infty$ or $t'_L = \infty$. The value Δ_M can be finite or infinite. If $\Delta_M = \infty$, without loss of generality (w.l.o.g.) all executions may be assumed to be infinite. Otherwise we assume that $\mathcal{S}$ is alive [29], i.e. for any discrete state $q \in Q$ there exists a discrete state q^+ and $\sigma \in \Sigma$ such that $(q, \sigma, q^+) \in E$, so that again all the executions may be assumed w.l.o.g. to be infinite.

We will use the following notation: $f^{-1}(\cdot)$ denotes the inverse image operator of $f(\cdot)$, $reach\,(Q_0)$ denotes the set of discrete states that can be reached from Q_0, i.e. such that there exists an execution, with initial discrete state in Q_0, which steers the discrete state in $reach\,(Q_0)$ in a finite number of switchings. We assume w.l.o.g. that $Q = reach(Q_0)$.

3 Observability Notions

A rather complete discussion on different definitions of observability for some subclasses of hybrid systems can be found in [7], [8]. In particular, our definition in [8] is based on the reconstructability of the hybrid state evolution from some instant of time on and after a finite number, namely k, of transitions for some suitable continuous input. However, in some important applications, as for example in Air Traffic Management, it is necessary to identify immediately and before a transition occurs, those discrete states – that we may call *critical* – that can lead to unsafe situations [14], [15]. In that case, even if the system is observable in the sense of [8], if a critical state is reached before k transitions take place, the corresponding critical situation is not identified. We therefore need to extend the definition of [8] by requiring, in addition to observability, the immediate detection of the critical states.

All the definitions presented here can be given for general hybrid systems. Let $Q_c \subset Q$ denote the set of critical states associated with the linear switching system $\mathcal{S}$. We assume w.l.o.g. that $Q_0 \subset reach^{-1}(Q_c)$.

Definition 3. *A linear switching system $\mathcal{S}$ is Q_c–observable if there exist a function $\hat{u} \in \mathcal{U}$, a function $\hat{\xi} : \mathcal{Y}^o \times \mathcal{U} \to \Xi$, a real $\Delta \in (0, \Delta_m)$ and for any $\xi_0 \in \Xi_0$ there exists $\hat{t} \in (t_0, \infty)$ such that for any execution of $\mathcal{S}$ with $u = \hat{u}$,*

$$\hat{\xi}\left(y^o|_{[t_0,t]}, \hat{u}|_{[t_0,t)}\right) = \xi(t, j),$$

for any j such that $q(j) \in Q_c, \forall t \in [t_j + \Delta, t'_j]$ and for any j such that $j \geq \min\{j : \hat{t} \in I_j\}$, $\forall t \in [\hat{t}, \infty) \cap [t_j + \Delta, t'_j]$.

Remark 1. The meaning of the above definition is that any hybrid evolution has to be reconstructed at any time but a finite interval after a transition occurs, and any current state belonging to a critical set has to be detected before the next switching. If Q_c is the empty set ($Q_c = \varnothing$), i.e. if there are no critical discrete states, Definition 3 of $\varnothing$–observability is equivalent to the notion of observability given in [8].

Remark 2. Definition 3 of observability is based on the existence of a control law that ensures the reconstruction of the hybrid state evolution. One could object that if a state is critical, it should be observable for all inputs. The results that we obtained in [11] answer this question: under the conditions of Theorem 1 (see Section 4), the class of control laws for which the hybrid state evolution cannot be reconstructed is a 'thin' set in the class of control laws $\mathcal{U}$. Consequently, our notion of observability is an 'almost everywhere' notion with respect to the chosen control law.

If one is interested in observing only the hybrid state related to the critical locations Q_c, Definition 3 can be relaxed as follows.

Definition 4. *A linear switching system $\mathcal{S}$ is Q_c–critically observable if there exist a function $\hat{u} \in \mathcal{U}$, a function $\hat{\xi}\colon \mathcal{Y}^o \times \mathcal{U} \rightarrow \Xi$ and a real $\Delta \in (0, \Delta_m)$ such that for any execution of $\mathcal{S}$ with $u = \hat{u}$,*

$$\hat{\xi}\left(y^o|_{[t_0,t]}\,, \hat{u}|_{[t_0,t)}\right) = \xi(t,j),$$

for any j such that $q(j) \in Q_c, \forall t \in [t_j + \Delta, t'_j]$.

The definition of Q_c–critical observability can be further relaxed, by requiring the reconstruction only of the discrete component of the critical states.

Definition 5. *A linear switching system $\mathcal{S}$ is Q_c–critically location observable if there exist a function $\hat{u} \in \mathcal{U}$, a function $\hat{q} : \mathcal{Y}^o \times \mathcal{U} \rightarrow Q$ and a real $\Delta \in (0, \Delta_m)$ such that for any execution of $\mathcal{S}$ with $u = \hat{u}$,*

$$\hat{q}\left(y^o|_{[t_0,t]}\,, \hat{u}|_{[t_0,t)}\right) = q(j),$$

for any j such that $q(j) \in Q_c, \forall t \in [t_j + \Delta, t'_j]$.

The relations among the different observability notions introduced above are summarized hereafter:

$$\begin{array}{c} Q_c\text{– observability} \\ \Downarrow \\ Q_c\text{– critical observability} \\ \Downarrow \\ Q_c\text{– critical location observability.} \end{array}$$

Moreover, as a direct consequence of the definitions, we have the following

Proposition 1. *A linear switching system $\mathcal{S}$ is Q_c–observable if and only if it is Q_c–critically observable and $\varnothing$–observable.*

4 Main Results

This section is devoted to the characterization of the observability notions introduced in the previous section, and in particular of Q_c–critical observability. In view of Proposition 1, we address first $\varnothing$–observability and then Q_c–critical observability. For the various observability notions of interest, a set of sufficient and, under some assumptions on the switching systems, necessary and sufficient conditions are given. Those conditions are sufficient also for the more general class of hybrid systems, where transitions can be forced by the value of the current continuous state (invariance transitions) or are enabled by appropriate conditions (guard conditions). In fact it is always possible to associate a switching system to a hybrid system, by replacing invariance transitions with switching transitions (i.e. due to external discrete uncontrollable input) and by removing guard conditions (see e.g. [10]). An observer (if it exists) for this switching system is also an observer for the original hybrid system.

4.1 Characterization of $\varnothing$–Observability

Given the semantics of linear switching systems and the definition of the observed output, the reconstruction of the discrete state evolution is based on both the discrete and the continuous components of the observed output. If the same discrete output is associated to two discrete states q_i and q_j of $\mathcal{S}$, i.e. $h(q_i) = h(q_j)$, then one may consider to discriminate q_i and q_j by means of the input–output behaviour of $S(q_i)$ and $S(q_j)$. In particular, if

$$\exists k \in \mathbb{N} \cup \{0\} : \ C_i A_i^k B_i \neq C_j A_j^k B_j, \tag{3}$$

there always exists a control law $u \in \mathcal{U}$, such that for any initial states of $S(q_i)$ and $S(q_j)$, the continuous outputs of $S(q_i)$ and $S(q_j)$ are different.

The following result gives a sufficient condition for $\varnothing$ –observability.

Theorem 1. *A linear switching system $\mathcal{S}$ is $\varnothing$–observable if the following conditions are satisfied*

$(i, 1)$ $\forall q_i, q_j \in Q_0$, $q_i \neq q_j$, *such that* $h(q_i) = h(q_j)$, *condition (3) holds;*

$(ii, 1)$ $\forall q_i, q_j \in reach\,(Q_0)$, $q_i \neq q_j$, *such that* $e = (q_i, \sigma, q_j) \in E$, $h(q_i) = h(q_j)$ *and* $\eta\,(e) = \epsilon$, *condition (3) holds;*

$(iii, 1)$ $\forall q_i \in Q$, $S(q_i)$ *is observable.*

The proof of the result above is a direct consequence of the results established in [11].

As already pointed out (see Remark 2), conditions of Theorem 1 guarantee the reconstruction of the hybrid state evolution not only for a particular control law but for 'almost all' control laws in the class $\mathcal{U}$.

It is easy to see that condition $(i, 1)$ ensures the reconstructability of the initial discrete state while condition $(ii, 1)$ ensures the reconstructability of the switching times: these two conditions guarantee that the discrete state evolution can be determined. The third condition $(iii, 1)$ ensures the reconstructability of the continuous component of the hybrid state, once the discrete state evolution is known.

If the space of initial conditions Ξ_0 coincides with the whole hybrid state space, i.e. $\Xi_0 = \Xi$, then condition $(i, 1)$ implies condition $(ii, 1)$. Moreover, if the system $\mathcal{S}$ is characterized by infinite maximum dwell time, i.e. $\Delta_M = +\infty$, then conditions $(i, 1)$ and $(iii, 1)$ are also necessary. Therefore, a consequence of Theorem 1 is

Corollary 1. *A linear switching system $\mathcal{S}$ with $\Xi_0 = \Xi$ and $\Delta_M = +\infty$, is $\varnothing$–observable if and only if conditions $(i, 1)$ and $(iii, 1)$ hold.*

In [8], the notion of $\varnothing$–observability was characterized for a linear switching system S with $\Xi_0 = \Xi$, $\Delta_M = +\infty$, and $\eta(e) = \epsilon$, $\forall e \in E$. The conditions given in [8] coincide with those of Corollary 1, since, if the maximum dwell time is infinite, the information that we get from the transitions plays no role.

4.2 Characterization of Q_c–Critical Observability

The characterization of the notion of Q_c–critical observability is addressed by abstracting the continuous outputs of a given switching system to a suitable discrete domain. More precisely, we embed the information coming from the continuous component of the observed output into the discrete component of the observed output.

For this reason, following [4], we introduce a so–called *signature generator*. We consider here a particular signature generator consisting of a system whose inputs are the continuous input and output of $\mathcal{S}$ and whose output is a 'signature' that can be considered as an additional discrete output $h_c(q)$ associated with a discrete state q of $\mathcal{S}$. The signature $h_c(q)$ has to be generated before the system leaves the discrete state q and therefore in a time interval $\Delta < \Delta_m$. Once this signature is generated, it remains constant until a new signature is generated.

If two dynamical systems $S(q_i)$ and $S(q_j)$ satisfy condition (3), there exists a control law $u \in \mathcal{U}$ such that different signatures can be associated with $S(q_i)$ and $S(q_j)$. Therefore, we assume that for any pair of distinct discrete states $q_i, q_j \in Q$, $h_c(q_i) = h_c(q_j)$ if and only if $C_i A_i^k B_i = C_j A_j^k B_j$, $\forall k \in \mathbb{N} \cup \{0\}$. This assumption allows stating *a priori* conditions for a signature to be generated, even if the information that we can collect from the continuous evolution could be richer. Indeed, even if $S(q_i)$ and $S(q_j)$ do not satisfy (3), there may exist initial conditions x_i^0 for $S(q_i)$ and x_j^0 for $S(q_j)$ such that, for any $u \in \mathcal{U}$, the continuous outputs of $S(q_i)$ and $S(q_j)$ are different. This is why the observability conditions presented in this section are in general sufficient, although there are cases in which they are also necessary, as shown later.

We now define, starting from the given switching system $\mathcal{S}$, a suitable switching system $\mathcal{S}_d$ whose discrete output gives also informations about the input–output behavior of the continuous systems associated with the discrete locations of $\mathcal{S}$. Formally, given $\mathcal{S} = (\Xi, \Xi_0, \Theta, S, E, R, \Upsilon)$, we define the following linear switching system:

$$\mathcal{S}_d = (\Xi, \Xi_0, \Theta, S_d, E, R, \Upsilon_d)\,,$$

where:

- S_d is a mapping that associates to any discrete state $q_i \in Q$, the following continuous–time linear system:

$$\dot{x}(t) = A_i x(t) + B_i u(t), \qquad y = \mathbf{0}, \qquad i \in J$$

  where $\mathbf{0}$ is the zero vector in $\mathbb{R}^p$ and the matrices A_i and B_i are as in (1);
- $\Upsilon_d = \bar{\Psi}_E \times \bar{\Psi}_Q \times \{0\}$, where:
  - $\bar{\Psi}_Q = \Psi_Q \times \Psi$ for some set Ψ such that $\Psi_Q \cap \Psi = \emptyset$ is the extended output space associated with the discrete states by means of the function $\bar{h}$: $Q \rightarrow \bar{\Psi}_Q$ such that

$$\bar{h}(q_i) = \bar{h}(q_j) \iff h(q_i) = h(q_j) \text{ and } h_c(q_i) = h_c(q_j);$$

  - $\bar{\Psi}_E = \Psi_E \cup \{\bar{\psi}_E\}$ such that $\bar{\psi}_E \notin \Psi_E$ and $\bar{\eta}$: $E \rightarrow \bar{\Psi}_E$ such that for any $e = (q_i, \sigma, q_j) \in E$,

$$\bar{\eta}(e) : = \begin{cases} \bar{\psi}_E & \text{if } \eta(e) = \epsilon \text{ and } \bar{h}(q_i) \neq \bar{h}(q_j), \\ \eta(e) & \text{otherwise.} \end{cases}$$

Two locations q_i and q_j of a switching system $\mathcal{S}$ may be distinguished either because $h(q_i) \neq h(q_j)$ or because condition (3) holds, i.e. equivalently, because $\bar{h}(q_i) \neq \bar{h}(q_j)$. Therefore,

Proposition 2. *Given a linear switching system $\mathcal{S}$, consider the associated linear switching system $\mathcal{S}_d$. Assume $0 \in X_i^0$ for any $q_i \in Q_0 \cap Q_c$. Then, $\mathcal{S}$ is Q_c–critically observable only if for any $q_c \in Q_0 \cap Q_c$,*

$(i, 2)$ $S(q_c)$ is observable;

$(ii, 2)$ for any $q_0 \in Q_0 \setminus \{q_c\}$, $\bar{h}(q_c) \neq \bar{h}(q_0)$.

Proof. *(i,2)* By definition of Q_c–critical observability, for any $q(0) = q_c \in Q_0 \cap Q_c$ it is necessary to reconstruct the continuous component of the hybrid state from the observed output, within the time interval I_0. Therefore $S(q_c)$ has to be observable. *(ii,2)* By contradiction, suppose $\bar{h}(q_c) = \bar{h}(q_0)$, for some $q_c \in Q_0 \cap Q_c$ and $q_0 \in Q_0 \setminus \{q_c\}$. Since the continuous component of the initial hybrid state can be zero, then, by definition of the function $\bar{h}$, it is not possible to distinguish q_c and q_0, and hence the system is not Q_c–critically observable.

Sufficient conditions for Q_c–critical observability can be given as follows:

Proposition 3. *The linear switching system $\mathcal{S}$ is Q_c–critically observable if:*

$(i,3)$ *$\mathcal{S}$ is Q_c–critically location observable;*

$(ii,3)$ *for any $q_c \in Q_c$, $S(q_c)$ is observable.*

By definition, condition $(i,3)$ is also necessary and condition $(ii,3)$ is necessary if $Q_c \subset Q_0$ for a switching system to be Q_c–critically observable.

Necessary and sufficient conditions for Q_c–critical observability may be given on the basis of an observer $\mathcal{O}$ for $\mathcal{S}_d$, which detects the critical states in the sense of Definition 5 whenever those critical states are reached. The construction of the observer $\mathcal{O}$ (see also [14], [15]) is inspired by [29], where a procedure was given for the construction of a finite state machine that, under appropriate conditions, allows an intermittent observation of the discrete state of $\mathcal{S}$, and by [4], where hybrid observers were proposed for reconstructing the hybrid state evolution of a hybrid system, in the sense of k–current state observability, namely after a certain fixed $k > 0$.

The observer $\mathcal{O}$ is a DES [20], that takes as inputs the observed output of $\mathcal{S}_d$ and gives back as outputs all and only the discrete states of $\mathcal{S}_d$ that match that observed output. The basic idea is as follows. Suppose the switching system $\mathcal{S}_d$ starts its evolution from a location $q_0 \in Q_0$. When the discrete output $\bar{h}(q_0)$ associated with q_0 is available, this output is captured as an input by the observer. This first piece of information allows the observer to discriminate among all the discrete states of Q_0 that are compatible with $\bar{h}(q_0)$. This actually implies that once this information is acquired, the observer gives back as output

$$Q_1 = \left\{ q \in Q_0 \colon \bar{h}(q) = \bar{h}(q_0) \right\} .$$

If a transition $e_1 \in E$ occurs, the system $\mathcal{S}_d$ provides a discrete output $\bar{\eta}(e_1)$ that will be an additional input for the observer. On the basis of $\bar{\eta}(e_1)$, the observer provides the set Q_2 of all discrete states that can be reached by a state in Q_1 through a transition e whose discrete output coincides with $\bar{\eta}(e_1)$. Therefore,

$$Q_2 = \Big\{ q \in Q \mid \exists q_1 \in Q_1, \exists \sigma \in \Sigma \colon e = (q_1, \sigma, q) \in E, \bar{\eta}(e) = \bar{\eta}(e_1) \Big\}.$$

By iterating this two–step procedure the observer can be built.

For later use, it is convenient to rewrite the discrete dynamics associated with $\mathcal{S}_d$ by means of a non–deterministic generator of formal language [31],

$$\begin{aligned} q(j+1) &\in \delta(q(j), \sigma(j)) \\ \sigma(j) &\in \phi(q(j)) \\ \psi_E(j) &= \eta(e_{j-1}), \quad \eta(e_{-1}) = \epsilon \\ \psi_Q(j) &= \bar{h}(q(j)) \end{aligned} \tag{4}$$

where $\delta\colon Q \times \Sigma \to 2^Q$ and $\phi\colon Q \to 2^\Sigma$ are respectively the transition and the input functions. Moreover, let $s_\epsilon \in \Sigma^*$ be the input strings whose output is a sequence of empty strings ϵ.

The following algorithm defines the observer

$$\mathcal{O} = (\hat{Q}, \hat{Q}_0, \hat{\Sigma}, \hat{\Psi}, \hat{\delta}, \hat{\phi}, \hat{h}),$$

where $\hat{Q} \subset 2^Q$ is the state space, $\hat{Q}_0 \subset \hat{Q}$ is the set of initial states, $\hat{\Sigma}$ is the set of inputs that coincides with the set of outputs of $\mathcal{S}_d$, $\hat{\Psi}$ is the set of outputs that coincides with $\hat{Q}$, $\hat{\delta} : \hat{Q} \times \hat{\Sigma} \to \hat{Q}$ is the transition function, $\hat{\phi}\colon \hat{Q} \to 2^{\hat{\Sigma}}$ is the input function and $\hat{h}\colon \hat{Q} \to \hat{\Psi}$ is the output function.

Algorithm 2

`Begin`

$\hat{q}_0\colon = Q_0 \cup \{\delta(q_0, s_\epsilon) \in Q \mid q_0 \in Q_0\}$
$\hat{Q}_0\colon = \{\hat{q}_0\}$
$\hat{Q}\colon = \hat{Q}_0$
$\hat{\Sigma}\colon = \left(\bar{\Psi}_E \setminus \{\epsilon\}\right) \cup \bar{\Psi}_Q$
$j\colon = 0$
`repeat`
  $\hat{Q}_{j+1} = \varnothing$
  `for any` $\hat{q} \in \hat{Q}_j$
    $\hat{\phi}(\hat{q})\colon = \left\{\psi_Q \in \bar{\Psi}_Q \mid \exists q \in \hat{q}\colon \bar{h}(q) = \psi_Q\right\}$
    `for any` $\psi_Q \in \hat{\phi}(\hat{q})$
      $\hat{\delta}(\hat{q}, \psi_Q)\colon = \left\{q \in \hat{q}\colon \bar{h}(q) = \psi_Q\right\} \neq \varnothing$
      `if` $\hat{\delta}(\hat{q}, \psi_Q) \notin \hat{Q}$
        $\hat{Q}_{j+1}\colon = \hat{Q}_{j+1} \cup \left\{\hat{\delta}(\hat{q}, \psi_Q)\right\}$
        $\hat{Q}\colon = \hat{Q} \cup \hat{Q}_{j+1}$
      `end if`
    `end for`
  `end for`
  `for any` $\hat{q} \in \hat{Q}_{j+1}$
    $\hat{\phi}(\hat{q})\colon = \left\{\psi_E \in \bar{\Psi}_E \setminus \{\epsilon\} \mid \exists q \in \hat{q},\ \exists \sigma \in \phi(q)\colon \eta_E((q, \sigma, q^+)) = \psi_E,\ \textit{for some } q^+ \in \delta(q, \sigma)\right\}$
    `for any` $\psi_E \in \hat{\phi}(\hat{q})$
      $\hat{\delta}(\hat{q}, \psi_E)\colon = \left\{ q \in Q \mid \exists \bar{q} \in \hat{q},\ \exists s \in \Sigma^*\colon q \in \delta(\bar{q}, s)!\ \textit{and}\ \eta_E(s) \in \psi_E \epsilon^* \right\}$
      `if` $\hat{\delta}(\hat{q}, \psi_E) \notin \hat{Q}$
        $\hat{Q}_{j+1}\colon = \hat{Q}_{j+1} \cup \left\{\hat{\delta}(\hat{q}, \psi_E)\right\}$
        $\hat{Q}\colon = \hat{Q} \cup \hat{Q}_{j+1}$
      `end if`
    `end for`
  `end for`

$j: = j + 1$
`until` $\hat{Q}_{j+1} = \varnothing$
$\hat{\Psi}: = \hat{Q}$
$\hat{h}(\hat{q}): = \hat{q}, \; \forall \, \hat{q} \in \hat{Q}$

`End`

The finite convergence of Algorithm 2 is guaranteed by the finiteness of the discrete state space Q of $\mathcal{S}_d$. The set of critical states Q_c of the system $\mathcal{S}$ induces a set of critical states $\hat{Q}_c$ on the observer $\mathcal{O}$, whose analysis is fundamental for assessing critical location observability. $\hat{Q}_c$ is formally defined as

$$\hat{Q}_c: = \{\hat{q} \in \hat{Q} \mid \hat{q} \cap Q_c \neq \varnothing \wedge \hat{\phi}(\hat{q}) \cap \bar{\Psi}_Q = \varnothing\}$$

The following result holds.

Theorem 2. *$\mathcal{S}_d$ is Q_c–critically location observable if and only if for any $\hat{q}_c \in \hat{Q}_c$, $card(\hat{q}_c) = 1$.*

The proof of the result above is a straightforward consequence of the definition of $\mathcal{O}$ and of the notion of Q_c–critical location observability. Moreover, Theorem 2 allows us also to give some sufficient conditions for characterizing Q_c–critical location observability of $\mathcal{S}$, as follows.

Theorem 3. *Consider the linear switching systems $\mathcal{S}$ and $\mathcal{S}_d$. The following statements hold:*

(i,3) $\mathcal{S}$ is Q_c-critically location observable if $\mathcal{S}_d$ is Q_c-critically location observable.

(ii,3) If $Q_c \subset Q_0$ and for any $q_i \in Q_c$, $0 \in X_i^0$, then $\mathcal{S}$ is Q_c-critically location observable only if $\mathcal{S}_d$ is Q_c-critically location observable.

Proof. (i,3) The statement follows by definition of system $\mathcal{S}_d$. (ii,3) By applying Proposition 2, if $Q_c \subset Q_0$ and for any $q_i \in Q_c$, $0 \in X_i^0$, then any two critical states q_i and q_j in Q_c can be distinguished only if $\bar{h}(q_i) \neq \bar{h}(q_j)$. Since this last condition implies the Q_c-critical location observability of $\mathcal{S}_d$, the result follows.

4.3 Example

We now analyze an example of application of the methodology proposed in the previous section for checking critical observability.

Consider a switching system

$$\mathcal{S} = (\Xi, \Xi_0, \Theta, S, E, R, \Upsilon),$$

where:

- $\Xi = Q \times \mathbb{R}^n$ where $Q = \{q_1, q_2, q_3, q_4\}$;
- $\Xi_0 = \{q_1, q_2, q_3\} \times \mathbb{R}^n$;
- $\Theta = \Sigma \times \mathbb{R}^m$ where $\Sigma = \{\sigma\}$;
- $S(q) = S$ for any $q \in Q$, where S is a linear dynamical system $\dot{x} = Ax + Bu$, $y = Cx$ that is supposed to be observable;
- $E = \{(q_1, \sigma, q_2), (q_1, \sigma, q_3), (q_3, \sigma, q_1), (q_2, \sigma, q_4), (q_4, \sigma, q_1), (q_4, \sigma, q_2), (q_4, \sigma, q_3)\}$;
- $R(e, (q_i, x)) = (q_j, x)$, $\forall e = (q_i, \sigma, q_j) \in E$, $\forall x \in \mathbb{R}^n$;
- $\Upsilon = \Psi_E \times \Psi_Q \times \mathbb{R}^p$, is the output space, where $\Psi_E = \{\epsilon, \alpha\}$, $\Psi_Q = \{a, b\}$ and

$$h(q) = \begin{cases} a, \text{ if } q \in \{q_1, q_3\} \\ b, \text{ if } q \in \{q_2, q_4\} \end{cases}$$

$$\eta(e) = \begin{cases} \epsilon, \text{ if } e = (q_1, \sigma, q_3) \\ \alpha, \text{ otherwise.} \end{cases}$$

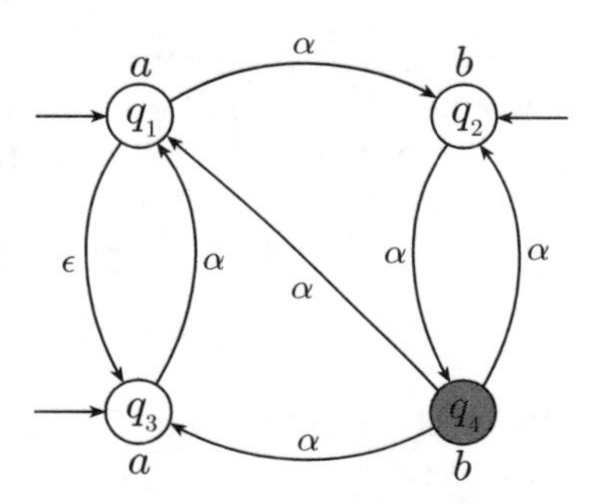

Fig. 1. DES associated with the switching system $\mathcal{S}$

The DES associated with $\mathcal{S}$ is depicted in Figure 1, where the discrete inputs driving the transitions are omitted and the arrows with no labels indicate the initial discrete states. We suppose that the set of critical states is $Q_c = \{q_4\}$. Since dynamical systems associated with each of the locations of $\mathcal{S}$ coincide, the signatures play no role and therefore the discrete dynamics of $\mathcal{S}_d$ coincide with the discrete dynamics of $\mathcal{S}$.

By applying Algorithm 2, the observer $\mathcal{O}$ depicted in Figure 2 is obtained. It is easily seen that $\hat{Q}_c = \{\{q_4\}\}$ and therefore the conditions of Theorem 2 are fulfilled: thus $\mathcal{S}_d$ is Q_c –critically location observable. By combining Proposition 3 and Theorem 3, we can conclude that $\mathcal{S}$ is Q_c –critically observable. For the sake of explanation, locations q_2 and q_4 are characterized by the same discrete output and the same continuous dynamics S, hence $\bar{h}(q_2) = \bar{h}(q_4)$. However, since the topological properties of the DES associated to $\mathcal{S}$ do not allow reaching q_4 before reaching q_2 and since the transitions connecting the states q_2 and q_4 have no unobservable output, the observer $\mathcal{O}$ is able to detect if the current location is q_2 or q_4.

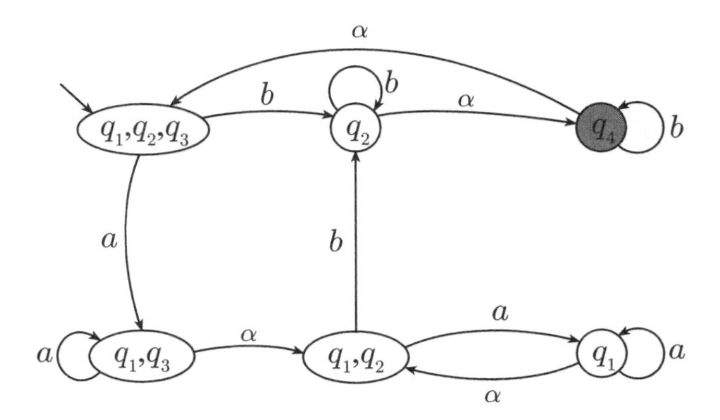

Fig. 2. Observer $\mathcal{O}$ associated with the switching system $\mathcal{S}$

5 A Case Study: The Active Runway Crossing System

In this section, we consider the example proposed in [33] and [23], and analyzed in [14], [16], of an active runway crossing with the intent of testing the applicability of the theoretical results on observers to a realistic ATM situation for the detection of situation awareness errors. This will be a sufficiently simple case study that summarizes the main difficulties in the formulation, analysis and control of a typical accident risk situation for ATM. The active runway crossing will be decomposed into various subsystems, each with hybrid dynamics modeling its specific operations.

The active runway crossing environment consists of a runway A (with holdings, crossings and exits), a maintenance area and aprons. The crossings connect the aprons and the maintenance area. Crossings (on both sides) and holdings have remotely controlled stopbars to access the runway, and each exit has a fixed stopbar (see Figure 3).

The following relevant areas can be defined:

$$\begin{aligned}
\Omega_{Ap} &= \{(x,y) \mid x > a_4, y \in [b_1, b_6]\} \\
\Omega_{AW_1} &= \{(x,y) \mid x \in [a_3, a_4], y \in [b_1, b_2]\} \\
\Omega_{AW_2} &= \{(x,y) \mid x \in [a_3, a_4], y \in [b_3, b_4]\} \\
\Omega_{AW_3} &= \{(x,y) \mid x \in [a_3, a_4], y \in [b_5, b_6]\} \\
\Omega_{S_1} &= \{(x,y) \mid x \in [a_2, a_3], y \in [b_1, b_2]\} \\
\Omega_{S_2} &= \{(x,y) \mid x \in [a_2, a_3], y \in [b_3, b_4]\}
\end{aligned}$$

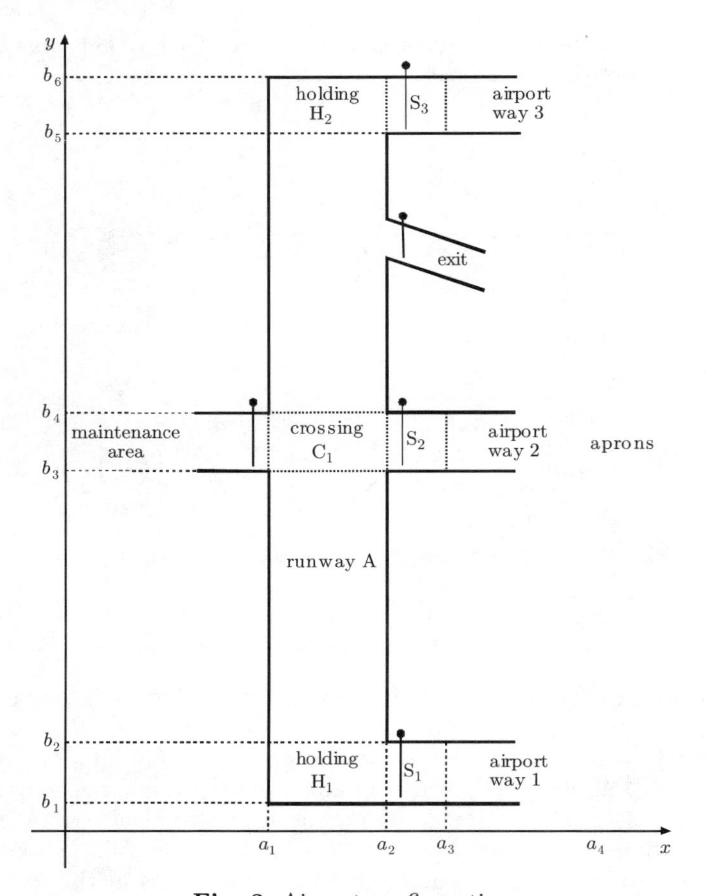

Fig. 3. Airport configuration

$$
\begin{aligned}
\Omega_{S_3} &= \{(x, y) \mid x \in [a_2, a_3], y \in [b_5, b_6]\} \\
\Omega_{H_1} &= \{(x, y) \mid x \in [a_1, a_2], y \in [b_1, b_2]\} \\
\Omega_{H_2} &= \{(x, y) \mid x \in [a_1, a_2], y \in [b_5, b_6]\} \\
\Omega_{C_1} &= \{(x, y) \mid x \in [a_1, a_2], y \in [b_3, b_4]\} \\
\Omega_{RW_A} &= \{(x, y) \mid x \in [a_1, a_2], y \in [b_1, b_6]\} \\
\Omega_M &= \{(x, y) \mid x < a_1, y \in [b_3, b_4]\}
\end{aligned}
$$

where 'Ap' stands for aprons, 'AW' for airport way, 'S' for stopbar, 'H' for holding, 'C' for crossing, 'RW_A' for runway A and 'M' for maintenance area.

Humans may not have a correct 'Situation Awareness' (SA) [19], [33] of the various elements in the environment:

Definition 6. *Situation Awareness (SA) is the perception of elements in the environment within a volume of time and space, the comprehension of their meaning, and the projection of their status in the near future. The projection in the near future of the perception of the actual environment is referred to as intent SA.*

The consequent errors can then evolve and create hazardous situations. Our goal is to identify these errors and possibly correct them before they may cause catastrophic events.

Within an ATM system, Stroeve *et al.* [33] define an *agent* as an entity, such as a human operator or a technical system, characterized by its *SA* of the environment. Following [33], *SA* can be incomplete or inaccurate, due to three different situations. An agent may:

1. wrongly perceive task–relevant information or miss them completely;
2. wrongly interpret the perceived information;
3. wrongly predict a future status.

An important source of error that has to be considered when analyzing multi–agent environments is the propagation of erroneous situation awareness due to agents interactions, e.g. via VHF communication.

5.1 Agents in an Active Runway Crossing

The runway crossing operation consists of various agents:

1. a pilot flying (P_t) directed to RW_A to perform a take off operation;
2. a pilot flying (P_c) directed to the M, taxiing through AW_2 and the runway crossing C_1;
3. a ground controller (C_g);
4. a tower controller (C_t);
5. the airport technical support system (ATS).

The pilot P_t proceeds towards the holding area (regular taxiway) with the intent of completing a take off operation, while the pilot P_c is approaching the crossing area. The tower controller C_t and ground controller C_g, with the aid of visual observation of the runway and VHF communication, respectively, are responsible of granting take off and crossing, avoiding the use of the runway by two aircraft simultaneously. Technical support systems help the pilots and the controllers to communicate (VHF) and detect dangerous situations (alerts).

The specific behavior of these agents in the runway crossing operation can be described as follows:

1. *Pilot flying of taking off aircraft* P_t. Initially P_t executes boarding and waits for start up grant by C_g. He begins taxiing on AW_1, stops at stopbar

S_1 and communicates with the C_t at the reserved frequency to obtain take off grant. Depending on the response, P_t waits for grant or executes take off immediately. Because of a *SA* error, the take off could be initiated without grant. For simplicity, we will not consider this kind of error here. When the aircraft is airborne, he confirms the take off has been completed to C_t . During take off operations, P_t monitors the traffic situation on the runway visually and via VHF. If a crossing aircraft is visible, or in reaction to an emergency braking command by the controller, the P_t starts a braking action and take off is rejected.

2. *Pilot Flying of crossing aircraft* P_c. When start up is granted by C_g, the P_c proceeds on the AW_2 and stops at stopbar S_2. He asks to C_g crossing permission and crosses when granted. While proceeding towards the AW_2, he may have the *intent SA* that the next airport way point is either a regular taxiway (erroneous *intent SA*) or a runway crossing. In the first case, P_c enters RW_A without waiting for crossing permission. In the second case, P_c could have the *SA* that crossing is allowed while it is not. Then, he would enter the runway performing an unauthorized runway crossing. The reaction of P_c to the detection of a collision risk, due to visual observation or a tower controller call, is an emergency braking action.
3. *Ground Controller* C_g. C_g is a human operator supported by visual observation and by the *ATS* system. He grants start up both to P_t and P_c, and handles crossing operations on RW_A. If C_g has *SA* of a collision risk, C_g specifies an emergency braking action to the crossing aircraft.
4. *Tower Controller* C_t. C_t is a human operator supported by visual observation and by the *ATS* system. The C_t handles take off operations on RW_A. If the C_t has *SA* of a collision risk, he specifies an emergency braking action to the taking off aircraft.
5. *ATS system.* This is the technical system supporting the decisions of the controllers, and consists of a communication system, a runway incursion alert and a stopbar violation alert.

5.2 Pilot Flying Observation Problem

The agents previously described can be modeled either as hybrid systems [26] or as DESs [16].

The pilot flying P_t can be modeled as a non–deterministic hybrid system $\mathcal{H}_{P_t}$ with

- $Q_1 = \{q_{1,1}, q_{1,2}, q_{1,3}, q_{1,4}, q_{1,5}, q_{1,6}, q_{1,7}, q_{1,8}\}$ the set of discrete states with $q_{1,1}$ the P_t communicating with C_g and waiting for start up grant, $q_{1,2}$ the P_t taxiing on AW_1, $q_{1,3}$ the P_t aborting taxi, $q_{1,4}$ the P_t at stopbar S_1, $q_{1,5}$ the P_t executing an authorized take off on RW_A, $q_{1,6}$ the P_t lined up and waiting for take off grant, $q_{1,7}$ the P_t executing an unauthorized take

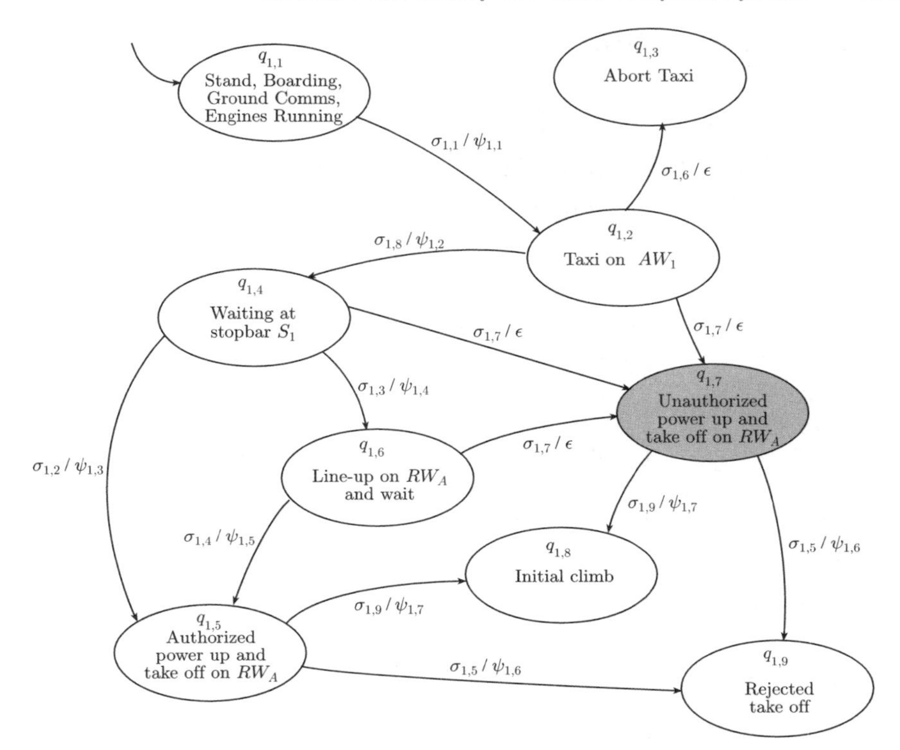

Fig. 4. Hybrid system $\mathcal{H}_{P_t}$ modelling P_t

off on RW_A, $q_{1,8}$ the P_t executing the initial climb, $q_{1,9}$ the P_t aborting take off (emergency braking);

- $\Sigma_1 = \{\sigma_{1,1}, \sigma_{1,2}, \sigma_{1,3}, \sigma_{1,4}, \sigma_{1,5}, \sigma_{1,6}, \sigma_{1,7}\}$ the set of discrete inputs, where $\sigma_{1,1}$ models the start up clearance by C_g, $\sigma_{1,2}$ the command for immediate take off by C_t, $\sigma_{1,3}$ the command to line up and wait by C_t, $\sigma_{1,4}$ the take off clearance by C_t , $\sigma_{1,5}$ an emergency braking command by C_t, $\sigma_{1,6}$ is a disturbance that causes a taxi abort, and $\sigma_{1,7}$ models a situation awareness error as a disturbance that causes an ungranted take off;
- $\Psi_1 = \{\psi_{1,1}, \psi_{1,2}, \psi_{1,3}, \psi_{1,4}, \psi_{1,5}, \psi_{1,6}, \psi_{1,7}, \psi_{1,8}\} \cup \{\epsilon\}$ the set of discrete outputs, with $\psi_{1,1}$ the start up confirmation to C_g, $\psi_{1,2}$ the take off request, $\psi_{1,3}$ the immediate take off confirmation, $\psi_{1,4}$ the line–up and wait confirmation, $\psi_{1,5}$ the take off confirmation, $\psi_{1,6}$ the emergency braking confirmation, $\psi_{1,7}$ the airborne confirmation;
- $X_1 = \{(s_1, v_1) : s_1 \in \mathbb{R}^2, v_1 \in \mathbb{R}^2\}$, is the set of the continuous state values, where s_1 indicates the position and v_1 the velocity of the agent;
- $U_1 = \mathbb{R}^2$, is the set of the continuous input u_1 values, $D_1 = \mathbb{R}^2$ is the set of the continuous disturbance d_1 values;
- The initial discrete state is $q_{1,1}$;

- The invariant conditions are defined as

$$
\begin{aligned}
I_{q_{1,1}} &= \{(s_1, v_1) \colon s_1 \in \Omega_{Ap}, \|v_1\| = 0\} \\
I_{q_{1,2}} &= \{(s_1, v_1) \colon s_1 \in \Omega_{AW_1} \cup \Omega_{S_1}, \|v_1\| > 0\} \\
I_{q_{1,3}} &= \{(s_1, v_1) \colon s_1 \in \Omega_{AW_1} \cup \Omega_{S_1}, \|v_1\| = 0\} \\
I_{q_{1,4}} &= \{(s_1, v_1) \colon s_1 \in \Omega_{S_1}, \|v_1\| = 0\} \\
I_{q_{1,5}} &= \{(s_1, v_1) \colon s_1 \in \Omega_{RW_A}, \|v_1\| > 0\} \\
I_{q_{1,6}} &= \{(s_1, v_1) \colon s_1 \in \Omega_{H_1}, \|v_1\| \geq 0\} \\
I_{q_{1,7}} &= \{(s_1, v_1) \colon s_1 \in \Omega_{RW_A} \cup \Omega_{S_1}, \|v_1\| > 0\} \\
I_{q_{1,8}} &= \{(s_1, v_1) \colon s_1 \in \Omega_{RW_A}, \|v_1\| > v_t\} \\
I_{q_{1,9}} &= \{(s_1, v_1) \colon s_1 \in \Omega_{RW_A}, \|v_1\| \geq 0\}
\end{aligned}
$$

  where v_t is the take off velocity;
- $S_{C_1} = \{f_{q_{j,1}} : q_{j,1} \in Q_1\}$, $f_{q_{j,1}} \colon X_1 \times U_1 \times D_1 \to T_{X_1}$, the continuous (simplified) dynamics $\dot{s}_1 = v_1$, $\dot{v}_1 = u_1 + d_1$, where d_1 represents possible disturbance forces acting on the aircraft (e.g. wind);
- $E_1 \subseteq Q_1 \times \Sigma_1 \times Q_1$ the set of transitions given by the graph in Figure 4;
- $\eta_1 : E_1 \to \Psi_1$ the discrete output function, defined by the graph in Figure 4, where the outputs corresponding to transitions due to situation awareness errors ($\{q_{1,2}, q_{1,7}\}$, $\{q_{1,4}, q_{1,7}\}$ and $\{q_{1,6}, q_{1,7}\}$) are unobservable (ϵ output);
- $R_1(e, (q_i, x)) = (q_j, x)$, $\forall (e, (q_i, x)) \in E_1 \times Q_1 \times X_1, e = (q_i, \sigma, q_j), \sigma \in \Sigma_1$ the reset mapping;
- The guard conditions are

$$
\begin{aligned}
G(q_{1,2}, q_{1,4}) &= \{(s_1, v_1) \colon s_1 \in S_1, \|v_1\| = 0\} \\
G(q_{1,5}, q_{1,8}) &= G(q_{1,7}, q_{1,8}) = \{(s_1, v_1) \colon s_1 \in RW_A, \|v_1\| > v_t\}.
\end{aligned}
$$

The hybrid system model $\mathcal{H}_{P_t}$ is more general than the switching system model defined in Section 2. However as already explained, it is possible to define an abstraction $\mathcal{H}'_{P_t}$ of $\mathcal{H}_{P_t}$ by replacing the invariance and guard sets with the whole continuous state space. The resulting system $\mathcal{H}'_{P_t}$ is a switching system in the sense of Definition 1, with linear continuous dynamics subject to a disturbance. An observer designed for $\mathcal{H}'_{P_t}$ is also an observer for the pilot flying model $\mathcal{H}_{P_t}$.

An observer $\mathcal{O}_{P_t}$ for $\mathcal{H}_{P_t}$ is given in Figure 5. It is clear that the system $\mathcal{H}_{P_t}$ is not Q_c–critically observable, if the set of critical states is $Q_c = \{q_{1,7}\}$. In fact, the states of the observer $\{q_{1,2}, q_{1,3}, q_{1,7}\}$, $\{q_{1,4}, q_{1,7}\}$, $\{q_{1,6}, q_{1,7}\}$ are critical and have cardinality greater than 1.

In this case study, the same continuous dynamics is associated to each discrete state. Therefore, it is not possible to discriminate the discrete states using the input-output behavior and no signature in the sense of Section 4 can be generated a priori. However, if the continuous output $y(t) = s_1(t)$ were available, then an additional output $h(q_{1,7})$ could be generated when $s_1 \in \Omega_{RW_A}$. In that case, the observer $\mathcal{O}'_{P_t}$ (see Figure 6) is obtained and the

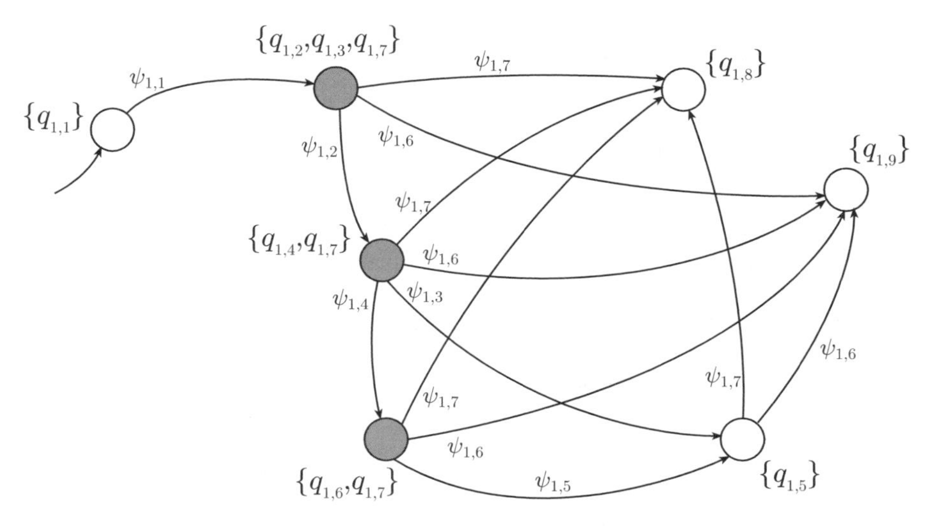

Fig. 5. Observer $\mathcal{O}_{P_t}$

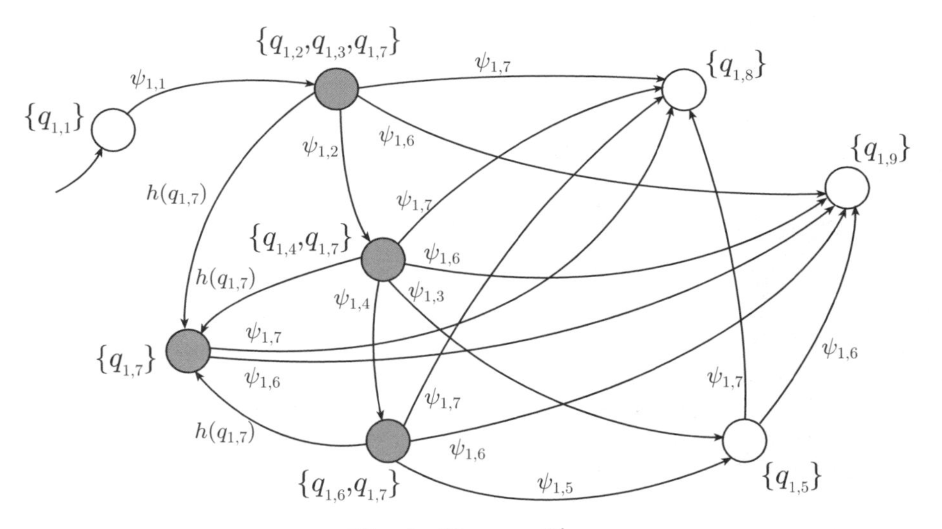

Fig. 6. Observer $\mathcal{O}'_{P_t}$

system $\mathcal{H}_{P_t}$ is critically observable. This shows how the observation problem for P_t can be solved.

An analogous model and a similar procedure can be followed for solving the observation problem for P_c (see Figure 7). P_c can be modeled by a hybrid system, where

- $Q_2 = \{q_{2,1}, q_{2,2}, q_{2,3}, q_{2,4}, q_{2,5}, q_{2,6}, q_{2,7}\}$, are the sets of discrete states where $q_{2,1}$ corresponds to P_c communicating with C_g and waiting for start

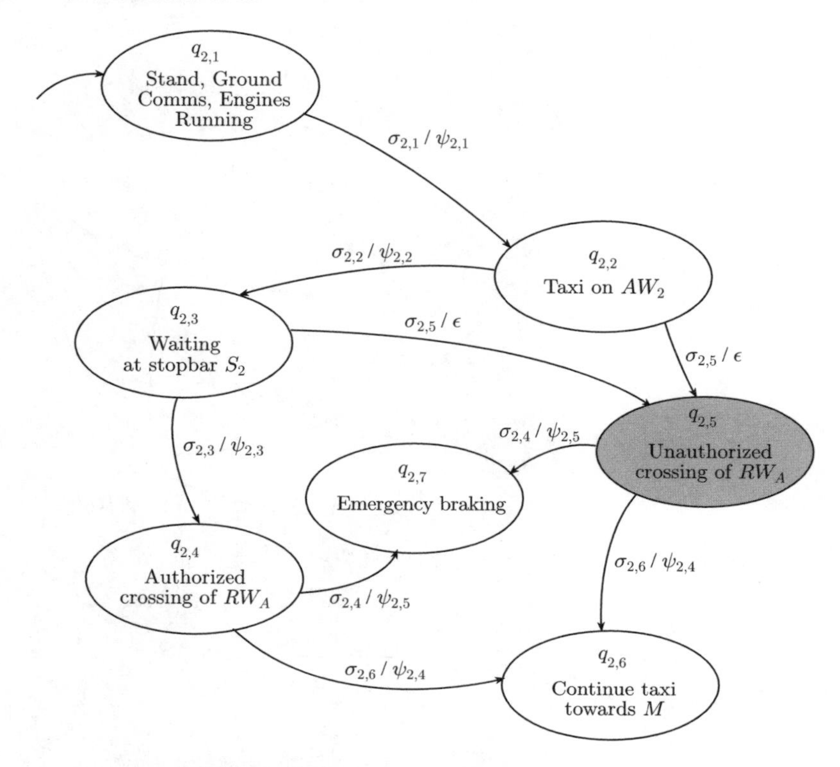

Fig. 7. Hybrid system $\mathcal{H}_{P_c}$ modelling P_c

up grant, $q_{2,2}$ to P_c taxiing on AW_2, $q_{2,3}$ to P_c waiting at stopbar S_2, $q_{2,4}$ to P_c executing an authorized crossing of RW_A, $q_{2,5}$ to P_c executing an unauthorized crossing of RW_A, $q_{2,6}$ to P_c taxiing towards M, $q_{2,7}$ to P_c performing an emergency braking operation;

- $\Sigma_2 = \{\sigma_{2,1}, \sigma_{2,2}, \sigma_{2,3}, \sigma_{2,4}, \sigma_{2,5}\}$, is the set of discrete inputs, where $\sigma_{2,1}$ models the start up clearance by the C_g, $\sigma_{2,2}$ the command by C_g to wait at stopbar S_2, $\sigma_{2,3}$ the crossing grant by C_g, $\sigma_{2,4}$ the emergency braking command by C_g, $\sigma_{2,5}$ models situation awareness error as a disturbance that causes an ungranted crossing;
- $\Psi_2 = \{\psi_{2,1}, \psi_{2,2}, \psi_{2,3}, \psi_{2,4}, \psi_{2,5}\} \cup \{\epsilon\}$, is the set of discrete outputs, with $\psi_{2,1}$ the start up confirmation, $\psi_{2,2}$ the crossing request, $\psi_{2,3}$ the RW_A crossing grant confirmation, $\psi_{2,4}$ the crossing complete confirmation, $\psi_{2,5}$ the emergency braking confirmation;
- $X_2 = \{(s_2, v_2) \colon s_2 \in \mathbb{R}^2, v_2 \in \mathbb{R}^2\}$, is the set of the continuous state values, where s_2 indicates the position and v_2 the velocity of the agent;
- $U_2 = \mathbb{R}^2$, is the set of the continuous input u_2 values, $D_2 = \mathbb{R}^2$ is that of the continuous disturbance d_2 values;
- The initial discrete state is $q_{2,1}$;

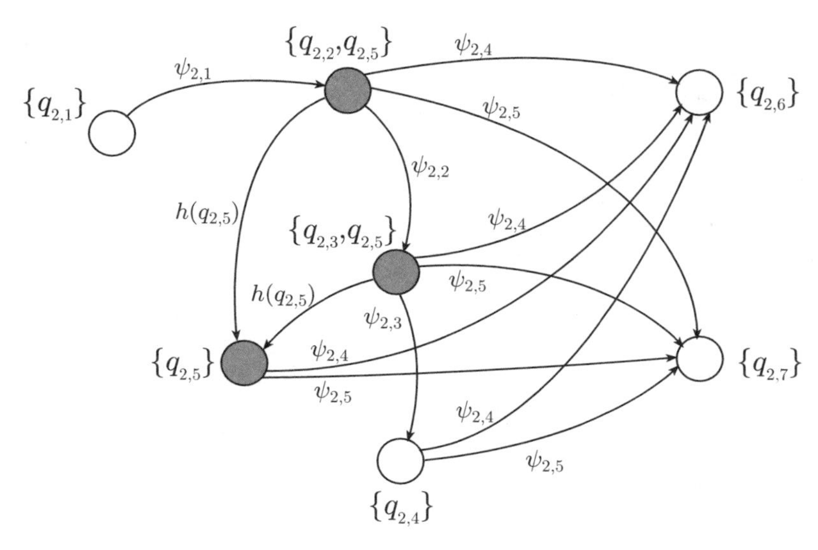

Fig. 8. Critical observer $\mathcal{O}'_{P_c}$

- The invariant conditions are defined as follows

$$
\begin{aligned}
I_{q_{2,1}} &= \{(s_2, v_2) \colon s_2 \in \Omega_{Ap}, \|v_2\| = 0\} \\
I_{q_{2,2}} &= \{(s_2, v_2) \colon s_2 \in \Omega_{AW} \cup \Omega_{S_2}, \|v_2\| > 0\} \\
I_{q_{2,3}} &= \{(s_2, v_2) \colon s_2 \in \Omega_{S_2}, \|v_2\| = 0\} \\
I_{q_{2,4}} &= \{(s_2, v_2) \colon s_2 \in \Omega_{C_1}, \|v_2\| > 0\} \\
I_{q_{2,5}} &= \{(s_2, v_2) \colon s_2 \in \Omega_{S_2} \cup \Omega_{C_1}, \|v_2\| > 0\} \\
I_{q_{2,6}} &= \{(s_2, v_2) \colon s_2 \in \Omega_{M}, \|v_2\| > 0\} \\
I_{q_{2,7}} &= \{(s_2, v_2) \colon s_2 \in \Omega_{C_1}, \|v_2\| \geq 0\}
\end{aligned}
$$

- $S_{C_2} = \{f_{q_{j,2}} \colon q_{j,2} \in Q_2\}$, $f_{q_{j,2}} \colon X_2 \times U_2 \times V_2 \to T_{X_2}$, $j = 1, 2$, are the continuous (simplified) dynamics $\dot{s}_2 = v_2$, $\dot{v}_2 = u_2 + d_2$, and d_2 represents possible disturbance forces acting on the aircraft (e.g. wind);
- $E_2 \subseteq Q_2 \times \Sigma_2 \times Q_2$ the set of transitions given by the graph in Figure 7;
- $\eta_2 : E_2 \to \Psi_2$ the discrete output function, defined by the graph in Figure 7, where the outputs corresponding to transitions due to situation awareness errors ($\{q_{2,2}, q_{2,5}\}$ and $\{q_{2,3}, q_{2,5}\}$) are unobservable, and are the source of the observability problems that we need to address;
- $R_2(e, (q_i, x)) = (q_j, x)$, $\forall (e, (q_i, x)) \in E_2 \times Q_2 \times X_2, e = (q_i, \sigma, q_j), \sigma \in \Sigma_2$ the reset mapping;
- The guard conditions are

$$G(q_{2,4}, q_{2,6}) = (q_{2,5}, q_{2,6}) = \{(s_2, v_2) \colon s_2 \in M, \|v_2\| > 0\}.$$

As done for $\mathcal{H}_{P_t}$, one can design an observer $\mathcal{O}_{P_c}$. The states $\{q_{2,2}, q_{2,5}\}$, $\{q_{2,3}, q_{2,5}\}$ with cardinality greater than 1 are critical, if the set of critical states is $Q_c = \{q_{2,5}\}$. If the continuous output $y(t) = s_2(t)$ were available, an additional discrete output $h(q_{2,5})$ generated when $s_2 \in \Omega_{C_1}$ would lead to the observer $\mathcal{O}'_{P_c}$. In that case, the system $\mathcal{H}_{P_c}$ is critically observable (see Figure 8).

More complicated observation problems involving the two pilots acting together can be formalized by considering the shuffle product of $\mathcal{H}_{P_t}$ and $\mathcal{H}_{P_c}$ [20], and determining the induced critical states on this new system $\mathcal{H}$. Indeed, in the case of the two pilots acting together, an emergency braking action may result into a halt of the aircraft on the runway, an unsafe situation to avoid. For the sake of shortness, we do not analyze this situation here, but in the next section we will show how our methods can take into account critical states arising from the composition of the behaviors of two agents, in particular the ground controller and the tower controller.

5.3 Controller Observation Problem

Consider now the observation problem of the controllers.

The ground controller C_g can be modeled by a DES $\mathcal{D}_{C_g}$ where:

- $Q_3 = \{q_{3,1}, q_{3,2}, q_{3,3}\}$ is the set of discrete states, with $q_{3,1}$ corresponding to C_g in miscellaneous monitoring operations, $q_{3,2}$ to C_g having granted crossing, $q_{3,3}$ to an emergency braking action on the runway;
- $\Sigma_3 = \{\sigma_{3,1}, \sigma_{3,2}, \sigma_{3,3}, \sigma_{3,4}, \sigma_{3,5}\}$ is the finite set of input symbols, with $\sigma_{3,1}$ the decision to give a crossing grant, $\sigma_{3,2} = \psi_{2,4}$ the crossing completed confirmation, $\sigma_{3,3}$ the stopbar violation alarm on, $\sigma_{3,4}$ the decision to give a start up, $\sigma_{3,5} = \psi_{2,2}$ the crossing request;
- $\Psi_3 = \{\psi_{3,1}, \psi_{3,2}, \psi_{3,3}, \psi_{3,4}\} \cup \{\varepsilon\}$ is the set of discrete outputs, with $\psi_{3,1} = \sigma_{2,3}$ the crossing grant, $\psi_{3,2} = \sigma_{2,4}$ the emergency braking command, $\psi_{3,3} = \sigma_{1,1} = \sigma_{2,1}$ the start up grant, $\psi_{3,4} = \sigma_{2,2}$ the command to wait for crossing grant at stopbar S_2;
- The set E_3 of transitions and the output function η_3 are defined by the graph in Figure 9.

The tower controller C_t can also be modeled by a DES $\mathcal{D}_{C_t}$ where:

- $Q_4 = \{q_{4,1}, q_{4,2}, q_{4,3}\}$ is the set of discrete states, with $q_{4,1}$ corresponding to C_t in miscellaneous operations, $q_{4,2}$ to C_t having granted take off, $q_{4,3}$ an emergency braking action on the runway;
- $\Sigma_4 = \{\sigma_{4,1}, \sigma_{4,2}, \sigma_{4,3}\}$ is the finite set of input symbols, with $\sigma_{4,1} = \psi_{1,2}$ the take off request, $\sigma_{4,2} = \psi_{1,5}$ the take off completed confirmation, $\sigma_{4,3}$ the runway incursion alert on;
- $\Psi_4 = \{\psi_{4,1}, \psi_{4,2}\} \cup \{\varepsilon\}$ is the set of discrete outputs, with $\psi_{4,1} = \sigma_{1,2}$ the take off grant, $\psi_{4,2} = \sigma_{1,5}$ emergency braking command;
- The set E_4 of transitions and the output function η_4 are defined by the graph in Figure 9.

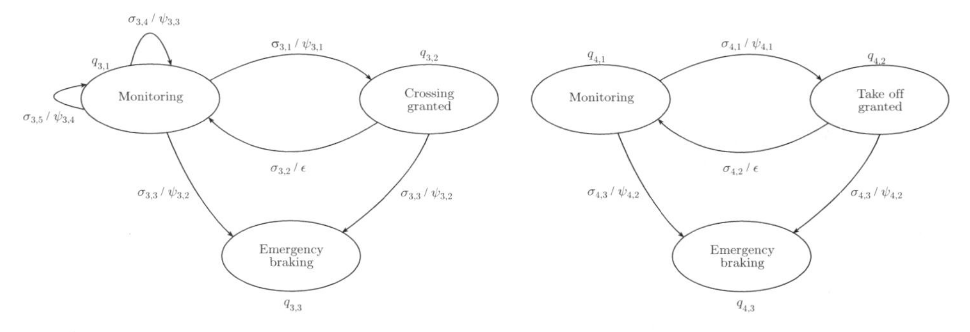

Fig. 9. DESs modelling $\mathcal{D}_{C_g}$ and $\mathcal{D}_{C_t}$

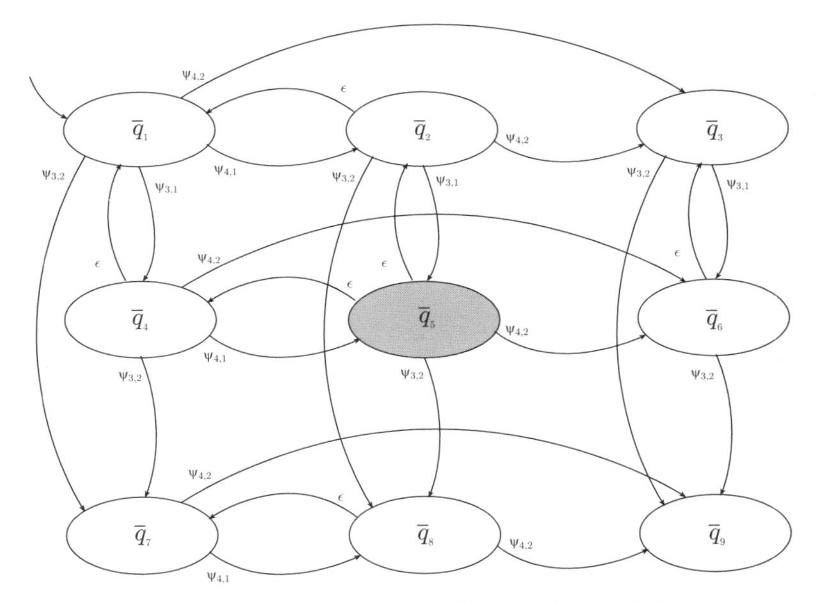

Fig. 10. Shuffle product $\mathcal{D}_{C_g}||\mathcal{D}_{C_t}$ of $\mathcal{D}_{C_g}$ and $\mathcal{D}_{C_t}$

The hazardous situation of a crossing grant given by C_g and a take–off grant simultaneously given by C_t should be detected. However, the DESs $\mathcal{D}_{C_g}$ and $\mathcal{D}_{C_t}$ have no critical states, because the hazardous situation arises when a crossing grant is given by C_g simultaneously with a take off grant given by C_t. Hence, the observation problem has to be considered for the composition (shuffle product [20]) $\mathcal{D}_{C_g}||\mathcal{D}_{C_t}$ of $\mathcal{D}_{C_g}$ and $\mathcal{D}_{C_t}$, represented in Figure 10.

Since we are dealing with a DES that can be viewed as a special case of switching system, the observability conditions presented in the previous sections can be applied to the system $\mathcal{D}_{C_g}||\mathcal{D}_{C_t}$. The observer associated with this system is illustrated in Figure 11. The state $\bar{q}_5 = \{q_{3,2}, q_{4,2}\}$ that corresponds to simultaneous crossing grant and take off grant, is critical. Then,

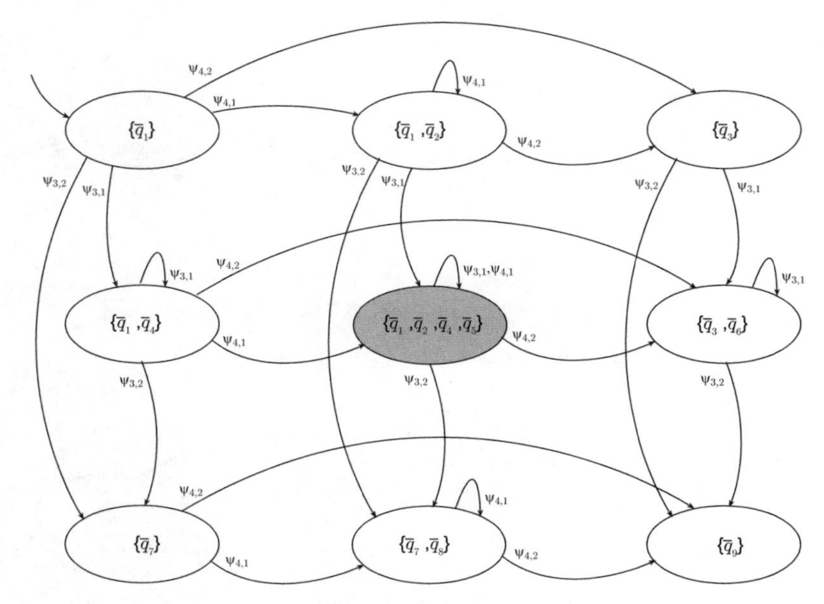

Fig. 11. Observer of $\mathcal{D}_{C_g} || \mathcal{D}_{C_t}$

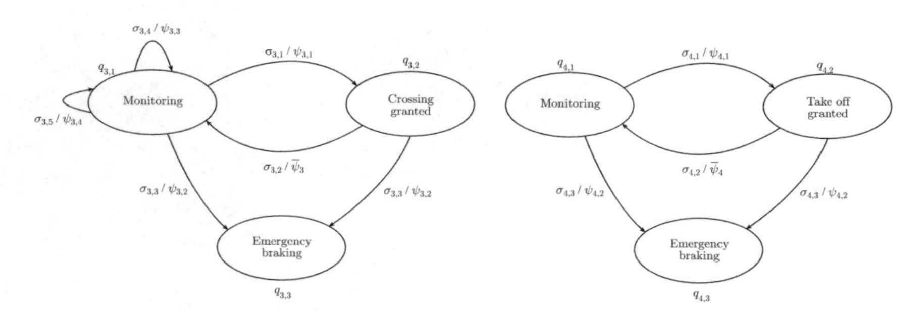

Fig. 12. DESs modelling $\bar{\mathcal{D}}_{C_g}$ and $\bar{\mathcal{D}}_{C_t}$

some additional information are needed to detect the critical state $\bar{q}_5$. However in a DES, no continuous information are available. Hence, the only way for solving the observability problem of the critical states is the introduction of new discrete outputs, e.g. the confirmation that crossing ($\bar{\psi}_3$) or take off ($\bar{\psi}_4$) are completed, as shown in Figure 12. This corresponds to a change in the procedure the controllers have to follow.

After the addition of new outputs, the observer of the shuffle product satisfies the critical observability criteria with respect to the critical state $\bar{q}_5$ (see Figure 13). In this case, the observer coincides with the original DES, because every transition has an observable discrete output.

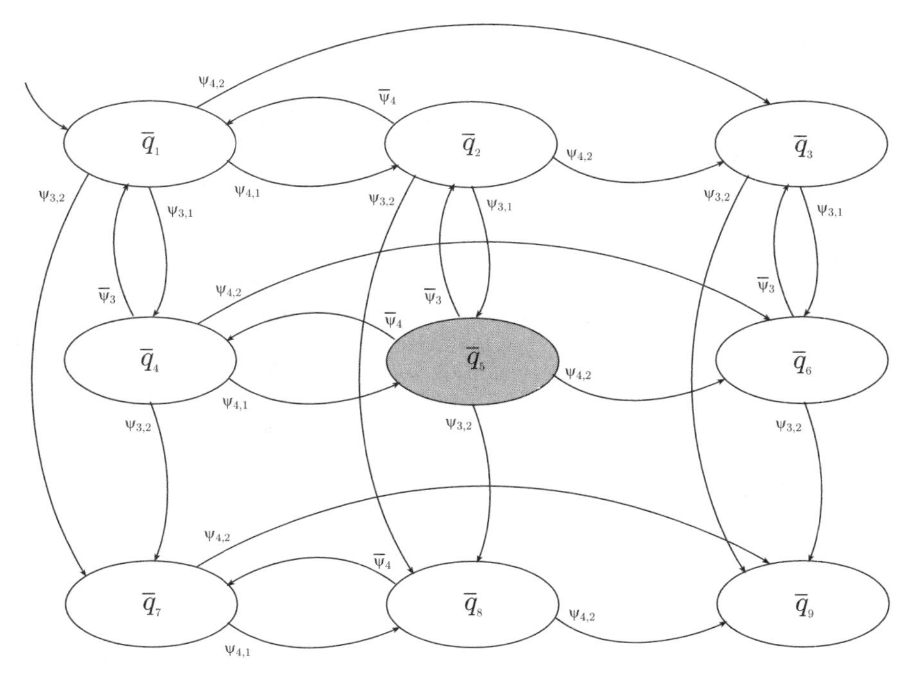

Fig. 13. Observer of $\bar{\mathcal{D}}_{C_g}||\bar{\mathcal{D}}_{C_t}$

6 Conclusions

We addressed the characterization of observability of linear switching systems. We derived some sufficient and some necessary conditions for assessing observability and critical observability, which can be checked by means of a computationally efficient procedure. We proposed an observer that under appropriate conditions is guaranteed to reconstruct the hybrid state evolution of a given switching system whenever a critical state is reached. We showed how critical observability can be used in the runway crossing problem where four human agents interact in a system consisting of five subsystems. The human agents are subject to errors that may lead to catastrophic situations and are modeled as hybrid systems. We developed a hybrid observer to detect the hazardous situations corresponding to critical states. Future work will focus on the analysis of the topology of the discrete event system associated with the linear switching system to find more efficient procedures for checking observability.

Acknowledgement

The authors are grateful to Ted Lewis and Derek Jordan who provided the scenario described in Section 5, which relies on the UK Radio Telephony (RT) procedures CAP 413 (2002).

References

1. A. Balluchi, M. D. Di Benedetto, C. Pinello, C. Rossi, A. L. Sangiovanni–Vincentelli, Hybrid Control in Automotive Applications: the Cut–off Control. *Automatica*, vol. 35, *Special Issue on Hybrid Systems*, March 1999, pp. 519–535.
2. A. Balluchi, L. Benvenuti, M. D. Di Benedetto, C. Pinello, A. L. Sangiovanni–Vincentelli, Automotive Engine Control and Hybrid Systems: Challenges and Opportunities. *Proceedings IEEE*, Invited Paper, vol. 88, no. 7, July 2000, pp. 888–912.
3. A. Balluchi, M. D. Di Benedetto, C. Pinello, A. L. Sangiovanni–Vincentelli, A Hybrid Approach to the Fast Positive Force Transient Tracking Problem in Automotive Engine Control. *Proceedings of the* 37th *IEEE Conference on Decision and Control* (CDC 98), Tampa, FL, December 98, pp. 3226–3231.
4. A. Balluchi, L. Benvenuti, M. D. Di Benedetto, A. L. Sangiovanni–Vincentelli, Design of Observers for Hybrid Systems. *Hybrid Systems: Computation and Control*, Claire J. Tomlin and Mark R. Greenstreet, Eds, vol. 2289 of Lecture Notes in Computer Science, Springer–Verlag, Berlin Heidelberg New York, 2002, pp. 76–89.
5. A. Balluchi, L. Benvenuti, M. D. Di Benedetto, A. L. Sangiovanni–Vincentelli, Observability for Hybrid Systems. *Proceedings of the* 42nd *IEEE Conference on Decision and Control* (CDC 03), Maui, Hawaii, USA, December 9–12, 2003.
6. A. Bemporad, G. Ferrari–Trecate, M. Morari, Observability and Controllability of Piecewise Affine and Hybrid Systems. *IEEE Transactions on Automatic Control*, vol. 45, no. 10, October 2000, pp. 1864–1876.
7. E. De Santis, M. D. Di Benedetto, S. Di Gennaro, G. Pola, Hybrid Observer Design Methodology. Public Deliverable D7.2, Project IST–2001–32460 HYBRIDGE, August 19, 2003. http://www.nlr.nl/public/hosted–sites/hybridge.
8. E. De Santis, M. D. Di Benedetto, G. Pola, On Observability and Detectability of Continuous–time Linear Switching Systems. *Proceedings of the* 42nd *IEEE Conference on Decision and Control* (CDC 03), Maui, Hawaii, USA, December 9–12, 2003, pp. 5777–5782 (extended version in www.diel.univaq.it/tr/web/web_search_tr.php).
9. E. De Santis, M. D. Di Benedetto, L. Berardi, Computation of Maximal Safe Sets for Switching Systems. *IEEE Transactions on Automatic Control*, vol. 41 no. 10, February 2004, pp. 184–195.
10. E. De Santis, M. D. Di Benedetto, G. Pola, Digital Idle Speed Control of Automotive Engines: A Safety Problem for Hybrid Systems. *International Journal of Hybrid Systems*, 6th Special Issue on Nonlinear Analysis: Hybrid Systems and Applications, 2006, to appear.

11. E. De Santis, M. D. Di Benedetto, G. Pola, Observability and Detectability of Linear Switching Systems: A Structural Approach. Technical Report no. R.05–82, Department of Electrical Engineering and Computer Science, University of L'Aquila, Italy, January 2006. (submitted) (also available from www.diel.univaq.it/tr/web/web_search_tr.php).
12. M. D. Di Benedetto and A. L. Sangiovanni–Vincentelli, Eds. *Hybrid Systems: Computation and Control*, Lecture Notes in Computer Science vol. 2034, Springer–Verlag, 2001.
13. M. D. Di Benedetto, S. Di Gennaro, A. D'Innocenzo, Situation Awareness Error Detection. Public Deliverable D7.3, Project IST–2001–32460 HYBRIDGE, August 18, 2004, http://www.nlr.nl/public/hosted–sites/hybridge.
14. M. D. Di Benedetto, S. Di Gennaro, A. D'Innocenzo, Critical Observability and Hybrid Observers for Error Detection in Air Traffic Management. *Proceedings of the 2005 International Symposium on Intelligent Control and* 13th *Mediterranean Conference on Control and Automation*, June 27–29, Limassol, Cyprus, 2005, pp. 1303–1308.
15. M. D. Di Benedetto, S. Di Gennaro, A. D'Innocenzo, Error Detection within a Specific Time Horizon and Application to Air Traffic Management, *Proceedings of the Joint Conference* 44th *IEEE Conference on Decision and Control* & European Control Conference (CDC–ECC 05), Seville, Spain, December 12–15, 2005, pp. 7472–7477.
16. M. D. Di Benedetto, S. Di Gennaro, A. D'Innocenzo, Error Detection within a Specific Time Horizon. Public Deliverable D7.4, Project IST–2001–32460 HYBRIDGE, January 26, 2005, http://www.nlr.nl/public/hosted–sites/hybridge.
17. S. Di Gennaro, Nested Observers for Hybrid Systems. *Proceedings of the Latin–American Conference on Automatic Control CLCA 2002*, Guadalajara, México, December 3–6, 2002.
18. S. Di Gennaro, Notes on the Nested Observers for Hybrid Systems. *Proceedings of the European Control Conference 2003* (ECC 03), Cambridge, UK, September 2003.
19. M. R. Endsley, Towards a Theory of Situation Awareness in Dynamic Systems. *Human Factors*, vol. 37, no. 1, 1995, pp. 32–64.
20. J. E. Hopcroft, J. D. Ullman, *Introduction to Automata Theory, Languages and Computation*, Addison–Wesley, Reading, MA, 1979.
21. I. Hwang, H. Balakrishnan, C. Tomlin, Observability Criteria and Estimator Design for Stochastic Linear Hybrid Systems. *Proceedings of European Control Conference 2003* (ECC 03), Cambridge, UK, September 2003.
22. R. E. Kalman, A New Approach to Linear Filtering and Prediction Problems. *Transactions of the ASME – Journal of Basic Engineering*, vol. D, 1960, pp. 35–45.
23. T. Lewis, D. Jordan, Personal communication, BAE Systems, 2004.
24. D. Liberzon, Switching in systems and control, Birkhauser 2003.
25. D. G. Luenberger, An Introduction to Observers. *IEEE Transactions on Automatic Control*, vol. 16, no. 6, December 1971, pp.596–602.
26. J. Lygeros, C. Tomlin, S. Sastry, Controllers for Reachability Specifications for Hybrid Systems. *Automatica*, Special Issue on Hybrid Systems, vol. 35, 1999.
27. A. S. Morse, Supervisory Control of Families of Linear Set–point Controllers–Part 1: Exact Matching. *IEEE Transactions on Automatic Control*, vol. 41, no. 10, October 1996, pp. 1413–1431.

28. M. Oishi, I. Hwang and C. Tomlin, Immediate Observability of Discrete Event Systems with Application to User–Interface Design. *Proceedings of the* 42nd *IEEE Conference on Decision and Control* (CDC 03), Maui, Hawaii, USA, December 9–12, 2003, pp. 2665–2672.
29. C. M. Özveren, and A. S. Willsky, Observability of Discrete Event Dynamic Systems. *IEEE Transactions on Automatic Control*, vol. 35, 1990, pp. 797–806.
30. P. J. Ramadge, Observability of Discrete Event Systems. *Proceedings of the* 25th *IEEE Conference on Decision and Control* (CDC 86), Athens, Greece, 1986, pp. 1108–1112.
31. P. J. Ramadge, W. M. Wonham, Supervisory Control of a Class of Discrete–Event Processes. *SIAM Journal of Control and Optimization*, vol. 25, no. 1, 1987, pp. 206–230.
32. E. D. Sontag, On the Observability of Polynomial Systems, I: Finite–time Problems. *SIAM Journal of Control and Optimization* , vol. 17, no. 1, 1979, pp. 139–151.
33. S. Stroeve, H. A. P. Blom, M. van der Park, Multi–Agent Situation Awareness Error Evolution in Accident Risk Modelling. FAA–Eurocontrol, ATM2003, June 2003, http://atm2003.eurocontrol.fr.
34. R. Vidal, A. Chiuso, S. Soatto, Observability and Identifiability of Jump Linear Systems. *Proceedings of the* 41st *IEEE Conference on Decision and Control*, Las Vegas, Nevada USA, December 2002, pp. 3614–3619.
35. R. Vidal, A. Chiuso, S. Soatto, S. Sastry, Observability of Linear Hybrid Systems. Lecture Notes in Computer Science vol. 2623, A. Pnueli and O. Maler Eds. (2003), Springer–Verlag Berlin Heidelberg, pp. 526–539.
36. T. Yoo and S. Lafortune, On The Computational Complexity Of Some Problems Arising In Partially–observed Discrete–Event Systems. *Proceedings of the 2001 American Control Conference* (ACC 01), Arlington, Virginia , June 25–27, 2001.

Multirobot Navigation Functions I

Savvas G. Loizou[1] and Kostas J. Kyriakopoulos[2]

[1] National Technical University of Athens, Athens, Greece,
sloizou@central.ntua.gr
[2] National Technical University of Athens, Athens, Greece,
kkyria@central.ntua.gr

Summary. This is the first of two chapters dealing with multirobot navigation. In this chapter a centralized methodology is presented for navigating a team of multiple robotic agents. The solution is a closed form feedback based navigation scheme. The considered robot kinematics include holonomic and non-holonomic constraints and are handled under the unifying framework of multirobot navigation functions. The derived methodology has theoretically guaranteed global convergence and collision avoidance properties. The feasibility of the proposed navigation scheme is verified through non-trivial computer simulations.

1 Introduction

Multi-Robot Navigation is a field of robotics that has recently gained increasing attention, due to the need to control more than one robot in the same workspace. The main motivation for our work initiated from the need to navigate concurrently several robotic agents sharing the same workspace. There are many application domains for multi-robot navigation ranging from navigation of teams of micro robots to conflict resolution in air traffic management systems.

The main focus of work on multi-robotic systems in the last few years has been on team formations [6, 24, 29, 9, 39, 28]. There have been several attempts to tackle multiagent navigation since the last two decades [43, 16, 15, 21, 42, 45, 44]. Most of them

- are based on heuristic approaches
- rely on simplifying assumptions i.e. point robots, convex obstacles, etc.
- do not possess theoretically guaranteed properties like stability, collision avoidance and global convergence
- are not applicable for online trajectory generation
- do not account for nonholonomic kinematics
- do not consider bounded inputs

H.A.P. Blom, J. Lygeros (Eds.): Stochastic Hybrid Systems, LNCIS 337, pp. 171–207, 2006.

In [43] the author defines separating planes at each moment and ensures that the robots stay in opposite half spaces but cannot guarantee that each robot will reach its goal since they may reach a deadlock state where one robot is blocking the other. In [16] a decoupled approach is presented, where first separate paths for the individual robots are computed and then possible conflicts of the generated paths are resolved in an off-line fashion. In [34] the authors consider an alternative problem in the domain of multirobot navigation, that is path coordination, where the robots paths are calculated off-line and a coordination scheme is executed in an off-line fashion. Although a large number of robots can be handled in this framework, it cannot handle inaccuracies in the executed trajectories, which are usually present in robotic systems due to the inability of the robots's hardware to follow exactly the pre-specified trajectory. In [3] a dynamic networks approach is adopted where a fast centralized planner is used to compute new coordinated trajectories on the fly. However this methodology does not have theoretically guaranteed global convergence properties.

A need for a unifying framework for robotic navigation, where one can perform analysis and establish theoretical guarantees for the properties of the system is apparent. Such a framework was proposed by Koditschek and Rimon [11] in their seminal work. This framework had all the sought qualities but could only handle single point-sized robot navigation. Two of the authors of this work in their previous work [38] had successfully extended the navigation function framework to take into account the volume of each robot and also to handle robots with non-holonomic kinematic constraints. In this work we present a provably correct way to extend the navigation function framework to the case of multiple robot navigation.

Of particular importance to multi-robot navigation is the case of systems possessing non-holonomic kinematic constraints. In [7] formation transitions of non-holonomic vehicle teams are studied using a graph theoretic approach. No general solutions have been proposed for closed loop navigation for multiple non-holonomic robots navigation, because of the problem's complexity and the fact that non-holonomic systems do not satisfy Brocket's necessary smooth feedback stabilization condition [2] hence no continuous static control law can stabilize a non-holonomic system to a point. Several motion planning strategies for non-holonomic systems are based on differential geometry [14, 13, 30, 26, 5, 23, 22]. Other strategies implement multi-rate [40] or time-varying controllers [27, 41]. Discontinuous control strategies are based on appropriately combining different controllers [12].

The main contributions of this work can be summarized as follows:

1. A new methodology for constructing provably correct Navigation Functions for multi-robot navigation
2. A provably correct way to implement dipolar potential fields in Multi-Robot Navigation Functions for application in mixed holonomic and non-holonomic systems

3. Development of a Multi-Robot control scheme, that takes into account bounds in the maximum achievable velocities of the system

The rest of the chapter is organized as follows: Section 2 presents the concept of Navigation Functions while section 3 introduces the considered system and presents the problem statement. Section 4 introduces the concept of Multi-Robot Navigation Functions, while in section 5 the controller synthesis is presented. In section 6 simulation results of the proposed methodology are presented and the chapter concludes with section 7.

2 Navigation Functions (NFs)

Navigation functions (NF's) are real valued maps realized through cost functions $\varphi(q)$, whose negated gradient field is attractive towards the goal configuration and repulsive wrt obstacles. It has been shown by Koditscheck and Rimon that strict global navigation (i.e. the system $\dot{q} = u$ under a control law of the form $u = -\nabla\varphi$ admits a globally attracting equilibrium state) is not possible, and a smooth vector field on any sphere world with a unique attractor, must have at least as many saddles as obstacles [11]. Figure 1 shows a navigation function in a workspace with three obstacles.

A navigation function can be defined as follows:

Definition 1. *[11] Let $\mathcal{F} \subset \mathcal{R}^{2N}$ be a compact connected analytic manifold with boundary. A map $\varphi : \mathcal{F} \rightarrow [0, 1]$ is a navigation function if:*

1. *Analytic on $\mathcal{F}$;*
2. *Polar on $\mathcal{F}$, with minimum at $q_d \in \mathring{\mathcal{F}}$;*
3. *Morse on $\mathcal{F}$;*
4. *Admissible on $\mathcal{F}$.*

Strictly speaking, the continuity requirements for the navigation functions are to be C^2. The first property of Definition 1 follows the intuition provided by the authors of [11], that is preferable to use closed form mathematical expressions to encode actuator commands instead of "patching together" closed form expressions on different portions of space, so as to avoid branching and looping in the control algorithm. Analytic navigation functions, through their gradient provide a direct way to calculate the actuator commands, and once constructed they provide a provably correct control algorithm for every environment that can be diffeomorphically transformed to a sphere world.

A function φ is called polar if it has a unique minimum on $\mathcal{F}$. By using smooth vector fields one cannot do better than have almost global navigation [11]. By using a polar function on a compact connected manifold with boundary, all initial conditions will either be brought to a saddle point or to the unique minimum: q_d.

A scalar valued function φ is called a Morse function if all its critical points (zero gradient vector field) are non-degenerate, that is the Hessian at the

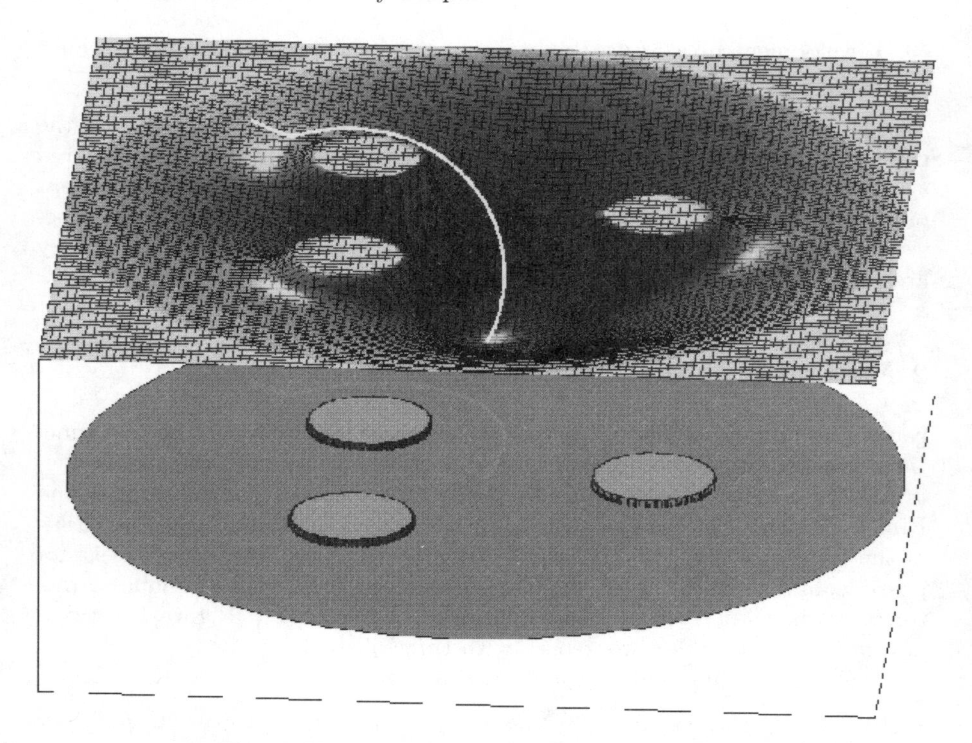

Fig. 1. Navigation Function with three obstacles

critical points is full rank. The requirement in Definition 1 that a navigation function must be a Morse function, establishes that the initial conditions that bring the system to saddle points are sets of measure zero [25]. In view of this property, all initial conditions away from sets of measure zero are brought to q_d.

The last property of Definition 1 guarantees that the resulting vector field is transverse to the boundary of $\mathcal{F}$. This establishes that the system will be safely brought to q_d, avoiding collisions.

3 System Description and Problem Statement

We assume that the n robots indexed from $1 \ldots n$ $(0 \leq n \leq m)$ are holonomic and the rest $z = m - n$ robots indexed from $(n+1) \ldots m$ are non-holonomic. Define the posture of each robot as $p_i = \left[q_i^T \theta_i\right]^T \in \mathbb{R}^2 \times (-\pi, \pi]$ with $i \in \{1 \ldots m\}$. The state vector of the holonomic robots is defined as $\mathbf{q_h} = \left[p_1^T \ldots p_n^T\right]^T$ and for the non-holonomic as $\mathbf{q_{nh}} = \left[p_{n+1}^T \cdots p_m^T\right]^T$. The state of the whole system is $\mathbf{p} = \left[\mathbf{q_h}^T \mathbf{q_{nh}}^T\right]^T$. We will also need the orientation vector $\theta = \left[\theta_{n+1}^T \dots \theta_m^T\right]^T$.

The kinematics of the holonomic subsystem can be described by the following model

$$\dot{\mathbf{q_h}} = M \cdot \mathbf{u_h} \tag{1}$$

and the kinematics of the non-holonomic subsystem by the model:

$$\dot{\mathbf{q_{nh}}} = C \cdot \mathbf{u_{nh}} \tag{2}$$

The augmented system we are considering can thus be described by the following kinematic model:

$$\dot{\mathbf{p}} = \begin{bmatrix} M_{3n\times 3n} & O_{3n\times 2z} \\ O_{3z\times 3n} & C_{3z\times 2z} \end{bmatrix} \cdot \begin{bmatrix} \mathbf{u_h} \\ \mathbf{u_{nh}} \end{bmatrix} \tag{3}$$

where $M = diag\left(u_{x_1}^{max},\ u_{y_1}^{max},\ w_1^{max},\ \dots\ w_n^{max}\right)$ contains the maximum velocities achievable by the holonomic subsystem, $\mathbf{u_h} = \left[\mathbf{u_{h_1}}^T \dots \mathbf{u_{h_n}}^T\right]^T$, $\mathbf{u_{h_i}} = \left[u_{x_i} u_{y_i} w_i\right]^T$, $\mathbf{u_{nh}} = \left[\mathbf{u_{nh_{n+1}}}^T \dots \mathbf{u_{nh_m}}^T\right]^T$, $\mathbf{u_{nh_i}} = \left[v_i w_i\right]^T$ and $C = \begin{bmatrix} C_{n+1} & & 0 \\ & \ddots & \\ 0 & & C_m \end{bmatrix}$ with $C_i = \begin{bmatrix} v_i^{max} cos(\theta_i) & 0 \\ v_i^{max} sin(\theta_i) & 0 \\ 0 & w_i^{max} \end{bmatrix}$ since we are modeling the non-holonomic systems as non-holonomic unicycles. $v_i^{\max}$ and $w_i^{\max}$ contained in C_i matrix are the maximum achievable linear and angular velocities by the non-holonomic subsystem. The considered upper bounds to the robots achievable velocities are reflected in the following restrictions over the inputs:

$$c\left|u_{x_i}\right| \le 1, \left|u_{y_i}\right| \le 1, \left|w_i\right| \le 1,\ i \in \{1 \dots n\} \tag{4}$$

$$\left|w_i\right| \le 1, \left|v_i\right| \le 1,\ i \in \{n+1 \dots m\} \tag{5}$$

The problem we are considering, of navigation of a mixed team of holonomic and non-holonomic agents, can be stated as follows:

Given the mixed holonomic and non-holonomic system (3) and the input constraints (4), derive a feedback kinematic control law that steers the system from any initial configuration to the goal configuration avoiding collisions. The environment is assumed perfectly known and stationary.

4 Multirobot Navigation Functions (MRNFs)

4.1 Preliminaries

Multi-Robot Navigation Functions(MRNFs), like NFs, are real valued maps realized through cost functions, whose negated gradient field is attractive towards the goal configuration and repulsive wrt obstacles. Considering a trivial system described kinematically as $\dot{x} = u$, the basic idea behind navigation functions is to use a control law of the form $u = -\nabla\varphi$, where φ is an MRNF, to drive the system to its destination. Our assumption that we have spherical robots and spherical obstacles does not constrain the generality of this methodology, since it has been proven [11] that navigation properties are invariant under diffeomorphisms. Methods for constructing analytic diffeomorphisms are discussed in ([32],[31]) for point robots and in [37] for rigid body robots. We should note here that a proper diffeomorphism for a multi-robot scenario must preserve the robot proximity relations discussed later in this section.

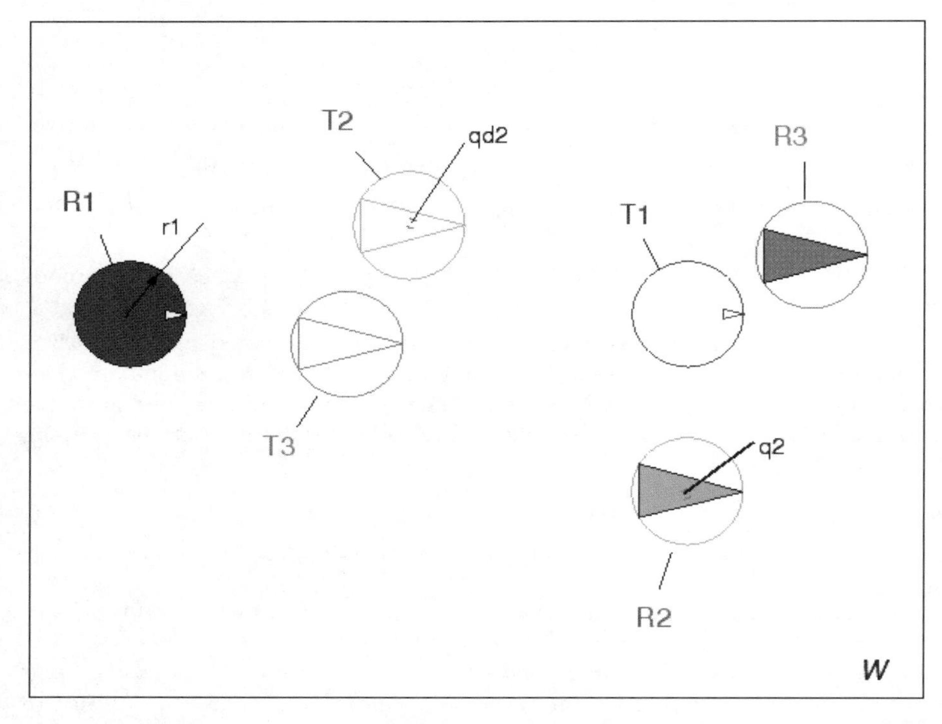

Fig. 2. Workspace populated with holonomic (filled disks) and non-holonomic (disks with filled triangle) robotic agents. Target configurations represented with non-filled disks

Let us assume the following situation: We have m mobile robots, and their workspace $\mathcal{W} \subset \mathbb{R}^r$ where r is the workspace dimension. Each robot R_i, $i = 1 \ldots m$ occupies a sphere in the workspace: $R_i = \{\mathbf{q} \in \mathbb{R}^r : \|\mathbf{q} - \mathbf{q}_i\| \leq r_i\}$ where $\mathbf{q}_i \in \mathbb{R}^r$ is the center of the sphere and r_i is the radius of the robot. The configuration of each robot is represented by $\mathbf{q}_i$ and the configuration space C is spanned by $\mathbf{q} = \left[\mathbf{q}_1^T \ldots \mathbf{q}_m^T\right]^T$. The destination configurations are denoted with the index d, i.e. $\mathbf{q}_d = \left[\mathbf{q}_{d_1}^T \dots \mathbf{q}_{d_m}^T\right]^T$. Figure 2 depicts a team of holonomic robots represented as filled disks and nonholonomic robots represented as disks with filled triangles in a spherical workspace. A multi-robot navigation function can be defined in an analogous manner to the navigation function definition [11] as follows:

Definition 2. *Let $\mathcal{F} \subset \mathbb{R}^{rm}$ be a compact connected analytic manifold with boundary. A map $\varphi : \mathcal{F} \to [0,1]$ is a multirobot navigation function if it is:*

1. *Analytic on $\mathcal{F}$,*
2. *Polar on $\mathcal{F}$, with minimum at $\mathbf{q}_d \in \overset{o}{\mathcal{F}}$,*
3. *Morse on $\mathcal{F}$*
4. $\lim_{q \to \partial\mathcal{F}} \varphi(q) = 1 > \varphi(q_{int}),\ \forall q_{int} \in \overset{\circ}{\mathcal{F}}$

Strictly speaking, the continuity requirements for the MRNFs are to be C^2. Analytic MRNFs, through their gradient provide a direct way to calculate the actuator commands, and once constructed they provide a provably correct control algorithm for every environment that can be diffeomorphically transformed to a sphere world.

The requirement in Definition 2 that an MRNF must be a Morse function, establishes that the initial conditions that bring the system to saddle points are sets of measure zero, hence all initial conditions away from sets of measure zero are brought to q_d.

The last property of definition 2 guarantees that the resulting vector field is transverse to the boundary of $\mathcal{F}$, hence a system inheriting the gradient field properties of the MRNF will be safely brought to q_d, avoiding collisions.

4.2 NFs vs MRNFs

The concept behind potential functions is that the system must be attracted toward the "good" sets and repelled away from "bad" sets. Multi-Robot Navigation functions are a special category of potential functions that have the properties defined in Definition 2. The navigation function proposed by Koditschek and Rimon [11] for single, point robot navigation, was a composition of three functions:

$$\varphi \triangleq \sigma_d \circ \sigma \circ \hat{\varphi} = \frac{\gamma_d}{\left(\gamma_d^k + \beta\right)^{1/k}} \tag{6}$$

where $\sigma_d(x) \triangleq x^{1/k}$ was used to render the destination point a non-degenerate critical point. $\sigma(x) \triangleq \frac{x}{1+x}$ was used to constrain the values of the navigation function in the range of $[0,1]$. Function $\hat{\varphi}$ was chosen to reflect this concept as $\hat{\varphi} = \frac{\gamma}{\beta}$ where $\gamma = \gamma_d^k = \|q - q_d\|^{2k}$ was a metric of the distance from the target - hence the good set was defined as $\gamma^{-1}(0)$ and the bad sets were defined as $\beta^{-1}(0)$. Now the essential difference between single point robot and multiple non-point robot navigation lies in the way of choosing the function β. For the single point robot case, this function was chosen as the product of the functions β_j that encoded class K_∞ functions of the distance of the robot from the obstacles and the workspace boundary. In initial attempts to tackle the non-point multi-robot navigation problem in the context of navigation functions, the authors of [45, 44] chose function β as the product of the functions $\beta_{i,j} = \frac{1}{2}\left(\|q_i - q_j\|^2 - (\rho_i + \rho_j)^2\right)$. They were able to theoretically establish that the resulting potential function attained a uniform maximum value on the configuration space boundary i.e. the resulting trajectories were collision free.

The major contribution of this work is in showing that an appropriate and more elaborate construction of the function β, first presented by the authors in [17], yields a provably correct multi-robot navigation function.

4.3 Terminology

Our intuition for developing this methodology was that in multi-robot scenarios, just avoiding the neighboring robots was not an adequate strategy. It makes more sense for a centralized controller to try to avoid any possible collision scheme. With this in mind we had to encode in β the "distance" of the system from every possible collision scheme. A key issue of this point of view is that collision schemes are categorized into discrete proximity relations.

The **robot proximity function**, which is a measure of the distance between two robots i and j is defined as $\beta_{i,j}(q) = q^T \cdot D_{i,j} \cdot q - (r_i + r_j)^2$ where the matrix $D_{i,j}$ is defined in Appendix A. We will use the term '**relation**' to describe the possible collision schemes that can be defined in a multi-robot scheme, possibly including obstacles. The '**set of relations**' between the members of the team can be defined as the set of all possible collision schemes between the members of the team. A '**binary relation**' is a relation between two robots. Any relation can be expressed as a set of binary relations. A '**relation tree**' is the set of robots-obstacles that form a linked team. Each **relation** may consist of more than one trees (figure 3). We will call the number of binary relations in a relation, the '**relation level**'. Figure (4) demonstrates several types of relations of a four – member team.

A **relation proximity function (RPF)** provides a measure of the distance between the robots involved in a relation. Each relation has its own RPF. An RPF is the sum of the robot proximity functions of a relation and assumes the value of zero whenever the related robots collide and increases wrt the distance of the related robots:

$$b_R = \mathbf{q}^T \cdot P_R \cdot \mathbf{q} - \sum_{\{i,j\} \in R} (r_i + r_j)^2 \tag{7}$$

where R is the set of binary relations (e.g. for the relation in figure (3.b) $R = \{\{A,B\},\{A,C\},\{B,C\},\{D,E\}\}$) and $P_R = \sum_{\{i,j\} \in R} D_{i,j}$ is the **rela-**

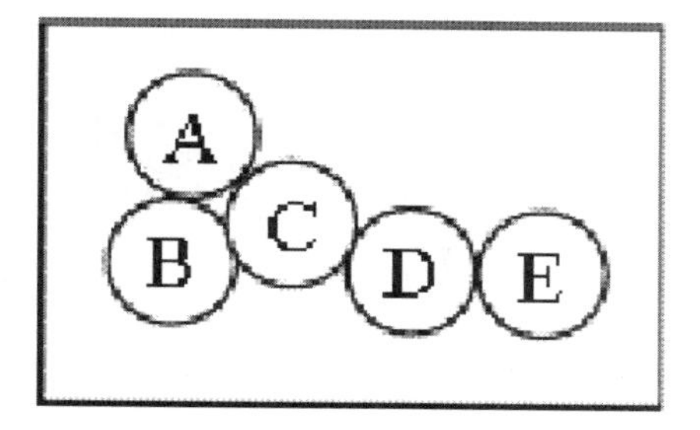

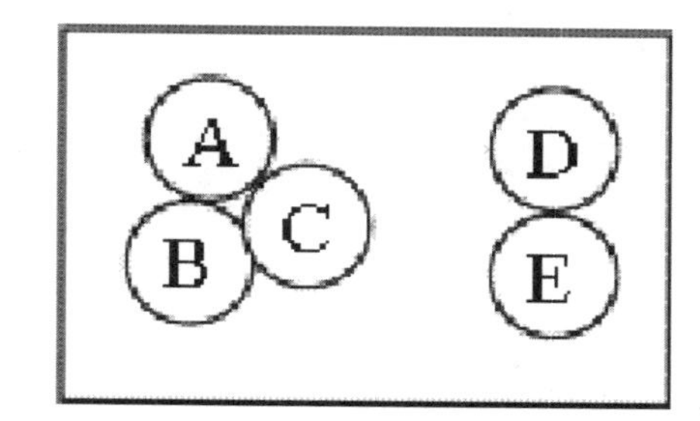

Fig. 3. (a) One – tree relation, (b) Two tree relation

tion matrix of *RPF*. A **Relation Verification Function (RVF)** is defined by:

$$g_{R_j}\left(b_{R_j}, B_{R_j^C}\right) = b_{R_j} + \frac{\lambda \cdot b_{R_j}}{b_{R_j} + B_{R_j^C}^{1/h}} \tag{8}$$

where $\lambda, h > 0$, R_j^C is the complementary to R_j set of *relations* in the same level, j is an index number defining the relation in the level and $B_{R_j^C} = \prod_{k \in R_j^C} b_k$. An *RVF* is zero if a relation holds while no other relation from the same level holds and has the properties: (a) $\lim_{x \to 0} \lim_{y \to 0} g_x(x, y) = \lambda$, (b) $\lim_{y \to 0} \lim_{x \to 0} g_x(x, y) = 0$.

Based on the above properties, in a robot proximity situation, one can verify that: if $\left(g_{R_j}\right)_k = 0$ at some *level* k then $(g_{R_i})_h \neq 0$ for any *level* h and $i \neq j$ in level k . It should be noted hereby that since in the highest relation level only one relation exists, there will be no complementary relations and the RVF will be identical to the RPF e.g. $\lambda = 0$ for this relation.

4.4 Construction of MRNFs

For the MRNFs, the β function used in eq. 6, is replaced with the G function defined as

$$G = \prod_{L=1}^{n_L} \prod_{j=1}^{n_{R,L}} \left(g_{R_j}\right)_L \tag{9}$$

with n_L the number of levels and $n_{R,L}$ the number of relations in level L.

The number of relation verification functions for a multirobot scenario with m robots, assuming that any relation is possible, is $2^{\frac{m \cdot (m-1)}{2}} - 1$. Hence the required computations for the construction of G in e.q. (9) increases exponentially wrt the number of robots in the workspace.

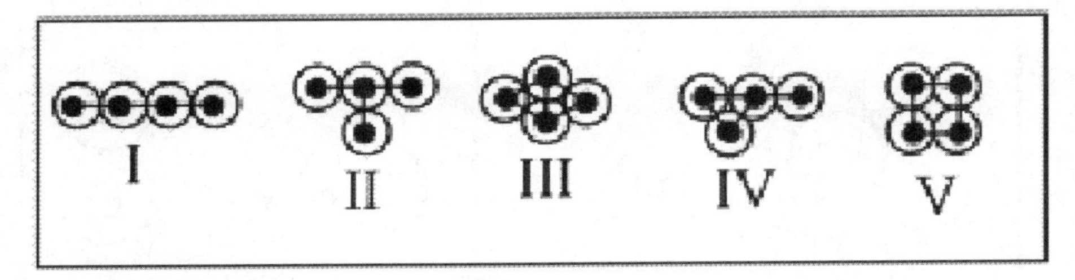

Fig. 4. I, II are level 3; IV, V are level 4 and III is a level 5 relation

Example

As an example, we will present the steps to construct an *MRNF* for a team of four robots. Assume the robots are indexed 1 through 4. We begin by dfining the for each relation j in every level k, the set of binary relations comprising the relation $(R_j)_k$. For each binary relation we calculate the robot proximity function . Knowing the members of each relation we can calculate the relation proximity functions of each relation, which are the sum of the robot proximity functions of the individual binary relations comprising the relation. Tables 1.a and 1.b. show the RPFs for several members of each level.

Table 1.a. Relation proximity functions in Levels 1 to 4

Relation	Level 1	Level 2	Level 3	Level 4
1	β_{12}	$\beta_{12}+\beta_{13}$	$\beta_{12}+\beta_{13}+\beta_{14}$	$\beta_{12}+\beta_{13}+\beta_{14}+\beta_{23}$
2	β_{13}	$\beta_{12}+\beta_{14}$	$\beta_{12}+\beta_{13}+\beta_{23}$	$\beta_{12}+\beta_{13}+\beta_{14}+\beta_{24}$
3	β_{14}	$\beta_{12}+\beta_{23}$	$\beta_{12}+\beta_{13}+\beta_{24}$	$\beta_{12}+\beta_{13}+\beta_{14}+\beta_{34}$
⋮	⋮	⋮	⋮	⋮
20	-	-	$\beta_{23}+\beta_{24}+\beta_{34}$	-

Table 1.b Relation proximity functions in Levels 5, 6

Relation	Level 5	Level 6
1	$\beta_{12}+\beta_{13}+\beta_{14}+\beta_{23}+\beta_{24}$	$\beta_{12}+\beta_{13}+\beta_{14}+\beta_{23}+\beta_{24}+\beta_{34}$
2	$\beta_{12}+\beta_{13}+\beta_{14}+\beta_{23}+\beta_{34}$	-
⋮	⋮	-
6	$\beta_{13}+\beta_{14}+\beta_{23}+\beta_{24}+\beta_{34}$	-

Notice that Levels 1 through 6 contain 6, 15, 20, 15, 6, 1 relations respectively. Once relation proximity functions have been defined for all levels, we can easily calculate the complements $B_{R_j^C}$ and then the RVFs through eqn. (8). G can then be calculated through eqn. (9) and the navigation function through eq. (6) with $\beta := G$. Parameter k in eq. (6), should be chosen to be large enough, as there exists a lower bound below which the function is not a navigation function. Such a lower bound is theoretically established in section 4.7 for a bounded workspace.

4.5 Assumptions

An assumption about the robot target configurations was needed in proving the navigation properties of our methodology. So for any valid workspace we need the destination configurations to be related with the robot radii through the following inequality:

$$\sum_{\{l,j\}\in R_H} \left\|\mathbf{q}_d^l - \mathbf{q}_d^j\right\|^2 > (m-1)^2 \cdot \sum_{\{l,j\}\in R_H} (r_l + r_j)^2 \tag{10}$$

where R_H is the highest level relation. It should be noted that as this is a requirement for the sphere world, it does not actually constrain the applicability of the methodology. This is due to navigation properties being invariant under diffeomorphisms. This means that when we are navigating robots in a diffeomorphic to a sphere world this requirement is equivalent to selecting target configurations in such a way that robots are not touching at their targets. In the equivalent diffeomorphic sphere world the robot radii can be chosen to be sufficiently small so eq. (10) is satisfied.

4.6 Characterization

With the above definitions and construction in place we can state the following:

Theorem 1. *For any valid workspace there exists $K, h_0 \in \mathbb{Z}^+$ such that for every $k > K$ and $h > h_0$ the function:*

$$\varphi = \sigma_d \circ \sigma \circ \hat{\varphi} = \frac{\gamma_d}{\left(\gamma_d^k + G\right)^{1/k}} \tag{11}$$

with G as defined in (9) is a Multi-Robot Navigation Function

Proof. Properties 1 and 4 of Definition 2 hold by construction. By Proposition 1, there exists a positive integer N_1 such that for every $k > N_1$, φ is polar on $\mathcal{F}$. By Proposition 6 there exist an ε_1 and an h_0, such that for every $k > N_2 = N(\varepsilon_1)$, with $N(\cdot)$ as defined in Proposition 4, and for every integer $h > h_0$, φ is Morse on $\mathcal{F}$. Choosing a K such that $K > max\{N_1, N_2\}$ completes the proof. □

4.7 Proof of Correctness

The following theorem allows us to reason for function φ by examining the simpler function $\hat{\varphi}$.

Theorem 2 ([11]). *Let $I_1, I_2 \subseteq R$ be intervals, $\hat{\varphi} : \mathcal{F} \to I_1$ and $\sigma : I_1 \to I_2$ be analytic. Define the composition $\varphi : \mathcal{F} \to I_2$, to be $\varphi = \sigma \circ \hat{\varphi}$. If σ is monotonically increasing on I_1, then the set of critical points of $\hat{\varphi}$ and φ coincide and the (Morse) index of each critical point is identical.*

Let $\varepsilon > 0$. Define $B_i^l(\varepsilon) = \{\mathbf{q} : 0 < (g_{R_i}(\mathbf{q}))_l < \varepsilon\}$. Following the reasoning inspired by that of [11], we can discriminate the following topologies:

1. The destination point $\mathbf{q}_d$
2. The free space boundary: $\partial\mathcal{F}(\mathbf{q}) = G^{-1}(\delta)$, $\delta \to 0$
3. The robot/obstacle proximity set: $\mathcal{F}_0(\varepsilon) = \bigcup\limits_{L=1}^{n_L} \bigcup\limits_{i=1}^{n_{R,L}} B_i^L(\varepsilon) - \{\mathbf{q}_d\}$, with n_L and $n_{R,L}$ as defined above.
4. The robot/obstacle distant set: $\mathcal{F}_1(\varepsilon) = \mathcal{F} - (\{\mathbf{q}_d\} \bigcup \mathcal{F}_0(\varepsilon))$

We can now state the following:

Proposition 1. *For any valid workspace, there exists a positive integer N_1 such that for every $k > N_1$, function (11) with $\beta = G$ as defined in (9) is polar on $\mathcal{F}$.*

Proof. By Proposition 2, $\mathbf{q}_d$ is a local minimum of φ. By Proposition 3 all critical points are in the interior of free space and by Proposition 4, choosing $k > N(\varepsilon)$ no critical points exist in $\mathcal{F}_1$. Proposition 5 establishes the existence of an ε_0, such that $N_1 = N(\varepsilon_0)$ is a lower bound for k for which the critical points in $\mathcal{F}_0$ are not local minima. □

Proposition 2. *The destination point $\mathbf{q}_d$ is a non – degenerate local minimum of φ.*

Proof. See Appendix B.1 □

Proposition 3. *All the critical points are in the interior of the free space.*

Proof. See Appendix B.2 □

Proposition 4. *For every $\varepsilon > 0$, there exists a positive integer* $\mathrm{N}(\varepsilon)$ *such that if $k > \mathrm{N}(\varepsilon)$ then there are no critical points of $\hat{\varphi}$ in $\mathcal{F}_1(\varepsilon)$.*

Proof. See Appendix B.3 □

Hence the set away from the obstacles is 'cleaned' from critical points. The workspace can be bounded with several obstacles prohibiting the motion of robots beyond them or by defining a world obstacle in the sense of robot proximity function: $\beta_{w,i} = (-1)\left(\mathbf{q}_i^T\mathbf{q}_i - (r_w - r_i)^2\right)$ where the index i refers to the robot and the index w refers to the world obstacle.

The following proposition establishes that the critical points of the proposed function except from the target are saddles.

Proposition 5. *There exists an* $\varepsilon_0 > 0$ *, such that* $\hat{\varphi}$ *has no local minimum in* $\mathcal{F}_0(\varepsilon)$*, as long as* $\varepsilon < \varepsilon_0$*.*

Proof. See Appendix B.4 □

The following proposition establishes that the proposed function is a Morse function [25].

Proposition 6. *There exists* $\varepsilon_1 > 0$ *and* $h_0 > 0$*, such that the critical points of* $\hat{\varphi}$ *are non-degenerate as long as* $\varepsilon < \varepsilon_1$ *and* $h > h_0$*.*

Proof. See Appendix B.5 □

5 Controller Synthesis

5.1 The Holonomic Case

In the holonomic case, we are considering system 1. In this case we can directly use the MRNF's negated gradient field to drive the system to it's destination from any feasible initial configuration, using a control law of the form:

$$\mathbf{u_h} = -K \cdot \nabla\varphi(\mathbf{q_h}) \tag{12}$$

where K is a positive gain. We can state the following:

Proposition 7. *System (1) under the control law (12), with* φ *a Multi Robot Navigation Function, is globally asymptotically stable, almost everywhere* [3]

Proof. See Appendix B.6 □

5.2 The Mixed Holonomic and Non-holonomic Case

We will now proceed with presenting a controller design methodology that handles the more general case of teams having both holonomic and non-holonomic members with additional input constraints. The two ends of this configuration is the purely holonomic case and the purely non-holonomic case both with input constraints, which is in accordance to the problem statement as posed in section 3.

[3] i.e. everywhere except from a set of initial conditions of measure zero

Dipolar MRNFs

As it was shown in [36] a navigation field with dipolar structure was particularly suitable for nonholonomic navigation. Based on [36] and [20], we apply the dipolar navigation methodology to the problem we are considering: To be able to produce a dipolar field, φ must be modified as follows:

$$\varphi = \frac{\|\mathbf{p} - \mathbf{p_d}\|^2}{\left(\|\mathbf{p} - \mathbf{p_d}\|^{2k} + H_{nh} \cdot G\right)^{1/k}}$$

where H_{nh} has the form of a pseudo - obstacle:

$$H_{nh} = \varepsilon_{nh} + \prod_{i=n+1}^{m} \eta_{nh_i}$$

Figure 5 shows a 2D dipolar Navigation Function. The navigation properties are not affected by this modification, as long as the workspace is bounded, η_{nh_i} can be bounded in the workspace and $\varepsilon_{nh} > \min\{\varepsilon_0, \varepsilon_1\}$ [19]. A possible choice of η_{nh_i} is:

$$\eta_{nh_i} = \left((\mathbf{q} - \mathbf{q}_d)^T \cdot \mathbf{n}_{d_i}\right)^2 \tag{13}$$

where $\mathbf{n}_{d_i} = \left[O_{1\times2(i-1)} \ \cos(\theta_{d_i}) \ \sin(\theta_{d_i}) \ O_{1\times2(m-i)}\right]^T$.

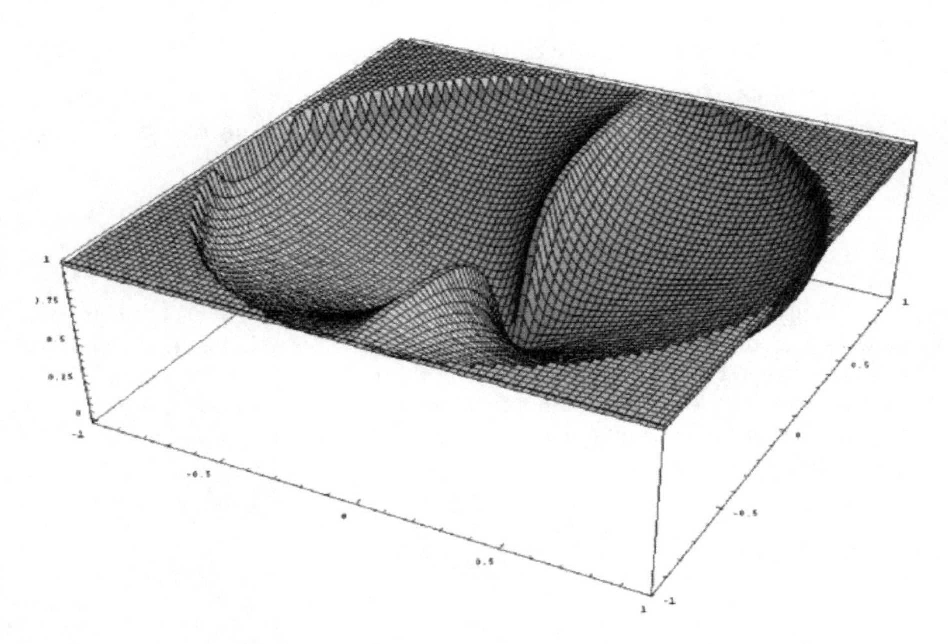

Fig. 5. 2D Dipolar Navigation Function

Design

In the following analysis we will use $V = \varphi(\mathbf{p})$, where φ an MRNF, as a Lyapunov function candidate.

Define $\mathcal{M} = \{n+1, \ldots, m\}$ and $\Omega = P(\mathcal{M})$ where P denotes the power set operator. Assuming that Ω is an ordered set, let N_j denote the j 'th element of Ω where $j \in \{1, \ldots, 2^z\}$. Then $N_j \subseteq \mathcal{M}$ with $N_1 = \{\emptyset\}$ and $N_{2^z} = \mathcal{M}$. We can now define:

$$\Delta_j = \delta_{\theta_{nh}}(j) - \delta_{V_q} - \delta_h \tag{14}$$

where $\delta_{\theta_{nh}}, \delta_{V_q}, \delta_{V_\theta}$ are defined as follows:

$$\delta_{\theta_{nh}}(j) = \sum_{i \in \{M \backslash N_j\}} \left[\frac{(\theta_{nh_i} - \theta_i) \cdot w_i^{max} \cdot V_{\theta_i}}{a_1 + |\theta_{nh_i} - \theta_i|} \right] - - \sum_{i \in \{N_j\}} \left[\frac{w_i^{max} \cdot V_{\theta_i}^2}{a_1 + |V_{\theta_i}|} \right]$$

$$\delta_{V_q} = \sum_{i=n+1}^{m} \left[|V_{x_i} \cdot cos(\theta_i) + V_{y_i} \cdot sin(\theta_i)| \cdot \frac{v_i^{max} \cdot Z_i}{a_2 + Z_i} \right]$$

$$\delta_h = \sum_{i=1}^{n} \frac{u_{x_i}^{max} \cdot V_{x_i}^2}{a_1 + |V_{x_i}|} + \frac{u_{y_i}^{max} \cdot V_{y_i}^2}{a_1 + |V_{y_i}|} + \frac{w_i^{max} \cdot V_{\theta_i}^2}{a_1 + |V_{\theta_i}|}$$

$$Z_i = a_3 \cdot \left(V_{x_i}^2 + V_{y_i}^2\right) + a_4 \left((x_i - x_{d_i})^2 + (y_i - y_{d_i})^2 \right)$$

$$\theta_{nh_i} = atan2\left(V_{y_i} \cdot side_i, V_{x_i} \cdot side_i\right)$$

with

$$side_i = sgn\left((\mathbf{q} - \mathbf{q}_d) \cdot \mathbf{n}_{d_i}\right)$$

$$sgn(x) = \begin{cases} -1 & x < 0 \\ 1 & x \geq 0 \end{cases}$$

and V_x, V_y, V_θ denotes the derivative of V along q_x, q_y, θ respectively. a_1, a_2, a_3, a_4 are positive constants. Define $H = \{j : \Delta_j < 0\}$ and $\rho = \min\{H \bigcup \{2^z\}\}$. Also define $s(x) = \frac{x}{a_1 + |x|}$. We can now state the following:

Proposition 8. *The system (3) under the control law:*

$$\begin{aligned} u_{x_t} &= -s\left(V_{x_t}\right) \\ u_{y_t} &= -s\left(V_{y_t}\right), \; t \in \{1, \ldots n\} \\ \omega_t &= -s\left(V_{\theta_t}\right) \end{aligned}$$

$$\begin{aligned} \omega_l &= -s\left(\theta_l - \theta_{nh_l}\right), \; l \in \{\mathcal{M} \backslash N_\rho\} \\ \omega_j &= -s\left(V_{\theta_j}\right), \qquad j \in \{N_\rho\} \end{aligned}$$

$$v_i = -\frac{Z_i}{a_2 + Z_i} \cdot \operatorname{sgn}\left(V_{x_i} \cdot \cos\left(\theta_i\right) + V_{y_i} \cdot \sin\left(\theta_i\right)\right), \qquad i \in \mathcal{M}$$

is globally asymptotically stable a.e.[4]

Proof. See Appendix B.7 □

Corollary 1. *The control law defined in Proposition 8 respects the input constraints defined in (4).*

Proof. Since the range of function $-1 \leq s(x) \leq 1, \forall x \in \mathbb{R}$ and $|u_i| = \frac{Z_i}{a_2+Z_i} \leq 1$, the constraints (4) are not violated. □

6 Simulations

To verify the effectiveness of our algorithms, we have set-up a simulation with 5 robots. The robots are represented as circles with an inscribed triangle indicating their current orientation. Holonomic robots were represented as filled disks and non-holonomic robots as disks with an inscribed filled triangle (figure 2)

In the first simulation we used only holonomic robots to demonstrate the effectiveness of the multirobot navigation functions. Shown in figure 6-a, are the initial robot configurations indicated with Ri and their target configurations Ti, with $i \in \{1 \ldots 5\}$. Robots 1 and 2 were initially placed at each others target, whereas robots $3 \ldots 5$ were initially placed at their destination configurations. The rest snapshots of figure 6 show the evolution of the system. Observe how robots $3 \ldots 5$ move away from their targets to allow for robots 1 and 2 to maneuver their way to their targets. Eventually all robots converge to their targets. In figure 7 the control effort for each robot is shown. Since initial and final angles are identical for the holonomic simulation, there is no control effort for the angular velocity. As can be seen from figure 7, the control effort for each actuation direction lies in the predefined velocity bounds indicated by the dotted red lines at ±100% control effort levels.

In the second simulation we used 2 holonomic robots ($R1$, $R2$) and 3 non-holonomic robots $R3 \ldots R5$ to show the effectiveness of dipolar multirobot navigation functions in scenarios with mixed holonomic - non-holonomic robot teams. Shown in figure 8-a, are the initial and final robot configurations indicated as Ri, Ti resp., with $i \in \{1 \ldots 5\}$. Figure 8-b - 8-i depict the robot trajectories and maneuvers to reach their targets. And in this mixed scenario

[4] a.e.: almost everywhere, i.e. everywhere except a set of initial conditions of measure zero that lead the holonomic subsystem to saddle points

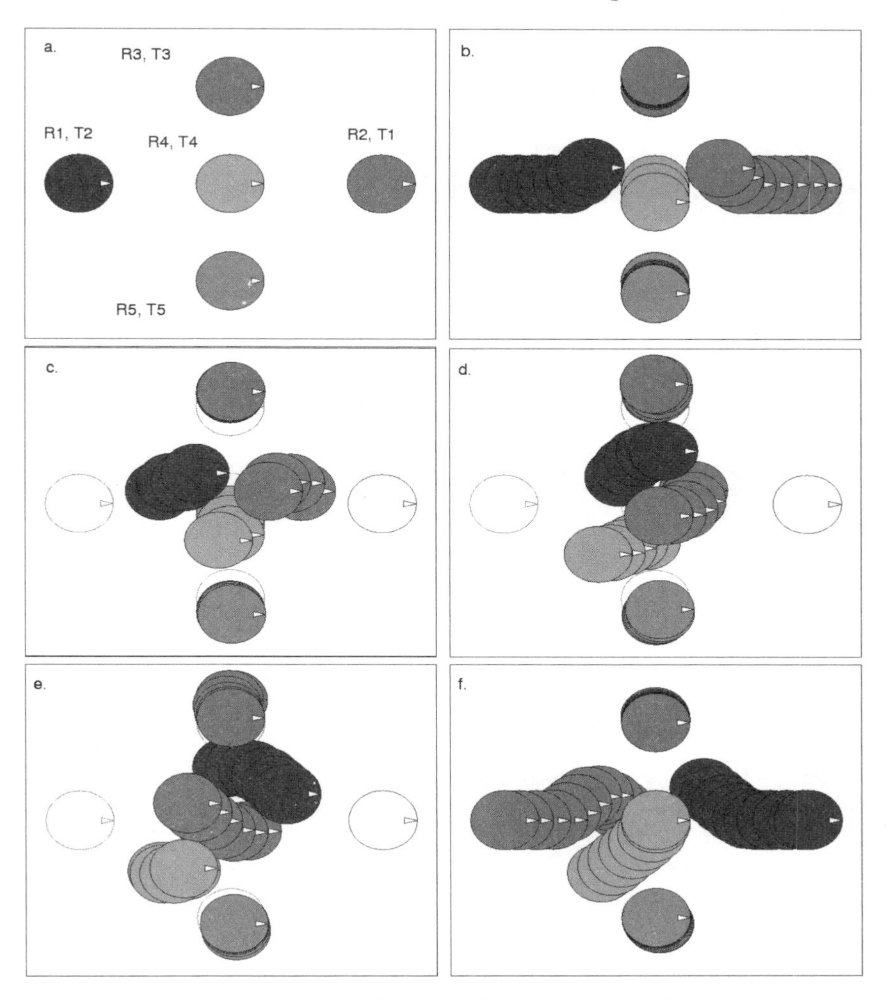

Fig. 6. First simulation with 5 holonomic robots

the multirobot navigation functions augmented with an appropriate dipolar structure succeeds in navigating the mixed robotic team to its destination. Figure 9 depicts the control effort for each robot. While the control signal for the holonomic robots (R1, R2) is absolutely continuous (fig. 9), the control signal for the non-holonomic robots (R3-5) exhibits at some time instants a high frequency switching known as chattering. This is expected due to the discontinuous controllers being used for the non-holonomic subsystem. In [35] it is shown that one can translate a discontinuous kinematic controller to a dynamic one using a non-smooth backstepping controller design technique, maintaining the kinematic controller's convergence properties, and at the same

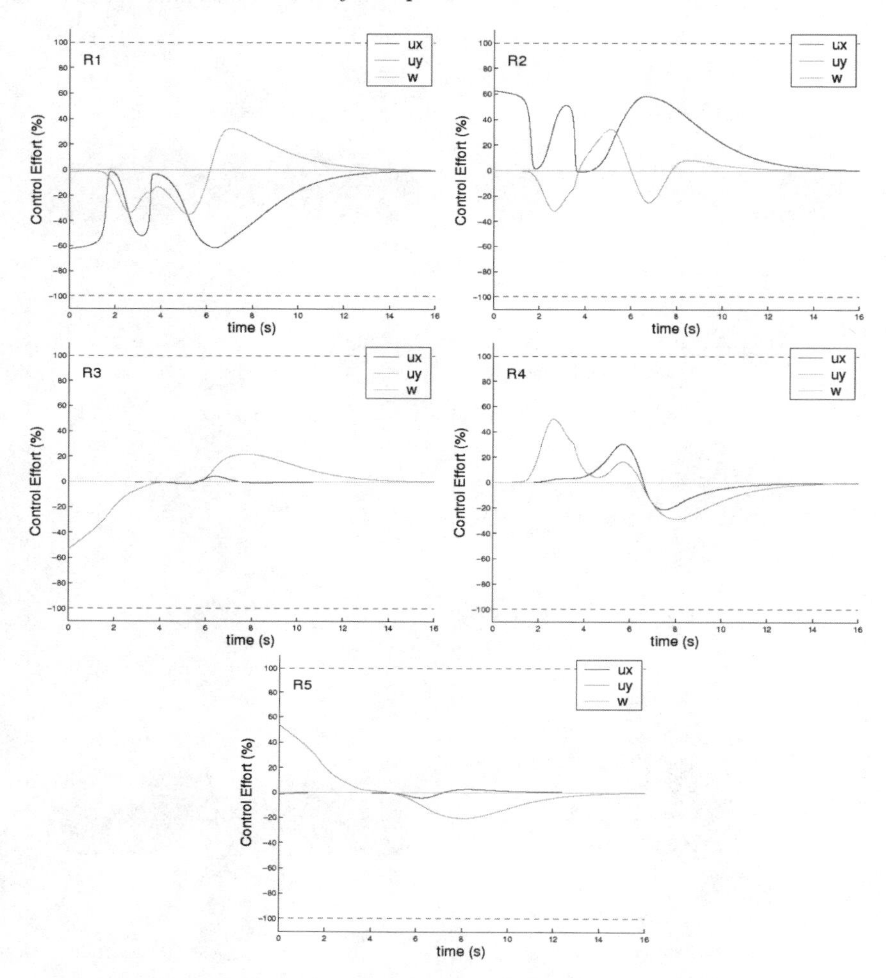

Fig. 7. Control effort for the first simulation for each robot

time smoothing out the chattering behavior through the backstepping integrator which acts as a low pass filter .

7 Conclusion

A new methodology for constructing provably correct multirobot navigation functions was presented in this chapter. The derived methodology can be applied to mixed holonomic - non-holonomic teams when augmented with an appropriate dipolar structure. The proposed controllers provide upper bounded inputs to the system, while maintaining the MRNF's global convergence and

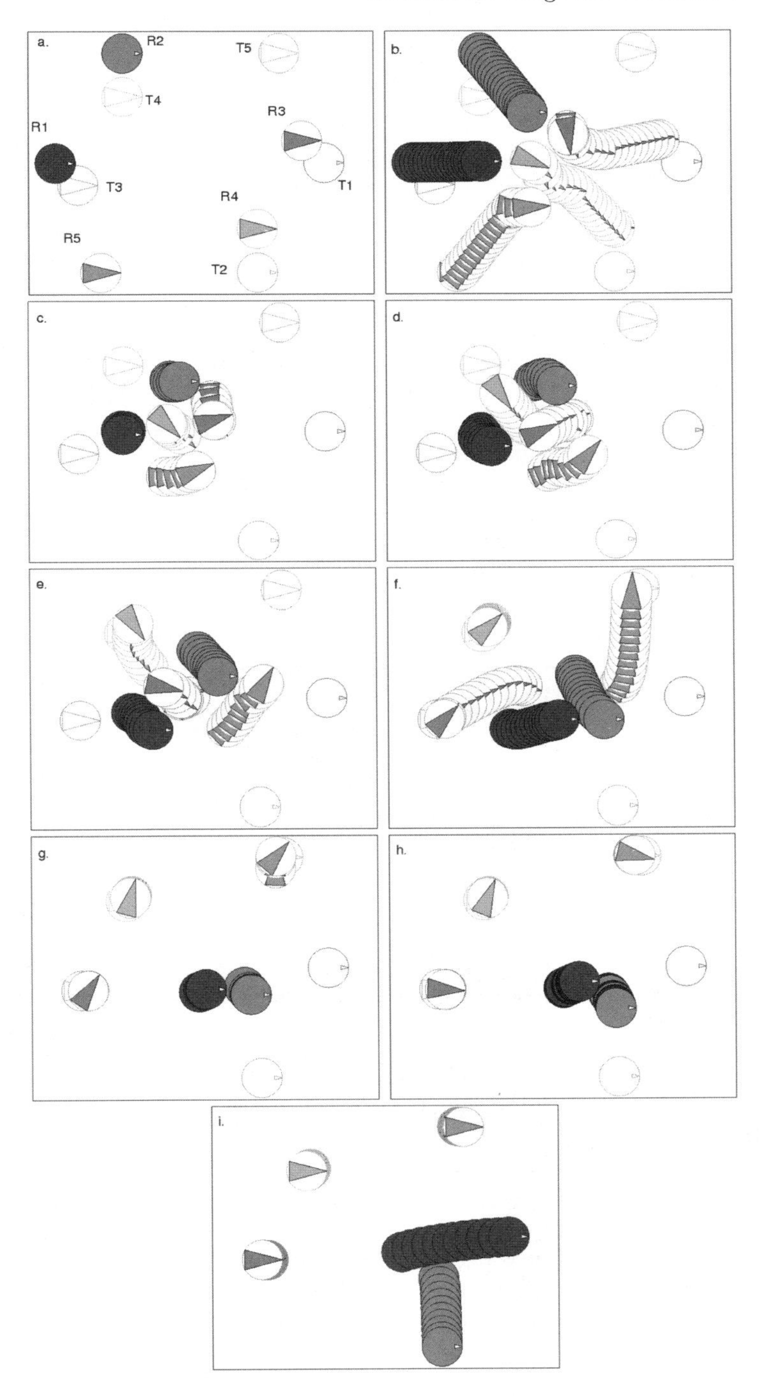

Fig. 8. Second simulation with 2 holonomic and 3 non-holonomic robots

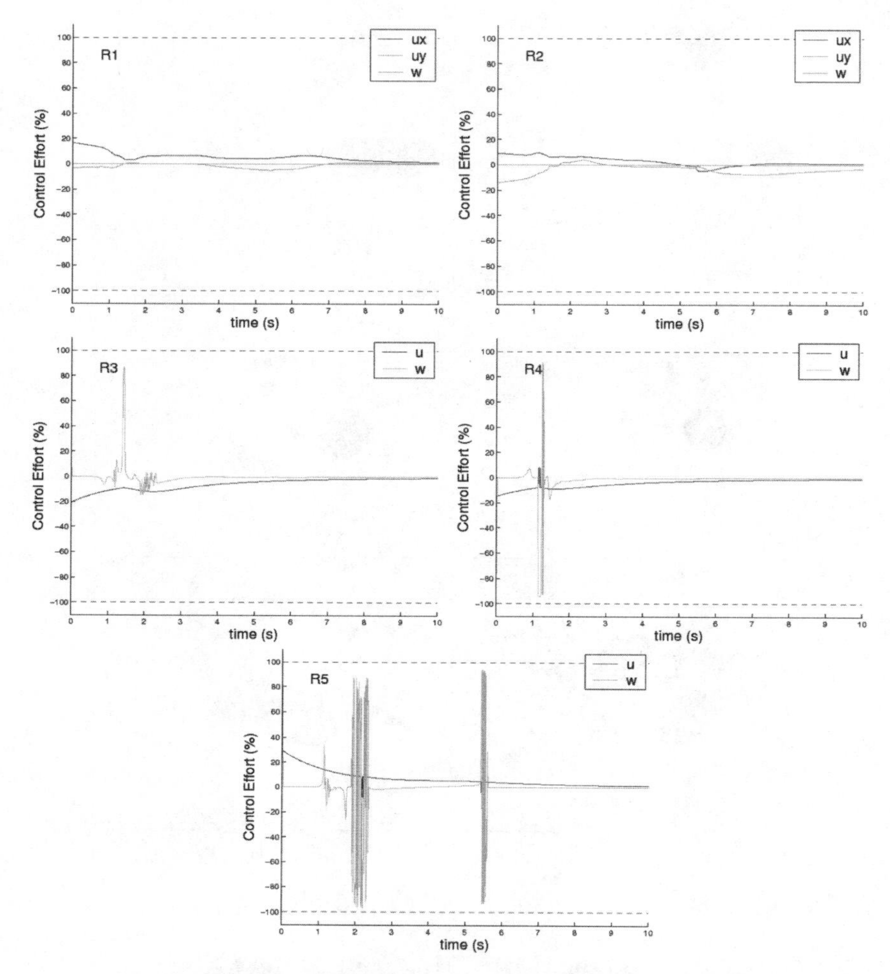

Fig. 9. Control effort for the second simulation for each robot

collision avoidance properties. The methodology due to its closed loop nature provides a robust navigation scheme with guaranteed collision avoidance and its global convergence properties guarantee that a solution will be found if one exists. The closed form control law and the analytic expression of the potential function and the derivatives provide fast feedback making the methodology suitable for real time applications. The methodology can be readily applied to a three dimensional workspace and through proper transformations to arbitrarily shaped robots. The complexity of the methodology, as discussed in section 4.4 increases exponentially wrt the number of robots.

Current research directions are towards reducing the methodology's complexity using a hybrid systems framework and hierarchical application of the methodology to robotic swarms.

In this chapter we discussed the centralized multiagent navigation problem basing our approach on the navigation functions concept. The next chapter extends the multiagent navigation functions concept to the domain of decentralized multiagent navigation.

A Definitions

This section contains several definitions used in this chapter.

$$D_{ij} = \tag{15}$$
$$\begin{bmatrix} & & O_{2(i-1)\times 2m} & & \\ O_{2\times 2(i-1)} & I_{2\times 2} & O_{2\times 2(j-i-1)} & -I_{2\times 2} & O_{2\times 2(m-j)} \\ & & O_{2(j-i-1)\times 2m} & & \\ O_{2\times 2(i-1)} & -I_{2\times 2} & O_{2\times 2(j-i-1)} & I_{2\times 2} & O_{2\times 2(m-j)} \\ & & O_{2(m-j)\times 2m} & & \end{bmatrix}$$

B Proofs

B.1 Proof of Proposition 2

Similar to this found in [11]. From eq. (11), we have:

$$\nabla\varphi\left(\mathbf{q}_d\right) = \frac{\left(\left(\gamma_d^k + G\right)^{1/k}\nabla\gamma_d - \gamma_d\nabla\left(\gamma_d^k + G\right)^{1/k}\right)}{\left(\gamma_d^k + G\right)^{2/k}} = 0$$

since at $\mathbf{q}_d$ both γ_d and $\nabla\gamma_d$ are zero. The Hessian at a critical point is:

$$\nabla^2\varphi = \frac{\left(\left(\gamma_d^k + G\right)^{1/k}\nabla^2\gamma_d - \gamma_d\nabla^2\left(\gamma_d^k + G\right)^{1/k}\right)}{\left(\gamma_d^k + G\right)^{2/k}}$$

but at qd, $\nabla^2\gamma_d = 2I$ and the Hessian reduces to:

$$\left(\nabla^2\varphi\right)\left(\mathbf{q}_d\right) = 2G^{-1/k}I$$

which is non – degenerate. □

B.2 Proof of Proposition 3

Let $\mathbf{q}_0$ be a point on $\partial\mathcal{F}$ and suppose that $\left(g_{R_j}\right)_\kappa(\mathbf{q}_0) = 0$ for the relation j of level k. Then $(g_{R_i})_h(\mathbf{q}_0) > 0$, for any *level* h and $i \neq j$ in level k, because only one *RVF* can hold at a time. Then at $\mathbf{q}_0$:

$$\nabla\varphi(\mathbf{q}_0) = \left(\frac{\left(\gamma_d^k + G\right)^{\frac{1}{k}} \nabla\gamma_d - \gamma_d \nabla\left(\gamma_d^k + G\right)^{\frac{1}{k}}}{\left(\gamma_d^k + G\right)^{\frac{2}{k}}} \right)_{|\mathbf{q}_0} =$$

$$-\frac{1}{k}\gamma_d^{-k} \left(\prod_{L=1}^{n_L} \prod_{\substack{i(L)=1 \\ i(k)\neq j}}^{n_{R,L}} (g_{R_i})_L \right) \cdot \nabla\left(g_{R_j}\right)_k \neq 0$$

□

B.3 Proof of Proposition 4

Similar to this found in [11]. From $\hat{\varphi} = \frac{\gamma_d}{G}$ it follows:

$$\nabla\hat{\varphi} = \frac{1}{G^2}\left(Gk\gamma_d^{k-1}\nabla\gamma_d - \gamma_d^k \nabla G\right)$$

At a critical point it will be: $\gamma_d \nabla G = Gk\nabla\gamma_d$ and taking the magnitude of both sides we get: $2\kappa G = \sqrt{\gamma_d}\,\|\nabla G\|$ since $\|\nabla\gamma_d\| = 2\sqrt{\gamma_d}$. A sufficient condition for the above equality not to hold is:

$$\kappa > \frac{1}{2}\frac{\sqrt{\gamma_d}\,\|\nabla G\|}{G}$$

for all

$$\mathbf{q} \in \mathcal{F}_1(\varepsilon)$$

An upper bound for the right side of the inequality can be derived, provided that the workspace (or configuration space) C is bounded and is given by:

$$\frac{1}{2}\frac{\sqrt{\gamma_d}\|\nabla G\|}{G} <$$

$$\frac{1}{2}\frac{1}{\varepsilon}\max_C\left\{\sqrt{\gamma_d}\right\} \sum_{L=1}^{n_L}\sum_{j=1}^{n_{R,L}} \max_C\left\{\left\|\nabla\left(g_{R_j}\right)_L\right\|\right\} \triangleq N(\varepsilon)$$

since $\left(g_{R_j}\right)_L > \varepsilon, j \in \{1..n_{R,L}\}, L \in \{1..n_L\}$. □

B.4 Proof of Proposition 5

If $\mathbf{q} \in \mathcal{F}_0(\varepsilon) \cap C_{\hat{\varphi}}$, where $C_{\hat{\varphi}}$ is the set of critical points, then $\mathbf{q} \in B_i^L(\varepsilon)$ for at least one set $\{L, i\}$, $i \in \{1..n_{R,L}\}$, $L \in \{1 \ldots n_L\}$ with n_L the number of *levels* and $n_{R,L}$ the number of *relations* in *level* L . We will use a unit vector as a test direction to demonstrate that $\left(\nabla^2 \hat{\varphi}\right)(\mathbf{q})$ has at least one negative eigenvalue. At a critical point,

$$\left(\nabla \hat{\varphi}\right)(\mathbf{q}) = \frac{1}{G^2}\left(k \cdot G \cdot \gamma_d^{k-1} \cdot \nabla \gamma_d - \gamma_d^k \cdot \nabla G\right) = 0$$

Hence

$$\gamma_d \nabla G = G k \nabla \gamma_d \tag{16}$$

The Hessian at a critical point is:

$$\left(\nabla^2 \hat{\varphi}\right)(\mathbf{q}) = \frac{1}{G^2}\left(G \cdot \nabla^2 \gamma_d^k - \gamma_d^k \cdot \nabla^2 G\right)$$

and expanding:

$$\begin{aligned} &\left(\nabla^2 \hat{\varphi}\right)(\mathbf{q}) = \\ &\tfrac{\gamma_d^{k-2}}{G^2}\left(kG\left[\gamma_d \cdot \nabla^2 \gamma_d + (k-1)\nabla \gamma_d \nabla \gamma_d^T\right] - \gamma_d^2 \cdot \nabla^2 G\right) \end{aligned} \tag{17}$$

Taking the outer product of both sides of eq. (16), we get:

$$(Gk)^2 \nabla \gamma_d \nabla \gamma_d^T = \gamma_d^2 \nabla G \cdot \nabla G^T \tag{18}$$

Substituting eq. (18) in eq. (17), we get:

$$\begin{aligned} &\left(\nabla^2 \hat{\varphi}\right)(\mathbf{q}) = \\ &\tfrac{\gamma_d^{k-1}}{G^2}\left(kG \cdot \nabla^2 \gamma_d + \left(1 - \tfrac{1}{k}\right)\tfrac{\gamma_d}{G}\nabla G \cdot \nabla G^T - \gamma_d \cdot \nabla^2 G\right) \end{aligned} \tag{19}$$

We choose the test vector to be: $\hat{\mathbf{u}} = P_{R_i} \cdot \mathbf{q}^{\perp} / \left\| P_{R_i} \cdot \mathbf{q}^{\perp} \right\|$ where P_{R_i} is the *relation matrix* of b_{R_i} and $\left(\mathbf{q}^{\perp}\right)^T = \left[\begin{array}{ccc} \mathbf{q}_1^{\perp} & \ldots & \mathbf{q}_m^{\perp} \end{array}\right]$. With $\nabla^2 \gamma_d = 2I$ we form the quadratic form:

$$\begin{aligned} &\frac{G^2}{\gamma_d^{k-1}} \hat{\mathbf{u}}^{\mathrm{T}} \left(\nabla^2 \hat{\varphi}\right)(\mathbf{q})\, \hat{\mathbf{u}} = \\ &2kG + \left(1 - \tfrac{1}{k}\right)\tfrac{\gamma_d}{G} \hat{\mathbf{u}}^{\mathrm{T}} \cdot \nabla G \cdot \nabla G^T \cdot \hat{\mathbf{u}} - \gamma_d \cdot \hat{\mathbf{u}}^{\mathrm{T}} \cdot \nabla^2 G \cdot \hat{\mathbf{u}} \end{aligned} \tag{20}$$

Taking the inner product of $\mathbf{u}$ and ∇b_{R_i} we have:

$$\left\langle \left(2P_{R_i} \cdot \mathbf{q}\right), \left(P_{R_i} \cdot \mathbf{q}^{\perp}\right) \right\rangle = 2\mathbf{q}^T P_{R_i}^T P_{R_i} \mathbf{q}^{\perp}$$

As is shown in [18], the product $P_{R_i}^T P_{R_i}$, is a linear combination of the matrices $D_{i,j}$, with $\{i, j\} \in \mathbf{P}_{R_i}^{\mathbf{2}}$ where $\mathbf{P^2}$ is the set of relations contained in the product of P matrices. Hence we can write:

$$P_{R_i}^T \cdot P_{R_i} = \sum_{\{i,j\}\in \mathbf{P^2}_{R_i}} a_{ij} D_{ij}$$

with $a_{i,j}$ integer constants (see [18]). So:

$$\mathbf{q}^T \cdot P_{R_i}^T \cdot P_{R_i} \cdot \mathbf{q}^{\perp} = 0$$

Hence $\hat{\mathbf{u}} \perp \nabla b_{R_i}$. In the following analysis we will use the subscript 'i' instead of 'R_i' to simplify notation.

After manipulation of the term $\hat{\mathbf{u}}^{\mathrm{T}} \cdot \nabla G \cdot \nabla G^T \cdot \hat{\mathbf{u}}$ in eq. (20), we get:

$$\left(1 - \frac{1}{k}\right) \frac{\gamma_d}{G} \hat{\mathbf{u}}^{\mathrm{T}} \cdot \nabla G \cdot \nabla G^T \cdot \hat{\mathbf{u}} = g_i \gamma_d \cdot \eta_i \tag{21}$$

where

$$\begin{aligned} \eta_i = \left(1 - \frac{1}{k}\right) &\left(\bar{g}_i^{-1} \hat{\mathbf{u}}^{\mathrm{T}} \cdot \nabla \bar{g}_i \cdot \nabla \bar{g}_i^T \cdot \hat{\mathbf{u}} + \cdots \right. \\ &+ \lambda^2 c_i^{-2} d_i^4 \cdot \bar{g}_i \cdot \hat{\mathbf{u}}^T \cdot \nabla \tilde{b}_i^{1/h} \cdot \nabla \tilde{b}_i^{1/h^T} - \ldots \\ &\left. - 2\lambda c_i^{-1} d_i^2 \hat{\mathbf{u}}^T \cdot \nabla \tilde{b}_i^{1/h} \cdot \nabla \bar{g}_i^T \cdot \hat{\mathbf{u}} \right) \end{aligned}$$

$$G = g_i \cdot \bar{g}_i, \; g_i = c_i \cdot b_i, \; c_i = 1 + \lambda d_i$$

$$\tilde{b}_i = B_{R_i^C}, \; d_i = \frac{1}{b_i + \tilde{b}_i^{1/h}}$$

After manipulation of the term $\hat{\mathbf{u}}^{\mathrm{T}} \cdot \nabla^2 G \cdot \hat{\mathbf{u}}$ (see [18] for details), we get:

$$\hat{\mathbf{u}}^{\mathrm{T}} \cdot \nabla^2 G \cdot \hat{\mathbf{u}} = g_i \cdot \xi_i + v_i \cdot \bar{g}_i \cdot c_i \tag{22}$$

where

$$\xi_i = \hat{\mathbf{u}}^{\mathrm{T}} \cdot \nabla^2 \bar{g}_i \cdot \hat{\mathbf{u}} + \frac{\bar{g}_i}{c_i} \cdot \hat{\mathbf{u}}^{\mathrm{T}} \cdot A_i \cdot \hat{\mathbf{u}} - 2\frac{\lambda}{c_i} d_i^2 \hat{\mathbf{u}}^{\mathrm{T}} \cdot \nabla \tilde{b}_i^{1/h} \cdot \nabla \bar{g}_i \cdot \hat{\mathbf{u}}$$

$$A_i = \lambda \left(2 d_i^3 \mathbf{f}_i \cdot \mathbf{f}_i{}^T - d_i^2 T_i\right), \; \mathbf{f_i} = \nabla b_i + \nabla \tilde{b}_i^{1/h}$$

$$T_i = \nabla^2 b_i + \nabla^2 \tilde{b}_i^{1/h}, \; v_i \geq 2$$

Using equations (21) and (22), eq. (20) becomes:

$$\begin{aligned} &\frac{G^2}{\gamma_d^{k-1}} \hat{\mathbf{u}}^{\mathrm{T}} \left(\nabla^2 \hat{\varphi}\right)(\mathbf{q}) \,\hat{\mathbf{u}} = \\ &(2kG - v_i \cdot \bar{g}_i \cdot \gamma_d \cdot c_i) + g_i \cdot (\gamma_d \cdot \eta_i - \gamma_d \cdot \xi_i) \end{aligned} \tag{23}$$

Taking the inner product of both sides of eq. (16) with $\nabla \gamma_d$ we get:

$$4Gk = \nabla \gamma_d \nabla G = \bar{g}_i \nabla g_i \cdot \nabla \gamma_d + g_i \nabla \bar{g}_i \cdot \nabla \gamma_d \tag{24}$$

Substituting $2Gk$ from eq. (24) in eq. (23) and expanding ∇g_i we get:

$$\frac{G^2}{\gamma_d^{k-1}} \hat{\mathbf{u}}^{\mathrm{T}} \left(\nabla^2 \hat{\varphi}\right)(\mathbf{q})\, \hat{\mathbf{u}} = \bar{g}_i \cdot c_i \left(\tfrac{1}{2}\nabla b_i \cdot \nabla \gamma_d - v_i \cdot \gamma_d\right) + \cdots + g_i \cdot \left(\gamma_d \cdot \eta_i - \gamma_d \cdot \xi_i - \sigma_i + \nabla \bar{g}_i \cdot \nabla \gamma_d\right) \tag{25}$$

where $\sigma_i = \frac{\lambda \bar{g}_i d_i^2}{2c_i} \mathbf{f_i} \cdot \nabla \gamma_d$. Setting $\mu_i = \frac{1}{2}\nabla b_{R_i} \cdot \nabla \gamma_d - v_i \cdot \gamma_d$, eq. (25) becomes:

$$\frac{G^2}{\gamma_d^{k-1}} \hat{\mathbf{u}}^{\mathrm{T}} \left(\nabla^2 \hat{\varphi}\right)(\mathbf{q})\, \hat{\mathbf{u}} = \bar{g}_i c_i \mu_i + g_{R_i} \left(\gamma_d \eta_i - \gamma_d \xi_i - \sigma_i + \nabla \bar{g}_i \nabla \gamma_d\right) \tag{26}$$

The second term of eq. (26) is proportional to g_{R_i} and can be made arbitrarily small by a suitable choice of ε but can still be positive, so the first term should be strictly negative. We will need the following lemma to proceed with our analysis:

Lemma 1.

$$\max_{\mathbf{q} \in \mathcal{F}_0} (\mu_i) = (x + a) \cdot (x - a/(m-1)) \cdot (m-1)/m$$

where $x = \sqrt{\varepsilon + \sum (r_i + r_j)^2}$ *and* $a = \sqrt{\mathbf{q}_d^T P_{R_i} \mathbf{q}_d}$

Proof.

$$\mu_i = \nabla b_{R_i} \cdot \nabla \gamma_d / 2 - v_i \cdot \gamma_d \leq 2 f(q)$$

where

$$f(q) = q^T \cdot P_{R_i} \cdot q - q^T \cdot P_{R_i} \cdot q_d - (q - q_d)^T \cdot (q - q_d)$$

Thus

$$\nabla f(q) = 2 P_{R_i} \cdot q - P_{R_i} \cdot q_d - 2(q - q_d)$$

If q_c is a critical point, then: $\nabla f(q_c) = 0$. Solving for q_c , we get:

$$q_c = 1/2 \cdot (P_{R_i} - I)^{-1} \cdot (P_{R_i} - 2I) \cdot q^d$$

But for the worst-case scenario (This is when the proximity relation is a complete graph)

$$(P_{R_i} - I)^{-1} = 1/(m-1) \cdot P_{R_i} - I$$

with

$$P_{R_i} = m \cdot I - \begin{bmatrix} 1 & \cdots & 1 \end{bmatrix}^T \begin{bmatrix} 1 & \cdots & 1 \end{bmatrix}$$

So

$$q_c = (I - P_{R_i} \cdot 1/(2m-2))\, q^d$$

and

$$f(q_c) = -m/(4m-4) < 0$$

The Hessian of $f(q)$ is:

$$\nabla^2 f(q) = 2(P_{R_i} - I)$$

It can be verified that $P_{R_i} - I$ has eigenvalues:

1. $\lambda = m-1$ of multiplicity $(m-1)D$, where D is the workspace dimension, and
2. $\lambda = -1$ of multiplicity D.

That means that $f(q)$ decreases only along D dimensions about q_c and increases along the $(m-1)D$ remaining (for some appropriate coordinate system), which in turn means that q_c is a saddle. We are interested in finding the maximum value that $f(q)$ may attain under the constraint that $b_{R_i} \leq \varepsilon$. We form the constraint function:

$$g(q) = q^T \cdot P_{R_i} \cdot q - \varepsilon - \sum_{\{l,j\} \in R_i} (r_l + r_j)^2 \leq 0$$

Since g is convex ($\nabla^2 g(q) = 2 \cdot P_{R_i} > 0$) and q_c is a saddle point of f , $f(q)$ will attain its maximum and minimum values over the constraint's boundary, $g(q) = 0$. This can be formulated as a nonlinear optimization problem:

$$\max_{q \in U} (f(q))$$

where

$$f(q) = q^T \cdot P_{R_i} \cdot q - q^T \cdot P_{R_i} \cdot q_d - (q - q_d)^T \cdot (q - q_d)$$

and

$$U = \left\{ q : g(q) = q^T \cdot P_{R_i} \cdot q - \varepsilon - \sum_{\{l,j\} \in R_i} (r_l + r_j)^2 \leq 0 \right\}$$

If

$$q^* = \arg\max_{q \in U} (f(q))$$

then, according to Kuhn Tucker conditions, there exists a $\rho \geq 0$ such that:

$$\nabla f(q^*) - \rho \nabla g(q^*) = 0 \tag{27}$$

$$\rho \cdot g(q^*) = 0 \tag{28}$$

$$g(q^*) \leq 0 \tag{29}$$

$$\rho \geq 0 \tag{30}$$

From eq. (27) we have:

$$2P_{R_i} \cdot q^* - P_{R_i} \cdot q_d - 2(q^* - q_d) - 2\rho \cdot P_{R_i} \cdot q^* = 0$$

Solving for q^*, we get

$$q^* = \frac{1}{2}\left(I + (\rho - 1) \cdot P_{R_i}\right)^{-1}\left(2I - P_{R_i}\right) q_d$$

One can easily verify that:

$$\left(I + (\rho - 1) \cdot P_{R_i}\right)^{-1} = \left(I - P_{R_i} \cdot (\rho - 1)/(1 + (\rho - 1)\, m)\right)$$

and

$$q^* = \frac{1}{2} \cdot \left(2I - P_{R_i} \cdot (2\rho - 1)/(1 + (\rho - 1)\, m)\right) q_d$$

As discussed above, the constraint should be activated, so $\rho > 0$ and from eq. (28) we get:

$$g\left(q^*\right) = 0$$

Solving for ρ we get:

$$\rho_{1,2} = \left(2\,(m - 1) \pm (m - 2)\, a/x\right)/(2m)$$

Both ρ_1 , ρ_2 could be made positive so by substituting in q^* we have:

$${}^{+}q^* = \left(I - P_{R_i}(a + x)/(ma)\right) q_d$$

and

$${}^{-}q^* = \left(I - P_{R_i}(a - x)/(ma)\right) q_d$$

where ${}^{+}q^*$, ${}^{-}q^*$ are the values of q^* for ρ_1, ρ_2 respectively. Examining the terms of $f\left(q\right)$, we have:

1. $q^T P_{R_i} q = x^2$ for both ${}^{+}q^*, {}^{-}q^*$
2. $q^T P_{R_i} q_d = -ax$ for ${}^{+}q^*$
3. $q^T P_{R_i} q_d = ax$ for ${}^{-}q^*$
4. $\left(q - q_d\right)^T\left(q - q_d\right) = \left(a + x\right)^2 \big/ m$ for ${}^{+}q^*$ and
5. $\left(q - q_d\right)^T\left(q - q_d\right) = \left(a - x\right)^2 \big/ m$ for ${}^{-}q^*$

After substituting in $f\left(q\right)$, we get:

$$f\left({}^{+}q^*\right) = x^2 + ax - \left(a + x\right)^2 \big/ m = (x + a)\left(x - a/(m - 1)\right)(m - 1)/m$$

and

$$f\left({}^{-}q^*\right) = x^2 - ax - \left(a - x\right)^2 \big/ m = \left(x + a/(m - 1)\right)(x - a)\,(m - 1)/m$$

Then $f\left({}^{+}q^*\right) < 0$ for

$$-a < x < a/(m - 1)$$

and $f\left({}^{-}q^*\right) < 0$ for

$$-a/(m - 1) < x < a$$

We can observe that $f\left({}^{+}q^{*}\right) = f\left({}^{-}q^{*}\right)$ for $x = 0$ and since we are interested for $x > 0$, it holds that $f\left({}^{+}q^{*}\right) > f\left({}^{-}q^{*}\right)$ since

$$f\left({}^{+}q^{*}\right) - f\left({}^{-}q^{*}\right) = 2a\left(m-2\right)x/m > 0, \forall x > 0, m > 2$$

Therefore, by choosing $f\left({}^{+}q^{*}\right)$ we have the result:

$$\max_{q \in F_0}\left(\mu_i\right) = \left(x+a\right)\cdot\left(x - a/(m-1)\right)\cdot(m-1)/m$$

and the proof of Lemma 1 is complete. ∎

So according to *Lemma 1*, for μ_i to be negative, it is sufficient to make sure that:

$$\varepsilon < \frac{1}{(m-1)^2}\cdot \mathbf{q}_d^T P_{R_H}\mathbf{q}_d - \sum_{\{l,j\}\in R_H}\left(r_l + r_j\right)^2 = \ldots$$
$$\frac{1}{(m-1)^2}\sum_{\{l,j\}\in R_H}\left\|\mathbf{q}_d^l - \mathbf{q}_d^j\right\|^2 - \sum_{\{l,j\}\in R_H}\left(r_l + r_j\right)^2 = \varepsilon_0$$

Another constraint arises from the fact that $\varepsilon > 0$. So for a valid workspace it will be:

$$\sum_{\{l,j\}\in R_H}\left\|\mathbf{q}_d^l - \mathbf{q}_d^j\right\|^2 > (m-1)^2\cdot\sum_{\{l,j\}\in R_H}\left(r_l + r_j\right)^2$$

where R_H is the highest level relation. □

B.5 Proof of Proposition 6

Following the line of thought presented in [11], to prove that $\hat{\varphi}$ is non-degenerate, we need to prove that the quadratic form associated to the orthogonal complement of $\mathcal{N}_q = span\left\{\hat{\mathbf{u}}\right\}$ is positive definite. Since $\nabla b_i \perp \hat{\mathbf{u}}$ we need to prove that $\tilde{\mathbf{u}}^T\left(\nabla^2\varphi\right)\tilde{\mathbf{u}} > 0$, where $\tilde{\mathbf{u}} = \widehat{\nabla b_i}$. At a critical point from eq. (16) we get: $\left(k\cdot G\right)^2\left\|\nabla\gamma_d\right\|^2 = \gamma_d^2\left\|\nabla G\right\|^2$

$$2kG = \frac{\gamma_d}{2kG}\left\|\nabla G\right\|^2$$

Multiplying eq. (19) from both sides with $\tilde{\mathbf{u}}$, we get:

$$\frac{G^2}{\gamma_d^{k-1}}\tilde{\mathbf{u}}^T\left(\nabla^2\hat{\varphi}\right)(\mathbf{q})\,\tilde{\mathbf{u}} =$$
$$2kG + \left(1-\tfrac{1}{k}\right)\tfrac{\gamma_d}{G}\tilde{\mathbf{u}}^T\cdot\nabla G\cdot\nabla G^T\cdot\tilde{\mathbf{u}} - \gamma_d\cdot\tilde{\mathbf{u}}^T\cdot\nabla^2 G\cdot\tilde{\mathbf{u}}$$

$$= L + M + N$$

where after replacing $2kG$:

$$L = \frac{\gamma_d}{2kG} \|\nabla G\|^2$$

$$M = \left(1 - \frac{1}{k}\right) \frac{\gamma_d}{G} \tilde{\mathbf{u}}^T \cdot \nabla G \cdot \nabla G^T \cdot \tilde{\mathbf{u}} \tag{31}$$

$$N = -\gamma_d \cdot \tilde{\mathbf{u}}^T \cdot \nabla^2 G \cdot \tilde{\mathbf{u}}$$

Expanding the term L we get:

$$L = \frac{\gamma_d}{2kG} \left(g_i^2 \|\nabla \bar{g}_i\|^2 + 2G \nabla g_i \cdot \nabla \bar{g}_i + \bar{g}_i^2 \|\nabla g_i\|^2 \right)$$

and denote $L_a = \frac{\gamma_d}{2kG} (2G \nabla g_i \nabla \bar{g}_i)$. Expanding the term M we get:

$$M = \left(1 - \frac{1}{k}\right) \frac{\gamma_d}{G} \cdot \left(g_i^2 (\tilde{\mathbf{u}} \cdot \nabla \bar{g}_i)^2 + \bar{g}_i^2 (\tilde{\mathbf{u}} \cdot \nabla g_i)^2 \right) +$$

$$+2 \frac{\gamma_d}{G} G (\tilde{\mathbf{u}} \cdot \nabla g_i) \cdot (\nabla \bar{g}_i \cdot \tilde{\mathbf{u}}) - 2 \frac{1}{k} \frac{\gamma_d}{G} G (\tilde{\mathbf{u}} \cdot \nabla g_i) \cdot (\nabla \bar{g}_i \cdot \tilde{\mathbf{u}})$$

and denote $M_a = 2\frac{\gamma_d}{G} G (\tilde{\mathbf{u}} \cdot \nabla g_i) \cdot (\nabla \bar{g}_i \cdot \tilde{\mathbf{u}})$ and $M_b = -2\frac{1}{k}\frac{\gamma_d}{G} G (\tilde{\mathbf{u}} \cdot \nabla g_i) \cdot (\nabla \bar{g}_i \cdot \tilde{\mathbf{u}})$. Let $M_1 = \tilde{\mathbf{u}} \cdot \nabla g_i$. Expanding M_1 we get:

$$M_1 = \|\nabla b_i\| + \lambda d_i^2 \widehat{\nabla b_i} \cdot \left(\tilde{b}_i^{1/h} \nabla b_i - b_i \nabla \tilde{b}_i^{1/h} \right)$$

For term: $\widehat{\nabla b_i} \cdot \left(\tilde{b}_i^{1/h} \nabla b_i - b_i \nabla \tilde{b}_i^{1/h} \right)$ we have:

$$\widehat{\nabla b_i} \cdot \left(\tilde{b}_i^{1/h} \nabla b_i - b_i \nabla \tilde{b}_i^{1/h} \right) \geq \tilde{b}_i^{1/h} \|\nabla b_i\| - b_i \left\| \nabla \tilde{b}_i^{1/h} \right\|$$

and since $\|\nabla b_i\| = 2\sqrt{b_i + \sum_{\{l,j\} \in R_i} (r_l + r_j)^2}$ we have:

$$\widehat{\nabla b_i} \cdot \left(\tilde{b}_i^{1/h} \nabla b_i - b_i \nabla \tilde{b}_i^{1/h} \right) \geq$$

$$\tilde{b}_i^{1/h} \left(2\sqrt{\sum_{\{l,j\} \in R_i} (r_l + r_j)^2} - \varepsilon \left\| \frac{\nabla \tilde{b}_i^{1/h}}{\tilde{b}_i^{1/h}} \right\| \right)$$

but after some manipulation we have that $\left\| \frac{\nabla \tilde{b}_i^{1/h}}{\tilde{b}_i^{1/h}} \right\| < \frac{1}{h \cdot \varepsilon} \sum_{\mu \in R_i^C} \|\nabla b_\mu\|$ so

$$\widehat{\nabla b_i} \cdot \left(\tilde{b}_i^{1/h} \nabla b_i - b_i \nabla \tilde{b}_i^{1/h} \right) >$$

$$\tilde{b}_i^{1/h} \left(2\sqrt{\sum_{\{l,j\} \in R_i} (r_l + r_j)^2} - \frac{1}{h} \sum_{\mu \in R_i^C} \|\nabla b_\mu\| \right)$$

For this to be positive, it must be:

$$h > \frac{1}{2} \cdot \frac{\max\left(\sum\limits_{\mu \in R_i^C} \|\nabla b_\mu\|\right)}{\sqrt{\sum\limits_{\{l,j\} \in R_i} (r_l + r_j)^2}} = h_1 \tag{32}$$

So choosing h according to (32) we have that: $M_1 = \widehat{\nabla b_i} \cdot \nabla g_i \geq \|\nabla b_i\|$ and of course:

$$\|\nabla g_i\| \geq \widehat{\nabla b_i} \cdot \nabla g_i \geq \|\nabla b_i\| \tag{33}$$

Examining the term:

$$\begin{aligned} &L_a + M_b = \\ &\quad = \tfrac{\gamma_d}{k}\left(\nabla g_i \nabla \bar{g}_i - 2\left(\tilde{\mathbf{u}} \cdot \nabla g_i\right) \cdot \left(\nabla \bar{g}_i \cdot \tilde{\mathbf{u}}\right)\right) \end{aligned}$$

but after manipulation

$$\begin{aligned} &\nabla g_i \nabla \bar{g}_i - 2\left(\tilde{\mathbf{u}} \cdot \nabla g_i\right) \cdot \left(\nabla \bar{g}_i \cdot \tilde{\mathbf{u}}\right) \geq \\ &- \|\nabla \bar{g}_i\| \left\|\left(2\left(\tilde{\mathbf{u}} \cdot \nabla g_i\right) \cdot \tilde{\mathbf{u}} - \nabla g_i\right)\right\| \end{aligned}$$

noticing that $\left\|\left(2\left(\tilde{\mathbf{u}} \cdot \nabla g_i\right) \cdot \tilde{\mathbf{u}} - \nabla g_i\right)\right\| = \|\nabla g_i\|$, we get

$$\nabla g_i \nabla \bar{g}_i - 2\left(\tilde{\mathbf{u}} \cdot \nabla g_i\right) \cdot \left(\nabla \bar{g}_i \cdot \tilde{\mathbf{u}}\right) \geq - \|\nabla \bar{g}_i\| \|\nabla g_i\|$$

Hence

$$L_a + M_b \geq -\frac{\gamma_d}{k} \|\nabla \bar{g}_i\| \|\nabla g_i\|$$

Hence examining the term $L + M_b$ we have:

$$L + M_b \geq \frac{\gamma_d}{2kG}\left(g_i \|\nabla \bar{g}_i\| - \bar{g}_i \|\nabla g_i\|\right)^2 \geq 0$$

which is non-negative and can be neglected.

For the term N: Expanding the $\tilde{\mathbf{u}}^T \cdot \nabla^2 G \cdot \tilde{\mathbf{u}}$ part we get:

$$\begin{aligned} &\tilde{\mathbf{u}}^T \cdot \nabla^2 G \cdot \tilde{\mathbf{u}} = \\ &\quad \tilde{\mathbf{u}}^T \cdot \left(g_i \nabla^2 \bar{g}_i + \bar{g}_i \nabla^2 g_i\right) \cdot \tilde{\mathbf{u}} + 2\left(\tilde{\mathbf{u}}^T \cdot \nabla \bar{g}_i\right) \cdot \left(\tilde{\mathbf{u}}^T \cdot \nabla g_i\right) \end{aligned}$$

Notice that the second term is canceled with M_a. Using equation (33), we can write (since $k > 1$):

$$\begin{aligned} &\frac{G^2}{\gamma_d^{k-1}} \tilde{\mathbf{u}}^T \left(\nabla^2 \hat{\varphi}\right)(\mathbf{q})\, \tilde{\mathbf{u}} \geq \\ &\quad \tfrac{\gamma_d}{g_i}\left(\left(1 - \tfrac{1}{k}\right) \bar{g}_i \|\nabla b_i\|^2 - g_i^2 \tilde{\mathbf{u}}^T \nabla^2 \bar{g}_i \tilde{\mathbf{u}} - g_i \bar{g}_i \tilde{\mathbf{u}}^T \nabla^2 g_i \tilde{\mathbf{u}}\right) \end{aligned}$$

We will now proceed by examining the term: $\tilde{\mathbf{u}}^T \cdot \nabla^2 g_i \cdot \tilde{\mathbf{u}}$. After expanding it, we get:

$$\tilde{\mathbf{u}}^T \cdot \nabla^2 g_i \cdot \tilde{\mathbf{u}} = c_i \cdot \tilde{\mathbf{u}}^T \cdot \nabla^2 b_i \cdot \tilde{\mathbf{u}} - \cdots$$
$$-s_i \cdot \tilde{\mathbf{u}}^T \cdot \nabla b_i \cdot \left(\nabla b_i + \nabla \tilde{b}_i^{1/h}\right)^T \cdot \tilde{\mathbf{u}} + b_i \cdot \tilde{\mathbf{u}}^T \cdot A_i \cdot \tilde{\mathbf{u}}$$

where $s_i = \frac{2\lambda}{b_i + \tilde{b}_i^{1/h}\ {}^2}$. After manipulation of the term $\tilde{\mathbf{u}}^T \cdot \nabla b_i \cdot \left(\nabla b_i + \nabla \tilde{b}_i^{1/h}\right)^T \cdot \tilde{\mathbf{u}}$ we get:

$$\tilde{\mathbf{u}}^T \cdot \nabla b_i \cdot \left(\nabla b_i + \nabla \tilde{b}_i^{1/h}\right)^T \cdot \tilde{\mathbf{u}} \geq \|\nabla b_i\|^2 - \|\nabla b_i\| \cdot \left\|\nabla \tilde{b}_i^{1/h}\right\|$$

Examining the term:

$$\|\nabla b_i\| - \left\|\nabla \tilde{b}_i^{1/h}\right\| =$$
$$2\sqrt{b_i + \sum_{\{l,j\} \in R_i} (r_l + r_j)^2} - \frac{\tilde{b}_i^{1/h}}{h \cdot \varepsilon} \sum_{\mu \in R_i^C} \|\nabla b_\mu\| \tag{34}$$

Requiring (34) to be positive, we need:

$$h \geq \frac{\max\left\{1,\ \tilde{b}_i\right\} \cdot \max\left\{\sum_{\mu \in R_i^C} \|\nabla b_\mu\|\right\}}{2 \cdot \varepsilon \cdot \min\left\{\sqrt{\sum_{\{l,j\} \in R_i} (r_l + r_j)^2}\right\}} = h_2(\varepsilon)$$

Hence the term $s_i \cdot \tilde{\mathbf{u}}^T \cdot \nabla b_i \cdot \left(\nabla b_i + \nabla \tilde{b}_i^{1/h}\right)^T \cdot \tilde{\mathbf{u}} > 0$ and can be neglected. From the expansion of the term $b_i \cdot \tilde{\mathbf{u}}^T \cdot A_i \cdot \tilde{\mathbf{u}}$ let us consider the following term:

$$\widehat{\nabla b_i}^T \mathbf{f}_i \mathbf{f}_i^T \widehat{\nabla b_i} = \left(\|\nabla b_i\| + \widehat{\nabla b_i} \cdot \nabla \tilde{b}_i^{1/h}\right)^2 < 4\|\nabla b_i\|^2$$

because of (34). For $\left(b_i + \tilde{b}_i^{1/h}\right)^3$, with $\varepsilon < 1$ we have:

$$\left(b_i + \tilde{b}_i^{1/h}\right)^3 > \tilde{b}_i^{3/h} > \varepsilon^{3n_R/h}$$

where $n_R + 1$ is the number of relations in the level with maximum relations. With

$$h > 3n_R = h_3$$

we have:

$$\left(b_i + \tilde{b}_i^{1/h}\right)^3 > \varepsilon$$

Hence:

$$\tilde{\mathbf{u}}^T \cdot A_i \cdot \tilde{\mathbf{u}} < \frac{8\lambda}{\varepsilon}\|\nabla b_i\|^2 - \frac{s_i}{2}\tilde{\mathbf{u}}^T \nabla^2 b_i \tilde{\mathbf{u}} - \frac{s_i}{2}\tilde{\mathbf{u}}^T \nabla^2 \tilde{b}_i^{1/h} \tilde{\mathbf{u}}$$

and

$$\begin{gathered}\tilde{\mathbf{u}}^T \cdot \nabla^2 g_i \cdot \tilde{\mathbf{u}} < c_i \cdot \tilde{\mathbf{u}}^T \cdot \nabla^2 b_i \cdot \tilde{\mathbf{u}} + \cdots \\ +8\lambda \left\| \nabla b_i \right\|^2 - b_i \tfrac{s_i}{2} \tilde{\mathbf{u}}^T \nabla^2 b_i \tilde{\mathbf{u}} - b_i \tfrac{s_i}{2} \tilde{\mathbf{u}}^T \nabla^2 \tilde{b}_i^{1/h} \tilde{\mathbf{u}}\end{gathered}$$

Hence

$$\begin{gathered}\tfrac{G^2}{\gamma_d^{k-1}} \tilde{\mathbf{u}}^T \left(\nabla^2 \hat{\varphi} \right) (\mathbf{q}) \, \tilde{\mathbf{u}} \geq \\ \tfrac{\gamma_d}{g_i} \Big(\left(1 - \tfrac{1}{k} \right) \bar{g}_i \left\| \nabla b_i \right\|^2 - g_i^2 \tilde{\mathbf{u}}^T \nabla^2 \bar{g}_i \tilde{\mathbf{u}} - \cdots \\ -g_i \bar{g}_i \left(c_i \cdot \tilde{\mathbf{u}}^T \cdot \nabla^2 b_i \cdot \tilde{\mathbf{u}} + \cdots \right. \\ \left. +8\lambda \left\| \nabla b_i \right\|^2 - b_i \tfrac{s_i}{2} \tilde{\mathbf{u}}^T \nabla^2 b_i \tilde{\mathbf{u}} - b_i \tfrac{s_i}{2} \tilde{\mathbf{u}}^T \nabla^2 \tilde{b}_i^{1/h} \tilde{\mathbf{u}} \right) \Big)\end{gathered}$$

From the term $\tilde{\mathbf{u}}^T \cdot \nabla^2 b_i \cdot \tilde{\mathbf{u}}$, following a similar analysis with the one used in the proof of Proposition 5 (see [18]), we get:

$$2 < \tilde{\mathbf{u}}^T \nabla^2 b_i \tilde{\mathbf{u}} < 2 + \frac{(m-2)\left(m^2-1\right)}{2m} \cdot \frac{r_{\max}}{r_{\min}}$$

Hence:

$$\begin{gathered}\tfrac{G^2}{\gamma_d^{k-1}} \tilde{\mathbf{u}}^T \left(\nabla^2 \hat{\varphi} \right) (\mathbf{q}) \, \tilde{\mathbf{u}} \geq \\ \geq \tfrac{\gamma_d}{g_i} \Big(\left(1 - \tfrac{1}{k} \right) \bar{g}_i \left\| \nabla b_i \right\|^2 - g_i^2 \tilde{\mathbf{u}}^T \nabla^2 \bar{g}_i \tilde{\mathbf{u}} - \\ -g_i \bar{g}_i \left(c_i \left(2 + \tfrac{(m-2)\left(m^2-1\right)}{2m} \cdot \tfrac{r_{\max}}{r_{\min}} \right) + 8\lambda \left\| \nabla b_i \right\|^2 - \cdots \right. \\ \left. -b_i \tfrac{s_i}{2} \tilde{\mathbf{u}}^T \nabla^2 \tilde{b}_i^{1/h} \tilde{\mathbf{u}} \right) \Big)\end{gathered}$$

For c_i with $h > 3n_R$, we have:

$$c_i = 1 + \frac{\lambda}{b_i + \tilde{b}_i^{1/h}} \leq 1 + \frac{\lambda}{\tilde{b}_i^{1/h}} < 1 + \frac{\lambda}{\varepsilon^{n_R/h}} \leq 1 + \frac{\lambda}{\varepsilon^{1/3}}$$

Hence after a term rearrangement we get:

$$\begin{gathered}\tfrac{G^2}{\gamma_d^{k-1}} \tilde{\mathbf{u}}^T \left(\nabla^2 \hat{\varphi} \right) (\mathbf{q}) \, \tilde{\mathbf{u}} \geq \\ \tfrac{\gamma_d}{g_i} \left(\bar{g}_i \sum_{j=1}^{3} \left(\tfrac{1}{6} \left(1 - \tfrac{1}{k} \right) \left\| \nabla b_i \right\|^2 - \varepsilon \cdot K_j \right) + \cdots \right. \\ \left. + \left(\tfrac{1}{2} \left(1 - \tfrac{1}{k} \right) \left\| \nabla b_i \right\|^2 - \varepsilon^2 \tilde{\mathbf{u}}^T \nabla^2 \bar{g}_i \tilde{\mathbf{u}} \right) \right)\end{gathered} \tag{35}$$

where

$$K_1 = \frac{1+\lambda}{\varepsilon^{\frac{1}{3}}} \left(2 + \frac{(m-2)\left(m^2-1\right)}{2m} \cdot \frac{r_{\max}}{r_{\min}} \right)$$

$$K_2 = 8\lambda \left\| \nabla b_i \right\|^2$$

$$K_3 = \lambda \varepsilon^{\frac{1}{3}} \left| \tilde{\mathbf{u}}^T \nabla^2 \tilde{b}_i^{1/h} \tilde{\mathbf{u}} \right|$$

Assuming that $k > 2$, and noting that $\min \left(\left\| \nabla b_i \right\|^2 \right) = 4 \sum_{\{l,j\} \in R_i} (r_l + r_j)^2$, then for both the right hand side terms of ineq. (35) to be positive, the sufficient conditions are:

$$\varepsilon < \left(\frac{2 \sum_{\{l,j\} \in R_i} (r_l + r_j)^2}{3(1+\lambda)\left(2 + \frac{(m-2)(m^2-1)}{2m} \cdot \frac{r_{\max}}{r_{\min}}\right)} \right)^{\frac{3}{2}} = \varepsilon_2$$

$$\varepsilon < \frac{1}{48\lambda} = \varepsilon_3$$

$$\varepsilon < \left(\frac{2 \sum_{\{l,j\} \in R_i} (r_l + r_j)^2}{3\lambda \cdot \max\left(\left| \tilde{\mathbf{u}}^T \nabla^2 \tilde{b}^{1/h} \tilde{\mathbf{u}} \right|\right)} \right)^{\frac{3}{2}} = \varepsilon_4$$

So to render (35) positive it is sufficient to choose an $\varepsilon < \varepsilon_1 = \min\{1, \varepsilon_2, \varepsilon_3, \varepsilon_4\}$ and an $h > h_0 = \max\{h_1, h_2(\varepsilon), h_3\}$. □

B.6 Proof of Proposition 7

Using φ as a Lyapunov function candidate, since $\varphi(q_d) = 0$ and $\varphi(q) > 0$, $\forall q \in F \setminus \{q_d\}$ by definition, and taking the time derivative of φ, we get:

$$\dot{\varphi} = \dot{q} \cdot \nabla \varphi(q) = -K \cdot \nabla \varphi(q) \cdot \nabla \varphi(q) = -K \cdot \left\| \nabla \varphi(q) \right\|^2 \leq 0$$

where the equality holds at the set of critical points $\mathcal{C} = \{q : \nabla \varphi(q) = \mathbf{0}\}$. By the definition of φ the set of critical points contains only one minimum, which is the target configuration q_d. The rest of critical points can be either maxima or saddles of φ. Obviously, a maximum point is the positive limit set of no initial condition other than itself. The 3rd property indicates that φ is a Morse function, hence its critical points are isolated [25]. Thus the set of initial conditions that lead to saddle points are sets of measure zero [11]. □

B.7 Proof of Proposition 8

Since the control scheme we are considering is discontinuous, the right hand side of (3) is discontinuous hence we need to consider the Filippov sets created over the switching regions. To this extend we will need the following results from non-smooth analysis:

Definition 3. *([8]) A vector function x is called a solution of $\dot{x} = f(x)$ if x is absolutely continuous and $\dot{x} \in K[f](x)$ where $K[f](x) \triangleq \overline{co}\{\lim f(x_i) | x_i \to x,\ x_i \notin N\}$, where N is a set of measure zero.*

Theorem 3. *[33] Let $x(\cdot)$ be a Filippov solution to $\dot{x} = f(x)$ and $V : \mathbb{R}^m \to \mathbb{R}$ be a Lipschitz and regular function. Then $V(x)$ is absolutely continuous, $\frac{d}{dt}V(x)$ exists almost everywhere and*

$$\frac{d}{dt}V(x) \in^{a.e.} \dot{\tilde{V}}$$

where $\dot{\tilde{V}} \triangleq \bigcap_{\xi \in \partial V(x)} \xi^T \cdot K[f](x)$ and ∂V is the Clarke's generalized gradient [4].

The following theorem is an extension to LaSalle's invariance principle for non-smooth systems:

Theorem 4. *[33] Let Ω be a compact set such that every Filippov solution to the autonomous system $\dot{x} = f(x)$, $x(0) = x(t_0)$ starting in Ω is unique and remains in in Ω for all $t > t_0$. Let $V : \Omega \to \mathbb{R}$ be a time independent regular function such that $v \leq 0$ for all $v \in \dot{\tilde{V}}$. (If $\dot{\tilde{V}}$ is the empty set then this is trivially satisfied). Define $S = \left\{x \in \Omega | 0 \in \dot{\tilde{V}}\right\}$. Then every trajectory in Ω converges to the largest invariant set, M in the closure of S.*

Function V is a regular function, since it's smooth. To reason about its time derivative, from Theorem 3, we need to examine:

$$\dot{\tilde{V}} = \bigcap_{\xi \in \partial V(x)} \xi^T \cdot \begin{bmatrix} M & O \\ O & C \end{bmatrix} \cdot K \begin{bmatrix} \mathbf{u_h} \\ \mathbf{u_{nh}} \end{bmatrix}$$

and since V is smooth,

$$\dot{\tilde{V}} = \nabla V^T \cdot \begin{bmatrix} M & O \\ O & C \end{bmatrix} \cdot K \begin{bmatrix} \mathbf{u_h} \\ \mathbf{u_{nh}} \end{bmatrix}$$

Substituting the control law from Proposition 8 we get:

$$\dot{\tilde{V}} \subset -v_h - v_{nh_u} + v_{nh_w} \tag{36}$$

where $v_h = \delta_h$, $v_{nh_u} = \sum_{i=n+1}^{m} \left[K\left[sgn\left(\nabla_{x_i,y_i} V \cdot \eta_i\right)\right] \cdot \left(\nabla_{x_i,y_i} V \cdot \eta_i\right) \cdot \frac{v_i^{max} \cdot Z_i}{a_2 + Z_i}\right] = \sum_{i=n+1}^{m} \left[\left|\nabla_{x_i,y_i} V \cdot \eta_i\right| \cdot \frac{v_i^{max} \cdot Z_i}{a_2 + Z_i}\right] = \delta_{V_q}$, where $\nabla_{x_i,y_i} V = \left[V_{x_i}, V_{y_i}\right]^T$ and $\eta_i = \left[\cos(\theta_i), \sin(\theta_i)\right]^T$. For v_{nh_w} we have $v_{nh_w} = K\left[\sum_{i \in \mathcal{M}} \left[w_i \cdot V_{\theta_i} \cdot w_i^{max}\right]\right]$. Then

(36) becomes: $\dot{\tilde{V}} \subset -v_h - v_{nh_u} - v_{nh_w} = -\delta_h - \delta_{V_q} + K\left[\sum_{i \in \mathcal{M}} \left[w_i \cdot V_{\theta_i} \cdot w_i^{max}\right]\right] =$ $-\delta_h - \delta_{V_q} + K\left[\delta_{\theta_{nh}}(\rho)\right] \subseteq [\Delta_\rho, 0]$ since the switchings occur between negative values of $\Delta(\cdot)$ away of the target, while at the target $\rho = 2^z$ and $\Delta_\rho = 0$. The eventual set is closed due to the closure of operator $K[\cdot]$.

Now let $E = \{\mathbf{x} : \dot{V}(\mathbf{x}) = 0\}$ and $E \supset S = \{\mathbf{p} : u_{x_t} = u_{y_t} = \omega_t = \omega_i = u_i = 0, \forall t \in \{1 \dots n\}, \forall i \in \mathcal{M}\}$ is an invariant set. From the proposed control law, it can be seen that $u_i = 0, \forall i \in \mathcal{M}$ only at the destination, and for all other configurations the controller provides a direction of movement and $\left\|u_{x_t}^2 + u_{y_t}^2\right\| > 0$ a.e. and vanishes at the origin. The set of initial conditions that lead the holonomic subsystem to saddle points is guaranteed to be of measure zero due to the Morse property (Proposition 6) of MRNFs. According to LaSalle's invariance principle for non-smooth systems (Theorem 4), the trajectories of the system converge asymptotically to the largest invariant set, which is the destination configuration □

References

1. D. Bertsekas. *Nonlinear Programming.* Athena Scientific, 1995.
2. R. W. Brockett. Control theory and singular riemannian geometry. In *New Directions in Applied Mathematics*, pages 11–27. Springer, 1981.
3. S. M. Rock C. M. Clark and J.-C. Latombe. Motion planning for multiple mobile robots using dynamic networks. *Proceedings of the IEEE International Conference on Robotics and Automation*, pages 4222–4227, 2003.
4. F. Clarke. *Optimization and Nonsmooth Analysis.* Addison - Wesley, 1983.
5. R. Murray D. Tilbury and S. Sastry. Trajectory generation for the n-trailer problem using goursat normal forms. *32rd IEEE Conference on Decision and Control*, pages 971–977, 1993.
6. J. P. Desai, J. Ostrowski, and V. Kumar. Controlling formations of multiple mobile robots. *Proc. of IEEE Int. Conf. on Robotics and Automation*, pages 2864–2869, 1998.
7. Jaydev P. Desai, James P. Ostrowski, and Vijay Kumar. Modeling and control of formations of non-holonomic mobile robots. *IEEE Transaction on Robotics and Automation*, 17(6):905–908, 2001.
8. A. Filippov. *Differential equations with discontinuous right-hand sides.* Kluwer Academic Publishers, 1988.
9. J. Hu and S. Sastry. Optimal collision avoidance and formation switching on riemannian manifolds. *IEEE Conf. on Decision and Control*, pages 1071–1076, 2001.
10. D. E. Koditschek. Robot planning and control via potential functions. In *The Robotics Review*, pages 349–368. MIT Press, 1989.
11. D. E. Koditschek and E. Rimon. Robot navigation functions on manifolds with boundary. *Advances Appl. Math.*, 11:412–442, 1990.
12. G. Laferriere and E. Sontag. Remarks on control lyapunov functions for discontinuous stabilizing feedback. *Proceedings of the 32nd IEEE Conference on Decision and Control*, pages 2398–2403, 1993.

13. G. Lafferrierre and H. Sussmann. Motion planning for controlable systems without drift. *Proceedings of the 1991 IEEE International Conference on Robotics and Automation*, 1991.
14. G. Lafferrierre and H. Sussmann. A differential geometric approach to motion planning. In *Nonholonomic Motion Planning, Z. Li and J. Canny, Eds.*, pages 235–270. Kluwer Academic Publishers, 1993.
15. J. C. Latombe. *Robot Motion Planning.* Kluwer Academic Publishers, 1991.
16. Y.H. Liu et al. A practical algorithm for planning collision free coordinated motion of multiple mobile robots. *Proc of IEEE Int. Conf. on Robotics and Automation*, pages 1427–1432, 1989.
17. S. G. Loizou and K. J. Kyriakopoulos. Closed loop navigation for multiple holonomic vehicles. *Proc. of IEEE/RSJ Int. Conf. on Intelligent Robots and Systems*, pages 2861–2866, 2002.
18. S. G. Loizou and K. J. Kyriakopoulos. Closed loop navigation for multiple holonomic vehicles. Tech. report, NTUA, http://users.ntua.gr/sloizou/academics/-TechReports/TR0102.pdf, 2002.
19. S. G. Loizou and K. J. Kyriakopoulos. Closed loop navigation for multiple non-holonomic vehicles. Tech. report, NTUA, http://users.ntua.gr/sloizou/-academics/TechReports/TR0202.pdf, 2002.
20. S.G. Loizou and K.J. Kyriakopoulos. Closed loop navigation for multiple non-holonomic vehicles. *IEEE Int. Conf. on Robotics and Automation*, pages 420–425, 2003.
21. V. J. Lumelsky and K. R. Harinarayan. Decentralized motion planning for multiple mobile robots: The cocktail party model. *Journal of Autonomous Robots*, 4:121–135, 1997.
22. P. Martin M. Fliess, J. Lévine and P. Rouchon. On differentially flat non-linear systems. *Proceedings of the 3rd IFAC Symposium on Nonlinear Control System Design*, pages 408–412, 1992.
23. P. Martin M. Fliess, J. Lévine and P. Rouchon. Flatness and defect of non-linear systems: Introductory theory and examples. *International Journal of Control*, 61(6):1327–1361, 1995.
24. M.Egerstedt and X. Hu. Formation constrained multi-agent control. *IEEE Trans. on Robotics and Automation*, 17(6):947–951, 2001.
25. J. Milnor. *Morse theory.* Annals of Mathematics Studies. Princeton University Press, Princeton, NJ, 1963.
26. R. Murray. Applications and extensions of goursat normal form o control f nonlinear systems. *32rd IEEE Conference on Decision and Control*, pages 3425–3430, 1993.
27. R. Murray and S. Sastry. Nonholonomic motion planning: Steering using sinusoids. *IEEE Transactions on Automatic Control*, pages 700–716, 1993.
28. P. Ogren and N. Leonard. Obstacle avoidance in formation. *EEE Int. Conf. on Robotics and Automation*, pages 2492–2497, 2003.
29. G. J. Pappas P. Tabuada and P. Lima. Feasible formations of multi-agent systems. *Proceedings of the American Control Conference*, pages 56–61, 2001.
30. M. Reyhanoglu. A general non-holonomic motion planning strategy for chaplygin systems. *33rd IEEE Conference on Decision and Control*, pages 2964–2966, 1994.
31. E. Rimon and D. E. Koditschek. The construction of analytic diffeomorphisms for exact robot navigation on star worlds. *Trans. of the American Mathematical Society*, 327(1):71–115, 1991.

32. E. Rimon and D. E. Koditschek. Exact robot navigation using artificial potential functions. *IEEE Trans. on Robotics and Automation*, 8(5):501–518, 1992.
33. D. Shevitz and B. Paden. Lyapunov stability theory of nonsmooth systems. *IEEE Trans. on Automatic Control*, 49(9):1910–1914, 1994.
34. S. Leroy T. Siméon and J.-P. Laumond. Path coordination for multiple mobile robots: A resolution-complete algorithm. *EEE Transactions On Robotics And Automation*, 18(1):41–49, 2002.
35. H. Tanner and K.J. Kyriakopoulos. Backstepping for nonsmooth systems. *Automatica*, 39:1259–1265, 2003.
36. H. G. Tanner and K. J. Kyriakopoulos. Nonholonomic motion planning for mobile manipulators. *Proc of IEEE Int. Conf. on Robotics and Automation*, pages 1233–1238, 2000.
37. H. G. Tanner, S. G. Loizou, and K. J. Kyriakopoulos. Nonholonomic stabilization with collision avoidance for mobile robots. *Proc. of IEEE/RSJ Int. Conf. on Intelligent Robots and Systems*, pages 1220–1225, 2001.
38. H. G. Tanner, S. G. Loizou, and K. J. Kyriakopoulos. Nonholonomic navigation and control of cooperating mobile manipulators. *IEEE Trans. on Robotics and Automation*, 19(1):53–64, 2003.
39. H. G. Tanner and G. J. Pappas. Formation input-to-state stability. *Proceedings of the 15th IFAC World Congress on Automatic Control*, pages 1512–1517, 2002.
40. D. Tilbury and A. Chelouah. Steering a three input non-holonomic system using multirate controls. *Proceedings of the European Control Conference*, pages 1993–1998, 1992.
41. D. Tilbury and A. Chelouah. Steering a three input non-holonomic system using multirate controls. *Proceedings of the European Control Conference*, pages 1432–1437, 1993.
42. E. Todt, G. Raush, and R. Suárez. Analysis and classification of multiple robot coordination methods. *Proc. of IEEE Int. Conf. on Robotics and Automation*, pages 3158–3163, 2000.
43. P. Tournassoud. A strategy for obstacle avoidance and its applications to multi - robot systems. *Proc. of IEEE Int. Conf. on Robotics and Automation*, pages 1224–1229, 1986.
44. L. Whitcomb and D. Koditschek. Automatic assembly planning and control via potential functions. *Proceedings of the IEEE/RSJ International Workshop on Intelligent Robots and Systems*, pages 17–23, 1991.
45. L. Whitcomb and D. Koditschek. Toward the automatic control of robot assembly tasks via potential functions: The case of 2-d sphere assemblies. *Proceedings of the IEEE International Conference on Robotics and Automation*, pages 2186–2191, 1992.

Multirobot Navigation Functions II: Towards Decentralization

Dimos V. Dimarogonas, Savvas G. Loizou and Kostas J. Kyriakopoulos

Control Systems Laboratory, National Technical University of Athens, 9 Heroon Polytechniou Street, Zografou 15780, Greece

Summary. This is the second part of a two part paper regarding Multirobot Navigation Functions. In this part, we discuss extensions of the centralized scheme presented in the first part, towards decentralization concepts. Both holonomic and nonholonomic kinematic models are considered and the limited sensing capabilities of each agent are taken into account. An extension to dynamic models of the agents' motion is also included. The conflict resolution as well as destination convergence properties are verified in each case through nontrivial computer simulations.

1 Introduction

This is the second part of a two part paper regarding Multirobot Navigation Functions. In this part, we discuss extensions of the centralized scheme presented in the first part, towards decentralization concepts.

Multi-agent Navigation is a field that has recently gained increasing attention both in the robotics and the control communities, due to the need for autonomous control of more than one mobile agents (vehicles/robots) in the same workspace. While most efforts in the past had focused on centralized planning, specific real-world applications have lead researchers throughout the globe to turn their attention to decentralized concepts. The basic motivation for this work comes from two application domains: (i) decentralized conflict resolution in air traffic management and (ii) the field of micro robotics, where a team of autonomous micro robots must cooperate to achieve manipulation precision in the sub micron level.

Decentralized navigation approaches are more appealing to centralized ones, due to their reduced computational complexity and increased robustness with respect to agent failures. The main focus of work in this domain has been cooperative and formation control of multiple agents, where so much effort has been devoted to the design of systems with variable degree of autonomy ([12],[14], [17], [41], [43]). There have been many different approaches to the decentralized motion planning problem. Open loop approaches use game

H.A.P. Blom, J. Lygeros (Eds.): Stochastic Hybrid Systems, LNCIS 337, pp. 209–253, 2006.

theoretic and optimal control theory to solve the problem taking the constraints of vehicle motion into account; see for example [2],[20],[35], [42] . On the other hand, closed loop approaches use tools from classical Lyapunov theory and graph theory to design control laws and achieve the convergence of the distributed system to a desired configuration both in the concept of cooperative ([13], [22],[23],[30]), and formation control ([1],[16],[24],[32] [33],[40]). A few approaches use computer science based tools to treat the problem;see for example [19],[28],[29]. However, the latter fail to guarantee convergence of the multi-agent system.

Closed loop strategies are apparently preferable to open loop ones, mainly because they provide robustness with respect to modelling uncertainties and agent failures and guaranteed convergence to the desired configurations. However, a common point of most work in this area is devoted to the case of point agents. Although this allows for variable degree of decentralization, it is far from realistic in real world applications. For example, in conflict resolution in Air Traffic Management, two aircraft are not allowed to approach each other closer than a specific "alert" distance. The construction of closed loop methods for distributed non-point multi-agent systems is both evident and appealing.

This chapter presents the first to the authors knowledge' extension of centralized multi-agent control using navigation functions, to a decentralized scheme. The level of decentralization depends on the knowledge each agent has for the state, objectives and actions of the rest of the team. A first step towards decentralization is discussed both for holonomic and for nonholonomic kinematics and allows each agent to ignore the desired destination of the others. In the process, we show how this scheme can be redefined in order to cope with the limited sensing capabilities of each agent, namely with the case when each agent has only partial knowledge of the state space.

The great advantages of the proposed scheme are (i) its relatively low complexity with respect to the number of agents, compared to centralized approaches to the problem and (ii) its application to non-point agents. The effectiveness of the methodology is verified through non-trivial computer simulations.

The rest of this chapter is organized as follows: section 2 refers to the case of decentralized conflict resolution for multiple holonomic kinematic agents with global sensing capabilities. The extension of the centralized approach to the decentralized case and the concept of decentralized navigation functions is encountered in section 3. Section 4 deals with the case of limited sensing capabilities for each agent. The nonholonomic counterparts of the previous sections are considered in section 5 while dynamic models of the agents' motion are taken into account in section 6. Section 7 includes some non-trivial computer simulations of the adopted theory and section 8 summarizes the results of this chapter and indicates current research. Sketches of the proofs of the propositions in section 3 are included in the Appendix.

2 Global Decentralized Conflict Resolution and Holonomic Kinematics

In this section, we present a decentralized conflict resolution algorithm for the case when the kinematics of each aircraft are considered purely holonomic. We first present the fundamental approach using Decentralized Navigation Functions (DNF's) for agents with global sensing capabilities.

For the case where of global sensing capabilities, the decentralization factor lies in the assumption that each agent does not need to know the desired destinations of the others in order to navigate to its goal configuration. A provable way to extend this method to the case of limited sensing capabilities is presented in the sequel.

Consider a system of N agents operating in the same workspace $W \subset \mathcal{R}^2$. Each agent i occupies a disc: $R = \{q \in \mathcal{R}^2 : \| q - q_i \| \leq r_i\}$ in the workspace where $q_i \in \mathcal{R}^2$ is the center of the disc and r_i is the radius of the agent. The configuration space is spanned by $q = [q_1, \ldots, q_N]^T$. The motion of each agent is described by the single integrator:

$$\dot{q}_i = u_i, i \in \mathcal{N} = [1, \ldots, N] \tag{1}$$

The desired destinations of the agents are denoted by the index d: $q_d = [q_{d1}, \ldots, q_{dN}]^T$. The following figure shows a three-agent conflict situation:

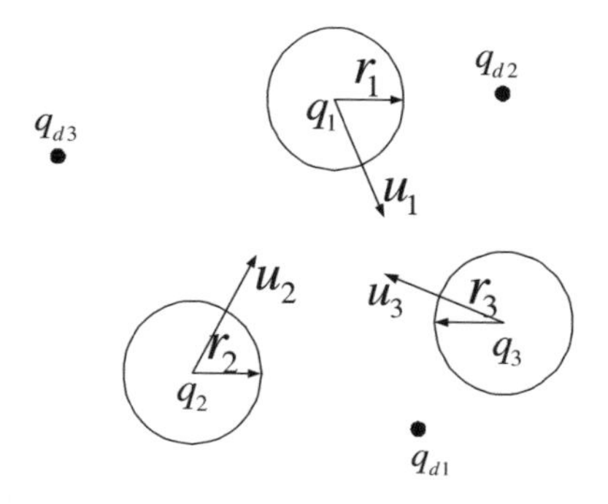

Fig. 1. A conflict scenario with three agents.

The multi agent navigation problem can be stated as follows: "*Derive a set of control laws (one for each agent) that drives the team of agents from any initial configuration to a desired goal configuration avoiding, at the same time, collisions.*"

We make the following assumptions:

- Each agent has global knowledge of the position of the others at each time instant.

- Each agent has knowledge only of its own desired destination but not of the others.
- We consider spherical agents.
- The workspace is bounded and spherical.

Our assumption regarding the spherical shape of the agents does not constrain the generality of this work since it has been proven that navigation properties are invariant under diffeomorphisms ([21]). Arbitrarily shaped agents diffeomorphic to spheres can be taken into account. Methods for constructing analytic diffeomorphisms are discussed in [39] for point agents and in [36] for rigid body agents.

The second assumption makes the problem decentralized. Clearly, in the centralized case a central authority has knowledge of everyones goals and positions at each time instant and it coordinates the whole team so that the desired specifications (destination convergence and collision avoidance) are fulfilled. In the current situation no such authority exists and we have to deal with the limited knowledge of each agent. This is of course the first step towards a variable degree of decentralization. The first assumption, regarding the global knowledge each agent has about the state space, is overcome in section 4, where we discuss how the methodology presented in the next subsections, can be extended to the case of limited sensing capabilities.

3 Decentralized Navigation Functions(DNF's)

3.1 DNF's Versus MRNF's

In the first part of this book chapter, it was shown how the Navigation Functions' method of [21] has been extended to the case of centralized control of multiple mobile agents with the use of Multi-Robot navigation functions (MRNF's).

In the form of a centralized setup [25], where a central authority has knowledge of the current positions and desired destinations of all agents, the sought control law is of the form: $u = -K\nabla\varphi(q)$ where K is a gain. In the decentralized case addressed in this chapter, each agent has knowledge of only the current positions of the others, and not of their desired destinations. Hence each agent has a different navigation law.

Following the procedure of [21],[25], we consider the following class of decentralized navigation functions(*DNF's*):

$$\varphi_i \triangleq \sigma_d \circ \sigma \circ \hat{\varphi}_i = \left(\frac{\gamma_i}{\gamma_i + G_i}\right)^{1/k} \tag{2}$$

which is a composition of $\sigma_d \triangleq x^{1/k}$, $\sigma \triangleq \frac{x}{1+x}$ and the cost function $\hat{\varphi}_i \triangleq \frac{\gamma_i}{G_i}$,where $\gamma_i^{-1}(0)$ denotes the desirable set(i.e. the goal configuration) and $G_i^{-1}(0)$ the set that we want to avoid(i.e. collisions with other agents).A suitable choice is:

$$\gamma_i = (\gamma_{di} + f_i)^k \tag{3}$$

where $\gamma_{di} = \| q_i - q_{di} \|^2$, is the squared metric of the current agent's configuration q_i from its destination q_{di}. The definition of the function f_i will be given later. Function G_i has as arguments the coordinates of all agents, i.e. $G_i = G_i(q)$, in order to express all possible collisions of agent i with the others. The proposed navigation function for agent i is

$$\varphi_i(q) = \frac{\gamma_{di} + f_i}{\left((\gamma_{di} + f_i)^k + G_i\right)^{1/k}} \tag{4}$$

By using the notation $\tilde{q}_i \triangleq [q_1, \ldots, q_{i-1}, q_{i+1}, \ldots, q_N]^T$, the decentralized NF can be rewritten as

$$\varphi_i = \varphi_i(q_i, \tilde{q}_i) = \varphi_i(q_i, t)$$

that is, the potential function in hand contains a *time-varying* element which corresponds to the movement in time of all the other agents apart from i. This element is neglected in the case of a single agent moving in an environment of static obstacles ([21]), but in this case the term $\frac{\partial \varphi_i}{\partial t}$ is nonzero.

3.2 Construction of the G Function

In the proposed decentralized control law, each agent has a different G_i which represents its relative position with all the other agents. In contrast to the centralized case, in which a central authority has global knowledge of the positions and desired destinations of the whole team and plans a global G function accordingly, in the decentralized case, each member i of the team has its own G_i function, which encodes the different proximity relations with the rest. The main difference of the DNF's and the MRNF's in [25] from the NF's introduced in [21] lies in the structure of the function G. While there were attempts to prove convergence and collision avoidance to the straightforward extension of [21] to the multiple moving agents case, only collision avoidance properties were established. Furthermore simulation results motivated us to consider a different approach to [25] for the decentralized setup. The basic difference with respect to the centralized case is that each G_i is constructed with respect to the specific agent i and not in a centralized fashion. Hence each G_i takes into account only the collision schemes in which i is involved.

We review now the construction of the "collision" function G_i for each agent i. The "Proximity Function" between agents i and j is given by

$$\beta_{ij} = \|q_i - q_j\|^2 - (r_i + r_j)^2 \tag{5}$$

Consider now the situation in figure 2. There are 5 agents and we proceed to define the function G_R for agent R.

Definition 1. *A* relation *with respect to agent R is every possible collision scheme that can occur in a multiple agents scene with respect R.*

Definition 2. *A* binary relation *with respect to agent R is a relation between agent R and another.*

Definition 3. *The* relation level *in the number of binary relations in a relation.*

We denote by $(R_j)_l$ the jth relation of level-l with respect to agent R. With this terminology in hand, the collision scheme of figure (2a) is a level-1 relation (one binary relation) and that of figure (2b) is a level-3 relation (three binary relations), always with respect to the specific agent R. We use the notation

$$(R_j)_l = \{\{R, A\}, \{R, B\}, \{R, C\}, \ldots\}$$

to denote the set of binary relations in a relation with respect to agent R, where $\{A, B, C, ...\}$ the set of agents that participate in the specific relation. For example, in figure (2b):

$$(R_1)_3 = \{\{R, O_1\}, \{R, O_2\}, \{R, O_3\}\}$$

where we have set arbitrarily $j = 1$.

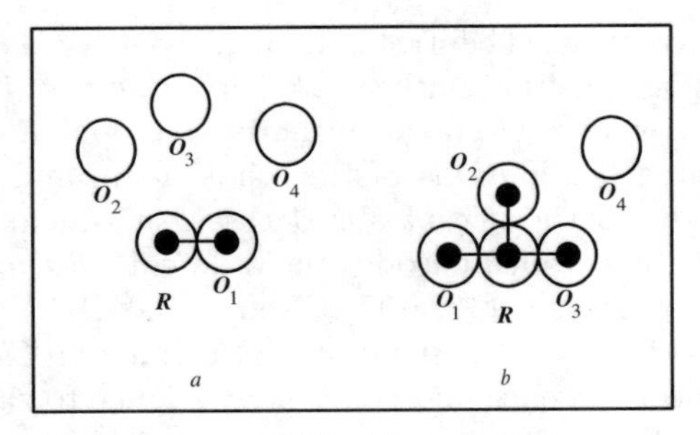

Fig. 2. Part a represents a level-1 relation and part b a level-3 relation wrt agent R.

The complementary set $(R_j^C)_l$ of relation j is the set that contains all the relations of the same level apart from the specific relation j. For example in figure (2b):

$$(R_1^C)_3 = \{(R_2)_3, (R_3)_3, (R_4)_3\}$$

where

$$(R_2)_3 = \{\{R, O_1\}, \{R, O_2\}, \{R, O_4\}\}$$
$$(R_3)_3 = \{\{R, O_1\}, \{R, O_3\}, \{R, O_4\}\}$$
$$(R_4)_3 = \{\{R, O_2\}, \{R, O_3\}, \{R, O_4\}\}$$

A "Relation Proximity Function" (RPF) provides a measure of the distance between agent i and the other agents involved in the relation. Each relation has its own RPF. Let R_k denote the kth relation of level l. The RPF of this relation is given by:

$$(b_{R_k})_l = \sum_{j \in (R_k)_l} \beta_{\{R,j\}} \tag{6}$$

where the notation $j \in (R_k)_l$ is used to denote the agents that participate in the specific relation of agent R. In the proofs, we also use the simplified notation $b_r = \sum_{j \in P_r} \beta_{ij}$ for simplicity, where r denotes a relation and P_r denotes the set of agents participating in the specific relation wrt agent i.

For example, in the relation of figure (2b) we have

$$(b_{R_1})_3 = \sum_{m \in (R_1)_3} \beta_{\{R,m\}} = \beta_{\{R,O_1\}} + \beta_{\{R,O_2\}} + \beta_{\{R,O_3\}}$$

A "Relation Verification Function" (RVF) is defined by:

$$(g_{R_k})_l = (b_{R_k})_l + \frac{\lambda (b_{R_k})_l}{(b_{R_k})_l + (B_{R_k^C})_l^{1/h}} \tag{7}$$

where λ, h are positive scalars and

$$(B_{R_k^C})_l = \prod_{m \in (R_k^C)_l} (b_m)_l$$

where as previously defined, $(R_k^C)_l$ is the complementary set of relations of level-l, i.e. all the other relations with respect to agent i that have the same number of binary relations with the relation R_k. Continuing with the previous example we could compute, for instance,

$$\left(B_{R_1^C}\right)_3 = (b_{R_2})_3 \cdot (b_{R_3})_3 \cdot (b_{R_4})_3$$

which refers to level-3 relations of agent R.

For simplicity we also use the notation $(B_{R_k^C})_l \equiv \tilde{b}_i = \prod_{m \in (R_k^C)_l} b_m$. The RVF can be written as $g_i = b_i + \frac{\lambda b_i}{b_i + \tilde{b}_i^{1/h}}$ It is obvious that for the highest level $l = n-1$ only one relation is possible so that $(R_k^C)_{n-1} = \emptyset$ and $(g_{R_k})_l = (b_{R_k})_l$ for $l = n-1$. The basic property that we demand from RVF is that it assumes the value of zero if a relation holds, while no other relations of the same or other levels hold. In other words it should indicate which of all possible relations holds. We have he following limits of RVF (using the simplified notation): (a) $\lim_{b_i \to 0} \lim_{\tilde{b}_i \to 0} g_i\left(b_i, \tilde{b}_i\right) = \lambda$ (b) $\lim_{\substack{b_i \to 0 \\ \tilde{b}_i \neq 0}} g_i\left(b_i, \tilde{b}_i\right) = 0$. These limits guarantee that RVF will behave in the way we want it to, as an indicator of a specific collision.

The function G_i is now defined as

$$G_i = \prod_{l=1}^{n_L^i} \prod_{j=1}^{n_{R_l}^i} (g_{R_j})_l \tag{8}$$

where n_L^i the number of levels and $n_{R_l}^i$ the number of relations in level-l with respect to agent i.

The definition of the G function in the multiple moving agents situation is slightly different than the one introduced by the authors in [21]. The collision scheme in that approach involved a single moving point agent in an environment with static obstacles. A collision with more than one obstacle was therefore impossible and the obstacle function was simply the product of the distances of the agent from each obstacle. In our case however, this is inappropriate, as can be seen in the next figure. The control law of agent A should

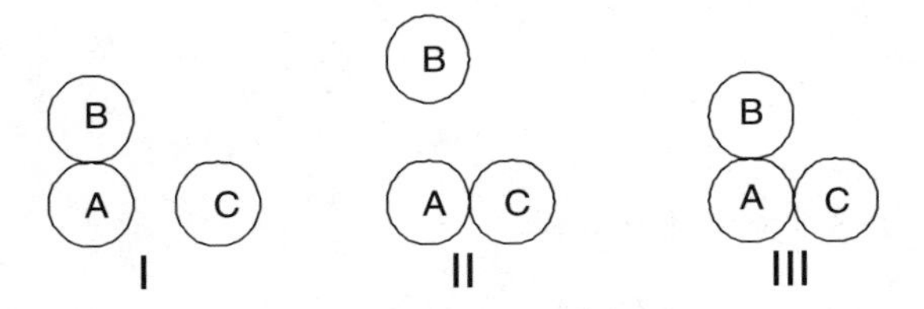

Fig. 3. I,II are level-1 relations with respect to A, while III is level-2. The RVFs of the level-1 relations are nonzero in situation III.

distinguish when agent A is in conflict with B, C, or B and C simultaneously. Mathematically, the first two situations are level-1 relations and the third a level-2 relation with respect to A. Whenever the latter occurs, the RVF of the level-2 relation tends to zero while the RVFs of the two separate level-1 relations (A,B and A,C) are nonzero. The key property of an RVF is that it tends to zero only when the corresponding relation holds. Hence it serves as an analytic switch that is activated (tends to zero) only when the relation it represents is realized.

3.3 An Example

As an example, we will present steps to construct the function G with respect to a specific agent in a team of 4 agents indexed 1 through 4. We construct the function G_1 wrt agent 1. We begin by defining the Relation Proximity Functions in every level (Table 1):

Table 1.

Relation	Level 1	Level 2	Level 3
1	$(b_1)_1 = \beta_{12}$	$(b_1)_2 = \beta_{12} + \beta_{13}$	$(b_1)_3 = \beta_{12}+$ $+\beta_{13} + \beta_{14}$
2	$(b_2)_1 = \beta_{13}$	$(b_2)_2 = \beta_{12} + \beta_{14}$	-
3	$(b_3)_1 = \beta_{14}$	$(b_3)_2 = \beta_{13} + \beta_{14}$	-

It is now easy to calculate the Relation Verification Functions for each relation based on equation (7). For example, for the second relation of level 2, the complement (term $(B_{R_k^C})_l$ in eq.(7)) is given by $(B_{2^C})_2 = (b_1)_2 \cdot (b_3)_2$ and substituting in (7), we have

$$(g_2)_2 = (b_2)_2 + \frac{\lambda\,(b_2)_2}{(b_2)_2 + ((b_1)_2 \cdot (b_3)_2)^{1/h}}$$

The function G_1 is then calculated as the product of the Relation Verification Functions of all relations.

3.4 The f Function

The key difference of the decentralized method with respect to the centralized case is that the control law of each agent ignores the destinations of the others. By using $\varphi_i = \frac{\gamma_{di}}{((\gamma_{di})^k + G_i)^{1/k}}$ as a navigation function for agent i, there is no potential for i to cooperate in a possible collision scheme when its initial condition coincides with its final destination. In order to overcome this limitation,we add a function f_i to γ_i so that the cost function φ_i attains positive values in proximity situations even when i has already reached its destination. A preliminary definition for this function was given in [11], [44]. Here, we modify the previous definitions to ensure that the destination point is a non-degenerate local minimum of φ_i with minimum requirements on assumptions. We define the function f_i by:

$$f_i(G_i) = \begin{cases} a_0 + \sum\limits_{j=1}^{3} a_j G_i^j, & G_i \leq X \\ 0, & G_i > X \end{cases} \tag{9}$$

where $X, Y = f_i(0) > 0$ are positive parameters the role of which will be made clear in the following. The parameters a_j are evaluated so that f_i is maximized when $G_i \to 0$ and minimized when $G_i = X$. We also require that f_i is continuously differentiable at X. Therefore we have:

$$a_0 = Y, a_1 = 0, a_2 = \frac{-3Y}{X^2}, a_3 = \frac{2Y}{X^3}$$

The parameter X serves as a sensing parameter that activates the f_i function whenever possible collisions are bound to occur. The only requirement we have for X is that it must be small enough to guarantee that f_i vanishes whenever the system has reached its equilibrium, i.e. when everyone has reached its destination. In mathematical terms:

$$X < G_i(q_{d1}, \ldots, q_{dN}) \; \forall i \tag{10}$$

That's the minimum requirement we have regarding knowledge of the destinations of the team.

The resulting navigation function is no longer analytic but merely C^1 at $G_i = X$. However, by choosing X large enough, the resulting function is analytic in a neighborhood of the boundary of the free space so that the characterization of its critical points can be made by the evaluation of its Hessian. Hence, the parameter X must be chosen small enough in order to satisfy (10) but large enough to include the region described above. Clearly, this is a tradeoff the control design has to pay in order to achieve decentralization. Intuitively, the destinations should be far enough from one another.

3.5 Control Strategy

The proposed feedback control strategy for agent i is defined as

$$u_i = -K_i \frac{\partial \varphi_i}{\partial q_i} \tag{11}$$

where $K_i > 0$ a positive gain.

3.6 Proof of Correctness

Let $\varepsilon > 0$. Define $B^i_{j,l}(\varepsilon) \equiv \{q : 0 < (g^i_{R_j})_l < \varepsilon\}$. Following [21],[25] we discriminate the following topologies for the function φ_i:

1. The destination point: q_{di}
2. The free space boundary: $\partial F(q) = G_i^{-1}(\delta), \delta \to 0$
3. The set near collisions: $F_0(\varepsilon) = \bigcup_{l=1}^{n^i_L} \bigcup_{j=1}^{n^i_{R,l}} B^i_{j,l}(\varepsilon) - \{q_{di}\}$
4. The set away from collisions: $F_1(\varepsilon) = F - (\{q_{di}\} \cup \partial F \cup F_0(\varepsilon))$

The following theorem allows us to derive results for the function φ_i by examining the simpler function $\hat{\varphi}_i(q) = \frac{\gamma_i}{G_i}$:

Theorem 1. *[21] Let I_1, I_2 be intervals, $\hat{\varphi} : F \to I_1$ and $\sigma : I_1 \to I_2$ be analytic. Define the composition $\varphi : F \to I_2$ to be $\varphi = \sigma \circ \hat{\varphi}$. If σ is monotonically increasing on I_1, then the set of critical points of φ and $\hat{\varphi}$ coincide and the (Morse) index of each critical point is identical.*

A key point in the discrimination between centralized and decentralized navigation functions is that the latter contain a time-varying part which depends on the movement of the other agents. Using the same procedure as in [21],[25] we first prove that the construction of each φ_i guarantees collision avoidance:

Proposition 1. *For each fixed t, the function $\varphi_i(q_i, \cdot)$ is a navigation function if the parameters h, k assume values bigger than a finite lower bound.*

Proof Sketch: For the complete proof see [7]. The set of critical points of φ_i is defined as $C_{\varphi_i} = \{q : \partial\varphi_i/\partial q_i = 0\}$. A critical point is non-degenerate if $\partial^2\varphi_i/\partial^2 q_i$ has full rank at that point.The statement of the proposition is guaranteed by the following Lemmas:

Lemma 1. *If the workspace is valid, the destination point q_{di} is a non-degenerate local minimum of φ_i.*

Lemma 2. *All critical points of φ_i are in the interior of the free space.*

Lemma 3. *For every $\varepsilon > 0$, there exists a positive integer $N(\varepsilon)$ such that if $k > N(\varepsilon)$ then there are no critical points of $\hat{\varphi}_i$ in $F_1(\varepsilon)$.*

Lemma 4. *There exists an $\varepsilon_0 > 0$ such that $\hat{\varphi}_i$ has no local minimum in $F_0(\varepsilon)$, as long as $\varepsilon < \varepsilon_0$.*

Lemma 5. *There exist $\varepsilon_1 > 0$ and $h_1 > 0$, such that the critical points of $\hat{\varphi}_i$ are non-degenerate as long as $\varepsilon < \varepsilon_1$ and $h > h_1$.*

The complete proofs of the Lemmas can be found in [7]. Sketches of the proofs are found in the Appendix. Lemmas 1-4 guarantee the polarity of the proposed DNF, whilst Lemma 5 guarantees the non-degeneracy of the critical points. By choosing k, h that satisfy the above Lemmas, the statement of Proposition 1 is proved.

This however does not guarantee global convergence of the system state to the destination configuration. This is achieved by using a Lyapunov function for the *whole* system which is *time invariant* that is a function that depends on the positions of all the agents. The candidate Lyapunov function that we use in this paper is simply the sum of the DNF's of all agents. Specifically we prove the following:

Proposition 2. *The time-derivative of $\varphi = \sum_{i=1}^{N} \varphi_i$ is negative definite across the trajectories of the system up to a set of initial conditions of measure zero if the parameters h, k assume values bigger than a finite lower bound.*

A detailed proof based on matrix calculus be found in [7] while a proof sketch in the Appendix.

4 The Case of Limited Sensing Capabilities

In the previous section, it was shown how with a suitable choice of the parameters h, k the proposed control law can satisfy the collision avoidance and destination convergence properties in a bounded workspace. The decentralization feature of the whole scheme lied in the fact that each agent didn't have knowledge of the desired destinations of the rest of the team. On the other hand, each one had global knowledge of the positions of the others at each time instant. This is far from realistic in real world applications.

In this section we provide the necessary machinery to take the limited sensing capabilities of each agent into account. Specifically, we alter the definition of inter-agent proximity functions in order to cope with the limited sensing range of each agent.

We consider a bounded workspace with n agents. Each agent has only local knowledge of the positions of the others at each time instant. Specifically, it only knows the position of agents which are in a cyclic neighborhood of specific radius d_C around its center. Therefore the Proximity Function between two agents has to be redefined in this case. We propose the following nonsmooth function:

$$\beta_{ij} = \begin{cases} \|q_i - q_j\|^2 - (r_i + r_j)^2, \text{for } \|q_i - q_j\| \leq d_C \\ d_C^2 - (r_i + r_j)^2, \text{for } \|q_i - q_j\| > d_C \end{cases} \tag{12}$$

The whole scheme is now modelled as a (deterministic) switched system in which switches occur whenever a agent enters or leaves the neighborhood of another. In the previous section, we have $\varphi = \sum_{i=1}^{n} \varphi_i$ as a Lyapunov function for the whole system. In this case this function is continuous everywhere, but nonsmooth whenever a switching occurs, i.e. whenever $\|q_i - q_j\| = d_c$ for some i, j. We define the *switching surface* as:

$$S = \{q : \exists i, j, i \neq j | \|q_i - q_j\| = d_c\} \tag{13}$$

We have proved that the system converges whenever $q \notin S$. On the switching surface the Lyapunov function is no longer smooth so classic stability theory for smooth systems is no longer adequate.

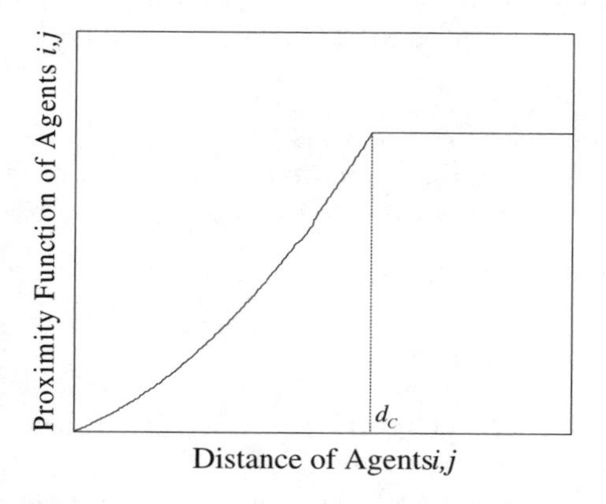

Fig. 4. The function β_{ij} for $r_i + r_j = 1, d_C = 4$.

In [6],[10] we prove the validity of Proposition 2 under the nonsmooth modification of the Proximity Functions. We make use of tools form nonsmooth stability theory ([5],[37]). It is shown than the nonsmooth alternative of the navigation function does not affect the stability and convergence properties of the system.

The prescribed control strategy is another step towards decentralization of the navigation functions' methodology. Although each agent must be aware of the *number* of agents in the entire workspace, it only has to know the *positions* of agents located in its neighborhood. The next step towards global decentralization is to consider the case where each agent is unaware of the global number of agents in the workspace, but only knows what is going on in its neighborhood.

5 Global Decentralized Conflict Resolution and Nonholonomic Kinematics

In this section, we present the decentralized conflict resolution algorithm for the case when the dynamics of each aircraft are considered nonholonomic. We first present the method of Decentralized Dipolar Navigation Functions (DDNF'sS) for agents with global sensing capabilities. We proceed by showing how this methodology can be extended to take into account the limited sensing capabilities of each agent.

5.1 Problem Statement

In this section, we consider the case where each agent has global knowledge of the positions and velocities of the others at each time instant. The decentralization factor lies in the assumption that each agent does not need to know the desired destinations of the others in order to navigate to its goal configuration. The means to extend this method to the case of limited sensing capabilities is presented in the sequel.

Consider the following system of N nonholonomic vehicles:

$$\begin{aligned} \dot{x}_i &= u_i \cos\theta_i \\ \dot{y}_i &= u_i \sin\theta_i \\ \dot{\theta}_i &= \omega_i \end{aligned} \tag{14}$$

with $i \in \{1 \ldots N\}$. (x_i, y_i, θ_i) are the position and orientation of each robot, u_i and w_i are the translational and rotational velocities respectively.

The problem we treat in this section can be now stated as follows: *"Given the N nonholonomic systems, derive a control law that steers every system from any feasible initial configuration to its goal configuration avoiding collisions."*

We make the following assumptions:

- Each agent has global knowledge of the position and velocity of the others at each time instant.
- Agents have no information about other agents targets.
- Around the target of each agent $\mathcal{A}$ there is a region called the agent's $\mathcal{A}$ safe region
- Agent's $\mathcal{A}$ safe region is only accessible by agent $\mathcal{A}$, while regarded as an obstacle by other agents.

5.2 Decentralized Dipolar Navigation Functions(DDNF's)

In this section, we show how the DNF's of the previous section have been redefined in [26] in order to provide trajectories suitable for nonholonomic navigation. This is accomplished by a enhancing a dipolar structure [38] to the navigation functions. Dipolar potential fields have been proven a very effective tool for stabilization [39] of nonholonomic systems as well as for centralized coordination of multiple agents with nonholonomic constraints [27]. The key advantage of this class of potential fields is that they drive the controlled agent to its destination with desired orientation.

The navigation function of the previous section is modified in the following manner in order to be able to produce a dipolar potential field:

$$\varphi_i = \frac{\gamma_{di}}{(\gamma_{di}^k + H_{nh_i} \cdot G_i \cdot b_{t_i})^{1/k}} \tag{15}$$

where $b_{t_i} = \prod_{j \neq i}(\|q_i - q_{d_j}\|^2 - (\varepsilon + r_i)^2)$. The term $\varepsilon > 0$ is the radius of the safe region of its agent. H_{nh_i} has the form of a pseudo-obstacle and is defined as

$$H_{nh_i} = \varepsilon_{nh} + \eta_{nh_i}$$

with $\varepsilon_{nh} > 0$, $\eta_{nh_i} = \|(q_i - q_{di}) \cdot n_{di}\|^2$ and $n_{di} = [\cos(\theta_{di}), \sin(\theta_{di})]^T$. Moreover $\gamma_{di} = \|q_i - q_{di}\|^2$, i.e. the heading angle is not incorporated in the distance to the destination metric. Figure 5 shows a 2D dipolar navigation function.

An important feature that should be noticed is the fact that this navigation function does not have to include the f_i function as each agent treats the other agents' targets as static obstacles.

5.3 Nonholonomic Control

We consider convergence of the multi-agent system as a two-stage process: In the first stage agents converge to a ball of radius ε called safe region, containing the desired destination of each agent. Each agent can get in its own safe region but not in others. The safe region of one agent is regarded as an obstacle from the other agents. Once an agent gets in its own safe region, it remains in the set and asymptotically converges to the origin.

Before defining the control we need some preliminary definitions: We define by $\frac{\partial^2}{\partial q_i^2}\varphi_i(q_i, t) = {}^i\nabla^2\varphi_i(q_i, t)$ the Hessian of φ_i. Let $\lambda_{\min}, \lambda_{\max}$ be the

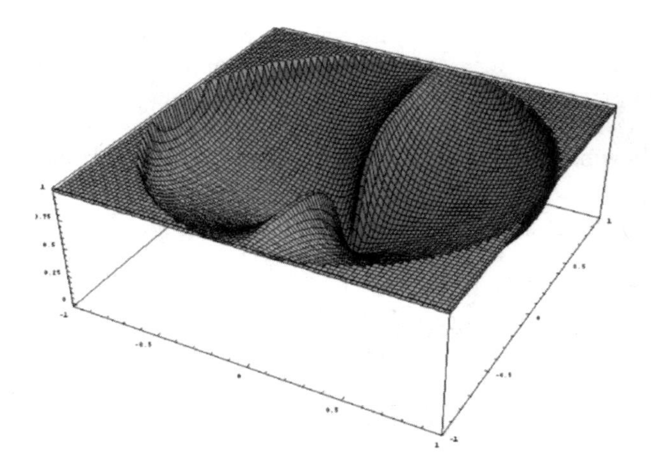

Fig. 5. A dipolar potential field

minimum and maximum eigenvalues of the Hessian and $\hat{\upsilon}_{\lambda_{\min}}, \hat{\upsilon}_{\lambda_{\max}}$ the unit eigenvectors corresponding to the minimum and maximum eigenvalues of the Hessian. Since navigation functions are Morse functions [31], their Hessian at critical points is never degenerate, i.e. their eigenvalues have always nonzero values.

As discussed before,φ_i is a dipolar navigation function. The flows of the dipolar navigation field provide feasible directions for nonholonomic navigation. What we need now is to extract this information from the dipolar function. To this extend we define the "nonholonomic angle":

$$\theta_{nh_i} = \begin{cases} \arg\left(\frac{\partial \varphi_i}{\partial x_i} \cdot s_i + \mathrm{i}\frac{\partial \varphi_i}{\partial y_i} \cdot s_i\right), & \neg P_1 \\ \arg\left(d_i \cdot s_i(\upsilon^x_{\lambda_{\min}} + \mathrm{i}\upsilon^y_{\lambda_{\min}}\right), & P_1 \end{cases}$$

where condition P_1 is used to identify sets of points that contain measure zero sets whose positive limit sets are saddle points:

$$P_1 = (\lambda_{\min} < 0) \wedge (\lambda_{\max} > 0) \wedge \left(\left|\hat{\upsilon}_{\lambda_{\min}} \cdot {}^i\nabla\varphi_i\right| < \varepsilon_1\right)$$

where

$$\varepsilon_1 < \min_{C=\{q_i:\|q_i - q_{di}\|=\varepsilon\}} \left(\left\|{}^i\nabla\varphi_i(C)\right\|\right), s_i = \mathrm{sgn}((q_i - q_{di}) \cdot \eta_{di})$$

$$d_i = \mathrm{sgn}(\upsilon_{\lambda_{\min_i}} \cdot {}^i\nabla\varphi_i), \eta_{di} = \left[\cos(\theta_{di})\ \sin(\theta_{di})\ \right]^T$$

$$\eta_i = \left[\cos(\theta_i)\ \sin(\theta_i)\ \right]^T$$

Before proceeding we need the following:

Lemma 6. *If* $\left|\hat{\upsilon}_{\lambda_{\min}} \cdot {}^i\nabla\varphi_i\right| = 0$ *then* P_1 *consists of the measure zero set of initial conditions that lead to saddle points.*

For a proof of this lemma the reader is referred to [26].

In view of Lemma 6, ε_1 can be chosen to be arbitrarily small so the sets defined by P_1 eventually consist of thin sets containing sets of initial conditions that lead to saddle points.

The following provides a suitable nonholonomic controller for the first stage:

Proposition 3. *The system under the control law*

$$\begin{array}{l} u_i = -\mathrm{sgn}\left(\frac{\partial\varphi_i}{\partial x_i}\cos\theta_i + \frac{\partial\varphi_i}{\partial y_i}\sin\theta_i\right)\cdot \\ \cdot\left(K_{u_i}K_{z_i} + c_i\frac{|\partial\varphi_i/\partial t|}{\frac{\partial\varphi_i}{\partial x_i}\cos\theta_i + \frac{\partial\varphi_i}{\partial y_i}\sin\theta_i}\tanh\left(\left|\frac{\partial\varphi_i}{\partial x_i}\cos\theta_i + \frac{\partial\varphi_i}{\partial y_i}\sin\theta_i\right|^2\right)\right) \\ \omega_i = \dot\theta_{nh_i} + (\theta_{nh_i} - \theta_i)\left(K_{\theta_i} + c_i\frac{|\partial\varphi_i/\partial t|}{2\left(\theta_{nh_i} - \theta_i\right)^2}\tanh\left(|\theta_{nh_i} - \theta_i|^3\right)\right) \end{array} \tag{16}$$

converges to the set $B_i = \{p_i : \|q_i - q_{di}\| \le \varepsilon - \delta, \theta_i \in (-\pi, \pi]\}$ *up to a set of measure zero of initial conditions where* $0 < \delta < \varepsilon$. *Here* $K_{z_i} = \left\|{}^i\nabla\varphi_i\right\|^2 + \|q_i - q_{di}\|^2$, K_{u_i}, K_{θ_i} *are positive constants,* $c_i > \frac{\varepsilon_2+1}{\varepsilon_2}$ *where*
$\varepsilon_2 = 2\pi^3\varepsilon_1^2\left(4\varepsilon_1 + \sqrt{2}\pi^{3/2}\right)^{-2}$ *and*

$$\frac{\partial\varphi_i}{\partial t} = \sum_{j\neq i}\left\{\left(\frac{\partial\varphi_i}{\partial x_j}\cos\theta_j + \frac{\partial\varphi_i}{\partial y_j}\sin\theta_j\right)\cdot u_j\right\} \tag{17}$$

.

Proof :We form the following Lyapunov function:

$$V_i = \varphi_i\left(x_i, y_i, t\right) + \left(\theta_{nh_i}\left(x_i, y_i, t\right) - \theta_i\right)^2$$

and take it's time derivative:

$$\begin{array}{l} \dot V_i = \frac{\partial\varphi_i}{\partial t} + u_i\eta_i{}^i\nabla\varphi_i + \\ \quad +2\left(\theta_{nh_i} - \theta_i\right)\left(-w_i + \frac{\partial\theta_{nh_i}}{\partial t} + u_i\eta_i\cdot{}^i\nabla\theta_{nh_i}\right) \end{array}$$

After substituting the control law u_i and w_i, we get:

$$\begin{array}{l} \dot V_i = \frac{\partial\varphi_i}{\partial t} - \left|{}^i\nabla\varphi_i\cdot\eta_i\right|\left(K_{z_i} + c_i\frac{|\partial\varphi_i/\partial t|}{|{}^i\nabla\varphi_i\cdot\eta_i|}\tanh\left(\left|{}^i\nabla\varphi_i\cdot\eta_i\right|^2\right)\right) \\ -2\left(\theta_{nh_i} - \theta_i\right)^2\left(K_{\theta_i} + c_i\frac{|\partial\varphi_i/\partial t|}{2\left(\theta_{nh_i} - \theta_i\right)^2}\tanh\left(|\theta_{nh_i} - \theta_i|^3\right)\right) \le \\ \le \frac{\partial\varphi_i}{\partial t} - c_i\left|\frac{\partial\varphi_i}{\partial t}\right|\left(\tanh\left(\left|{}^i\nabla\varphi_i\cdot\eta_i\right|^2\right) + \tanh\left(|\theta_{nh_i} - \theta_i|^3\right)\right) \end{array}$$

Since the set P_1 is by construction repulsive for ε_1 sufficiently small, we only need to consider the set $\neg P_1$. Then: $\left|{}^i\nabla\varphi_i\cdot\eta_i\right|^2 = \left\|{}^i\nabla\varphi_i\right\|^2\cos^2\left(\theta_{nh_i} - \theta_i\right)$. Let $\Delta\theta = |\theta_{nh_i} - \theta_i|$. After substituting we get:

$$\dot V_i \le \frac{\partial\varphi_i}{\partial t} - c_i\left|\frac{\partial\varphi_i}{\partial t}\right|\left(\tanh\left(\left\|{}^i\nabla\varphi_i\right\|^2\cos^2(\Delta\theta)\right) + \tanh\left(\Delta\theta^3\right)\right)$$

Before proceeding we need the following:

Lemma 7. *The following inequalities hold:*

1. $\tanh(x) \geq \frac{x}{x+1}$, $x \geq 0$
2. $\frac{x}{x+1} + \frac{y}{y+1} \geq \frac{x+y}{x+y+1}$, $x, y \geq 0$
3. $\cos^2 \Delta\theta \geq \frac{8}{\pi^3}\left(\left|\Delta\theta - \frac{\pi}{2}\right|\right)^3$ $\quad \Delta\theta \in \left[0, \frac{\pi}{2}\right]$

Proof:

1. For $x \geq 0$ we have that $e^{2x} - 1 - 2x \geq 0$. Hence $(x+1)(e^x - e^{-x}) \geq x(e^x + e^{-x})$ and we get the result: $\tanh(x) \geq \frac{x}{x+1}$. The equality holds at $x = 0$.
2. With $x, y \geq 0$ we have : $\frac{x}{x+1} + \frac{y}{y+1} = \frac{2xy+x+y}{xy+x+y+1} \geq \frac{xy+x+y}{xy+x+y+1} \geq \frac{x+y}{x+y+1}$ and the equality holds at $x = y = 0$
3. Denote $A(\Delta\theta) = \cos^2 \Delta\theta$ and $B(\Delta\theta) = \frac{8}{\pi^3}\left(\left|\Delta\theta - \frac{\pi}{2}\right|\right)^3$. Solving $A(\Delta\theta) = B(\Delta\theta)$, for $\Delta\theta \in \left[0, \frac{\pi}{2}\right]$ we get $\Delta\theta = 0$ for $A = B = 1$ and $\Delta\theta = \frac{\pi}{2}$ for $A = B = 0$. But at $\frac{\partial A}{\Delta\theta}_{|\Delta\theta=0} = 0 > -\frac{6}{\pi} = \frac{\partial \mathrm{B}}{\Delta\theta}_{|\Delta\theta=0}$ and since A and B have no other intersection for $\Delta\theta \in \left[0, \frac{\pi}{2}\right]$ it follows that $A(\Delta\theta) \geq B(\Delta\theta)$, for $\Delta\theta \in \left[0, \frac{\pi}{2}\right]$.

By use of Lemma 7.1 we get: $\dot{V}_i \leq \frac{\partial\varphi_i}{\partial t} - \left|{\partial\varphi_i}/{\partial t}\right| \left(c_i \frac{\left\|{}^i\nabla\varphi_i\right\|^2 \cos^2 \Delta\theta}{\left\|{}^i\nabla\varphi_i\right\|^2 \cos^2 \Delta\theta + 1} + c_i \frac{\Delta\theta^3}{\Delta\theta^3 + 1} \right)$.
By use of Lemma 7.2 we get: $\dot{V}_i \leq \frac{\partial\varphi_i}{\partial t} - \left|{\partial\varphi_i}/{\partial t}\right| c_i \left(\frac{\left\|{}^i\nabla\varphi_i\right\|^2 \cos^2 \Delta\theta + \Delta\theta^3}{\left\|{}^i\nabla\varphi_i\right\|^2 \cos^2 \Delta\theta + \Delta\theta^3 + 1} \right)$
and from Lemma 7.3 we get: $\dot{V}_i \leq \frac{\partial\varphi_i}{\partial t} - \left|{\partial\varphi_i}/{\partial t}\right| c_i \frac{\left\|{}^i\nabla\varphi_i\right\|^2 \frac{8}{\pi^3}\left(\left|\Delta\theta - \frac{\pi}{2}\right|\right)^3 + \Delta\theta^3}{\left\|{}^i\nabla\varphi_i\right\|^2 \frac{8}{\pi^3}\left(\left|\Delta\theta - \frac{\pi}{2}\right|\right)^3 + \Delta\theta^3 + 1}$.
In view of the fact that the function $\frac{f(x)}{f(x)+1}$ has the same extremal points with $f(x) \geq 0$ (see [21] for a proof), the minimum of $[\frac{\left\|{}^i\nabla\varphi_i\right\|^2 \frac{8}{\pi^3}\left(\left|\Delta\theta - \frac{\pi}{2}\right|\right)^3 + \Delta\theta^3}{\left\|{}^i\nabla\varphi_i\right\|^2 \frac{8}{\pi^3}\left(\left|\Delta\theta - \frac{\pi}{2}\right|\right)^3 + \Delta\theta^3 + 1}$ coincides with the minimum of $m = \left\|{}^i\nabla\varphi_i\right\|^2 \frac{8}{\pi^3}\left(\left|\Delta\theta - \frac{\pi}{2}\right|\right)^3 + \Delta\theta^3$. Trying to minimize m, we get: $\frac{\partial m}{\partial \left\|{}^i\nabla\varphi_i\right\|} = \frac{16}{\pi^3}\left\|{}^i\nabla\varphi_i\right\|\left(\left|\Delta\theta - \frac{\pi}{2}\right|\right)^3 \geq 0$ which means that m is strictly increasing in the direction of $\left\|{}^i\nabla\varphi_i\right\|$. Examining $\frac{\partial m}{\partial \Delta\theta} = 3 \cdot \Delta\theta^2 + \frac{24}{\pi^3}\left\|{}^i\nabla\varphi_i\right\|^2 \cdot \left(\Delta\theta - \frac{\pi}{2}\right)^2 \cdot sign\left(\Delta\theta - \frac{\pi}{2}\right)$ and requiring $\frac{\partial m}{\partial \Delta\theta} = 0$ for an extremum in the direction of $\Delta\theta$, we get:

$$\Delta\theta = \begin{cases} \frac{2\left\|{}^i\nabla\varphi_i\right\|\pi}{4\left\|{}^i\nabla\varphi_i\right\| \pm \sqrt{2} \cdot \pi^{3/2}} & \Delta\theta \leq {\pi}/{2} \\ \frac{2\left\|{}^i\nabla\varphi_i\right\|\pi}{4\left\|{}^i\nabla\varphi_i\right\| \pm \mathbf{i}\sqrt{2} \cdot \pi^{3/2}} & \Delta\theta > {\pi}/{2} \end{cases}$$

The only feasible solution is: $\Delta\theta = \frac{2\left\|{}^i\nabla\varphi_i\right\|\pi}{4\left\|{}^i\nabla\varphi_i\right\| + \sqrt{2} \cdot \pi^{3/2}}$. Substituting the solution in m we get: $\min_{\Delta\theta}(m) = \frac{2\left\|{}^i\nabla\varphi_i\right\|^2 \pi^3}{\left(4\left\|{}^i\nabla\varphi_i\right\| + \sqrt{2} \cdot \pi^{3/2}\right)^2}$. Minimizing the last we get:

$\frac{\partial \min_{\Delta\theta}(m)}{\partial\|^i\nabla\varphi_i\|} = \frac{4\sqrt{2}\|^i\nabla\varphi_i\|\pi^{9/2}}{\left(4\|^i\nabla\varphi_i\|+\sqrt{2}\cdot\pi^{3/2}\right)^3} \geq 0$. Activating the constraint $\|^i\nabla\varphi_i\| \geq \varepsilon_1$ we get: $\varepsilon_2 = \min(m) = \frac{2\varepsilon_1^2\pi^3}{\left(4\varepsilon_1+\sqrt{2}\cdot\pi^{3/2}\right)^2}$. Substituting in the time derivative of the Lyapunov function, we have that: $\dot{V}_i \leq \frac{\partial\varphi_i}{\partial t} - \left|\partial\varphi_i/\partial t\right| c\frac{\varepsilon_2}{\varepsilon_2+1}$, so choosing $c > \frac{\varepsilon_2+1}{\varepsilon_2}$ we get that $\dot{V}_i \leq \left|\partial\varphi_i/\partial t\right|\left(sign\left(\partial\varphi_i/\partial t\right) - k\right) \leq 0$ since $k = c\frac{\varepsilon_2}{\varepsilon_2+1} > 1$. The equality holds when $(q_i = q_{d_i}) \wedge \left(\partial\varphi_i/\partial t = 0\right)$. We assume that the system's initial conditions are in the set $\mathcal{W}_i \backslash \mathcal{S}_i$ where the set $\mathcal{S}_i = \left\{p_i : \|^i\nabla\varphi_i\| < \varepsilon_1\right\}$. ε_1 can be chosen to be arbitrarily small such that the set $\mathcal{S}_i$ includes arbitrarily small regions only around the saddle points and the target. Since we are considering convergence to the set B_i, we have that

$$\dot{V}_i < 0, \forall q_i \in W_{free} \backslash \left\{\bar{B}_i \cup \left\{q_i : \|^i\nabla\varphi_i(q_i)\| < \varepsilon_1\right\}\right\}$$

where the bar denotes the set internal. $\diamondsuit$

For the second stage each agent is isolated from the rest of the system. The dipolar navigation function for this case becomes:

$$\varphi_{int_i}(x_i, y_i, \theta_i) = \frac{\gamma_{d,\theta_i}}{\left(\gamma_{d,\theta_i}^k + H_{nh_i} \cdot \beta_{int_i}\right)^{1/k}} \tag{18}$$

where $\beta_{int_i} = \varepsilon^2 - \|q_i - q_{d_i}\|^2$, and $\gamma_{d,\theta_i} = \|q_i - q_{d_i}\|^2 + (\theta - \theta_{d_i})^2$. Define

$$\Delta_i = K_{\theta_i} \cdot \partial\varphi_{\text{int}_i}/\partial\theta_i \cdot (\theta_{inh_i} - \theta_i) - K_{u_i} \cdot K_{z_i} \cdot \left|^i\nabla\varphi_{\text{int}_i} \cdot \eta_i\right|$$

and

$$\theta_{inh_i} = \arg\left(\frac{\partial\varphi_{\text{int}_i}}{\partial x_i} \cdot s_i + \mathbf{i}\frac{\partial\varphi_{\text{int}_i}}{\partial y_i} \cdot s_i\right)$$

Then for each aircraft in isolation we have the following:

Proposition 4. *Each subsystem under the control law*

$$\begin{aligned} u_i &= -\text{sgn}\left(\tfrac{\partial\varphi_{\text{int}_i}}{\partial x_i}\cos\theta_i + \tfrac{\partial\varphi_{\text{int}_i}}{\partial y_i}\sin\theta_i\right) K_{u_i} K_{z_i} \\ \omega_i &= K_{\theta_i}(\theta_{inh_i} - \theta_i), \Delta_i < 0 \\ \omega_i &= -K_{\theta_i}\tfrac{\partial\varphi_{\text{int}_i}}{\partial\theta_i}, \Delta_i \geq 0 \end{aligned} \tag{19}$$

converges to p_{d_i}

Proof: Taking $V_i = \varphi_{\text{int}_i}$ as a Lyapunov function candidate, we have for the time derivative:

$$\dot{V}_i = \dot{\mathbf{x}} \cdot \nabla\varphi_{\text{int}_i} = u_i\left(^i\nabla\varphi_{\text{int}_i} \cdot \eta_i\right) + w_i\partial\varphi_{\text{int}_i}/\partial y_i \quad .$$

We can now discriminate two cases, depending on the level of Δ_i:

1. $\Delta_i < 0$. Then $\dot{V}_i = -K_{u_i} K_{z_i} \left| {}^i\nabla\varphi_{\text{int}_i} \cdot \eta_i \right| + K_{\theta_i} \left(\theta_{inh_i} - \theta_i\right) {}^{\partial\varphi_{\text{int}_i}}/_{\partial y_i} = \Delta_i < 0$
2. $\Delta_i \geq 0$. Then $\dot{V}_i = -K_{u_i} K_{z_i} \left| {}^i\nabla\varphi_{\text{int}_i} \cdot \eta_i \right| - K_{\theta_i} \left({}^{\partial\varphi_{\text{int}_i}}/_{\partial y_i}\right)^2 \leq 0$, with the equality holding only at the origin.

◇

The fact that each agent remains in its safe region after the first stage is established by the following lemma which is a direct application of the properties of the navigation function:

Lemma 8. *For each subsystem i under the control law (19)the set*

$$B_{\text{int}_i} = \{p_i : \|q_i - q_{di}\| \leq \varepsilon, \theta_i \in (-\pi, \pi]\}$$

is positive invariant.

Proof : The boundary of (18) is the set $\mathcal{B}_{int_i} = \{p_i : \beta_{int_i}(q_i) = 0\} = \{p_i : \|q_i - q_{d_i}\| = \varepsilon\} = \partial B_{\text{int}_i}$, i.e. the workspace boundary, which is positive invariant for a navigation function [21],[7]. ◇

5.4 The Case of Limited Sensing Capabilities

In the previous section, we presented the nonholonomic control scheme for multiple agents with global sensing capabilities. In this section we modify this in order to cope with the limited sensing range of each agent.

It is obvious that each agent takes into account the other agents only on the first stage. The inter-agent proximity functions are modified according to (12). However each agent has also only local knowledge of the velocities of the rest of the team. Therefore the term $\frac{\partial\varphi_i}{\partial t}$ must be modified according to:

$$\frac{\partial\varphi_i}{\partial t} = \sum_{j:\|q_i - q_{di}\| \leq d_C} \left\{ \left(\frac{\partial\varphi_i}{\partial x_j} \cos\theta_j + \frac{\partial\varphi_i}{\partial y_j} \sin\theta_j \right) \cdot u_j \right\} \tag{20}$$

where d_C is again the radius of the sensing zone of each agent. Hence each agent has to take into account only the positions and velocities of agents that are within each sensing zone at each time instant.

This modification of the control law (16) does not affect the stability results of the previous section as the nodes of the deterministic switched system admit a common Lyapunov function. Using arguments from established results on stability for hybrid systems([3],[34]) the convergence in the first stage is guaranteed for each agent in this case as well. The interested reader can refer to [10] for more details.

6 Dynamic Models

The mathematical models of the moving vehicles/agents in the previous sections were considered purely kinematic. In practice however, real mechanical systems and in particular moving vehicles are controlled through their acceleration. It is therefore evident that second order models are considered as well in the navigation functions' approach. The next two sections present the extension of the DNF's approach of the previous paragraphs to the cases of decentralized dynamic models for holonomic and nonholonomic systems, respectively.

6.1 Holonomic Dynamics

In this section, we present the decentralized control scheme for a multi-agent system with double integrator dynamics. The following discussion is based on [9].

We consider the following system of N agents with double integrator dynamics:

$$\begin{array}{l} \dot{q}_i = v_i \\ \dot{v}_i = u_i \end{array}, i \in \{1, \ldots, N\} \tag{21}$$

We will show that the system is asymptotically stabilized under the control law

$$u_i = -K_i \frac{\partial \varphi_i}{\partial q_i} + \theta_i \left(v_i, \frac{\partial \varphi_i}{\partial t} \right) - g_i v_i \tag{22}$$

where $K_i, g_i > 0$ are positive gains,

$$\theta_i \left(v_i, \frac{\partial \varphi_i}{\partial t} \right) \triangleq - \frac{c v_i}{\tanh \left(\|v_i\|^2 \right)} \left| \frac{\partial \varphi_i}{\partial t} \right|$$

and

$$\frac{\partial \varphi_i}{\partial t} = \sum_{j \neq i} \frac{\partial \varphi_i}{\partial q_j} \dot{q}_j$$

The first term of equation (22) corresponds to the potential field (decentralized navigation function described in section 2. The second term exploits the knowledge each agent has of the velocities of the others, and is designed to guarantee convergence of the whole team to the desired configurations. The last term serves as a damping element that ensures convergence to the destination point by suppressing oscillatory motion around it.

By using the notation $x = \left[x_1^T, \ldots, x_N^T \right]^T$, $x_i^T = \left[\begin{array}{cc} q_i^T & v_i^T \end{array} \right]$ the closed loop dynamics of the system can be rewritten as

$$\dot{x} = \xi(x) = \left[\xi_1^T(x), \ldots, \xi_N^T(x) \right]^T \tag{23}$$

with

$$\xi_i(x) = \begin{bmatrix} v_i \\ -K_i \frac{\partial \varphi_i}{\partial q_i} - \frac{c v_i}{\tanh(\|v_i\|^2)} \left| \frac{\partial \varphi_i}{\partial t} \right| - g_i v_i \end{bmatrix}$$

We will use the function $V = \sum_i K_i \varphi_i + \frac{1}{2} \sum_i \|v_i\|^2$ as a candidate Lyapunov function to show that the agents converge to their destinations points . We will check the stability of the multi-agent system with LaSalle's Invariance Principle. Specifically, the following theorem holds:

Theorem 2. *The system (23) is asymptotically stabilized to* $\begin{bmatrix} q_d^T & 0 \end{bmatrix}$,$q_d = [q_{d1}, \ldots, q_{dN}]^T$ *up to a set of initial conditions of measure zero if the exponent* k *assumes values bigger than a finite lower bound and* $c > \max_i(K_i)$.

Proof: The candidate Lyapunov Function we use is $V = \sum_i K_i \varphi_i + \frac{1}{2} \sum_i \|v_i\|^2$ and by taking its derivative we have

$$\begin{aligned}
& V = \sum_i K_i \varphi_i + \tfrac{1}{2} \sum_i \|v_i\|^2 \Rightarrow \\
& \dot{V} = \sum K_i \dot{\varphi}_i + \sum v_i^T \dot{v}_i = \sum K_i \left(\tfrac{\partial \varphi_i}{\partial t} + v_i^T \tfrac{\partial \varphi_i}{\partial q_i} \right) \\
& + \sum v_i^T \left(-K_i \tfrac{\partial \varphi_i}{\partial q_i} + \theta_i \left(v_i, \tfrac{\partial \varphi_i}{\partial t} \right) - g_i v_i \right) \\
& \Rightarrow \dot{V} = \sum \left(K_i \tfrac{\partial \varphi_i}{\partial t} + v_i^T \theta_i \left(v_i, \tfrac{\partial \varphi_i}{\partial t} \right) - g_i \|v_i\|^2 \right)
\end{aligned}$$

Using the notation $B_i \triangleq K_i \frac{\partial \varphi_i}{\partial t} + v_i^T \theta_i \left(v_i, \frac{\partial \varphi_i}{\partial t} \right)$ we first show that $\sum_i B_i \leq 0$ if $c > \max_i(K_i)$:

$$\begin{aligned}
& \tfrac{\partial \varphi_i}{\partial t} > 0 : \\
& c > \max_i (K_i) \Rightarrow c > K_i \tfrac{\tanh(\|v_i\|^2)}{\|v_i\|^2} \\
& \Rightarrow K_i > \tfrac{c\|v_i\|^2}{\tanh(\|v_i\|^2)} \operatorname{sgn} \left(\tfrac{\partial \varphi_i}{\partial t} \right) \\
& \Rightarrow K_i \tfrac{\partial \varphi_i}{\partial t} + v_i^T \theta_i \left(v_i, \tfrac{\partial \varphi_i}{\partial t} \right) < 0 \forall i : \tfrac{\partial \varphi_i}{\partial t} > 0 \\
& \tfrac{\partial \varphi_i}{\partial t} < 0 : \\
& c > 0 \Rightarrow c > -K_i \tfrac{\tanh(\|v_i\|^2)}{\|v_i\|^2} \\
& \Rightarrow K_i > \tfrac{c\|v_i\|^2}{\tanh(\|v_i\|^2)} \operatorname{sgn} \left(\tfrac{\partial \varphi_i}{\partial t} \right) \\
& \Rightarrow K_i \tfrac{\partial \varphi_i}{\partial t} + v_i^T \theta_i \left(v_i, \tfrac{\partial \varphi_i}{\partial t} \right) < 0 \forall i : \tfrac{\partial \varphi_i}{\partial t} < 0
\end{aligned}$$

Of course, $K_i \frac{\partial \varphi_i}{\partial t} + v_i^T \theta_i \left(v_i, \frac{\partial \varphi_i}{\partial t} \right) = 0$ for $\frac{\partial \varphi_i}{\partial t} = 0$. In the preceding equations we used the fact that $0 \leq \frac{\tanh(x)}{x} \leq 1 \forall x \geq 0$. So we have $\sum_i B_i \leq 0$ with equality holding only when $\frac{\partial \varphi_i}{\partial t} = 0 \forall i$. We have $\dot{V} = \sum_i B_i - \sum_i g_i \|v_i\|^2 \leq 0$. Hence, by LaSalle's Invariance Principle, the state of the system converges to the largest invariant set contained in the set

$$S = \left\{ q, v : \left(\frac{\partial \varphi_i}{\partial t} = 0 \right) \wedge (v_i = 0) \, \forall i \right\} = \\ = \{ q, v : (v_i = 0) \, \forall i \}$$

because by definition the set $\left\{ q, v : \left(\frac{\partial \varphi_i}{\partial t} = 0 \right) \forall i \right\}$ is contained in the set $\{ q, v : (v_i = 0) \, \forall i \}$. For this subset to be invariant we need $\dot{v}_i = 0 \Rightarrow \frac{\partial \varphi_i}{\partial q_i} = 0 \forall i$. The analysis of section 2 revealed that this situation occurs whenever the potential functions either reach the destination or a saddle point. By bounding the parameters k, h from below by a finite number, φ_i becomes a navigation function, hence its critical points are isolated ([21]). Thus the set of initial conditions that lead to saddle points are sets of measure zero ([31]). Hence the largest invariant set contained in the set $\frac{\partial \varphi_i}{\partial q_i} = 0 \forall i$ is simply $q_d \diamondsuit$

6.2 Nonholonomic Dynamics

In section 5, we presented the decentralized navigation functions methodology for multiple agents with nonholonomic kinematics. Although each agent had no specific knowledge about the destinations of the others, it treated a spherical region around the target of each agent as a static obstacle. In this section we modify the proposed control law in order to allow each agent to neglect the destinations of the others. Furthermore, the control inputs are the acceleration and rotational velocity of each vehicle, coping in this way with realistic classes of mechanical systems. The following discussion is based on [8].

We consider the following system of N nonholonomic agents with the following dynamics

$$\begin{array}{r} \dot{x}_i = v_i \cos \theta_i \\ \dot{y}_i = v_i \sin \theta_i \\ \dot{\theta}_i = \omega_i \\ \dot{v}_i = u_i \end{array} \quad , i \in \{1, \ldots, N\} \tag{24}$$

where v_i, ω_i are the translational and rotational velocities of agent i respectively, and u_i its acceleration.

The problem we treat in this paper can be now stated as follows:" *Given the N nonholonomic agents (24),consider the rotational velocity ω_i and the acceleration u_i as control inputs for each agent and derive a control law that steers every agent from any feasible initial configuration to its goal configuration avoiding, at the same, collisions.*"

We make the following assumptions:

- Each agent has global knowledge of the position of the others at each time instant.
- Each agent has knowledge only of its own desired destination but not of the others.
- We consider spherical agents.
- The workspace is bounded and spherical.

To be able to produce a dipolar potential field and cope with the prescribed assumptions, φ_i in this case is defined as follows:

$$\varphi_i = \frac{\gamma_{di} + f_i}{\left((\gamma_{di} + f_i)^k + H_{nh_i} \cdot G_i\right)^{1/k}} \tag{25}$$

where H_{nh_i} has been defined in section 5 and f_i in section 3.

Elements from Nonsmooth Analysis

In this section, we review some elements from nonsmooth analysis and Lyapunov theory for nonsmooth systems that we use in the stability analysis of the next section.

We consider the vector differential equation with discontinuous right-hand side:

$$\dot{x} = f(x) \tag{26}$$

where $f : R^n \to R^n$ is measurable and essentially locally bounded.

Definition 4. [15]*: In the case when n is finite, the vector function $x(.)$ is called a solution of (26) in $[t_0, t_1]$ if it is absolutely continuous on $[t_0, t_1]$ and there exists $N_f \subset R^n, \mu(N_f) = 0$ such that for all $N \subset R^n, \mu(N) = 0$ and for almost all $t \in [t_0, t_1]$*

$$\dot{x} \in K[f](x) \equiv \overline{co}\{\lim_{x_i \to x} f(x_i) | x_i \notin N_f \cup N\}$$

Lyapunov stability theorems have been extended for nonsmooth systems in [37],[4]. The authors use the concept of *generalized gradient* which for the case of finite-dimensional spaces is given by the following definition:

Definition 5. [5]*: Let $V : R^n \to R$ be a locally Lipschitz function. The generalized gradient of V at x is given by*

$$\partial V(x) = \overline{co}\{\lim_{x_i \to x} \nabla V(x_i) | x_i \notin \Omega_V\}$$

where Ω_V is the set of points in R^n where V fails to be differentiable.

Lyapunov theorems for nonsmooth systems require the energy function to be *regular*. Regularity is based on the concept of *generalized derivative* which was defined by Clarke as follows:

Definition 6. [5]*: Let f be Lipschitz near x and v be a vector in R^n. The generalized directional derivative of f at x in the direction v is defined*

$$f^0(x; v) = \lim_{y \to x} \sup_{t \downarrow 0} \frac{f(y + tv) - f(y)}{t}$$

Definition 7. [5]*: The function* $f : R^n \to R$ *is called regular if*
1) $\forall v$*, the usual one-sided directional derivative* $f'(x;v)$ *exists and*
2) $\forall v, f'(x;v) = f^0(x;v)$

The following chain rule provides a calculus for the time derivative of the energy function in the nonsmooth case:

Theorem 3. [37]*: Let* x *be a Filippov solution to* $\dot{x} = f(x)$ *on an interval containing* t *and* $V : R^n \to R$ *be a Lipschitz and regular function. Then* $V(x(t))$ *is absolutely continuous,* $(d/dt)V(x(t))$ *exists almost everywhere and*

$$\frac{d}{dt}V(x(t)) \in^{a.e.} \dot{\tilde{V}}(x) := \bigcap_{\xi \in \partial V(x(t))} \xi^T K[f](x(t))$$

We shall use the following nonsmooth version of LaSalle's invariance principle to prove the convergence of the prescribed system:

Theorem 4. [37] *Let* Ω *be a compact set such that every Filippov solution to the autonomous system* $\dot{x} = f(x), x(0) = x(t_0)$ *starting in* Ω *is unique and remains in* Ω *for all* $t \geq t_0$*. Let* $V : \Omega \to R$ *be a time independent regular function such that* $v \leq 0 \forall v \in \dot{\tilde{V}}$ *(if* $\dot{\tilde{V}}$ *is the empty set then this is trivially satisfied). Define* $S = \{x \in \Omega | 0 \in \dot{\tilde{V}}\}$*. Then every trajectory in* Ω *converges to the largest invariant set,*M*, in the closure of* S*.*

Nonholonomic Control and Stability Analysis

We will show that the system is asymptotically stabilized under the control law

$$\begin{array}{c} u_i = -v_i\{|\nabla_i \varphi_i \cdot \eta_i| + M_i\} - g_i v_i - \frac{v_i}{\tanh(|v_i|)} K_{v_i} K_{z_i} \\ \omega_i = -K_{\theta_i}(\theta_i - \theta_{di} - \theta_{nh_i}) + \dot{\theta}_{nh_i} \end{array} \tag{27}$$

where $K_{v_i}, K_{\theta_i}, g_i > 0$ are positive gains, $\theta_{nh_i} = \arg(\frac{\partial \varphi_i}{\partial x_i} \cdot s_i + \mathbf{i}\frac{\partial \varphi_i}{\partial y_i} \cdot s_i)$, $s_i = \mathrm{sgn}((q_i - q_{di}) \cdot \eta_{di})$, $\eta_i = \left[\begin{array}{cc} \cos\theta_i & \sin\theta_i \end{array}\right]^T$, $\eta_{di} = \left[\begin{array}{cc} \cos\theta_{di} & \sin\theta_{di} \end{array}\right]^T$, $K_{zi} = \|\nabla_i \varphi_i\|^2 + \|q_i - q_{di}\|^2$, $M_i > |\sum_{j \neq i} \nabla_i \varphi_j \cdot \eta_i|_{max}$ and $\nabla_i \varphi_j = \left[\begin{array}{cc} \frac{\partial \varphi_j}{x_i} & \frac{\partial \varphi_j}{y_i} \end{array}\right]$. In particular, we prove the following theorem:

Theorem 5. *Under the control law (27), the system is asymptotically stabilized to* $p_d = [p_{d1}, \ldots, p_{dN}]^T$*.*

Proof: Let us first consider the case $|v_i| > 0 \forall i$. We use

$$V = \sum V_i, V_i = \varphi_i + |v_i| + \frac{1}{2}(\theta_i - \theta_{di} - \theta_{nhi})^2$$

as a Lyapunov function candidate. For $|v_i| > 0$ we have

$$\dot{V} = \sum_i \dot{V}_i = \sum_i \left\{ \begin{array}{c} \sum_j v_j \left(\nabla_j \varphi_i\right) \cdot \eta_j + \operatorname{sgn}(v_i)\dot{v}_i + \\ + \left(\theta_i - \theta_{di} - \theta_{nhi}\right)\left(\dot{\theta}_i - \dot{\theta}_{nhi}\right) \end{array} \right\}$$

and substituting

$$\begin{array}{l} \dot{V} = \sum\limits_i \left\{ \sum\limits_j v_j \left(\nabla_j \varphi_i\right) \cdot \eta_j - |v_i| \left(\left| \left(\nabla_i \varphi_i\right) \cdot \eta_i \right| + M_i \right) \right\} \\ - \sum\limits_i \frac{|v_i|}{\tanh(|v_i|)} K_{v_i} K_{z_i} - \sum\limits_i g_i |v_i| \\ - \sum\limits_i K_{\theta_i} \left(\theta_i - \theta_{di} - \theta_{nhi}\right)^2 \end{array}$$

The first term of the right hand side of the last equation can be rewritten as

$$\begin{array}{l} \sum\limits_i \left\{ \sum\limits_j v_j \left(\nabla_j \varphi_i\right) \cdot \eta_j - |v_i| \left(\left| \left(\nabla_i \varphi_i\right) \cdot \eta_i \right| + M_i \right) \right\} = \\ = \sum\limits_i \left\{ \begin{array}{c} v_i \left(\nabla_i \varphi_i\right) \cdot \eta_i + v_i \sum\limits_{j \neq i} \left(\nabla_i \varphi_j\right) \cdot \eta_i - \\ - |v_i| \left(\left| \left(\nabla_i \varphi_i\right) \cdot \eta_i \right| + M_i \right) \end{array} \right\} \leq 0 \end{array}$$

so that $\dot{V} \leq -\sum\limits_i K_{v_i} K_{z_i} - \sum\limits_i g_i |v_i| - \sum\limits_i K_{\theta_i} \left(\theta_i - \theta_{di} - \theta_{nhi}\right)^2$ where the inequality $\frac{x}{\tanh x} \geq 1$ for $x \geq 0$.

The candidate Lyapunov function is nonsmooth whenever $v_i = 0$ for some i. The generalized gradient of V and the Filippov set of the closed loop system by are respectively given by

$$\partial V = \left[\begin{array}{c} \sum\limits_i \nabla_1 \varphi_i \\ \vdots \\ \sum\limits_i \nabla_N \varphi_i \\ \partial |v_1| \\ \vdots \\ \partial |v_N| \\ \frac{1}{2} \nabla_{\theta_1} \left(\theta_1 - \theta_{d1} - \theta_{nh1}\right)^2 \\ \vdots \\ \frac{1}{2} \nabla_{\theta_N} \left(\theta_N - \theta_{dN} - \theta_{nhN}\right)^2 \\ \frac{1}{2} \nabla_{\theta_{nh1}} \left(\theta_1 - \theta_{d1} - \theta_{nh1}\right)^2 \\ \vdots \\ \frac{1}{2} \nabla_{\theta_{nhN}} \left(\theta_N - \theta_{dN} - \theta_{nhN}\right)^2 \end{array} \right], K[f] = \left[\begin{array}{c} v_1 \cos\theta_1 \\ v_1 \sin\theta_1 \\ \vdots \\ v_N \cos\theta_N \\ v_N \sin\theta_N \\ u_1 \\ \vdots \\ u_N \\ \omega_1 \\ \vdots \\ \omega_N \\ \dot{\theta}_{nh1} \\ \vdots \\ \dot{\theta}_{nhN} \end{array} \right] = \left[\begin{array}{c} v_1 \cos\theta_1 \\ v_1 \sin\theta_1 \\ \vdots \\ v_N \cos\theta_N \\ v_N \sin\theta_N \\ K[u_1] \\ \vdots \\ K[u_N] \\ \omega_1 \\ \vdots \\ \omega_N \\ \dot{\theta}_{nh1} \\ \vdots \\ \dot{\theta}_{nhN} \end{array} \right]$$

We denote by $D \triangleq \{x : \exists i \in \{1, \ldots N\} \, \text{s.t.} v_i = 0\}$ the "discontinuity surface" and $D_S \triangleq \{i \in \{1, \ldots N\} \, \text{s.t.} v_i = 0\}$ the set of indices of agents that participate in D. We then have

$$\begin{aligned}
\dot{\tilde{V}} &= \bigcap_{\xi \in \partial V} \xi^T K\left[f\right] = \\
&v_1 \left(\sum_i \nabla_1 \varphi_i \right) \cdot \eta_1 + \ldots + v_N \left(\sum_i \nabla_N \varphi_i \right) \cdot \eta_N \\
&+ \bigcap_{\xi \in \partial |v_1|} \xi^T K\left[u_1\right] + \ldots + \bigcap_{\xi \in \partial |v_N|} \xi^T K\left[u_N\right] \\
&+ \sum_i \left(\theta_i - \theta_{di} - \theta_{nhi}\right) \left(\omega_i - \dot{\theta}_{nhi}\right) \Rightarrow
\end{aligned}$$

$$\begin{aligned}
\dot{\tilde{V}} &= \sum_{i \notin D_S} \left\{ v_i \left(\sum_i \nabla_i \varphi_j \right) \cdot \eta_i + \operatorname{sgn}\left(v_i\right) u_i \right\} \\
&+ \sum_{i \in D_S} \bigcap_{\xi \in \partial |v_i|} \xi^T K\left[u_i\right] - \sum_i K_{\theta_i} \left(\theta_i - \theta_{di} - \theta_{nhi}\right)^2
\end{aligned}$$

For $i \in D_S$ we have $\partial \left|v_i\right|_{v_i=0} = [-1,1]$ and $K\left[u_i\right]\Big|_{v_i=0} = \left[-\left|K_{vi}K_{zi}\right|, \left|K_{vi}K_{zi}\right|\right]$ so that $\bigcap_{\xi \in \partial |v_i|} \xi^T K\left[u_i\right] = 0$. From the previous analysis we also derive that

$$\begin{aligned}
&\sum_{i \notin D_S} \left\{ v_i \left(\sum_i \nabla_i \varphi_j \right) \cdot \eta_i + \operatorname{sgn}\left(v_i\right) u_i \right\} \leq \\
&- \sum_{i \notin D_S} \left\{ K_{vi} K_{zi} + g_i \left|v_i\right| \right\}
\end{aligned}$$

Going back to Theorem 5 it is easy to see that $v \leq 0 \forall v \in \dot{\tilde{V}}$. Each function V_i is regular as the sum of regular functions ([37]) and V is regular for the same reason. The level sets of V are compact so we can apply this theorem. We have that $S = \{x | 0 \in \dot{\tilde{V}}\} = \{x : (v_i = 0 \forall i) \bigwedge (\theta_i - \theta_{di} = \theta_{nhi} \forall i)\}$. The trajectory of the system converges to the largest invariant subset of S. For this subset to be invariant we must have

$$\dot{v}_i = 0 \Rightarrow K_{vi} K_{zi} = 0 \Rightarrow \left(\nabla_i \varphi_i = 0\right) \wedge \left(q_i = q_{di}\right) \forall i$$

For $\nabla_i \varphi_i = 0$ we have $\theta_{nhi} = 0$ so that $\theta_i = \theta_{di}$. $\diamondsuit$

7 Simulations

To demonstrate the navigation properties of our decentralized approach, we present a series of simulations of multiple agents that have to navigate from an initial to a final configuration, avoiding collision with each other. The chosen configurations constitute non-trivial setups since the straight-line paths connecting initial and final positions of each agent are obstructed by other agents. In the first screenshot of each figure $A - i, T - i$ denote the initial condition and desired destination of agent i respectively.

The first simulation in figure 6 involves 8 holonomic agents with global sensing capabilities. This is a case of decentralized conflict resolution of multiple holonomic agents with global sensing capabilities (see section 2). The guaranteed convergence and collision avoidance properties, as well as the cooperative nature of the proposed strategy, are easily verified. While all agents begin to navigate towards their desired goals, 4 agents return back towards their initial positions and allow the conflict resolution of the rest. Once the workspace is clear, the remaining four agents perform a conflict resolution manoeuver to converge to their final destinations.

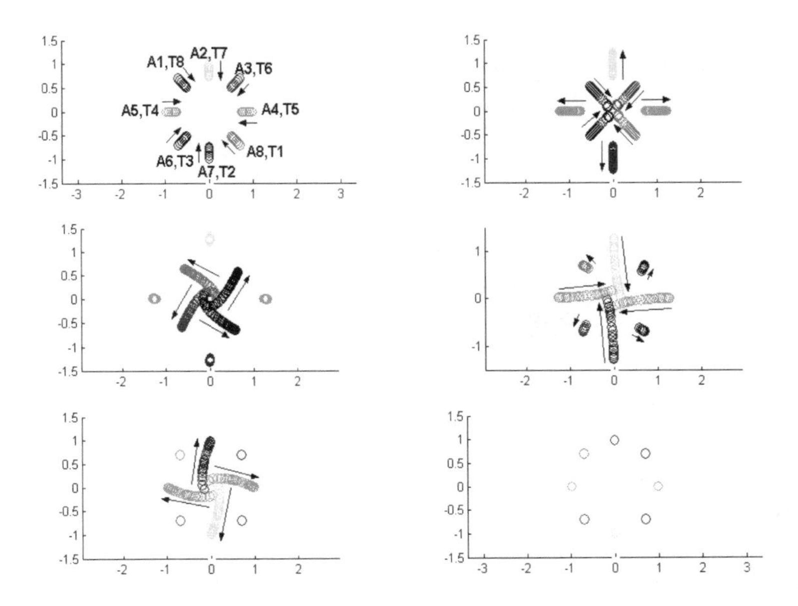

Fig. 6. Decentralized Conflict Resolution for 8 holonomic agents with Global Sensing Capabilities

The second simulation (fig. 7) involves four agents with local sensing capabilities. This is a case of decentralized conflict resolution of multiple holonomic agents with global sensing capabilities (see section 4). Each agent has no knowledge of the positions of agents outside its sensing zone, which is the big circle around its center of mass.

Figure 8 verifies the collision avoidance and global convergence properties of our algorithm in the nonholonomic case encountered in section 5 as well. In the first screenshot of this figure the ring around each target represents the corresponding transition guard where the transition from the first to the second stage takes place. In the second and third screenshot of this figure the four nonholonomic agents are outside their safe set and perform a conflict

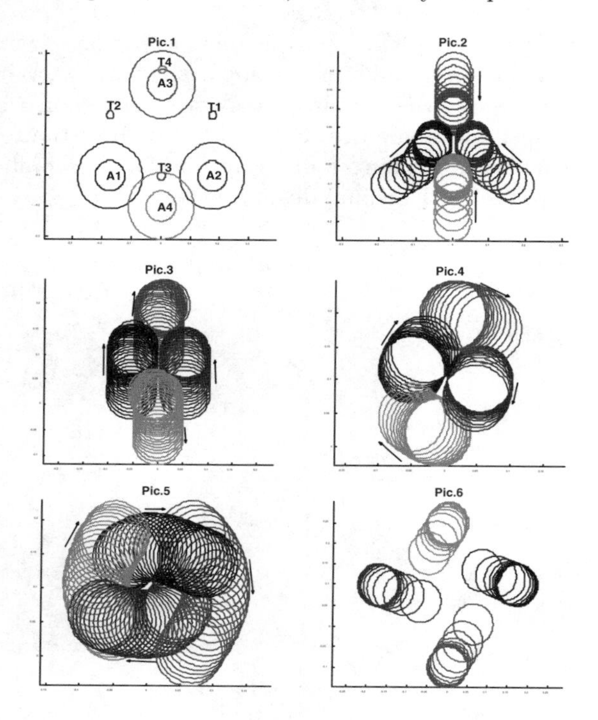

Fig. 7. Decentralized Conflict Resolution for 4 holonomic agents with Limited Sensing Capabilities

resolution maneuver, while in the last two screenshots each agent has entered its safe set surrounding its target, and it converges to its desired configuration.

The navigation properties of the proposed control scheme are verified in the dynamic case as well through the non-trivial simulations in figures 9,10 involving four holonomic and nonholonomic agents respectively. Figure 9 is an illustration of the control scheme developed in subsection 6.1 while figure 10 refers to the control scheme presented in subsection 6.2. The global convergence and collision avoidance properties are verified in this case as well.

The simulations presented in this section highlight the importance of this method as a feedback control strategy that guarantees satisfaction of the imposed specifications, namely collision avoidance and destination convergence, for multiple non-point agents. The results are significant as they deal both with holonomic and nonholonomic mathematical models of vehicle movement. The simulations of dynamic models of figures 9 and 10 have their own importance as the deal with mathematical models of real world applications, such as aircraft and mechanical systems.

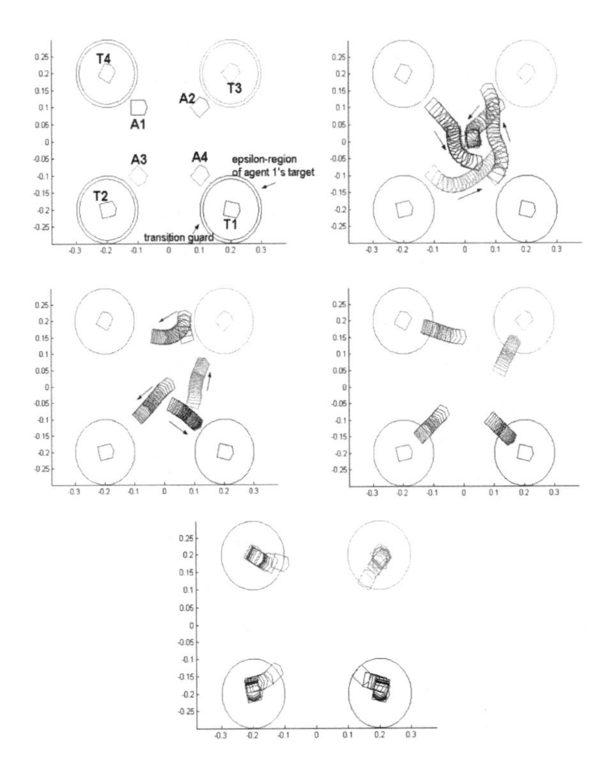

Fig. 8. Decentralized Conflict Resolution for 4 nonholonomic agents

8 Conclusion

In this work, a decentralized methodology for multiple mobile agent navigation has been presented. The methodology extends the centralized multi-agent navigation scheme of the previous chapter to a decentralized approach to the problem. The decentralization factor lies in the fact that each agent requires no knowledge of the desired destinations of the others, and also has limited sensing capabilities with respect to the whereabouts of agents located outside its sensing zone at each time instant. Dynamic models have also been taken into account in the sequel. This is the first to the authors' knowledge extension of centralized multi-agent control using navigation functions, to a decentralized scheme.

Current research includes extending the decentralization scheme to the case where no knowledge of the exact number of agents in the workspace is required as well as coping with three-dimensional models.

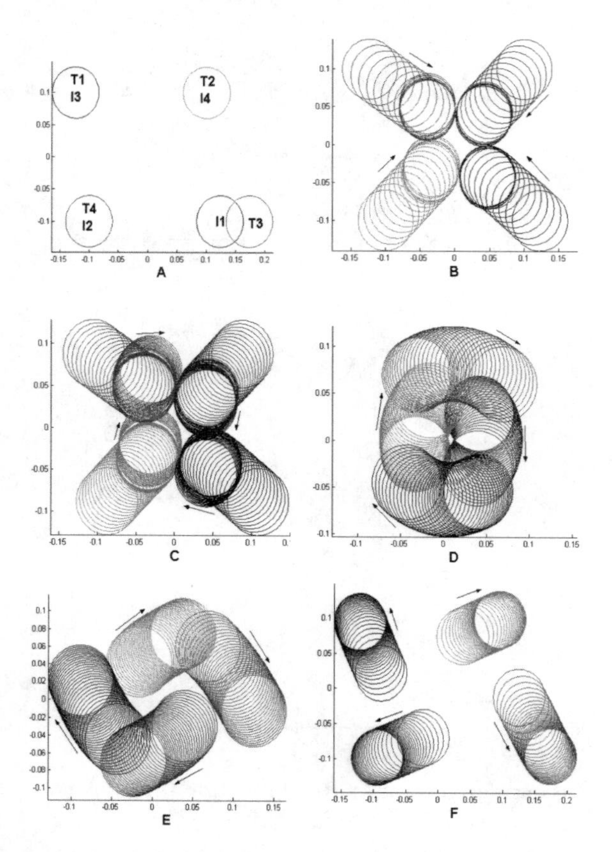

Fig. 9. Decentralized Conflict Resolution for 4 dynamic holonomic agents

A Proofs of Lemmas 1-5

Before proceeding with our proof, we introduce some simplifications concerning terminology. To simplify notation we denote by q instead of q_i the current agent configuration, by q_d instead of q_{di} its goal configuration, by G instead of its "G" function and by q_j the configurations of the other agents. In the proof sketches of Lemmas 1-5 we use the notation $\frac{\partial}{\partial q_i}(\cdot) \triangleq \nabla(\cdot)$ and $\frac{\partial^2}{\partial q_i^2}(\cdot) \triangleq \nabla^2(\cdot)$

A.1 Proof of Lemma 1

At steady state, the function f vanishes due to the constraint $X < G_i(q_{d1}, \ldots, q_{dN})\ \forall i$. Taking the gradient of the definition of φ we have:

$$\nabla\varphi(q_d) = \frac{\left(\gamma_d^k + G\right)^{1/k} \nabla\gamma_d - \gamma_d \nabla\left(\gamma_d^k + G\right)^{1/k}}{\left(\gamma_d^k + G\right)^{2/k}} = 0$$

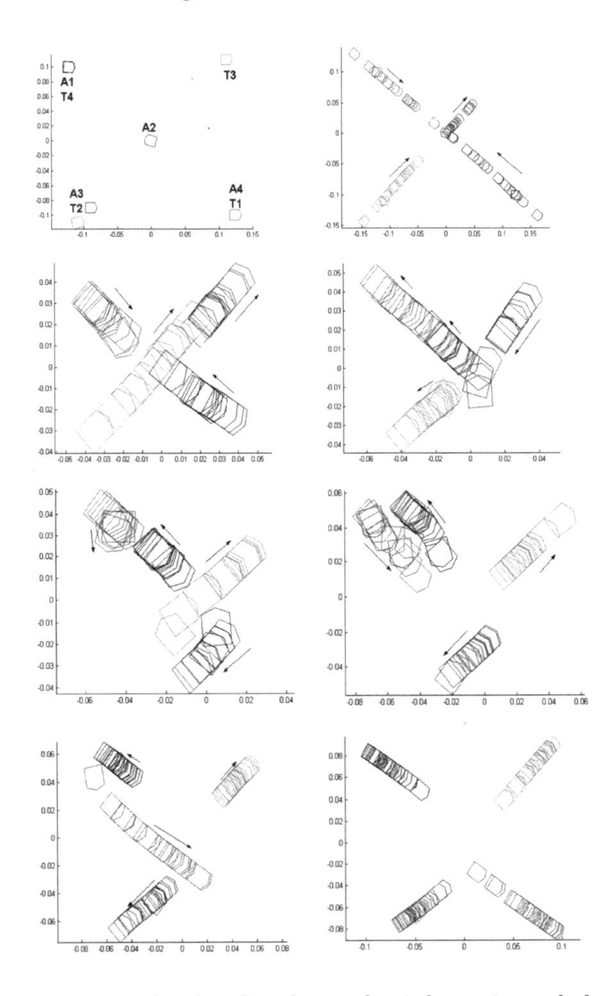

Fig. 10. Decentralized Conflict Resolution for 4 dynamic nonholonomic agents

since both γ_d and $\nabla(\gamma_d)$ vanish by definition at q_d. The Hessian at q_d is

$$\nabla^2\varphi(q_d) = \frac{\left(\gamma_d^k+G\right)^{1/k}\nabla^2\gamma_d - \gamma_d\nabla^2\left(\gamma_d^k+G\right)^{1/k}}{\left(\gamma_d^k+G\right)^{2/k}} =$$
$$= G^{-1/k}\cdot\nabla^2(\gamma_d) = 2G^{-1/k}I$$

which is non-degenerate.$\diamondsuit$

A.2 Proof of Lemma 2

Let q_0 be a point in ϑF and suppose that $(g_{R_a})_b(q_0) = 0$ for some relation a of level b. If the workspace is valid: $\left(g_{R_j}\right)_l(q_0) > 0$ for any level-l and $j \neq a$ since only one RVF can hold at a time. Using the terminology previously defined, and setting $g_i \equiv (g_{R_a})_b (q_0) = 0$, it follows that $\bar{g}_i > 0$. Taking the gradient of φ at q_0 , we obtain:

$$\begin{aligned}
\nabla\varphi(q_0) &= \left.\frac{\left((\gamma_d+f)^k+G\right)^{1/k}\nabla(\gamma_d+f)-(\gamma_d+f)\nabla\left((\gamma_d+f)^k+G\right)^{1/k}}{\left((\gamma_d+f)^k+G\right)^{2/k}}\right|_{q_0} \\
&\overset{G(q_0)=0}{=} \frac{(\gamma_d+f)\nabla(\gamma_d+f)-(\gamma_d+f)\nabla(\gamma_d+f)-\frac{1}{k}(\gamma_d+f)^{2-k}\nabla G}{(\gamma_d+f)^2} = \\
&= -\tfrac{1}{k}(\gamma_d+f)^{-k}\nabla G = -\tfrac{1}{k}(\gamma_d+f)^{-k}\,\bar{g}_i\nabla g_i \neq 0
\end{aligned}$$

A.3 Proof of Lemma 3

At a critical point $q \in C_{\hat{\varphi}} \bigcap F_1(\varepsilon)$ we have:

$$\begin{aligned}
&\hat{\varphi} = \tfrac{\gamma}{G} \Rightarrow \nabla\hat{\varphi} = \tfrac{1}{G^2}(G\nabla\gamma - \gamma\nabla G) \\
&\overset{\nabla\hat{\varphi}=0}{\Rightarrow} G\nabla\gamma = \gamma\nabla G \Rightarrow G\nabla(\gamma_d+f)^k = (\gamma_d+f)^k\nabla G \\
&\Rightarrow kG\nabla(\gamma_d+f) = (\gamma_d+f)\nabla G
\end{aligned}$$

Taking the magnitude of both sides yields:

$$kG\,\|\nabla(\gamma_d+f)\| = (\gamma_d+f)\,\|\nabla G\|$$

A sufficient condition for the above equality not to hold is given by:

$$\frac{(\gamma_d+f)\,\|\nabla G\|}{G\,\|\nabla(\gamma_d+f)\|} < k, \forall q \in F_1(\varepsilon)$$

An upper bound for the left side is given by:

$$\begin{aligned}
&\tfrac{(\gamma_d+f)\|\nabla G\|}{G\|\nabla(\gamma_d+f)\|} < \tfrac{(\gamma_d+f)}{\|\nabla(\gamma_d+f)\|}\cdot\sum_{l=1}^{n_L}\sum_{j=1}^{n_{R,l}}\tfrac{\overline{G}_{j,l}}{G}\left\|\nabla\left(g_{R_j}\right)_l\right\| < \\
&< \tfrac{1}{\varepsilon}\cdot\frac{\max\limits_W\{\gamma_d\}+\max\limits_W\{f\}\cdot\sum_{l=1}^{n_L}\sum_{j=1}^{n_{R,l}}\max\limits_W\nabla\left(g_{R_j}\right)_l}{\min\limits_W\|\nabla(\gamma_d+f)\|} = \\
&= \tfrac{1}{\varepsilon}\cdot\frac{\max\limits_W\{\gamma_d\}+Y\cdot\sum_{l=1}^{n_L}\sum_{j=1}^{n_{R,l}}\max\limits_W\nabla\left(g_{R_j}\right)_l}{\min\limits_W\|\nabla(\gamma_d+f)\|}
\end{aligned}$$

since: $\left(g_{R_j}\right)_l \geq \varepsilon.\Diamond$

A.4 Proof of Lemma 4

If $q \in F_0(\varepsilon) \cap C_{\hat{\varphi}}$, where $C_{\hat{\varphi}}$ is the set of critical points, then $q \in B_i^L(\varepsilon)$ for at least one set $\{L, i\}, i \in \{1...n_{R,L}\}, L \in \{1...n_L\}$, with n_L the number of levels and $n_{R,L}$ the number of relations in level L. We will use a unit vector as a test direction to demonstrate that $\left(\nabla^2\hat{\varphi}\right)(q)$ has at least one negative eigenvalue. At a critical point,

$$(\nabla\hat{\varphi})(q) = \frac{kG(\gamma_d+f)^{k-1}\nabla(\gamma_d+f)-(\gamma_d+f)^k\nabla G}{G^2} = 0$$

Hence,

$$k\cdot G\cdot\nabla(\gamma_d+f) = (\gamma_d+f)\cdot\nabla G \tag{28}$$

The Hessian at a critical point is:

$$(\nabla^2\hat{\varphi})(q) = \frac{1}{G^2}\left(G\cdot\nabla^2(\gamma_d+f)^k-(\gamma_d+f)^k\cdot\nabla^2 G\right)$$

and expanding

$$\begin{array}{l}(\nabla^2\hat{\varphi})(q) = \frac{(\gamma_d+f)^{k-2}}{G^2}\cdot \\ \cdot\left\{kG\left[\begin{array}{l}(\gamma_d+f)\nabla^2(\gamma_d+f)+\\(k-1)\nabla(\gamma_d+f)\nabla(\gamma_d+f)^T\end{array}\right]-(\gamma_d+f)^2\nabla^2 G\right\}\end{array}$$

Taking the outer product of both sides of equation 28, we get:

$$(kG)^2\nabla(\gamma_d+f)\nabla(\gamma_d+f)^T = (\gamma_d+f)^2\nabla G\nabla G^T \tag{29}$$

Substituting equation 29 in equation 28, we get:

$$(\nabla^2\hat{\varphi})(q) = \frac{(\gamma_d+f)^{k-1}}{G^2}\left\{\begin{array}{l}kG\nabla^2(\gamma_d+f)+\\+\left(1-\frac{1}{k}\right)\frac{(\gamma_d+f)}{G}\nabla G\nabla G^T-\\-(\gamma_d+f)\nabla^2 G\end{array}\right\}$$

We choose the test vector (unit magnitude) to be:$\hat{u} = \frac{\nabla b_i(q_c)^\perp}{\left\|\nabla b_i(q_c)^\perp\right\|}$. By its definition $\hat{u}$ is orthogonal to ∇b_i at a critical point q_c, and so the following properties hold:$\hat{u}^T\cdot\nabla b_i = 0$ and $\nabla b_i^T\cdot\hat{u} = 0$. With $\nabla^2(\gamma_d+f) = 2\cdot\mathrm{I}+\nabla^2 f$, we form the quadratic form:

$$\begin{array}{l}\frac{G^2}{(\gamma_d+f)^{k-1}}\hat{u}^T(\nabla^2\hat{\varphi})(q)\hat{u} = 2kG+kG\hat{u}^T\nabla^2 f\hat{u}\\+\left(1-\frac{1}{k}\right)\frac{(\gamma_d+f)}{G}\hat{u}^T\nabla G\nabla G^T\hat{u}-(\gamma_d+f)\hat{u}^T\nabla^2 G\hat{u}\end{array}$$

After many nontrivial calculation we get

$$\begin{array}{l}\frac{G^2}{(\gamma_d+f)^{k-1}}\hat{u}^T(\nabla^2\hat{\varphi})(q)\hat{u} = \\ \bar{g}_i c_i\left(1+\frac{a_0}{\gamma_d}\right)\left(\frac{1}{2}\nabla b_i^T\nabla\gamma_d-\upsilon_i\gamma_d\right)\\ +g_i\left\{\begin{array}{l}k\bar{g}_i\hat{u}^T\nabla^2 f\hat{u}+(\gamma_d+f)\eta_i-(\gamma_d+f)\psi_i+\frac{z_2}{2\gamma_d}\\ -\upsilon_i\bar{g}_i c_i\left(\sum\limits_{j=1}^{3}a_j g_i^{j-1}\bar{g}_i^j\right)-\zeta_i\end{array}\right\}\end{array} \tag{30}$$

where $c_i = 1+\frac{\lambda}{b_i+\tilde{b}_i^{1/h}}$,$\upsilon_i = 2\cdot l$,$l$ the relation level,

$$\eta_i = \left(1 - \frac{1}{k}\right) \left[\begin{array}{l} \frac{\hat{u}^T \nabla \bar{g}_i \nabla \bar{g}_i^T \hat{u}}{\bar{g}_i} - 2\lambda \frac{\hat{u}^T \nabla \bar{g}_i \left(\nabla \tilde{b}_i^{1/h}\right)^T \hat{u}}{c_i \left(b_i + \tilde{b}_i^{1/h}\right)^2} + \\ + \lambda^2 \bar{g}_i \frac{\hat{u}^T \nabla \tilde{b}_i^{1/h} \left(\nabla \tilde{b}_i^{1/h}\right)^T \hat{u}}{c_i^2 \left(b_i + \tilde{b}_i^{1/h}\right)^4} \end{array} \right]$$

$$\begin{array}{l} \psi_i = \hat{u}^T \cdot \nabla^2 \bar{g}_i \cdot \hat{u} + \frac{\bar{g}_i}{c_i} \cdot \hat{u}^T \cdot B_i \cdot \hat{u} - \\ -2 \frac{\lambda}{c_i \left(b_i + \tilde{b}_i^{1/h}\right)^2} \cdot \hat{u}^T \cdot \nabla \tilde{b}_i^{1/h} \cdot \nabla \bar{g}_i \cdot \hat{u} \end{array}$$

$$B_i = \lambda \left[\begin{array}{l} 2 \frac{\left(\nabla b_i + \nabla \tilde{b}_i^{1/h}\right) \left(\nabla b_i + \nabla \tilde{b}_i^{1/h}\right)^T}{\left(b_i + \tilde{b}_i^{1/h}\right)^3} - \\ - \frac{\nabla^2 b_i + \nabla^2 \tilde{b}_i^{1/h}}{\left(b_i + \tilde{b}_i^{1/h}\right)^2} \end{array} \right]$$

$$\begin{array}{l} z_2 \left(g_i, \bar{g}_i, \nabla g_i, \nabla \bar{g}_i\right) = \gamma_d \nabla \bar{g}_i^T \nabla \gamma_d + f \nabla \bar{g}_i^T \nabla \gamma_d + \ldots \\ -k \bar{g}_i \left(2 \nabla \gamma_d^T \cdot \nabla f - \nabla f^T \cdot \nabla f\right) \end{array}$$

$$\zeta_i = \frac{\lambda \bar{g}_i}{2 c_i \left(b + \tilde{b}^{1/h}\right)^2} \left(\nabla b + \nabla \tilde{b}^{1/h}\right)^T \cdot \nabla \gamma_d$$

Setting: $\tilde{\mu}_i = \left(1 + \frac{a_0}{\gamma_d}\right) \cdot \mu_i$ where $\mu_i = \frac{1}{2} \nabla b_i^T \nabla \gamma_d - \upsilon_i \cdot \gamma_d$ equation (30) becomes:

$$\begin{array}{l} \frac{G^2}{(\gamma_d + f)^{k-1}} \hat{u}^T \left(\nabla^2 \hat{\varphi}\right)(q) \hat{u} = \bar{g}_i \cdot c_i \cdot \tilde{\mu}_i + \\ g_i \left\{ \begin{array}{l} k \cdot \bar{g}_i \cdot \hat{u}^T \cdot \nabla^2 f \cdot \hat{u} + (\gamma_d + f) \cdot \eta_i - (\gamma_d + f) \cdot \psi_i \\ + \frac{z_2}{2\gamma_d} - \upsilon_i \bar{g}_i c_i \left(\sum\limits_{j=1}^{3} a_j g_i^{j-1} \bar{g}_i^j \right) - \zeta_i \end{array} \right\} \end{array} \tag{31}$$

The second term is proportional to g_i and can be made arbitrarily small by a suitable choice of ε but can still be positive, so the first term should be strictly negative.

From the result of Lemma 7 in [7], we have:

$$\begin{array}{l} \max\limits_{q \in F_0} \{\mu_i\} = \\ = \frac{2}{l} \left(\begin{array}{l} \frac{1}{l} \sqrt{\left\|\sum q_j\right\|^2 - l \sum \left\|q_j\right\|^2 + l \left(\sum (r + r_j)^2 + \varepsilon\right)} \\ - \left\|l q_d - \sum q_j\right\| \end{array} \right). \\ \cdot \left\|l q_d - \sum q_j\right\| \end{array}$$

For ε small enough, $\max\limits_{q \in F_0} \{\mu_i\}$ is negative. Moreover, the term $\left(1 + \frac{a_0}{\gamma_d}\right)$ is always greater than one, since we have assumed that $a_0 > 0$, and $\gamma_d > 0$ for $q \in F_0(\varepsilon)$. Thus for ε small enough, $\tilde{\mu}_i$ is also negative. So, for $\tilde{\mu}_i$, according to Lemma 1, it is sufficient to make sure that:

$$\begin{aligned}&\tfrac{1}{l}\cdot\sqrt{\left\|\sum q_j\right\|^2 - l\cdot\sum\left\|q_j\right\|^2 + l\cdot\left(\sum\left(r+r_j\right)^2+\varepsilon\right)} < \\ &< \left\|l\cdot q_d - \sum q_j\right\| \Rightarrow \varepsilon < l\cdot\left\|l\cdot q_d - \sum q_j\right\|^2 + \sum\left\|q_j\right\|^2 - \\ &-\tfrac{1}{l}\cdot\left\|\sum q_j\right\|^2 - \sum\left(r+r_j\right)^2 \equiv \varepsilon_0\end{aligned}$$

An other constraint arises from the fact that $\varepsilon > 0.$. So for a valid workspace it will be:

$$l\cdot\left\|l\cdot q_d - \sum q_j\right\|^2 + \sum\left\|q_j\right\|^2 - \frac{1}{l}\cdot\left\|\sum q_j\right\|^2 > \sum\left(r+r_j\right)^2$$

◇

A.5 Proof of Lemma 5

From the proof of the previous Lemma, we have at a critical point

$$\begin{aligned}&\tfrac{G^2}{(\gamma_d+f)^{k-1}}\left(\nabla^2\hat{\varphi}\right) = kG\nabla^2\left(\gamma_d+f\right) + \\ &\left(1-\tfrac{1}{k}\right)\tfrac{\gamma_d+f}{G}\nabla G\nabla G^T - \left(\gamma_d+f\right)\nabla^2 G\end{aligned}$$

We also have $\nabla f = \left(\underbrace{\sum_{j=1}^{3} j a_j G_i^{j-1}}_{\sigma(G)}\right)\nabla G$ and $\nabla^2 f = \sigma\nabla^2 G + \sigma^*\nabla G\nabla G^T, \sigma^* = \sum_{j=2}^{3} j(j-1)a_j G^{j-2}$. At a critical point:

$$\begin{aligned}&kG\nabla\left(\gamma_d+f\right) = \left(\gamma_d+f\right)\nabla G \Rightarrow \\ &kG\nabla\gamma_d = \left(\gamma_d+f\right)\nabla G - kG\nabla f \Rightarrow \\ &kG\nabla\gamma_d = \left(\gamma_d+f-kG\sigma(G)\right)\nabla G \Rightarrow \\ &G\nabla\gamma_d = \left\{\underbrace{\frac{\gamma_d+f}{k} - G\sigma(G)}_{-\sigma_i}\right\}\nabla G\end{aligned}$$

Taking the magnitude from both sides we have $2kG = \frac{k|\sigma_i|^2}{2G\gamma_d}\left\|\nabla G\right\|^2$. Choosing $\tilde{u} = \widehat{\nabla b_i}$ as a test direction and after some manipulation we have

$$\begin{aligned}&\tfrac{G^2}{k(\gamma_d+f)^{k-1}}\tilde{u}^T\left(\nabla^2\hat{\varphi}\right)\tilde{u} = \underbrace{\frac{|\sigma_i|^2}{2G\gamma_d}\left\|\nabla G\right\|^2}_{L} + \\ &\underbrace{\xi\tilde{u}^T\nabla G\nabla G^T\tilde{u}}_{M} + \underbrace{\sigma_i\tilde{u}^T\nabla^2 G\tilde{u}}_{N}\end{aligned}$$

where

$$\xi = \left(1-\frac{1}{k}\right)\frac{\gamma_d + Y}{kG} + \sum_{j=2}^{3}\left\{kj(j-1) + \left(1-\frac{1}{k}\right)\right\}\frac{a_j}{k}G^{j-1}$$

After some manipulation, we have

$$\begin{array}{l} L+M+N \geq \frac{|\sigma_i|^2}{2G\gamma_d}\left\{\begin{array}{l} g_i^2\left\|\nabla\bar{g}_i\right\|^2 + \bar{g}_i^2\left\|\nabla g_i\right\|^2 - \\ 2G\left\|\nabla\bar{g}_i\right\|\left\|\nabla g_i - 2\left(\tilde{u}^T\nabla g_i\right)\tilde{u}\right\| \end{array}\right\} \\ +2\left(\frac{|\sigma_i|^2}{\gamma_d} + \xi G + \sigma_i\right)\left(\tilde{u}^T\nabla g_i\right)\left(\nabla\bar{g}_i\tilde{u}\right) \\ +\xi\bar{g}_i^2\left(\tilde{u}^T\nabla g_i\right)^2 + \sigma_i\tilde{u}^T\left(g_i\nabla^2\bar{g}_i + \bar{g}_i\nabla^2 g_i\right)u \end{array}$$

But $\left\|\nabla g_i - 2\left(\tilde{u}^T\nabla g_i\right)\tilde{u}\right\|^2 = \left\|\nabla g_i\right\|^2$ so that

$$\begin{array}{l} g_i^2\left\|\nabla\bar{g}_i\right\|^2 + \bar{g}_i^2\left\|\nabla g_i\right\|^2 - \\ 2G\left\|\nabla\bar{g}_i\right\|\left\|\nabla g_i - 2\left(\tilde{u}^T\nabla g_i\right)\tilde{u}\right\| = \left(g_i\left\|\nabla\bar{g}_i\right\| - \bar{g}_i\left\|\nabla g_i\right\|\right)^2 \end{array}$$

so that

$$\begin{array}{l} L+M+N \geq 2\left(\frac{|\sigma_i|^2}{\gamma_d} + \xi G + \sigma_i\right)\left(\tilde{u}^T\nabla g_i\right)\left(\nabla\bar{g}_i\tilde{u}\right) \\ +\xi\bar{g}_i^2\left(\tilde{u}^T\nabla g_i\right)^2 + \sigma_i\tilde{u}^T\left(g_i\nabla^2\bar{g}_i + \bar{g}_i\nabla^2 g_i\right)u \end{array}$$

It is shown in [7] that the second term, which is strictly positive, dominates the third and the first term for sufficiently small ε.

B Proof of Proposition 2

In the proof sketch of Proposition 2, the terms $\nabla(\cdot)$, $\nabla^2(\cdot)$ have their usual meaning and refer to the whole state space and not a single agent, namely $\nabla(\cdot) \triangleq \left[\frac{\partial}{\partial q_1}(\cdot), \ldots, \frac{\partial}{\partial q_N}(\cdot)\right]^T$ and $\nabla^2(\cdot) \triangleq \left[\frac{\partial^2}{\partial q_{ij}}(\cdot)\right]$.

We immediately note that the following proof is existential rather than computational. We show that a finite k that renders the system almost everywhere asymptotically stable *exists*, but we do not provide an analytical expression for this lower bound. However, practical values of k have been provided in the simulation section.
Let us recall that the Proximity function between agents i and j is given by:

$$\beta_{ij}(q) = \left\|q_i - q_j\right\|^2 - (r_i + r_j)^2 = q^T D_{ij} q - (r_i + r_j)^2$$

where the $2N \times 2N$ matrix D_{ij} is given by:

$$D_{ij} = \left[\begin{array}{ccccc} & & O_{2(i-1)\times 2N} & & \\ O_{2\times 2(i-1)} & I_{2\times 2} & O_{2\times 2(j-i-1)} & -I_{2\times 2} & O_{2\times 2(N-j)} \\ & & O_{2(j-i-1)\times 2N} & & \\ O_{2\times 2(i-1)} & -I_{2\times 2} & O_{2\times 2(j-i-1)} & I_{2\times 2} & O_{2\times 2(N-j)} \\ & & O_{2(N-j)\times 2N} & & \end{array}\right]$$

We can also write $b_r^i = q^T P_r^i q - \sum_{j \in P_r} (r_i + r_j)^2$,where $P_r^i = \sum_{j \in P_r} D_{ij}$, and P_r denotes the set of binary relations in relation r. It can easily be seen that $\nabla b_r^i = 2P_r^i q, \nabla^2 b_r^i = 2P_r^i$. We also use the following notation for the r-th relation wrt agent i:

$$g_r^i = b_r^i + \frac{\lambda b_r^i}{b_r^i + (\tilde{b}_r^i)^{1/h}}, \tilde{b}_r^i = \prod_{\substack{s \in S_r \\ s \neq r}} b_s^i,$$

$$\nabla \tilde{b}_r^i = \sum_{\substack{s \in S_r \\ s \neq r}} \underbrace{\prod_{\substack{t \in S_r \\ t \neq s,r}} b_i^t}_{\tilde{b}_{s,r}^i} \cdot 2P_s^i q$$

where S_r denotes the set of relations in the same level with relation r. An easy calculation shows that

$$\nabla g_r^i = \ldots = 2\left[d_r^i P_r^i - w_r^i \tilde{P}_r^i\right] q \triangleq Q_r^i q, \tilde{P}_r^i \triangleq \sum_{\substack{s \in S_r \\ s \neq r}} \tilde{b}_{s,r}^i P_s^i$$

where $d_r^i = 1 + (1 - \frac{b_r^i}{b_r^i + (\widetilde{b_r^i})^{1/h}}) \frac{\lambda}{b_r^i + (\widetilde{b_r^i})^{1/h}}$, $w_r^i = \frac{\lambda b_r^i (\widetilde{b_r^i})^{\frac{1}{h}-1}}{h(b_r^i + (\widetilde{b_r^i})^{1/h})^2}$. The gradient of the G_i function is given by:

$$G_i = \prod_{r=1}^{N_i} g_r^i \Rightarrow \nabla G_i = \sum_{r=1}^{N_i} \underbrace{\prod_{\substack{l=1 \\ l \neq r}}^{N_i} g_l^i}_{\tilde{g}_r^i} \nabla g_r^i = \sum_{r=1}^{N_i} \tilde{g}_r^i Q_r^i q \triangleq Q_i q$$

We define $\nabla G \triangleq \begin{bmatrix} \nabla G_1 \\ \vdots \\ \nabla G_N \end{bmatrix} = \begin{bmatrix} Q_1 \\ \vdots \\ Q_N \end{bmatrix} q \triangleq Qq$

Remembering that $u_i = -K_i \frac{\partial \varphi_i}{\partial q_i}$ and that $\varphi_i = \frac{\gamma_{di} + f_i}{((\gamma_{di} + f_i)^k + G_i)^{1/k}}, f_i = \sum_{j=0}^{3} a_i G_i^j$ the closed loop dynamics of the system are given by:

$$\dot{q} = \begin{bmatrix} -K_1 A_1^{-(1+1/k)} \left\{ G_1 \frac{\partial \gamma_{d1}}{\partial q_1} + \sigma_1 \frac{\partial G_1}{\partial q_1} \right\} \\ \vdots \\ -K_N A_N^{-(1+1/k)} \left\{ G_N \frac{\partial \gamma_{dN}}{\partial q_N} + \sigma_N \frac{\partial G_N}{\partial q_N} \right\} \end{bmatrix} = \ldots$$
$$= -A_K G (\partial \gamma_d) - A_K \Sigma Q q$$

where $(\partial \gamma_d) = \left[\frac{\partial \gamma_{d1}}{\partial q_1} \ldots \frac{\partial \gamma_{dN}}{\partial q_N} \right]^T, \sigma_i = G_i \sigma(G_i) - \frac{\gamma_{di} + f_i}{k}, \sigma(G_i) = \sum_{j=1}^{3} j a_j G_i^{j-1}, A_i = (\gamma_{di} + f_i)^k + G_i$ and the matrices

$$G \triangleq \underbrace{diag\left(G_1, G_1, \ldots, G_N, G_N\right)}_{2N \times 2N}$$

$$A_K \triangleq \underbrace{diag\left(\begin{array}{c} K_1 A_1^{-(1+1/k)}, K_1 A_1^{-(1+1/k)}, \ldots, \\ K_N A_N^{-(1+1/k)}, K_N A_N^{-(1+1/k)} \end{array}\right)}_{2N \times 2N}$$

$$\Sigma \triangleq \underbrace{\left[\underbrace{\Sigma_1}_{2N \times 2N}, \ldots, \underbrace{\Sigma_N}_{2N \times 2N}\right]}_{2N \times 2N^2},$$

$$\Sigma_i = diag\left(0, 0, \ldots, \underbrace{\sigma_i, \sigma_i}_{2i-1, 2i}, \ldots, 0, 0\right)$$

By using $\varphi = \sum_i \varphi_i$ as a candidate Lyapunov function we have

$$\varphi = \sum_i \varphi_i \Rightarrow \dot{\varphi} = \left\{\sum_i (\nabla \varphi_i)^T\right\} \dot{q},$$
$$\nabla \varphi_i = A_i^{-(1+1/k)} \left\{G_i \nabla \gamma_{di} + \sigma_i \nabla G_i\right\}$$

and after some trivial calculation $\sum_i (\nabla \varphi_i)^T = \ldots = (\partial \gamma_d)^T A_G + q^T Q^T A_\Sigma$ where

$$A_G = \underbrace{diag\left(\begin{array}{c} G_1 A_1^{-(1+1/k)}, G_1 A_1^{-(1+1/k)}, \ldots, \\ G_N A_N^{-(1+1/k)}, G_N A_N^{-(1+1/k)} \end{array}\right)}_{2N \times 2N}$$

$$A_\Sigma = \underbrace{\left[\begin{array}{c} \underbrace{A_{\Sigma_1}}_{2N \times 2N} \\ \vdots \\ \underbrace{A_{\Sigma_N}}_{2N \times 2N} \end{array}\right]}_{2N^2 \times 2N}, A_{\Sigma_i} = \underbrace{diag\left(\begin{array}{c} A_i^{-(1+1/k)} \sigma_i, \ldots, \\ A_i^{-(1+1/k)} \sigma_i \end{array}\right)}_{2N \times 2N}$$

So we have

$$\dot{\varphi} = \left\{\sum_i (\nabla \varphi_i)^T\right\} \dot{q} = \ldots =$$
$$= -\left[\begin{array}{cc} (\partial \gamma_d)^T & q^T \end{array}\right] \underbrace{\left[\begin{array}{cc} M_1 & M_2 \\ M_3 & M_4 \end{array}\right]}_{M} \left[\begin{array}{c} \partial \gamma_d \\ q \end{array}\right]$$

where $M_1 = A_G A_K G$, $M_2 = A_G A_K \Sigma Q$, $M_3 = Q^T A_\Sigma A_K G$, $M_4 = Q^T A_\Sigma A_K \Sigma Q$.

In [7], we provide an analytic expression for the elements of the matrix Q.

We examine the positive definiteness of the matrix M by use of the following theorems:

Theorem 6. [18]*: Given a matrix* $A \in \Re^{n\times n}$ *then all its eigenvalues lie in the union of n discs:*

$$\bigcup_{i=1}^{n}\left\{z : |z-a_{ii}| \le \sum_{\substack{j=1\\ j\neq i}}^{n} |a_{ij}|\right\} \triangleq \bigcup_{i=1}^{n} R_i(A) \triangleq R(A)$$

Each of these discs is called a Gersgorin disc of A.

Corollary 1. [18]*: Given a matrix* $A \in \Re^{n\times n}$ *and n positive real numbers* $p_1,\ldots,p_n$ *then all its eigenvalues of A lie in the union of n discs:*

$$\bigcup_{i=1}^{n}\left\{z : |z-a_{ii}| \le \frac{1}{p_i}\sum_{\substack{j=1\\ j\neq i}}^{n} p_j\,|a_{ij}|\right\}$$

A key point of Corollary 1 is that if we bound the first $n/2$ Gersgorin discs of a matrix A sufficiently away from zero, then an appropriate choice of the numbers $p_1,\ldots,p_n$ renders the remaining $n/2$ discs sufficiently close to the corresponding diagonal elements. Hence, by ensuring the positive definiteness of the eigenvalues of the matrix M corresponding to the first $n/2$ rows, then we can render the remaining ones sufficiently close to the corresponding diagonal elements. This fact will be made clearer in the analysis that follows.

Some useful bounds are obtained by the following lemma:

Lemma 9. *: The following bounds hold for the terms* $Q_{ii}^i, Q_{ii}^j, \sigma_i$

$$\sigma_i(\varepsilon) \in \begin{cases} \left[-Y\left(\frac{1}{k}+\frac{8}{9}\right)-\frac{\gamma_{di}}{k}, \underbrace{-\frac{Y}{k}-\frac{\gamma_{di}}{k}}_{\sigma_i(0)}\right], 0 \le \varepsilon \le \varepsilon^* \\ \left[-Y\left(\frac{1}{k}+\frac{8}{9}\right)-\frac{\gamma_{di}}{k}, \underbrace{-\frac{\gamma_{di}}{k}}_{\sigma_i(X)}\right], X \ge \varepsilon \ge \varepsilon^* \end{cases}$$

$$0 < Q_{ii}^i < \left|Q_{ii}^i\right|_{\max} < \infty$$

and

$$0 < Q_{ii}^j < \left|Q_{ii}^j\right|_{\max} < \infty$$

Proof: See [7].

Let us examine the Gersgorin discs of the first half rows of the matrix M. We denote this procedure as $M_1 - M_2$, as the main diagonal elements of M_1 are "compared" with the corresponding raw elements of M_2. Note that the submatrices M_1, M_2 are both diagonal, therefore the only nonzero elements of raw i of the $4N \times 4N$ matrix M are the elements $M_{ii}, M_{i,2N+i}$ where of course $1 \le i \le 2N$ as we calculate the Gersgorin discs of the first half rows of the matrix M. We have:

$$\begin{array}{l} |z - M_{ii}| \le \frac{1}{p_i} \sum\limits_{j \ne i} p_j \left| M_{ij} \right|, 1 \le i \le 2N \Rightarrow \\ \left| z - A_i^{-2(1+1/k)} K_i G_i^2 \right| \le \frac{p_{2N+i}}{p_i} \left| A_i^{-2(1+1/k)} \sigma_i K_i G_i Q_{ii}^i \right| \Rightarrow \\ \Rightarrow z \ge A_i^{-2(1+1/k)} K_i G_i^2 - \frac{p_{2N+i}}{p_i} \left| A_i^{-2(1+1/k)} \sigma_i K_i G_i Q_{ii}^i \right| \end{array}$$

We examine the following three cases:

- $G_i < \varepsilon$ At a critical point in this region, the corresponding eigenvalue tends to zero, so that the derivative of the Lyapunov function could achieve zero values. However, the result of Lemma 5 indicates that φ_i is a Morse function, hence its critical points are isolated [21]. Thus the set of initial conditions that lead to saddle points are sets of measure zero [31].
- $G_i > X$ The corresponding eigenvalue is guaranteed to be positive as long as:
$$\begin{array}{l} z > 0 \Leftarrow A_i^{-2(1+1/k)} K_i \left(G_i - \frac{p_{2N+i}}{p_i} \left| \sigma_i Q_{ii}^i \right| \right) > 0 \Leftarrow \\ G_i \ge X > \frac{p_{2N+i}}{p_i} \left| \sigma_i Q_{ii}^i \right| = \frac{\gamma_{di}}{k} \frac{p_{2N+i}}{p_i} \left| Q_{ii}^i \right| \Leftarrow \\ \Leftarrow k > \frac{(\gamma_{di})_{\max}}{X} \frac{p_{2N+i}}{p_i} \left| Q_{ii}^i \right|_{\max} \end{array}$$
- $0 < \varepsilon \le G_i \le X$
$$\begin{array}{l} z > 0 \Leftarrow \varepsilon > \left\{ Y \left| \frac{1}{k} + \frac{8}{9} \right| + \left| \frac{\gamma_{di}}{k} \right| \right\} \frac{p_{2N+i}}{p_i} \left| Q_{ii}^i \right|_{\max} \Leftarrow \\ \Leftarrow \varepsilon > 2 \max \left\{ \begin{array}{c} 2 \max \left\{ \frac{Y}{k}, \frac{8Y}{9} \right\}, \\ \left| \frac{(\gamma_{di})_{\max}}{k} \right| \end{array} \right\} \frac{p_{2N+i}}{p_i} \left| Q_{ii}^i \right|_{\max} \\ \stackrel{Y \le \frac{\Theta_1}{k}}{\Longleftarrow} k > 2 \max \left\{ \begin{array}{c} 2\sqrt{\frac{\Theta_1}{\varepsilon}}, \frac{16\Theta_1}{9\varepsilon}, \\ \frac{(\gamma_{di})_{\max}}{\varepsilon} \end{array} \right\} \frac{p_{2N+i}}{p_i} \left| Q_{ii}^i \right|_{\max} \end{array}$$

A key point is that there is no restriction on how to select the terms $\frac{p_{2N+i}}{p_i}$. This will help us in deriving bounds that guarantee the positive definiteness of the matrix M.

Let us examine the Gersgorin discs of the second half rows of the matrix M. Likewise, we denote this procedure as $M_3 - M_4$. The discs of Corollary 1 are evaluated:

$$\begin{array}{l} |z - M_{ii}| \le \sum\limits_{j \ne i} \frac{p_j}{p_i} \left| M_{ij} \right|, 2N + 1 \le i \le 4N, 1 \le j \le 4N \Rightarrow \\ \Rightarrow \left| z - (M_4)_{ii} \right| \le R_i(M_3) + R_i(M_4) \end{array}$$

where

$$(M_4)_{ii} = \sum_j K_i A_i^{-(1+1/k)} A_j^{-(1+1/k)} \sigma_j \sigma_i Q_{ii}^i Q_{ii}^j$$

and

$$\begin{aligned} R_i(M_3) &= \sum_{j=1}^{2N} \tfrac{p_j}{p_i} \left|(M_3)_{ij}\right| = \\ &= \sum_{j=1}^{2N} \tfrac{p_j}{p_i} \left| \sum_l A_l^{-(1+1/k)} \sigma_l A_j^{-(1+1/k)} K_j G_j Q_{ij}^l \right| \\ R_i(M_4) &= \sum_{\substack{j=2N+1 \\ j\neq i}}^{4N} \tfrac{p_j}{p_i} \left|(M_4)_{ij}\right| = \\ &= \sum_{j\neq i} \tfrac{p_j}{p_i} \left| \sum_l (A_l A_j)^{-(1+1/k)} \sigma_l \sigma_j K_j Q_{ij}^l Q_{jj}^j \right| \end{aligned}$$

A sufficient condition for the positive definiteness of the corresponding eigenvalue for raw i is then:

$$\begin{aligned} &(M_4)_{ii} > R_i(M_3) + R_i(M_4) \Leftarrow \\ &\Leftarrow (M_4)_{ii} > \max\{2R_i(M_3), 2R_i(M_4)\} \end{aligned}$$

We first show that we always have $R_i(M_3) \geq R_i(M_4)$. By taking into account the relations $Q_{jk}^i = Q_{kj}^i = 0, Q_{ij}^i = -Q_{jj}^i, j \neq i \neq k \neq j$ and expanding it is easy to see that

$$\begin{aligned} R_i(M_3) &= -\tfrac{1}{p_i} \sum_{j=1}^{2N} p_j \left\{ \begin{array}{l} A_j^{-2(1+1/k)} \sigma_j K_j G_j Q_{ii}^j + \\ (A_j A_i)^{-(1+1/k)} \sigma_i K_j G_j Q_{jj}^i \end{array} \right\} = \\ &= -\sum_{\substack{j=1 \\ j\neq i}}^{2N} \tfrac{p_j}{p} \left\{ \begin{array}{l} \underbrace{A_j^{-2(1+1/k)} \sigma_j K_j G_j Q_{ii}^j +}_{(I)} \\ \underbrace{(A_j A_i)^{-(1+1/k)} \sigma_i K_j G_j Q_{jj}^i}_{(II)} \end{array} \right\} \\ &-2\tfrac{p_i}{p} A_i^{-2(1+1/k)} \sigma_i K_i G_i Q_{ii}^i \end{aligned}$$

where without loss of generality we choose $p_i = p, 2N+1 \leq i \leq 4N$.We also have

$$R_i(M_4) = \sum_{j\neq i} \left\{ \begin{array}{l} \underbrace{A_j^{-2(1+1/k)} \sigma_j^2 K_j Q_{ii}^j Q_{jj}^j +}_{(I)} \\ \underbrace{(A_i A_j)^{-(1+1/k)} \sigma_i \sigma_j K_j Q_{jj}^i Q_{jj}^j}_{(II)} \end{array} \right\}$$

By comparing the terms (I) and (II) in the last two equations we have:

$$\begin{array}{l}
(I): -\frac{p_j}{p} A_j^{-2(1+1/k)} \sigma_j K_j G_j Q_{ii}^j \geq A_j^{-2(1+1/k)} \sigma_j^2 K_j Q_{ii}^j Q_{jj}^j \\
\Leftarrow -\frac{p_j}{p} \sigma_j G_j \geq \sigma_j^2 Q_{jj}^j \Leftrightarrow \sigma_j \left(\sigma_j Q_{jj}^j + \frac{p_j}{p} G_j \right) \leq 0 \\
\overset{\sigma_j < 0}{\Leftrightarrow} \sigma_j Q_{jj}^j + \frac{p_j}{p} G_j \geq 0 \\
(II): -\frac{p_j}{p} \left(A_j A_i \right)^{-(1+1/k)} \sigma_i K_j G_j Q_{jj}^i \geq \\
\geq \left(A_i A_j \right)^{-(1+1/k)} \sigma_i \sigma_j K_j Q_{jj}^i Q_{jj}^j \\
\Leftarrow -\frac{p_j}{p} \sigma_i G_j \geq \sigma_i \sigma_j Q_{jj}^j \Leftrightarrow \sigma_i \left(\sigma_j Q_{jj}^j + \frac{p_j}{p} G_j \right) \leq 0 \\
\overset{\sigma_i < 0}{\Leftrightarrow} \sigma_j Q_{jj}^j + \frac{p_j}{p} G_j \geq 0
\end{array}$$

Thus, the condition $\sigma_j Q_{jj}^j + \frac{p_j}{p} G_j \geq 0$ guarantees that $R_i(M_3) \geq R_i(M_4) \forall i$. Hence it suffices to show that $(M_4)_{ii} > 2R_i(M_3)$. The fact that $\sigma_j Q_{jj}^j + \frac{p_j}{p} G_j \geq 0$ is a direct conclusion of the results of procedure $M_1 - M_2$. For example, by the last bound on k we have:

$$\begin{array}{l}
k > 2 \max \left\{ 2\sqrt{\frac{\Theta_1}{\varepsilon}}, \frac{16\Theta_1}{9\varepsilon}, \frac{(\gamma_{dj})_{\max}}{\varepsilon} \right\} \frac{p}{p_j} \left| Q_{jj}^j \right|_{\max} \\
\underset{G_j \geq \varepsilon}{\overset{Y \leq \frac{\Theta_1}{k}}{\Rightarrow}} G_j > 2 \max \left\{ 2 \max \left\{ \frac{Y}{k}, \frac{8Y}{9} \right\}, \left| \frac{(\gamma_{dj})_{\max}}{k} \right| \right\} \frac{p}{p_j} \left| Q_{jj}^j \right|_{\max} \\
\Rightarrow G_j > \left\{ Y \left| \frac{1}{k} + \frac{8}{9} \right| + \left| \frac{\gamma_{dj}}{k} \right| \right\} \frac{p}{p_j} \left| Q_{jj}^j \right|_{\max} \\
\Rightarrow \frac{p_j}{p} G_j > \left| \sigma_j \right|_{\max} \left| Q_{jj}^j \right|_{\max} \Rightarrow \sigma_j Q_{jj}^j + \frac{p_j}{p} G_j > 0
\end{array}$$

The fact that $(M_4)_{ii} > 0$ is guaranteed by Lemma 9. This lemma also guarantees that there is always a finite upper bound on the terms

$$\left| (M_3)_{ij} \right| = \left| \sum_l A_l^{-(1+1/k)} \sigma_l A_j^{-(1+1/k)} K_j G_j Q_{ij}^l \right|$$

We have

$$\begin{array}{l}
(M_4)_{ii} > 2R_i(M_3) = 2 \sum\limits_{j=1}^{2N} \frac{p_j}{p} \left| (M_3)_{ij} \right| \Leftarrow \\
p > \frac{4N}{(M_4)_{ii}} \max\limits_j \left\{ p_j \left| (M_3)_{ij} \right| \right\}, \\
2N + 1 \leq i \leq 4N, 1 \leq j \leq 2N
\end{array}$$

◇

References

1. C. Belta and V. Kumar. Abstraction and control of groups of robots. *IEEE Transactions on Robotics*, 20(5):865–875, 2004.
2. A. Bicchi and L. Pallottino. On optimal cooperative conflict resolution for air traffic management systems. *IEEE Transactions on Intelligent Transportation Systems*, 1(4):221–232, 2000.

3. M.S. Branicky. Multiple lyapunov functions and other analysis tools for switched and hybrid systems. *IEEE Trans. on Automatic Control*, 43(4):475–482, 1998.
4. F. Ceragioli. *Discontinuous Ordinary Differential Equations and Stabilization*. PhD thesis, Dept. of Mathematics, Universita di Firenze, 1999.
5. F. Clarke. *Optimization and Nonsmooth Analysis*. Addison - Wesley, 1983.
6. D. V. Dimarogonas and K. J. Kyriakopoulos. Decentralized stabilization and collision avoidance of multiple air vehicles with limited sensing capabilities. *2005 American Control Conference, to appear.*
7. D. V. Dimarogonas, S. G. Loizou, K.J. Kyriakopoulos, and M. M. Zavlanos. Decentralized feedback stabilization and collision avoidance of multiple agents. Tech. report, NTUA, http://users.ntua.gr/ddimar/TechRep0401.pdf, 2004.
8. D.V. Dimarogonas and K.J. Kyriakopoulos. A feedback stabilization and collision avoidance scheme for multiple independent nonholonomic non-point agents. *2005 ISIC-MED, to appear.*
9. D.V. Dimarogonas and K.J. Kyriakopoulos. Decentralized motion control of multiple agents with double integrator dynamics. *16th IFAC World Congress, to appear*, 2005.
10. D.V. Dimarogonas and K.J. Kyriakopoulos. Decentralized navigation functions for multiple agents with limited sensing capabilities. *in preparation*, 2005.
11. D.V. Dimarogonas, M.M. Zavlanos, S.G. Loizou, and K.J. Kyriakopoulos. Decentralized motion control of multiple holonomic agents under input constraints. *42nd IEEE Conference on Decision and Control*, pages 3390–3395, 2003.
12. M. Egerstedt and X. Hu. A hybrid control approach to action coordination for mobile robots. *Automatica*, 38:125–130, 2002.
13. J. Feddema and D. Schoenwald. Decentralized control of cooperative robotic vehicles. *IEEE Transactions on Robotics*, 18(5):852–864, 2002.
14. R. Fierro, A. K. Das, V. Kumar, and J. P. Ostrowski. Hybrid control of formations of robots. *2001 IEEE International Conference on Robotics and Automation*, pages 3672–3677, 2001.
15. A. Filippov. *Differential equations with discontinuous right-hand sides*. Kluwer Academic Publishers, 1988.
16. V. Gazi and K.M. Passino. Stability analysis of swarms. *IEEE Transactions on Automatic Control*, 48(4):692–696, 2003.
17. V. Gupta, B. Hassibi, and R.M. Murray. Stability analysis of stochastically varying formations of dynamic agents. *42st IEEE Conf. Decision and Control*, pages 504–509, 2003.
18. R.A. Horn and C. R. Johnson. *Matrix Analysis*. Cambridge University Press, 1996.
19. D. Hristu-Varsakelis, M. Egerstedt, and P. S. Krishnaprasad. On the complexity of the motion description language mdle. *42st IEEE Conf. Decision and Control*, pages 3360–3365, 2003.
20. G. Inalhan, D.M. Stipanovic, and C.J. Tomlin. Decentralized optimization, with application to multiple aircraft coordination. *41st IEEE Conf. Decision and Control*, pages 1147–1155, 2002.
21. D. E. Koditschek and E. Rimon. Robot navigation functions on manifolds with boundary. *Advances Appl. Math.*, 11:412–442, 1990.
22. J.R. Lawton, R.W. Beard, and B.J. Young. A decentralized approach to formation maneuvers. *IEEE Transactions on Robotics and Automation*, 19(6):933–941, 2003.

23. J. Lin, A.S. Morse, and B. D. O. Anderson. The multi-agent rendezvous problem. *42st IEEE Conf. Decision and Control*, pages 1508–1513, 2003.
24. Y. Liu and K.M. Passino. Stability analysis of swarms in a noisy environment. *42st IEEE Conf. Decision and Control*, pages 3573–3578, 2003.
25. S. G. Loizou and K. J. Kyriakopoulos. Closed loop navigation for multiple holonomic vehicles. *Proc. of IEEE/RSJ Int. Conf. on Intelligent Robots and Systems*, pages 2861–2866, 2002.
26. S.G. Loizou, D.V. Dimarogonas, and K.J. Kyriakopoulos. Decentralized feedback stabilization of multiple nonholonomic agents. *2004 IEEE International Conference on Robotics and Automation*, pages 3012–3017, 2004.
27. S.G. Loizou and K.J. Kyriakopoulos. Closed loop navigation for multiple nonholonomic vehicles. *IEEE Int. Conf. on Robotics and Automation*, pages 420–425, 2003.
28. V. J. Lumelsky and K. R. Harinarayan. Decentralized motion planning for multiple mobile robots: The cocktail party model. *Journal of Autonomous Robots*, 4:121–135, 1997.
29. V. Manikonda, P.S. Krishnaprasad, and J. Hendler. Languages, behaviors, hybrid architectures and motion control. In *Mathematical Control Theory, special volume in honor of the 60th birthday of Roger Brockett, (eds. John Baillieul and Jan C. Willems)*, pages 199–226. Springer, 1998.
30. M. Mazo, A.Speranzon, K. H. Johansson, and X.Hu. Multi-robot tracking of a moving object using directional sensors. *2004 IEEE International Conference on Robotics and Automation*, pages 1103–1108, 2004.
31. J. Milnor. *Morse theory.* Annals of Mathematics Studies. Princeton University Press, Princeton, NJ, 1963.
32. P. Ogren, M.Egerstedt, and X. Hu. A control lyapunov function approach to multiagent coordination. *IEEE Transactions on Robotics and Automation*, 18(5):847–851, 2002.
33. R. Olfati-Saber and R.M. Murray. Flocking with obstacle avoidance: Cooperation with limited communication in mobile networks. *42st IEEE Conf. Decision and Control*, pages 2022–2028, 2003.
34. S. Pettersson and B. Lennartson. Stability and robustness for hybrid systems. *35th IEEE Conf. Decision and Control*, 1996.
35. G. Ribichini and E.Frazzoli. Efficient coordination of multiple-aircraft systems. *42st IEEE Conf. Decision and Control*, pages 1035–1040, 2003.
36. E. Rimon and D. E. Koditschek. Exact robot navigation using artificial potential functions. *IEEE Trans. on Robotics and Automation*, 8(5):501–518, 1992.
37. D. Shevitz and B. Paden. Lyapunov stability theory of nonsmooth systems. *IEEE Trans. on Automatic Control*, 49(9):1910–1914, 1994.
38. H. G. Tanner and K. J. Kyriakopoulos. Nonholonomic motion planning for mobile manipulators. *Proc of IEEE Int. Conf. on Robotics and Automation*, pages 1233–1238, 2000.
39. H. G. Tanner, S. G. Loizou, and K. J. Kyriakopoulos. Nonholonomic navigation and control of cooperating mobile manipulators. *IEEE Trans. on Robotics and Automation*, 19(1):53–64, 2003.
40. H.G. Tanner, A. Jadbabaie, and G.J. Pappas. Stable flocking of mobile agents. *42st IEEE Conf. Decision and Control*, pages 2010–2021, 2003.
41. C. Tomlin, G.J. Pappas, and S. Sastry. Conflict resolution for air traffic management: A study in multiagent hybrid systems. *IEEE Transactions on Automatic Control*, 43(4):509–521, 1998.

42. J.P. Wangermann and R.F. Stengel. Optimization and coordination of multiagent systems using principled negotiation. *Jour.Guidance Control and Dynamics*, 22(1):43–50, 1999.
43. H. Yamaguchi and J. W. Burdick. Asymptotic stabilization of multiple nonholonomic mobile robots forming group formations. *1998 IEEE International Conference on Robotics and Automation*, pages 3573–3580, 1998.
44. M.M. Zavlanos and K.J. Kyriakopoulos. Decentralized motion control of multiple mobile agents. *11th Mediterranean Conference on Control and Automation*, 2003.

Part III

Complexity and Randomization

Monte Carlo Optimisation for Conflict Resolution in Air Traffic Control

Andrea Lecchini[1], William Glover[1], John Lygeros[2], and Jan Maciejowski[1]

[1] Department of Engineering,
University of Cambridge, Cambridge, CB2 1PZ, UK,
{al394, wg214, jmm}@eng.cam.ac.uk

[2] Department of Electrical and Computer Engineering,
University of Patras, Rio, Patras, GR-26500, Greece,
lygeros@ee.upatras.gr

Summary. The safety of the flights, and in particular separation assurance, is one of the main tasks of Air Traffic Control. Conflict resolution refers to the process used by air traffic controllers to prevent loss of separation. Conflict resolution involves issuing instructions to aircraft to avoid loss of safe separation between them and, at the same time, direct them to their destinations. Conflict resolution requires decision making in the face of the considerable levels of uncertainty inherent in the motion of aircraft. We present a framework for conflict resolution which allows one to take into account such levels of uncertainty through the use of a stochastic simulator. The conflict resolution task is posed as the problem of optimizing an expected value criterion. Optimization of the expected value resolution criterion is carried out through an iterative procedure based on Markov Chain Monte Carlo. Simulation examples inspired by current air traffic control practice in terminal maneuvering areas and approach sectors illustrate the proposed conflict resolution strategy.

1 Introduction

In the current organization of the Air Traffic Management (ATM) system the centralized Air Traffic Control (ATC) is in complete control of air traffic and ultimately responsible for safety. Before take off, aircraft file flight plans which cover the entire flight. During the flight, ATC sends additional instructions to them, depending on the actual traffic, to improve traffic flow and avoid dangerous encounters. The primary concern of ATC is to maintain safe separation between the aircraft. The level of accepted minimum safe separation may depend on the density of air traffic and the region of the airspace. For example, a largely accepted value for horizontal minimum safe separation between two aircraft at the same altitude is 5 nmi in general en-route airspace; this is reduced to 3 nmi in approach sectors for aircraft landing and departing. A conflict is defined as a situation of loss of minimum safe separation

H.A.P. Blom, J. Lygeros (Eds.): Stochastic Hybrid Systems, LNCIS 337, pp. 257–276, 2006.

between two aircraft. If safety is not at stake, ATC also tries to fulfill the (possibly conflicting) requests of aircraft and airlines; for example, desired paths to avoid turbulence, or desired time of arrivals to meet schedule. To improve the performance of ATC, mainly in anticipation of increasing levels of air traffic, research effort has been devoted over the last decade on creating tools to assist ATC with conflict detection and resolution tasks. A review of research work in this area of ATC is presented in [15].

Uncertainty is introduced in air traffic by the action of wind, incomplete knowledge of the physical coefficients of the aircraft and unavoidable imprecision in the execution of ATC instructions. To perform conflict detection one has to evaluate the possibility of future conflicts given the current state of the airspace and taking into account uncertainty in the future position of aircraft. For this task, one needs a model to predict the future. In a probabilistic setting, the model could be either an empirical distribution of future aircraft positions [18], or a dynamical model, such as a stochastic differential equation (see, for example, [1, 12, 19]), that describes the aircraft motion and defines implicitly a distribution for future aircraft positions. On the basis of the prediction model one can evaluate metrics related to safety. An example of such a metric is conflict probability over a certain time horizon. Several methods have been developed to estimate different metrics related to safety for a number of prediction models, e.g [1, 12, 13, 18, 19]. Among other methods, Monte Carlo methods have the main advantage of allowing flexibility in the complexity of the prediction model since the model is used only as a simulator and, in principle, it is not involved in explicit calculations. In all methods a trade off exists between computational effort (simulation time in the case of Monte Carlo methods) and the accuracy of the model. Techniques to accelerate Monte Carlo methods especially for rare event computations are under development, see for example [14].

For conflict resolution, the objective is to provide suitable maneuvers to avoid a predicted conflict. A number of conflict resolution algorithms have been proposed in the deterministic setting, for example [7, 11, 21]. In the stochastic setting, the research effort has concentrated mainly on conflict detection, and only a few simple resolution strategies have been proposed [18, 19]. The main reason for this is the complexity of stochastic prediction models which makes the quantification of the effects of possible control actions intractable.

In this contribution we present a Markov Chain Monte Carlo (MCMC) framework [20] for conflict resolution in a stochastic setting. The aim of the proposed approach is to extend the advantages of Monte Carlo techniques, in terms of flexibility and complexity of the problems that can be tackled, to conflict resolution. The approach is motivated from Bayesian statistics [16, 17]. We consider an expected value resolution criterion that takes into account separation and other factors (e.g. aircraft requests). Then, the MCMC optimization procedure of [16] is employed to estimate the resolution maneuver that optimizes the expected value criterion. The proposed approach is illustrated in simulation, on some realistic benchmark problems, inspired by current ATC practice. The benchmarks were implemented in an air traffic simulator developed in previous work [8, 9, 10].

The material is organized in 5 sections. Section 2 presents the formulation of conflict resolution as an optimization problem. The randomized optimization procedure that we adopt to solve the problem is presented in Section 3. Section 4 is devoted to the benchmark problems used to illustrate our approach. Section 4.1 introduces the problems associated with ATC in terminal and approach sectors and Section 4.2 provides a brief overview of the simulator used to carry out the experiments. Sections 4.3 and 4.4 present results on benchmark problems in terminal and approach sectors respectively. Conclusions and future objectives are discussed in Section 5.

2 Conflict Resolution with an Expected Value Criterion

We formulate conflict resolution as a constrained optimization problem. Given a set of aircraft involved in a conflict, the conflict resolution maneuver is determined by a parameter ω which defines the nominal paths of the aircraft. From the point of view of the ATC, the execution of the maneuver is affected by uncertainty, due to wind, imprecise knowledge of aircraft parameters (e.g. mass) and Flight Management System (FMS) settings, etc. Therefore, the sequence of actual positions of the aircraft (for example, the sequence of positions observed by ATC every 6 seconds, which is a typical time interval between two successive radar sweeps) during the resolution maneuver is, a-priori of its execution, a random variable, denoted by X. A conflict is defined as the event that two aircraft get too close during the execution of the maneuver. The goal is to select ω to maximize the expected value of some measure of performance associated to the execution of the resolution maneuver, while ensuring a small probability of conflict. In this section we introduce the formulation of this problem in a general framework.

Let X be a random variable whose distribution depends on some parameter ω. The distribution of X is denoted by $p_\omega(x)$ with $x \in \mathbf{X}$. The set of all possible values of ω is denoted by $\mathbf{\Omega}$. We assume that a constraint on the random variable X is given in terms of a feasible set $\mathbf{X_f} \subseteq \mathbf{X}$. We say that a realization x, of random variable X, violates the constraint if $x \notin \mathbf{X_f}$. The probability of satisfying the constraint for a given ω is denoted by $\mathrm{P}(\omega)$

$$\mathrm{P}(\omega) = \int_{x \in \mathbf{X_f}} p_\omega(x) dx \,.$$

The probability of violating the constraint is denoted by $\bar{\mathrm{P}}(\omega) = 1 - \mathrm{P}(\omega)$.

For a realization $x \in \mathbf{X_f}$ we assume that we are given some definition of performance of x. In general performance can depend also on the value of ω, therefore performance is measured by a function $\mathrm{perf}(\cdot,\cdot) : \mathbf{\Omega} \times \mathbf{X_f} \to [0,1]$. The expected performance for a given $\omega \in \mathbf{\Omega}$ is denoted by $\textsc{Perf}(\omega)$, where

$$\textsc{Perf}(\omega) = \int_{x \in \mathbf{X_f}} \mathrm{perf}(\omega, x) p_\omega(x) dx \,.$$

Ideally one would like to select ω to maximize the performance, subject to a bound on the probability of constraint satisfaction. Given a bound $\bar{\mathbf{P}} \in [0, 1]$, this corresponds to solving the constrained optimization problem

$$\textsc{Perf}_{\max|\bar{\mathrm{P}}} = \sup_{\omega \in \boldsymbol{\Omega}} \textsc{Perf}(\omega) \tag{1}$$

$$\text{subject to } \bar{\mathrm{P}}(\omega) < \bar{\mathbf{P}}. \tag{2}$$

Clearly, for feasibility we must assume that there exists $\omega \in \boldsymbol{\Omega}$ such that $\bar{\mathrm{P}}(\omega) < \bar{\mathbf{P}}$, or, equivalently,

$$\bar{\mathrm{P}}_{\min} = \inf_{\omega \in \boldsymbol{\Omega}} \bar{\mathrm{P}}(\omega) < \bar{\mathbf{P}}.$$

The optimization problem (1)-(2) is generally difficult to solve, or even to approximate by randomized methods. Here we approximate this problem by an optimization problem with penalty terms. We show that with a proper choice of the penalty term we can enforce the desired maximum bound on the probability of violating the constraint, provided that such a bound is feasible, at the price of sub-optimality in the resulting expected performance.

We introduce a function $u(\omega, x)$ defined on the entire $\mathbf{X}$ by

$$u(\omega, x) = \begin{cases} \mathrm{perf}(\omega, x) + \Lambda & x \in \mathbf{X_f} \\ 1 & x \notin \mathbf{X_f}, \end{cases}$$

with $\Lambda > 1$. The parameter Λ represents a reward for constraint satisfaction. For a given $\omega \in \Omega$, the expected value of $u(\omega, x)$ is given by

$$U(\omega) = \int_{x \in \mathbf{X}} u(\omega, x) p_\omega(x) dx \qquad \omega \in \boldsymbol{\Omega} \,.$$

Instead of the constrained optimisation problem (1)–(2) we solve the unconstrained optimisation problem:

$$U_{\max} = \sup_{\omega \in \boldsymbol{\Omega}} U(\omega). \tag{3}$$

Assume the supremum is attained and let $\bar{\omega}$ denote the optimum solution, i.e. $U_{\max} = U(\bar{\omega})$. The following proposition introduces bounds on the probability of violating the constraints and the level of suboptimality of $\textsc{Perf}(\bar{\omega})$ over $\textsc{Perf}_{\max|\bar{\mathrm{P}}}$.

Proposition 1. *The maximiser, $\bar{\omega}$, of $U(\omega)$ satisfies*

$$\bar{\mathrm{P}}(\bar{\omega}) \;\leq\; \frac{1}{\Lambda} + \left(1 - \frac{1}{\Lambda}\right) \bar{\mathrm{P}}_{\min} \,, \tag{4}$$

$$\textsc{Perf}(\bar{\omega}) \;\geq\; \textsc{Perf}_{\max|\bar{\mathrm{P}}} - (\Lambda - 1)(\bar{\mathbf{P}} - \bar{\mathrm{P}}_{\min}) \,. \tag{5}$$

Proof. The optimisation criterion $U(\omega)$ can be written in the form

$$U(\omega) = \text{PERF}(\omega) + \Lambda - (\Lambda - 1)\bar{\mathrm{P}}(\omega)\,.$$

By the definition of $\bar{\omega}$ we have that $U(\bar{\omega}) \geq U(\omega)$ for all $\omega \in \boldsymbol{\Omega}$. We therefore can write

$$\text{PERF}(\bar{\omega}) + \Lambda - (\Lambda - 1)\bar{\mathrm{P}}(\bar{\omega}) \geq \text{PERF}(\omega) + \Lambda - (\Lambda - 1)\bar{\mathrm{P}}(\omega) \qquad \forall\omega$$

which can be rewritten as

$$\bar{\mathrm{P}}(\bar{\omega}) \leq \frac{\text{PERF}(\bar{\omega}) - \text{PERF}(\omega)}{\Lambda - 1} + \bar{\mathrm{P}}(\omega) \qquad \forall\omega\,. \tag{6}$$

Since $0 < \text{perf}(\omega, x) \leq 1$, $\text{PERF}(\omega)$ satisfies

$$0 < \text{PERF}(\omega) \leq P(\omega)\,. \tag{7}$$

Therefore we can use (7) to obtain an upper bound on the right-hand side of (6) from which we obtain

$$\bar{\mathrm{P}}(\bar{\omega}) \leq \frac{1}{\Lambda} + \left(1 - \frac{1}{\Lambda}\right)\bar{\mathrm{P}}(\omega) \qquad \forall\omega \in \boldsymbol{\Omega}.$$

We eventually obtain (4) by taking a minimum to eliminate the quantifier on the right-hand side of the above inequality.

In order to obtain (5) we proceed as follows. By definition of $\bar{\omega}$ we have that $U(\bar{\omega}) \geq U(\omega)$ for all $\omega \in \boldsymbol{\Omega}$. In particular, we know that

$$\text{PERF}(\bar{\omega}) \geq \text{PERF}(\omega) - (\Lambda - 1)\left[\bar{\mathrm{P}}(\omega) - \bar{\mathrm{P}}(\bar{\omega})\right] \qquad \forall\omega : \bar{\mathrm{P}}(\omega) \leq \bar{\mathbf{P}}\,.$$

Taking a lower bound of the right-hand side, we obtain

$$\text{PERF}(\bar{\omega}) \geq \text{PERF}(\omega) - (\Lambda - 1)\left[\bar{\mathbf{P}} - \bar{\mathrm{P}}_{\min}\right] \qquad \forall\omega : \bar{\mathrm{P}}(\omega) \leq \bar{\mathbf{P}}\,.$$

Taking the maximum and eliminating the quantifier on the right-hand side we obtain the desired inequality.

Proposition 1 suggests a method for choosing Λ to ensure that the solution $\bar{\omega}$ of the optimisation problem will satisfy $\bar{\mathrm{P}}(\bar{\omega}) \leq \bar{\mathbf{P}}$. In particular it suffices to know $\bar{\mathrm{P}}(\omega)$ for some $\omega \in \boldsymbol{\Omega}$ with $\bar{\mathrm{P}}(\omega) < \bar{\mathbf{P}}$ to obtain a bound. If there exists $\omega \in \boldsymbol{\Omega}$ for which $\hat{\mathrm{P}} = \bar{\mathrm{P}}(\omega)$ is known, then any

$$\Lambda \geq \frac{1 - \hat{\mathrm{P}}}{\bar{\mathbf{P}} - \hat{\mathrm{P}}}$$

ensures that $\bar{\mathrm{P}}(\bar{\omega}) \leq \bar{\mathbf{P}}$. If we know that there exists a parameter $\omega \in \boldsymbol{\Omega}$ for which the constraints are satisfied almost surely, a tighter (and potentially more useful) bound can be obtained. If there exists $\omega \in \mathbf{\Omega}$ such that $\bar{\mathrm{P}}(\omega) = 0$, then any

$$\Lambda \geq \frac{1}{\bar{\mathbf{P}}} \tag{8}$$

ensures that $\bar{\mathrm{P}}(\bar{\omega}) \leq \bar{\mathbf{P}}$. Clearly to minimise the gap between the optimal performance and the performance of $\bar{\omega}$ we need to select Λ as small as possible. Therefore the optimal choices of Λ that ensure the bounds on constraint satisfaction and minimise the suboptimality of the solution are $\Lambda = \frac{1-\hat{\mathrm{P}}}{\bar{\mathbf{P}}-\hat{\mathrm{P}}}$ and $\Lambda = \frac{1}{\bar{\mathbf{P}}}$ respectively.

3 Monte Carlo Optimisation

In this section we describe a simulation-based procedure, to find approximate optimizers of $U(\omega)$. The only requirement for applicability of the procedure is to be able to obtain realizations of the random variable X with distribution $p_\omega(x)$ and to evaluate $u(\omega, x)$ point-wise. This optimization procedure is in fact a general procedure for the optimization of expected value criteria. It has been originally proposed in the Bayesian statistics literature [16].

The optimisation strategy relies on extractions of a random variable Ω whose distribution has modes which coincide with the optimal points of $U(\omega)$. These extractions are obtained through Markov Chain Monte Carlo (MCMC) simulation [20]. The problem of optimising the expected criterion is then reformulated as the problem of estimating the optimal points from extractions concentrated around them. In the optimisation procedure, there exists a tunable trade-off between estimation accuracy of the optimiser and computational effort. In particular, the distribution of Ω is proportional to $U(\omega)^J$ where J is a positive integer which allows the user to increase the "peakedness" of the distribution and concentrate the extractions around the modes at the price of an increased computational load. If the tunable parameter J is increased during the optimisation procedure, this approach can be seen as the counterpart of Simulated Annealing for a stochastic setting. Simulated Annealing is a randomised optimisation strategy developed to find tractable approximate solutions to complex deterministic combinatorial optimisation problems, [22]. A formal parallel between these two strategies has been derived in [17].

The MCMC optimisation procedure can be described as follows. Consider a stochastic model formed by a random variable Ω, whose distribution has not been defined yet, and J conditionally independent replicas of random variable X with distribution $p_\Omega(x)$. Let us denote by $h(\omega, x_1, x_2, \ldots, x_J)$ the joint distribution of $(\Omega, X_1, X_2, X_3, \ldots, X_J)$. It is straightforward to see that if

$$h(\omega, x_1, x_2, \ldots, x_J) \propto \prod_j u(\omega, x_j) p_\omega(x_j) \tag{9}$$

then the marginal distribution of Ω, also denoted by $h(\omega)$ for simplicity, satisfies

$$h(\omega) \propto \left[\int u(\omega, x) p_\omega(x) dx \right]^J = U(\omega)^J . \tag{10}$$

This means that if we can extract realisations of $(\Omega, X_1, X_2, X_3, \ldots, X_J)$ then the extracted Ω's will be concentrated around the optimal points of $U(\Omega)$ for a sufficiently high J. These extractions can be used to find an approximate solution to the optimisation of $U(\omega)$.

Realisations of the random variables $(\Omega, X_1, X_2, X_3, \ldots, X_J)$, with the desired joint probability density given by (9), can be obtained through Markov Chain Monte Carlo simulation. The algorithm is presented below. In the algorithm, $g(\omega)$ is known as the instrumental (or *proposal*) distribution and is freely chosen by the user; the only requirement is that $g(\omega)$ covers the support of $h(\omega)$.

MCMC Algorithm

initialization:

`Set` $k = 0$

`Generate` $\Omega_0 \sim g(\omega)$

`Generate` $X_0^j \sim p_{\Omega_0}(x)\,, \quad j = 1, \ldots, J$

`Set` $U_0 = \prod_{j=1}^J u(\Omega_0, X_0^j)$

repeat:

`Set` $k = k + 1$

`Generate` $\tilde{\Omega} \sim g(\omega)$

`Generate` $\tilde{X}^j \sim p_{\tilde{\Omega}}(x)\,, \quad j = 1, \ldots, J$

`Set` $\tilde{U} = \prod_{j=1}^J u(\tilde{\Omega}, \tilde{X}^j)$

`Set` $\rho_k = \min \left\{ \frac{\tilde{U}}{g(\tilde{\Omega})} \frac{g(\Omega_{k-1})}{U_{k-1}}, 1 \right\}$

`Set` $[\Omega_k\,, U_k] = \begin{cases} [\tilde{\Omega}\,, \tilde{U}] & \texttt{with probability } \rho_k \\ [\Omega_{k-1}\,, U_{k-1}] & \texttt{with probability } 1 - \rho_k \end{cases}$

until `True`

This algorithm is a formulation of the Metropolis-Hastings algorithm for a desired distribution given by $h(\omega, x_1, x_2, \ldots, x_J)$ and proposal distribution given by

$$g(\omega) \prod_j p_\omega(x_j).$$

In this case, the acceptance probability for the standard Metropolis-Hastings algorithm is

$$\frac{h(\tilde{\omega}, \tilde{x}_1, \tilde{x}_2, \ldots, \tilde{x}_J)}{h(\omega, x_1, x_2, \ldots, x_J)} \frac{g(\omega) \prod_j p_\omega(x_j)}{g(\tilde{\omega}) \prod_j p_\omega(\tilde{x}_j)} .$$

By inserting (9) in the above expression one obtains $\rho(\omega, u_J, \tilde{\omega}, \tilde{u}_J)$. Under minimal assumptions, the Markov Chain generated by the $\Omega(k)$ is uniformly ergodic with stationary distribution $h(\omega)$ given by (10). Therefore, after a burn in period, the extractions $\Omega(k)$ accepted by the algorithm will concentrate around the modes of $h(\omega)$, which, by (10) coincide with the optimal points of $U(\omega)$. Results that characterize the convergence rate to the stationary distribution can be found, for example, in [20].

In the initialization step the state $[\Omega(0), U_J(0)]$ is always accepted. In subsequent steps the new extraction $[\tilde{\Omega}, \tilde{U}_J]$ is accepted with probability ρ otherwise it is rejected and the previous state of the Markov chain $[\Omega(k), U_J(k)]$ is maintained. Practically, the algorithm is executed until a certain number of extractions (say 1000) have been accepted. Because we are interested in the stationary distribution of the Markov chain, the first few (say 10%) of the accepted states are discarded to allow the chain to reach its stationary distribution ("burn in period").

A general guideline to obtain faster convergence is to concentrate the search distribution $g(\omega)$ where $U(\omega)$ assumes nearly optimal values. The algorithm represents a trade-off between computational effort and the "peakedness" of the target distribution. This trade-off is tuned by the parameter J which is the power of the target distribution and also the number of extractions of X at each step of the chain. Increasing J concentrates the distribution more around the optimizers of $U(\omega)$, but also increases the number of simulations one needs to perform at each step. Obviously if the peaks of $U(\omega)$ are already quite sharp, this implies some advantages in terms of computation, since there is no need to increase further the peakedness of the criterion by running more simulations. For the specific $U(\omega)$ proposed in the previous section, a trade-off exists between its peakedness and the parameter Λ, which is related to probability of constraint violation. In particular, the greater Λ is the less peaked the criterion $U(\omega)$ becomes, because the relative variation of $u(\omega, x)$ is reduced, and therefore more computational effort is required for the optimization of $U(\omega)$.

4 ATC in Terminal and Approach Sectors

4.1 Current Practice

Terminal Maneuvering Area (TMA) and Approach Sectors are perhaps the most difficult areas for ATC. The management of traffic, in this case, includes tasks such as determining landing sequences, issuing of "vector" maneuvers to avoid collisions, holding the aircraft in "stacks" in case of congested traffic, etc. Here, we give a schematic representation of the problem as described in [2, 6].

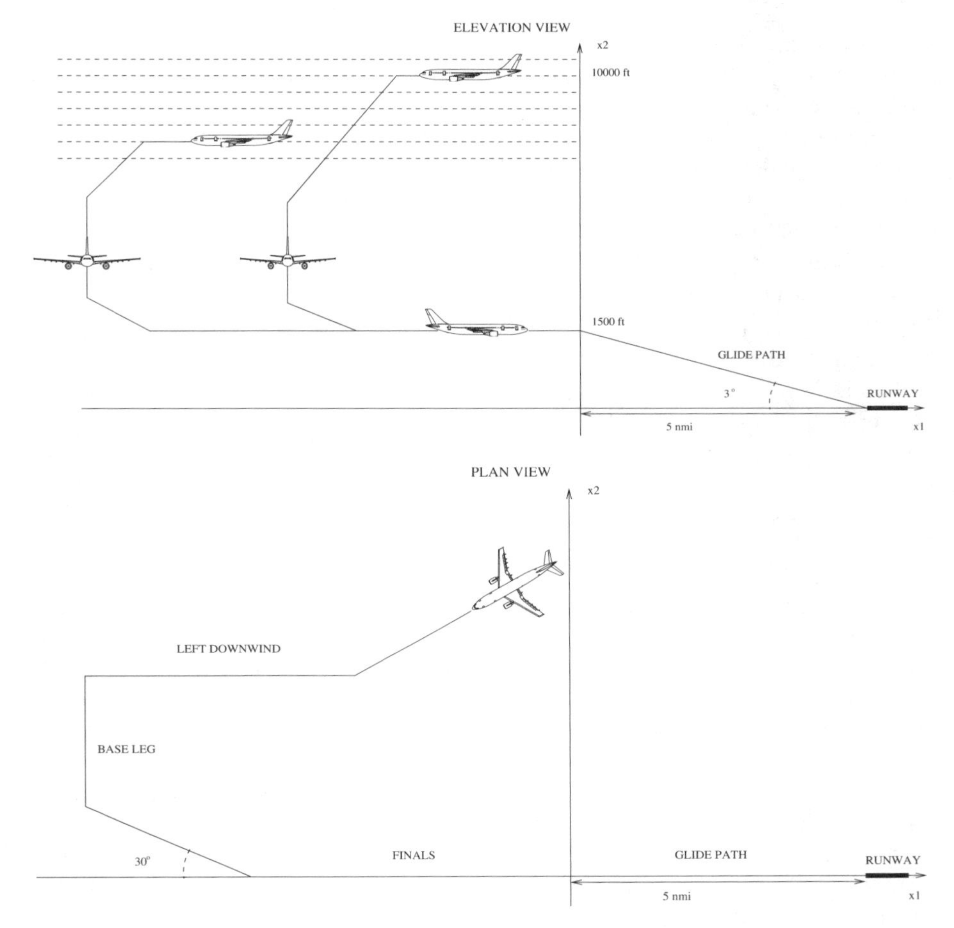

Fig. 1. Schematic representation of approach maneuver

During most of the flight, aircraft stay at cruising altitudes, above 30000 ft. In the current organization, the traffic at these altitudes has an en-route structure, which facilitates the action of ATC. Aircraft follow pre-specified corridors at different *flight levels*. Flight levels are given in 3 digit numbers, representing hundreds of feet; for example, the altitude of 30000 ft is denoted by FL300.

Towards the end of the flight, aircraft enter the TMA where ATC guides them from cruising altitudes to the entry points of the Approach Sector, which are typically between FL50 and FL150. Ideally, aircraft should enter the Approach Sector in a sequence properly spaced in time. The controllers of the Approach Sector are then responsible for guiding the aircraft towards the runway for landing. The tasks of ATC in the Approach Sectors include:

1) Maintain safe separation between aircraft. This is the most important requirement for safety, in any sector and during all parts of the flight. Aircraft must always maintain a minimum level of separation. A conflict between two aircraft is defined as the situation of loss of minimum safe separation between them. Safe separation is defined by a protected zone centered around each aircraft. The level of accepted minimum separation can vary with the density of the traffic and the region of the airspace. A largely accepted shape of the protected zone is defined by a vertical cylinder, centered on the aircraft with radius 5nmi and height 2000ft (i.e. aircraft which do not have 5nmi of horizontal separation must have 1000ft of vertical separation).

2) Take aircraft from entry altitude down to intercept the localizer. Once aircraft have entered the Approach Sector, ATC must guide them from the entry altitude (FL50 to FL150) to FL15. This is the altitude at which they can intercept the *localizer*, i.e. the radio beacons which will guide them onto the runway. The point at which the aircraft will actually start the descent towards the runway is an important variable which has to be carefully chosen since it can affect the rest of the maneuver and the coordination with other aircraft. The reason is that aircraft fly following pre-specified speed profiles which depend on the altitude; they fly faster at high altitudes and slower at low altitudes. This implies that aircraft, flying at lower altitudes, are slower in joining the landing queue.

3) Sequence aircraft towards the runway. The air traffic controllers must direct the aircraft towards the runway in a properly spaced queue. This is done by adjusting the way-points (corners) of a standard approach route (STAR) — see Figure 1. Typically the route is composed of four legs. During their descent, aircraft are first aligned, on one of the two sides of the runway, in the direction of the runway but with opposite heading. This leg is called the *left/right downwind leg*, since aircraft are expected to land against the wind. Aircraft then perform a turn of approximately 90°, to approach the localizer. This second segment is called the *base leg*. Aircraft perform an additional turn in order to intercept the plane of the localizer with an angle of incidence of approximately 30°. The reason is that 30° is a suitable angle for pilots to perform the final turn in the direction of the runway as soon as possible when the localizer has been intercepted. It is required that aircraft intercept the localizer plane at least 5 nmi from the beginning of the runway and at an altitude of 1000 – 1500 ft, so that they can follow a 3° – 5° glide path to the runway.

This approach geometry (which is referred to as the "trombone maneuver") is advantageous to air traffic controllers as it allows them great flexibility in spacing aircraft by adjusting the length of the downwind leg (hence the name trombone).

4.2 Simulation Procedure

A necessary step in the implementation of the MCMC algorithm is the extraction of the $\tilde{X}_j$ for a given value of ω. In the ATC case studies this corresponds to the extraction of aircraft trajectories for a given airspace configuration (current aircraft positions and given flight plans). The extraction of multiple such trajectories (J to be precise) will be necessary in the algorithm implementation. The trajectories will be in general different from one another because of the uncertainty that enters the process (due to wind, aircraft parameters unknown to ATC, etc.)

In earlier work we developed an air traffic simulator to simulate adequately the behavior of a set of aircraft from the point of view of ATC [8, 9, 10]. The simulator implements realistic models of current commercial aircraft described in the Base of Aircraft Data (BADA) [5]. The simulator contains also realistic stochastic models of the wind disturbance [3]. The aircraft models contain continuous dynamics, arising from the physical motion of the aircraft, discrete dynamics, arising from the logic embedded in the Flight Management System, and stochastic dynamics, arising from the effect of the wind and incomplete knowledge of physical parameters (for example, the aircraft mass, which depends on fuel, cargo and number of passengers). The simulator has been coded in Java and can be used in different operation modes, either to generate accurate data for validation of the performance of conflict detection and resolution algorithm, or to run faster simulations of simplified models. The nominal path for each aircraft is entered in the simulator as a sequence of way-points, either by reading data files, or manually. The simulator can also be used in a so called interactive mode, where the user is allowed to manipulate the flight plans on-line (move way-points, introduce new way points, etc.)

The actual trajectories of the aircraft generated by the simulator are a perturbed version of the flight plan that depends on the particular realizations of wind disturbances and uncertain parameters. The trajectories of the aircraft can be displayed on the screen (Figure 2) and/or stored in files for post-processing. The reader is referred to [9, 10] for a more detailed description of the simulator.

The air traffic simulator has been used to produce the examples presented in this section. The full accurate aircraft, FMS and wind models have been used both during the Monte Carlo optimization procedure and to obtain Monte Carlo estimates of post-resolution conflict probabilities. The simulator was invoked from Matlab on a Linux workstation with a Pentium 4 3GHz processor. Under these conditions the simulation of the flight of two aircraft for 30 minutes, which is approximately the horizon considered in the examples, took 0.2 seconds on the average. Notice that this simulation speed (5 simulations/sec) is quite low for a Monte Carlo framework. This is mainly due to the fact that no attempt has been made to optimize the code at this stage. For example, executing the Java simulator from a Matlab environment introduces unnecessary and substantial computational overhead. The reader is requested to evaluate the computation times reported in the following examples keeping this fact in mind.

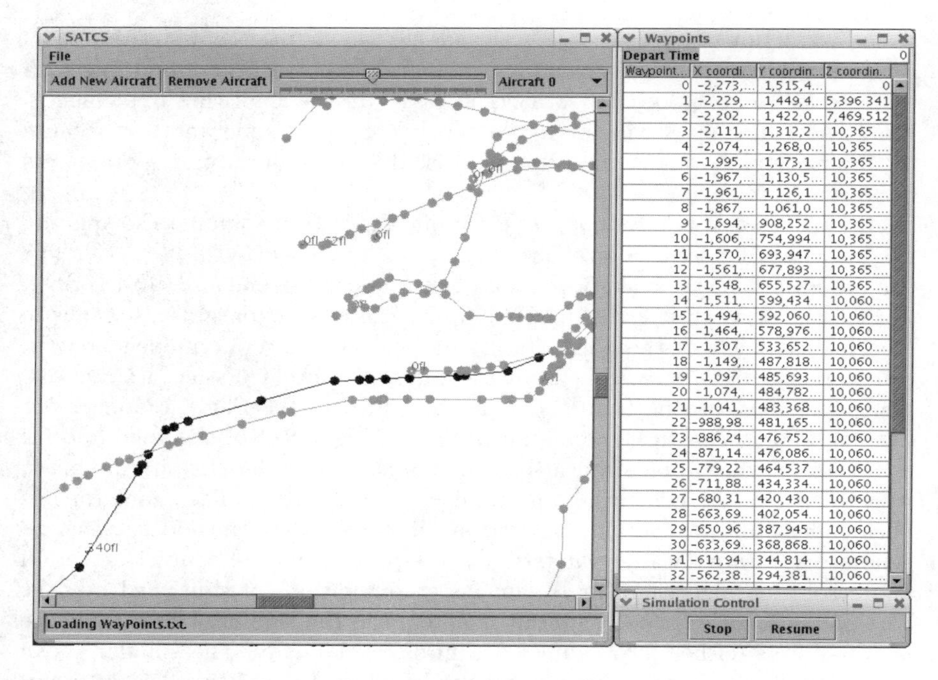

Fig. 2. Screen shot of a simulation of all flights using the Barcelona airport on a given day.

4.3 Sequencing Aircraft in Terminal Maneuvering Areas

We consider the problem of sequencing two aircraft. This is a typical task of ATC in TMA where aircraft descend from cruising altitude and need to be sequenced and separated by a certain time interval before entering in the final Approach Sector. In Figure 3 several possible trajectory realizations of a descending aircraft corresponding to the same nominal path are displayed. In this figure, the aircraft descends from FL350 to FL100. In addition to stochastic wind terms, uncertainty about the mass of the aircraft is introduced as a uniform distribution between two extreme values. The figure suggests that the resulting uncertainty in the position of aircraft is of the order of magnitude of some kilometers.

We consider the problem of sequencing two descending aircraft as illustrated in Figure 4(a). The initial position of the first aircraft (A1) is $(-100000, 100000)$ (where coordinates are expressed in meters) and FL350. The path of this aircraft is fixed: The aircraft proceeds to way-point $(-90000, 90000)$ where it will start a descent to FL150. The trajectory of A1, while descending, is determined by an intermediate way-point in $(0, 0)$ and a final way-point in $(100000, 0)$, where the aircraft is assumed to exit the TMA and enter the approach sector. The second aircraft (A2) is ini-

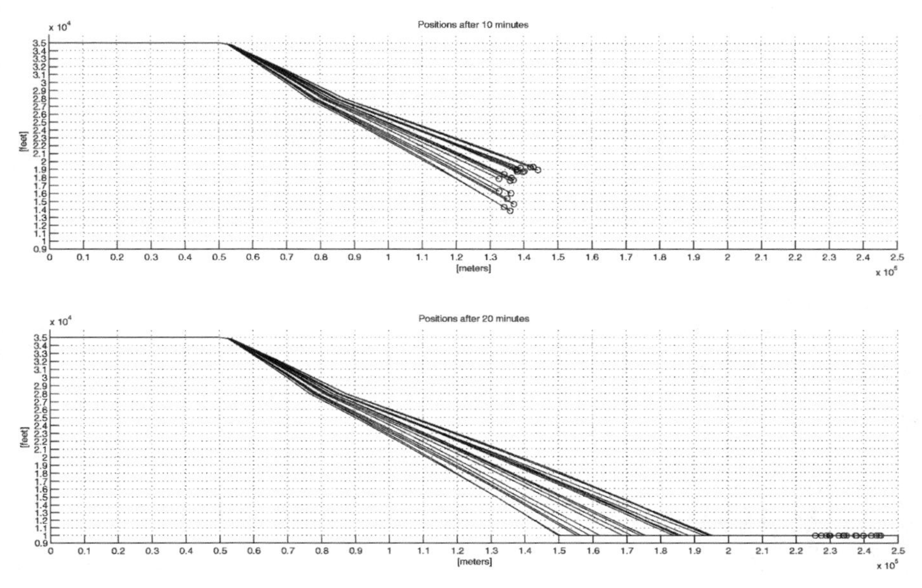

Fig. 3. Several trajectory realisations of aircraft descent

tially at $(-100000, -100000)$ and FL350. This aircraft proceeds to way-point $(-90000, -90000)$ where it will start its descent to FL150. The intermediate way-point $\omega = (\omega_1, \omega_2)$ must be selected in the range $\omega_1, \omega_2 \in [-90000, 90000]$. The aircraft will then proceed to way-point $(90000, 0)$ and then to the exit way-point $(100000, 0)$.

We assume that the objective is to obtain a time separation of 300 seconds between the arrivals of the two aircraft at the exit way-point, $(10000, 0)$. Performance in this sense is measured by perf $= e^{-a \cdot ||T_1 - T_2| - 300|}$ where T_1 and T_2 are the arrival times of A1 and A2 at the exit way-point and $a = 5 \cdot 10^{-3}$. The constraint is that the trajectories of the two aircraft should not be conflicting. In our simulations we define a conflict as the situation in which two aircraft have less than 5nmi of horizontal separation and less than 1000ft of vertical separation[3]. We optimize initially with an upper bound on the probability of constraint violation of $\bar{\mathbf{P}} = 0.1$. It is easy to see that there exists a maneuver in the set of optimization parameters that gives negligible conflict probability. Therefore, based on inequality (8), we select $\Lambda = 10$ in the optimization criterion.

The results of the optimization procedure are illustrated in Figures 4(b-d). Each figure shows the scatter plot of the accepted parameters during MCMC simulation for different choices of J and search distribution g. In all cases the

[3] In the TMA of large airports horizontal minimal separation is sometimes reduced to 3nmi, but this fact is ignored here.

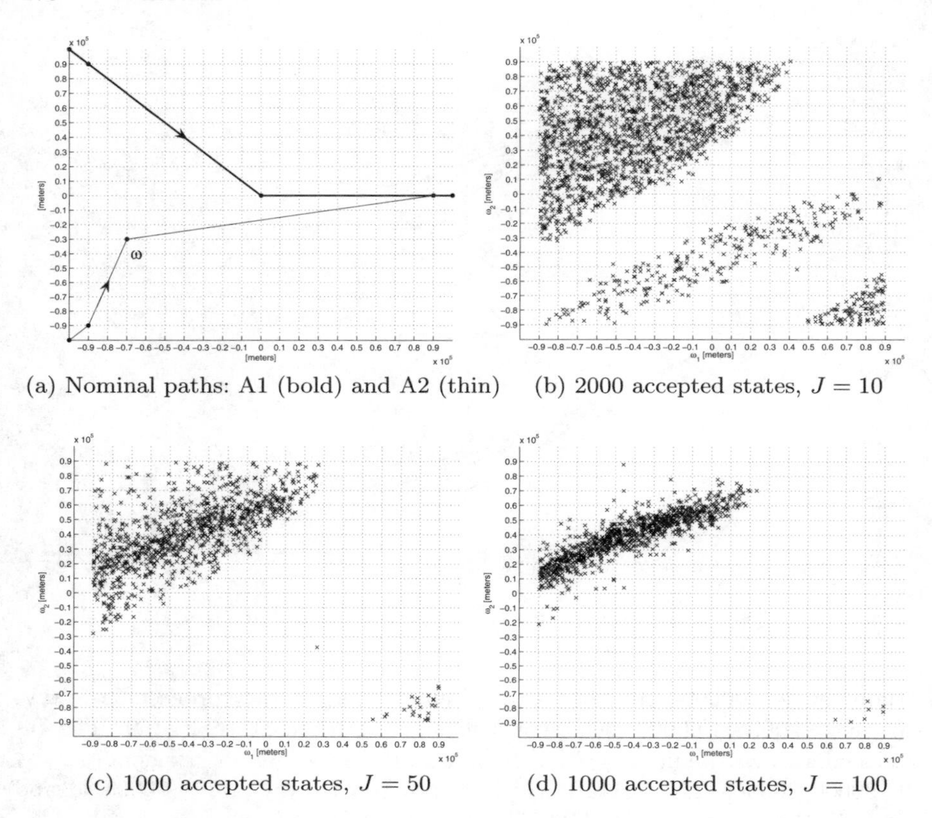

(a) Nominal paths: A1 (bold) and A2 (thin)

(b) 2000 accepted states, $J = 10$

(c) 1000 accepted states, $J = 50$

(d) 1000 accepted states, $J = 100$

Fig. 4. Accepted states during MCMC simulation

first 10% of accepted parameters was discarded as a burn in period, to allow convergence of the Markov chain to its stationary distribution.

Figure 4(b) illustrates the case $J = 10$. Regions characterized by a low density of accepted parameters can be clearly seen in the figure. These are parameters which correspond to nominal paths with high probability of conflict. The figure also shows distinct "clouds" of accepted maneuvers. They correspond to different sequences of arrivals at the exit point: either A1 arrives before A2 (top left and bottom right clouds) or A1 arrives after A2 (middle cloud). In this case the proposal distribution g was uniform over the parameter space and the ratio of accepted/proposed states was 0.27. This means that approximately $1100 \cdot 10/0.27 = 40740$ simulations were needed to obtain 1000 accepted states. At the average simulation speed of 5 simulations/second, the required computational time to obtain 1000 accepted states was then approximately 2 hours. In this simulation we actually extracted 5100 states. Figure 4(b) displays the last 2000 extracted states.

Figure 4(c) illustrates the case $J = 50$. In this case the proposal distribution g was a sum of 2000 Gaussian distributions $N(\mu, \sigma^2 I)$ with variance $\sigma^2 = 10^7\, m^2$. The means of Gaussian distributions were 2000 parameters randomly chosen from those accepted in the MCMC simulation for $J = 10$. The choice of this proposal distribution gives clear computational advantages since less computational time is spent searching over regions of non optimal parameters. In this case the ratio accepted/proposed states was 0.34. This means that approximately $1100 \cdot 50/0.34 = 161764$ simulations were needed to obtain 1000 accepted states. At an average of 5 simulations/second, the required computational time to obtain 1000 states was approximately 9 hours.

Figure 4(d) illustrates the case $J = 100$ and a proposal distribution constructed as before from states accepted for $J = 50$. Here the ratio accepted/proposed states was 0.3. This means that approximately $1100 \cdot 100/0.3 = 366666$ simulations were needed to obtain 1000 accepted states. At an average of 5 simulations/second, the required computational time to obtain 1000 states was approximately 20 hours. Figure 4(d) indicates that a nearly optimal maneuver is $\omega_1 = -40000$ and $\omega_2 = 40000$. The probability of conflict for this maneuver, estimated by 1000 Monte Carlo runs, was zero. The estimated expected time separation between arrivals was 283 seconds.

4.4 Coordination of Approach Maneuvers

In this section, we optimize an approach maneuver with coordination between two aircraft. In Figure 5 several trajectory realizations, of an aircraft performing the approach maneuver described in Section 4.4, are displayed. Here, the aircraft is initially at FL100 and descends to 1500ft during the approach. Uncertainty in the trajectory is due to the action of the wind and randomness in the aircraft mass. In the final leg a function that emulates the localizer and so eliminates cross-track error is implemented.

The problem formulation is illustrated in Figure 6(a). We consider Aircraft One (A1) and Aircraft Two (A2) approaching the runway. The glide path towards the runway starts at the origin of the reference frame and coordinates are expressed in meters. The aircraft are initially in level flight. The parameters of the approach maneuver are the distance, from initial position, of the start of the final descent (ω_1) and the length of the downwind leg (ω_2).

The initial position of A1 is $(0, 50000)$ and FL100. The approach maneuver of this aircraft is fixed to $\bar{\omega}_1 = 30000$ and $\bar{\omega}_2 = 50000$. The initial position of A2 is $(0, 50000)$ and altitude FL100. The parameters of its approach maneuver will be selected using the optimization algorithm. The range of the optimization parameters is $\omega_2 \in [35000, 60000]$ and $\omega_1 \in [0, \omega_2]$. We assume that the performance of the approach maneuver is measured by the arrival time of A2 at the start of the glide path (T_2). The measure of performance is given by perf $= e^{-a \cdot T_2}$ with $a = 5 \cdot 10^{-4}$. The constraint is that the trajectories of the two aircraft are not in conflict. In addition, aircraft 2 must also reach the

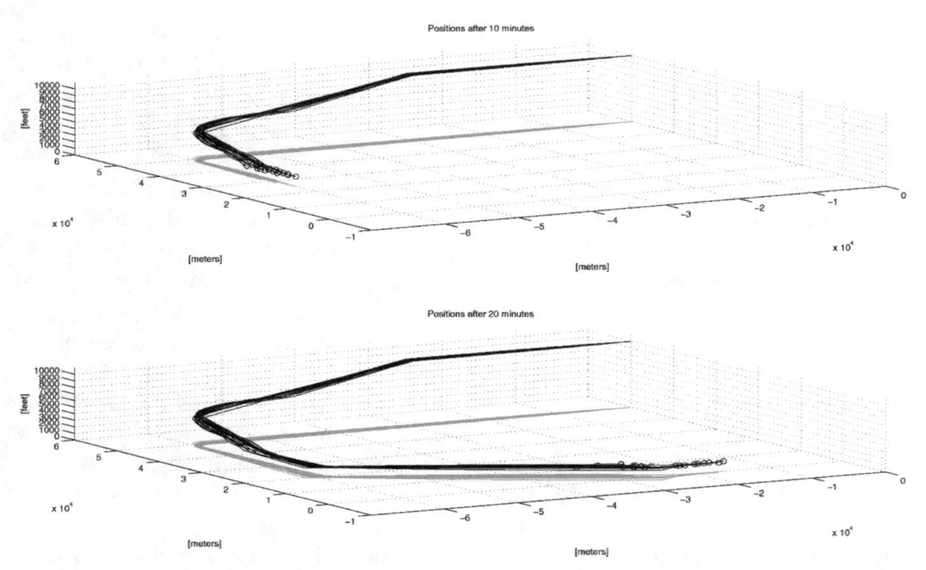

Fig. 5. Several trajectory realisations of aircraft descent

altitude of 1500 ft before the start of the glide path, to ensure that it intercepts the localizer. We optimize initially with an upper bound on probability of constraint violation given by $\bar{\mathbf{P}} = 0.1$. Since there exists a maneuver in the set of optimization parameters that gives negligible conflict probability, we select $\Lambda = 10$ in the optimization criterion according to inequality (8).

The results of the optimization procedure are illustrated in Figures 6(b-d) for different values of J and proposal distribution g. Each figure shows the scatter plot of the accepted parameters during MCMC simulation. Again, the first 10% of accepted parameters were discarded in each case as a burn in period. Figure 6(b) illustrates the case $J = 10$ and proposal distribution g uniform over the parameter space. In this case, the ratio between accepted and proposed parameters during MCMC simulation was 0.23. This means that approximately $1100 \cdot 10/0.23 = 47826$ simulations were needed to obtain 1000 accepted states, which took approximately 2.6 hours. A region characterized by a low density of accepted parameters can be clearly seen in the figure. These are parameters which correspond to a conflicting maneuver where the aircraft are performing an almost symmetrical approach. The figure also shows two distinct "clouds" of accepted maneuvers. They correspond to a discrete choice that the air traffic controller has to make: either land A2 before A1 (bottom right cloud) or land A1 before A2 (top left cloud).

Figure 6(c) illustrates the case $J = 50$. In this case the proposal distribution g was a sum of 100 Gaussian distributions $N(\mu, \sigma^2 I)$ with means selected randomly among the maneuvers accepted for $J = 10$ and variance $\sigma^2 = 10^5\, m^2$. In this case the ratio between accepted and proposed parameters

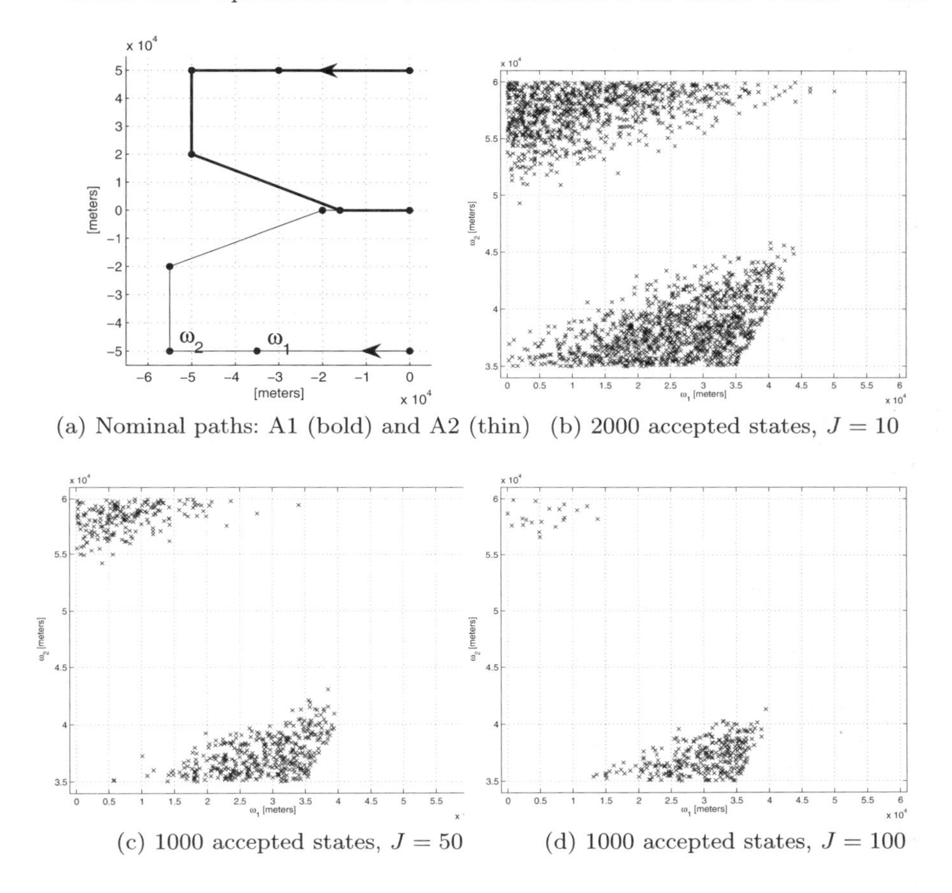

(a) Nominal paths: A1 (bold) and A2 (thin) (b) 2000 accepted states, $J = 10$

(c) 1000 accepted states, $J = 50$ (d) 1000 accepted states, $J = 100$

Fig. 6. Accepted states during MCMC simulation

was 0.25. This means that approximately $1100 \cdot 50/0.25 = 220000$ simulations were needed to obtain 1000 accepted states, which required approximately 12 hours of computation.

Figure 6(d) illustrates the case $J = 100$ and proposal distribution constructed as before from states accepted for $J = 50$. In this case the ratio between accepted and proposed parameters was 0.3. This means that approximately $1100 \cdot 100/0.3 = 366666$ simulations were needed to obtain 1000 accepted states (approximately 20 hours). Figure 6(d) indicates that a nearly optimal maneuver is $\omega_1 = 35000$ and $\omega_2 = 35000$. The probability of conflict for this maneuver, estimated by 1000 Monte Carlo runs, was zero.

5 Conclusions

In this contribution we illustrated an approach to air traffic conflict resolution in a stochastic setting based on Monte Carlo methods. The main motivation for our approach is to enable the use of realistic stochastic hybrid models of aircraft flight; Monte Carlo methods appear to be the only ones that allow such models. We have formulated conflict resolution as the optimization of an expected value criterion with probabilistic constraints. Here, a penalty formulation of the problem was developed, which guarantees constraint satisfaction but delivers suboptimal solutions. A side effect of the optimization procedure is that structural differences between maneuvers are highlighted as "clouds" of maneuvers accepted by the algorithm, with the density of the clouds related to the cost of the particular maneuver.

We presented the application of this method to two realistic scenarios inspired by terminal area and final approach maneuvering respectively. The solutions proposed by our algorithm were tested in Monte Carlo simulations and gave very good performance in terms of post-resolution conflicts. Computations times were high, partly because no attempt was made to optimize the implementation of the algorithms. Note also that the algorithm can provide a fairly accurate separation between "good" and "bad" maneuvers (i.e. maneuvers that meet, or violate the constraints) relatively cheaply, using a low J. It is only when we try to find an optimal maneuver among these that the computational load really kicks in.

Current research concentrates on overcoming the sub-optimality imposed by the penalty formulation of the constrained optimization problem considered in Section 2. A possible way is to use the Markov Chain Monte Carlo procedure presented in Section 3 to obtain optimization parameters that satisfy the constraints and then to optimize over this set in a successive step. We are also working on sequential Monte Carlo implementations of the optimization algorithm [4]. This will allow considerable computational savings, since it will enable the re-use of simulations from one step of the procedure to the next. It will also introduce feedback to the process, since it will make it possible to repeat the optimization on line in a receding horizon manner. Finally, we are continuing to work on modeling and implementation in the simulator of typical ATC situation with a realistic parameterization of control actions and control objectives.

References

1. H.A.P. Blom and G.J. Bakker. Conflict probability and incrossing probability in air traffic management. In *IEEE Conference on Decision and Control*, Las Vegas, Nevada, U.S.A., December 2002.
2. EUROCONTROL Experimental Centre. Analysis of the current situation in the paris, london and frankfurt tma. Technical Report Work Package 1: Final Project Report, FALBALA, April 27, 2004.

3. R.E. Cole, C. Richard, S. Kim, and D. Bailey. An assessment of the 60 km rapid update cycle (ruc) with near real-time aircraft reports. Technical Report NASA/A-1, MIT Lincoln Laboratory, July 15, 1998.
4. A. Doucet, N de Freitas, and N. Gordon (eds). *Sequential Monte Carlo Methods in Practice.* Springer-Verlag, 2001.
5. EUROCONTROL Experimental Centre. *User Manual for the Base of Aircraft Data (BADA) — Revision 3.3.* 2002. Available from World Wide Web: `http://www.eurocontrol.fr/projects/bada/`.
6. EUROCONTROL Experimental Centre. *Air-Traffic Control Familiarisation Course.* 2004.
7. E. Frazzoli, Z.H. Mao, J.H. Oh, and E. Feron. Aircraft conflict resolution via semi-definite programming. *AIAA Journal of Guidance, Control, and Dynamics*, 24(1):79–86, 2001.
8. W. Glover and J. Lygeros. A stochastic hybrid model for air traffic control simulation. In R. Alur and G. Pappas, editors, *Hybrid Systems: Computation and Control*, number 2993 in LNCS, pages 372–386. Springer Verlag, 2004.
9. W. Glover and J. Lygeros. A multi-aircraft model for conflict detection and resolution algorithm evaluation. Technical Report WP1, Deliverable D1.3, HYBRIDGE, February 18, 2004. http://www.nlr.nl/public/hosted-sites/hybridge/.
10. W. Glover and J. Lygeros. Simplified models for conflict detection and resolution algorithms. Technical Report WP1, Deliverable D1.4, HYBRIDGE, June 17, 2004. http://www.nlr.nl/public/hosted-sites/hybridge/.
11. J. Hu, M. Prandini, and S. Sastry. Optimal Coordinated Maneuvers for Three-Dimensional Aircraft Conflict Resolution. *AIAA Journal of Guidance, Control and Dynamics*, 25(5), 2002.
12. J. Hu, M. Prandini, and S. Sastry. Aircraft conflict detection in presence of spatially correlated wind perturbations. In *AIAA Guidance, Navigation and Control Conf.*, Austin, Texas, USA, 2003.
13. R. Irvine. A geometrical approach to conflict probability estimation. In *4th USA/Europe Air Traffic Management R&D Seminar*, Budapest, Hungary, 2001.
14. J. Krystul and H.A.P. Blom. Monte Carlo simulation of rare events in hybrid systems. Technical Report WP8, Deliverable D8.3, HYBRIDGE, 2004. Available from World Wide Web: http://www.nlr.nl/public/hosted-sites/ hybridge/.
15. J.K. Kuchar and L.C. Yang. A review of conflict detection and resolution methods. *IEEE Transactions on Intelligent Transportation Systems*, 1(4):179–189, 2000.
16. P. Mueller. Simulation based optimal design. In *Bayesian Statistics 6*, pages 459–474. J.O. Berger, J.M. Bernardo, A.P. Dawid and A.F.M. Smith (eds.), Oxford University Press, 1999.
17. P. Mueller, B. Sanso, and M. De Iorio. Optimal Bayesian design by inhomogeneous Markov chain simulation. Technical report, 2003. Available from World Wide Web: http://www.ams.ucsc.edu.
18. R.A. Paielli and H. Erzberger. Conflict probability estimation for free flight. *Journal of Guidance, Control and Dynamics*, 20(3):588–596, 1997.
19. M. Prandini, J. Hu, J. Lygeros, and S. Sastry. A probabilistic approach to aircraft conflict detection. *IEEE Transactions on Intelligent Transportation Systems*, 1(4), 2000.
20. C.P. Robert and G. Casella. *Monte Carlo Statistical Methods.* Springer-Verlag, 1999.

21. C. Tomlin, G. Pappas, and S. Sastry. Conflict resolution for Air Traffic Management: a case study in multi-agent hybrid systems. *IEEE Transactions on Automatic Control*, 43(4):509–521, 1998.
22. P.J.M. van Laarhoven and E.H. Aarts. *Simulated Annealing: Theory and Applications.* D.Reidel Publishing Company, 1987.

Branching and Interacting Particle Interpretations of Rare Event Probabilities

Pierre Del Moral[1] and Pascal Lezaud[2]

[1] Université de Nice Sophia Antipolis-06108 Nice Cedex 02, France
delmoral@math.unice.fr
[2] Centre d'Etudes de la Navigation Aérienne-31055 Toulouse Cedex, France
pascal.lezaud@recherche.enac.fr

Summary. This article focuses on branching particle interpretations of rare events. We connect importance sampling techniques with interacting particle algorithms, and multi-splitting branching models. These Monte Carlo methods are illustrated with a variety of examples arising in particle trapping analysis, as well as in ruin type estimation problems. We also provide a rather detailed presentation of the asymptotic theory of these particle algorithms, including exponential extinction probabilities, $\mathbb{L}_p$-mean error bounds, central limit theorem, and fluctuation variance comparaisons.

1 Introduction

The study of rare events is an important and very active area in a variety of scientific disciplines. In particle physics, rare event problems are often related to the estimation of non absorption probabilities of a particle evolving in a trapping medium. These quantities are also connected to the estimation of the Lyapunov exponent of Schrödinger type operators. In engineering sciences, these rare event problems arise in the analysis and prediction of major risks, as such earthquakes, floods, air collision risks, nuclear radiation dispersions. Studying major risks can undertaken utilising two main approaches, the statistical analysis of collected data and the probabilistic modelling of the processes leading to the incident. The statistical analysis of extreme values often needs an extended observation period, due to the very low occurrence probability of rare events. They are often based on the standard extreme value distributions, like the Gumbel, the Fréchet and the Weibull laws (see for instance [11, 7], and references therein). The probabilistic approach firstly consists in modelling the randomness of the underlying system, and secondly in using some mathematical, or simulation tools, to obtain an accurate estimate.

The use of analytical, and numerical approaches are often based on simplified, and ad hoc assumptions. On the other hand, the Monte Carlo simulation

H.A.P. Blom, J. Lygeros (Eds.): Stochastic Hybrid Systems, LNCIS 337, pp. 277–323, 2006.

is a practical alternative when the analysis calls for fewer simplifying assumptions. Nevertheless, obtaining accurate estimates of rare event probabilities, say about 10^{-9} to 10^{-12}, using traditional techniques really requires a huge amount of computing time.

Many techniques for reducing the number of trials in Monte Carlo simulation have been proposed, the more promising are based on importance sampling. Importance sampling consists in modifying the underlying probability distribution in such a way that rare events occur much more frequently. To use importance sampling, we need to have a deep knowledge of the system under study; and even in such a case, importance sampling may not reduce the number of trials. In addition, for large time scale problems, the importance weightings are also degenerate with respect to the time parameter. This degeneration reduces considerably the performance and the accuracy of the Monte Carlo approximation.

An alternative way to increase the relative number of visits to the rare event, is to use interactive evolution models and trajectory splitting techniques. These two approaches are based on the fact that there exist some physical potential functions reflecting the rare event regime, or there exist some intermediate levels that are visited much more often than the rare level. These intermediate events act as gateways to reach the desired rare level set. Particle methodologies were first introduced as heuristic algorithms in the beginning of the 1950's, in biology by M.N. Rosenbluth and A. W. Rosenbluth [12] for macromolecular simulations, as well as in physics with the article of T.E. Harris and H. Kahn [8] for particle transmission simulations. Since this period, the range of applications of these interactive particle ideas have increased, revealing unexpected connections between a variety of domains, including signal processing, financial mathematics, particle physics, biology and engineering sciences. The application in rare event simulation was firstly introduced in [1] and shortly after has been adapted to the hybrid systems [9]. For a detailed description of these applications models, and a precise mathematical analysis of these particle methods, the reader is referred to the research monograph [2], and references therein.

In the present review article, we focus on branching particle interpretations of rare events. A short description of the paper is as follows. Section 2 sets out a brief description of different types of branching particle methodologies. We connect importance sampling techniques with interacting particle algorithms, and multi-splitting branching models. We illustrate these Monte Carlo methods with a variety of examples arising in particle trapping analysis, as well as in ruin type estimation problems. Finally, we end this section with a description of a more refined interacting particle analysis of rare event probabilities based on multi-level decompositions of the state-space regions. The capture of the behavior of the Markov chain between each level needs the introduction of the random excursions models. This is provided by the Section 3, in which Markov chain models in abstract path are briefly introduced. The law of the excursions in the rare regime are described by functional representations which belong to the class of Feynman-Kac formulas.

Section 4 describes the Feynman-Kac models which are at the corner of diverses disciplines. The Feynman-Kac models, in general nonhomogeneous state spaces, are built with two ingredients: A Markov chain associated with a reference probability measure, and a sequence of potential functions. From the pure mathematical point of view, they correspond to a change of probability on path space associated with a given sequence of potential functions. From the point of view of physics, they represent for instance the path distribution of a single particle evolving in absorbing and disordered media (see for instance [13]). In this interpretation, the potential function represents a "killing or creation" rate related to the absorbing nature of the medium. In engineering sciences the use of Feynman-Kac models is of course not restricted to rare event modelling and analysis. They are commonly used in non linear filtering to represent the conditional distribution of a given random signal with respect to a sequence of noisy observations delivered by some sensors. Different physical interpretations of the Feynman-Kac models are also provided in this section. These models have natural particle interpretations in terms of genealogical tree-based evolutions. The Section 5 is devoted to particle interpretations of Feynman-Kac models. These particle models can be sought in many different ways depending on the application we have in mind. For the analysis of rare events, we have chosen to describe these models as an abstract stochastic linearization technique for solving nonlinear and measure-valued equations. The basic idea is to associate to a given nonlinear dynamical structure, a sequence of N Markov processes, in such a way that the N-empirical measures of the configurations converge, as $N \to \infty$, to the desired distribution. The parameter N represents the precision parameter, as well as the size of the systems. In some sense, these particle models can be regarded as a new approximation simulation technique. All these particle models are built on the same paradigm: When exploring a state space with many particles, we duplicate better fitted individuals at the expense of having light particles with poor fitness die.

Finally, Section 6 is concerned with the asymptotic behavior of the particle methods when the size of the systems tends to infinity. We provide a rather detailed presentation of the asymptotic theory, including exponential extinction probabilities, $\mathbb{L}_p$-mean error bounds, central limit theorem, and fluctuation variance comparaisons between these particles algorithms.

2 Branching Particle Methodologies

This section sets out a brief description of four different types of branching particle methodologies and an interacting particle algorithms for estimating rare events. The first concerns an original genetic type interpretation of importance sampling representations of rare event probabilities. The second one is concerned with a class of rare event problems arising in physics, and more particularly in nuclear engineering. We design an interacting particle interpretation for the evolution of a Markov chain in an absorbing medium. Special attention is paid to the study of the genealogical structure of these interacting jump particle models. We also connect these trapping problems with the estimation of the Lyapunov exponent of a Schrödinger type semigroup. The third and fourth are devoted respectively to an elementary Bernoulli branching method, and to a more sophisticated branching splitting variant.

2.1 Importance Sampling Branching Models

Let X_n be a Markov chain with transition probabilities $M_n(x_{n-1}, dx_n) = \mathbb{P}(X_n \in dx_n | X_{n-1} = x_{n-1})$. Suppose we want to estimate the probability $\mathbb{P}(A)$, that X_n reaches some rare region A of the state space. To fix the ideas, we can think of a simple random walk starting at $X_0 = 0$ and evolving to the right with a small probability

$$\mathbb{P}(X_n = X_{n-1} + 1) = p = 1 - \mathbb{P}(X_n = X_{n-1} - 1) < 1/2 . \tag{1}$$

In this situation for large values of M, the probabilities $\mathbb{P}(X_n \geq M)$ are extremely small. As we mentioned in the introduction, the importance sampling methodology consists in changing the whole distribution of the chain X_n so that to deal with a new random process X'_n which is "attracted" by the rare event. If we let $M'_n(x_{n-1}, dx_n)$ be the Markov transitions of X'_n, and assuming that M_n and M'_n are mutually absolutely continuous, then we have the rare event probability representation.

$$\mathbb{P}(X_n \in A) = \mathbb{E}\left(1_A(X'_n) \prod_{k=1}^{n} G_k(X'_{k-1}, X'_k)\right) , \tag{2}$$

with

$$G_k(X'_{k-1}, X'_k) = \frac{\mathrm{d}M_k(X'_{k-1}, \cdot)}{\mathrm{d}M_k(X'_{k-1}, \cdot)}(X'_k) ,$$

and for any bounded function f

$$\mathbb{E}(f(X_n) | X_n \in A) = \frac{\mathbb{E}\left(f(X'_n) 1_A(X'_n) \prod_{k=1}^{n} G_k(X'_{k-1}, X'_k)\right)}{\mathbb{E}\left(1_A(X'_n) \prod_{k=1}^{n} G_k(X'_{k-1}, X'_k)\right)} .$$

In the simple random walk example, we can exchange the role of p and $q = 1-p$. In this case, X'_n tends to move to the right, and the change of probability formula (2) holds true with $G_k(x, x+1) = p/q < 1$ and $G_k(x, x-1) = q/p > 1$.

The importance sampling method consists in evolving N independent copies X'^i of X', and taking the weighted Monte Carlo estimates

$$\frac{1}{N}\sum_{i=1}^{N} 1_A(X'^i_n)\left[\prod_{k=1}^{n} G_k(X'^i_{k-1}, X'^i_k)\right] \xrightarrow[N\to\infty]{} \mathbb{P}(X_n \in A)$$

$$\sum_{i=1}^{N} f(X'^i_n)\frac{1_A(X'^i_n)\prod_{k=1}^{n} G_k(X'^i_{k-1}, X'^i_k)}{\sum_{j=1}^{N} 1_A(X'^j_n)\prod_{k=1}^{n} G_k(X'^j_{k-1}, X'^j_k)} \xrightarrow[N\to\infty]{} \mathbb{E}(f(X_n)|X_n \in A) \ .$$

This Monte Carlo method works rather well when the so-called twisted process X'_n is well identified and the time parameter n is not too large, but it cannot be interpreted in any way as a simulation methodology of the process in the rare event regime. A complementary methodology is to interpret, at each stage, the local Radon-Nikodym potential functions G_n as birth rates. These favour the particle transitions $X'_{n-1} \to X'_n$ moving too slowly towards the rare level set. The corresponding algorithm consists in evolving N-particles according to a genetic type mutation/selection method:

$$(\widehat{X}'^i_{n-2})_{1\leq i\leq N} \xrightarrow{\text{Mutat.}} (X'^i_{n-1})_{1\leq i\leq N} \xrightarrow{\text{Select.}} (\widehat{X}'^i_{n-1})_{1\leq i\leq N} \xrightarrow{\text{Mutat.}} (X'^i_n)_{1\leq i\leq N} \ .$$

- During the selection mechanism, we examine the potential value of each past transition $(\widehat{X}'^i_{n-2}, X'^i_{n-1})_{1\leq i\leq N}$ and we select randomly N states $\widehat{X}'^i_{n-1}$ according to the discrete distribution

$$\sum_{i=1}^{N} \frac{G_{n-1}(\widehat{X}'^i_{n-2}, X'^i_{n-1})}{\sum_{j=1}^{N} G_{n-1}(\widehat{X}'^j_{n-2}, X'^j_{n-1})}\delta_{X'^i_{n-1}} \ .$$

- During the mutation mechanism, we simply evolve each selected particle $\widehat{X}'^i_{n-1}$ with a random elementary transition $\widehat{X}'^i_{n-1} \rightsquigarrow X'^i_n \sim M'_n(\widehat{X}'^i_{n-1}, \cdot)$.

The particle approximation models are now given by the occupation measures:

$$1_{|I^N_n|>0} \times \frac{1}{|I^N_n|}\sum_{i\in I^N_n} f(\widehat{X}'^i_n) \xrightarrow[N\to\infty]{} \mathbb{E}(f(X_n)|X_n \in A) \ ,$$

and the product formula

$$\frac{|I^N_n|}{N}\left[\prod_{k=1}^{n} \frac{1}{N}\sum_{i=1}^{N} G_k(\widehat{X}'^i_{k-1}, X'^i_k)\right] \xrightarrow[N\to\infty]{} \mathbb{P}(X_n \in A) \ ,$$

where $|I^N_n|$ represents the cardinality of the set of indices of the particles having succeeded to enter in A at time n. Furthermore, if we trace back the complete genealogy of the particles having succeeded to reach the level A at time n, then we have for any test function f_n on the path space

$$1_{|I^N_n|>0} \times \frac{1}{|I^N_n|}\sum_{i\in I^N_n} f_n(\widehat{X}'^i_{0,n}, \cdots, \widehat{X}'^i_{n,n}) \xrightarrow[N\to\infty]{} \mathbb{E}(f_n(X_0, \cdots, X_n)|X_n \in A) \ ,$$

where $(\widehat{X}'^i_{k,n})_{0\leq k\leq n}$ represents the ancestral line of the end-time particle $\widehat{X}'^i_{n,n} = \widehat{X}'^i_n$. Although, we can prove that $\mathbb{P}(I_n^N = \emptyset)$ decreases to 0 exponentially fast, as $N \to \infty$, in practice we still need to choose a sufficiently large number of particles to ensure that a reasonably large proportion arrives to the target set. The propagation of chaos properties of the interactive particle models ensure that the random variables $\widehat{X}'^i_n$ behaves asymptotically as independent copies of X'_n in the rare event regime.

2.2 Interacting Trapping Models

This section is concerned with rare event estimation problems arising in particle trapping analysis, and nuclear engineering. These probabilistic models also provide interesting physical interpretations of rare events in terms of interactive trapping particles, and the associated genealogical structure. We also connect these rare event estimations with the analysis of Lyapunov exponents of Schrödinger operators.

We consider a physical particle X_n evolving in an absorbing medium E, related to a given potential function $G : E \to [0,1]$. In the state space regions, where $G = 1$, the particle evolves randomly, and freely, according to a given Markov transition kernel $M(x, dy)$. When it enters in other regions, where $G < 1$, its life time decreases, and it is instantly absorbed when it visits the subset of null potential values. For indicator potential function, $G = 1_A$, $A \subset E$, this model reduces to a particle evolution killed on the complementary set $A^c = E \setminus A$. To visualize these models, Fig. 1 shows a particle evolution on $E = \mathbb{Z}$ killed outside an interval A at a random time T, and Fig. 2 illustrates the evolution of an absorbed particle in a lattice.

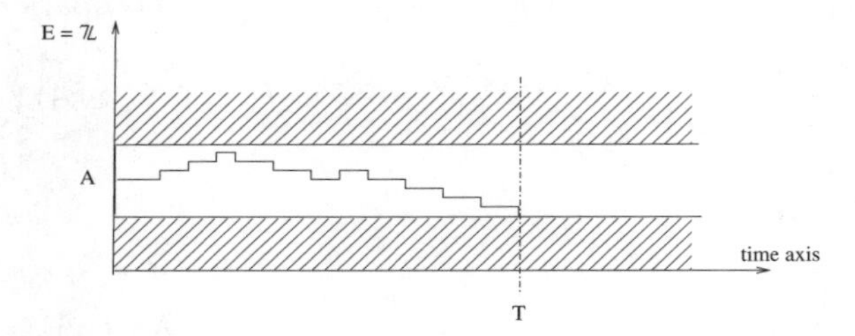

Fig. 1. Evolution of a particle in $E = \mathbb{Z}$ killed outside of A.

These probabilistic models arise in particle physics, such as in neutron collision/absorption analysis [8], as well as in nuclear engineering such as in the risk analysis of radiation containers shields. In this situation, the radiation source emits particles, which evolve in an absorbing shielding environment. In this context, the particle desintegrates when it visits the obstacles. The precise probabilistic model associated to these physical evolutions are discussed in Section 4.2.

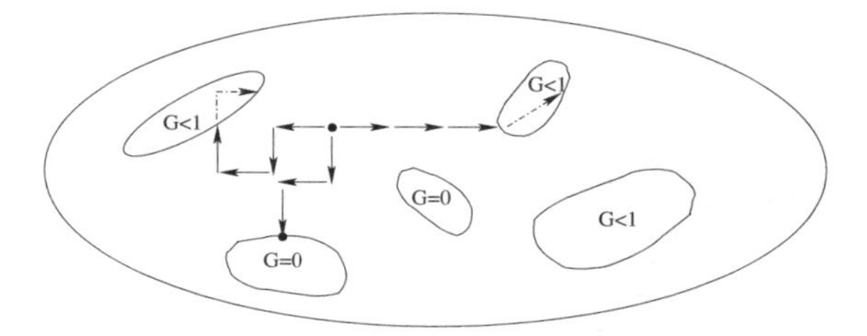

Fig. 2. Evolution of a particle in an absorbing medium.

If we let T be the first time the particle is absorbed, then we are interested in the rare event probabilities

$$\mathbb{P}(T > n) = \int_{E^{n+1}} \eta_0(dx_0)G(x_0)M(x_0, dx_1)G(x_1)\cdots M(x_{n-1}, dx_n)G(x_n) .$$

In the above formula, we integrate over all the particle paths $(x_0, \cdots, x_n) \in E^{n+1}$. The distribution η_0 represents the initial law of X_0, $G(x_0)$ is the probability that the particle at x_0 is not killed, $M(x_0, dx_1)$ is the distribution of the transition from x_0 to x_1, $G(x_1)$ is the probability that the particle at x_1 is not killed, and so on. For large values of the time parameter n, these probabilities are extremely small. In some sense, we have that

$$\mathbb{P}(T > n) = \mathbb{P}(T > 0) \prod_{i=1}^{n} \mathbb{P}(T > p | T > p-1) \approx e^{-n\lambda} , \tag{3}$$

for some constant $\lambda > 0$, which reflects the strength of the obstacles. This constant corresponds to the logarithmic Lyapunov exponent of the integral Schrödinger type semigroup, $G(x, dy) = G(x)M(x, dy)$. For more details, the reader is referred to [5].

To estimate these constants, and these rare event probabilities, we evolve N interacting particles, $\xi_n = (\xi_n^i)_{1 \le i \le N} \in E^N$, according to the following rules

$$\xi_n = (\xi_n^i)_{1 \le i \le N} \xrightarrow{\text{trapping/selection}} \widehat{\xi}_n = (\widehat{\xi}_n^i)_{1 \le i \le N} \xrightarrow{\text{evolution}} \xi_{n+1} = (\xi_{n+1}^i) .$$

During the trapping transition, each particule ξ_n^i survives with a probability $G(\xi_n^i)$, and in this case we set $\widehat{\xi}_n^i = \xi_n^i$. Otherwise, with a probability $1 - G(\xi_n^i)$, the particle is absorbed, and instantly another randomly chosen particle in the current configuration duplicates. More precisely, when the particle ξ_n^i is absorbed, we chose randomly a new particle ξ_n^i according to the discrete Gibbs measure

$$\sum_{j=1}^{N} \frac{G(\xi_n^j)}{\sum_{k=1}^{N} G(\xi_n^k)} \delta_{\xi_n^j} .$$

During the evolution step, each selected particule $\widehat{\xi}_n^i$ evolves randomly according to the Markov transition M. The rare event probabilities are approxi-

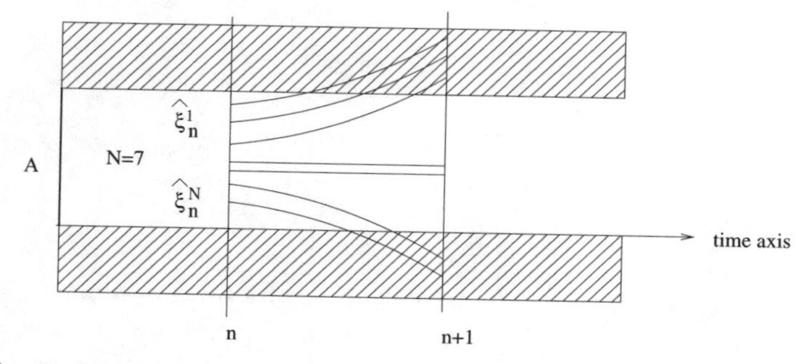

Fig. 3. Interacting particle with indicator potential function $G = 1_A$.

mated by the product formula

$$\mathbb{P}^N(T > n) = \prod_{p=0}^{n} \left(\frac{1}{N} \sum_{i=1}^{N} G(\xi_p^i) \right) \to \mathbb{P}(T > n)$$

$$= \mathbb{P}(T > 0) \prod_{i=1}^{n} \mathbb{P}(T > p | T > p-1) .$$

In the case of indicator potential function $G = 1_A$, we notice that the empirical mean potentials corresponds to the population of evolving transitions which have not been absorbed. In Fig. 3, we illustrate an example with $N = 7$ and $N^{-1} \sum_{i=1}^{N} 1_A(\xi_{n+1}^i) = 2/7$.

For long time horizon, we also have a particle interpretation of the Lyapunov exponent λ, previously introduced in (3)

$$-\frac{1}{n+1} \sum_{p=0}^{n} \log \left(\frac{1}{N} \sum_{i=1}^{N} G(\xi_p^i) \right) \approx \lambda.$$

In the birth and death interpretation, we can trace back the complete genealogy of a given particle ξ_n^i. If we let

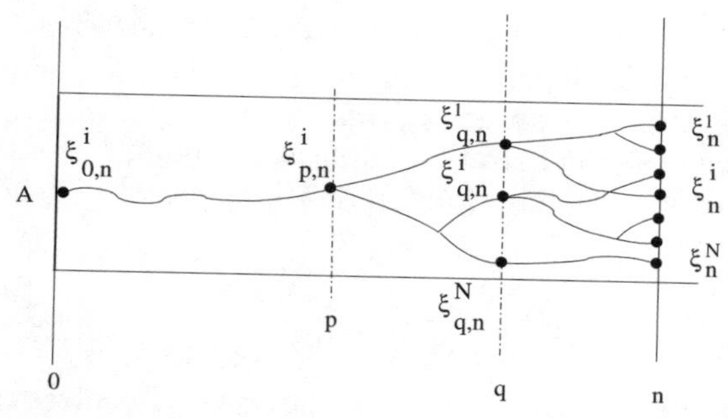

Fig. 4. Genealogical tree associated with the interactif trapping model.

$$\xi_{0,n}^{i} \leftarrow \xi_{1,n}^{i} \leftarrow \cdots \leftarrow \xi_{n-1,n}^{i} \leftarrow \xi_{n,n}^{i} = \xi_{n}^{i}$$

be the ancestral line of the particle with label i, at time n, then we have for any test function f_n on the state space E^{n+1},

$$\frac{1}{N}\sum_{i=1}^{N} f_n(\xi_{0,n}^{i}, \xi_{1,n}^{i}, \cdots, \xi_{n,n}^{i}) \xrightarrow[N\to\infty]{} \mathbb{E}\left(f(X_0, \cdots, X_n)|T \geq n\right) .$$

In some sense, the genealogical tree, associated with interaction trapping model, represents the path strategy used by the Markov particle to stay alive up to time n. Returning to the indicator potential function example, a model of a random tree is represented in Fig. 4.

In the lattice example, the genealogical tree models correspond to a spider web type strategy, as such illustrated in Fig. 5

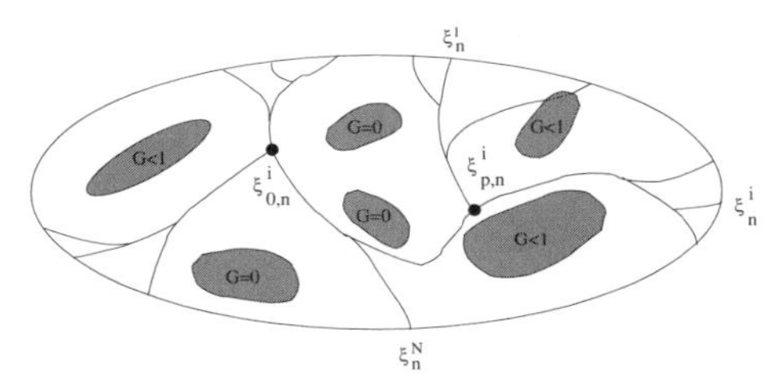

Fig. 5. Genealogical tree model in the lattice example.

2.3 A Bernoulli Splitting Technique

In contrast to importance sampling type algorithms, in the trajectory splitting methodology, the step-by-step evolution of the system follows the original probability measure. Entering the intermediate states, which is usually characterized by crossing a threshold by a control parameter, triggers the splitting of the trajectory. The current system state is held, and a number of independent subtrajectories are simulated from that state.

For example, let us consider $(m+1)$ sets B_i such that

$$B_{m+1} \subset B_m \subset \cdots \subset B_1 .$$

When the rare event A coincide with B_{m+1}, we have the product formula

$$\mathbb{P}(A) = \mathbb{P}(A|B_m)\mathbb{P}(B_m|B_{m-1})\cdots\mathbb{P}(B_2|B_1)\mathbb{P}(B_1). \tag{4}$$

On the right hand side of (4), each conditioning event is "not rare". The branching splitting technique proceeds as follows. Make a $\{0,1\}$ Bernoulli trial to check whether or not the set event B_1 has occured. If B_1 has occured, then we split this trial in R_1 Bernoulli additional trials, and for each of them we check again wether or not the event B_2 has occured. This procedure is repeated at each level. More precisely, each time the event B_i has occured, we sample R_i trials and we repeat this splitting technique each time B_{i+1} has occured. If an event level is not reached, neither is A, then we stop the current retrial. Using R_0 independent replications of this procedure, we have then considered $R_0 R_1 \cdots R_m$ trials, taking into account for example, that if we have failed to reach a level B_i at the i-th step, the $R_i \cdots R_m$ possible retrials have failed. An unbiased estimator of $\mathbb{P}(A)$ is clearly given by the quantity

$$\widehat{P} = \frac{N_A}{R_0 \prod_{i=1}^{m} R_i},$$

where N_A is the total number of trajectories having reached the set A. It can be proven [10] that in some sense the optimal simulation is obtained if

$$m = \lfloor -0.6275 \log P(A) - 1 \rfloor, \quad \mathbb{P}(B_i | B_{i-1}) \approx 1/5, \quad R_i = 5.$$

Nevertheless, in practice the trajectory splitting method may be difficult to apply. For example, the case of the estimation of the probability of a rare event in dynamical system is more complex, since the difficulty to find theoretically the optimal B_i, and R_i for each level i. Furthermore, the probability to reach B_i varies generally with the state of entrance in level B_{i-1}. Finally, but not the least, the conditional probabilities $\mathbb{P}(B_i | B_{i-1})$ are of course generally unknown! In this sense, this rather crude splitting strategy is of pure academic interest.

2.4 Branching Splitting Models

The branching strategy, we are about to describe, is rather close in spirit to the Bernoulli splitting method described above. The essential difference is that it enters the random evolution of the process in the rare event region. To be more precise, we consider a Markov process evolving in some state space, in such a way that a given region, or a particular site, say O, is visited infinitly often. Our objective is to estimate the probability $\mathbb{P}(A)$ to reach a rare level A before returning to O. For instance, if T_A represents the first hitting time of A, and T_O the first return time to O, then $\mathbb{P}(A) = \mathbb{P}(T_A < T_O)$. We proceed as shown in Fig. 6. If the first level B_1 is reached before going back to O, then we split the path into R_1 trials; otherwise, if we are back to O, we stop the exploration. At the next step we evolve each of these R_1 paths, starting from its entrance state in B_1. If a path hits B_2, before returning to O, then it is again splitted into R_2 trials; otherwise we stop its exploration. We repeat the branching transition at each level B_k. Finally, a path from level B_m that succeed to reach A (before returning to O) is considered as a success, and it is stopped. An implicit, and technical assumption is that the level $(i+1)$ cannot be reached from level $(i-1)$ without entering previously the intermediate level

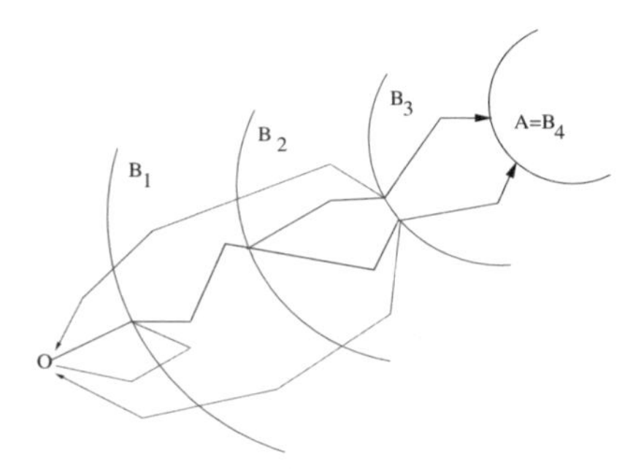

Fig. 6. Example with $R_i = 2$ at each level

i . This condition is clearly met, if we consider a decreasing sequence of level set $A = B_{m+1} \subset B_m \subset \cdots \subset B_1$.

In practice, the simulation time is limited to a given value T , so that we estimate

$$\mathbb{P}(T_A < \min(T_O, T)) .$$

Moreover, going back from B_i to O may take a long time, so we can reduce the computations by freezing the path exploration when it goes back to one, two or more levels down. Nevertheless, this rather crude strategy induces a bias, which is difficult to estimate. An alternative approach is the RESTART method, introduced in [15, 16].

2.5 Interacting Particle Systems Algorithm

In this section we design a genetic, and genealogical tree base model for estimating a rather general class of rare events, following [1]. The main idea behind this evolutionary type algorithm is again to decompose the state space into a judicious choice of threshold levels. This decomposition reflects the successive levels the stochastic process needs to cross before entering into the rare event. More precisely, we consider a strong Markov chain X_n , which is assumed to start in some set O with a given initial probability distribution. We associate to a given target set A, the first time the process hits the set A, namely

$$T_A = \inf\{n \geq 0 \ : \ X_n \in A\} .$$

We use the classical convention $\inf \emptyset = \infty$. We would like to estimate the quantities

$$\mathbb{P}(T_A \leq T) \quad \text{and} \quad \mathrm{Law}((X_n), 0 \leq n \leq T_A | T_A \leq T) . \tag{5}$$

In the above formulas, T is either a deterministic finite horizon time, or the (finite) entrance time into a recurrent set R when $R \cap O = \emptyset$ (or the first return time to O , if $R = O$).

As previously, before visiting R, or entering into A, the random process passes through a decreasing sequence of level sets

$$A = B_{m+1} \subset B_m \subset \cdots \subset B_1 ,$$

with $B_1 \cap (O \cup R) = \emptyset$. To capture the precise behavior of X between the different levels, we consider the random excursions $\mathcal{X}_n$ of X between the successive random times T_{n-1} and T_n ,where $(T_n)_{n=1,\cdots,m+1}$ represent the entrance times of the level sets $(B_n)_{n=1,\cdots,m+1}$. A synthetic picture of these excursions is given in the Fig. 7. We observe that these excursions may have different random lengths, and we have the decomposition formula

$$(T_A \leq T) = (T_{m+1} \leq T) = (T_1 \leq T, \cdots, T_{m+1} \leq T) .$$

To check wether or not a given path $(x_k)_{p \leq k \leq q}$, starting at $x_p \in B_{n-1}$ at time p, has succeeded to reach the level B_n at time q, it is convenient to introduce the indicator potential functions G_n defined on the excursion space by

$$G_n((x_k)_{p \leq k \leq q}) = \mathbf{1}_{\{x_q \in B_n\}} .$$

With this notation, and for each n we have

$$(T_n \leq T) = (G_1(\mathcal{X}_1) = 1, \cdots, G_n(\mathcal{X}_n) = 1) = \left(\prod_{p=1}^{n} G_p(\mathcal{X}_p) = 1 \right) .$$

Integrating over all the process excursions, we obtain the following formula

$$\mathbb{P}(T_n \leq T) = \mathbb{E}\left(\prod_{p=1}^{n} G_p(\mathcal{X}_p) \right) . \tag{6}$$

More generally, the law of the excursions in the rare event regime are described by the following formulas

$$\mathbb{E}(f(\mathcal{X}_0, \cdots, \mathcal{X}_n) | T_n \leq T) = \frac{\mathbb{E}\left(f(\mathcal{X}_0, \cdots, \mathcal{X}_n) \prod_{p=0}^{n} G_p(\mathcal{X}_p) \right)}{\mathbb{E}\left(\prod_{p=0}^{n} G_p(\mathcal{X}_p) \right)} \tag{7}$$

for any bounded test function f.

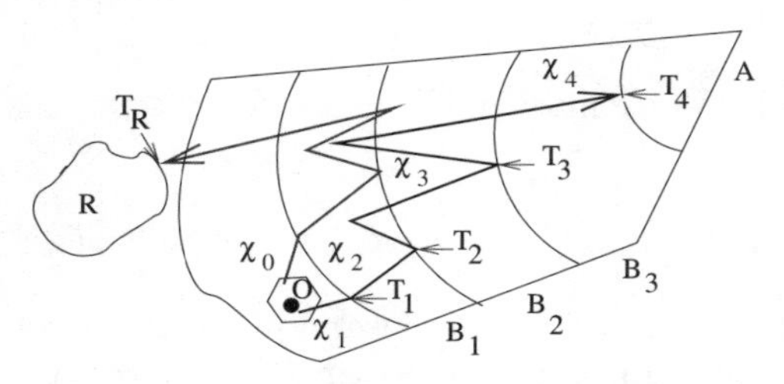

Fig. 7. Embedded Markov Chain

Functional representation of the form (6), and (7), belong to the class of Feynman-Kac formulas. A detailed account on these models can be found in [2], and the references therein. These models have natural particle interpretations, in terms of genealogical tree-based evolutions. An elementary genetic type approximation model is briefly described as follows: When a particle, starting at some level, does not succeed to enter into the next one, it is killed. Otherwise, each time it enters into a closer level of the rare set, it splits into several offsprings. Between the levels, these offsprings evolve as independent copies of the stochastic process $\mathcal{X}_n$, until they reach (or not) an even closer level, and so on.

To be more precise, we evolve N particles according to a two-steps, and genetic type mechanism:

$$(\xi_n^1, \cdots, \xi_n^N) \xrightarrow{\text{selection}} (\widehat{\xi}_n^1, \cdots, \widehat{\xi}_n^N) \xrightarrow{\text{mutation}} (\xi_{n+1}^1, \cdots, \xi_{n+1}^N) \tag{8}$$

- During the selection, each particle with label i having succeeded to reach the n-th level is held, and we set $\widehat{\xi}_n^i = \xi_n^i$. The others $\widehat{\xi}_n^j$ are chosen randomly (and uniformly) in the set of those having succeeded to reach the level B_n . If N_n denotes the number of particle which succeeded to reach the level B_n , then the estimate of the conditional probability $\mathbb{P}(B_n|B_{n-1})$ is simply given by the proportion ratio N_n/N .
- During the mutation, each particle $\widehat{\xi}_n^i$ evolves to a new location ξ_{n+1}^i , randomly chosen according to the transition probability of the chain $\mathcal{X}_n$ in the excursion space.

From previous consideration, a natural unbiased estimate of $\mathbb{P}(T_A \leq T)$ is simply given by the product $\prod_{p=1}^{m+1}(N_p/N)$. Using the propagation of chaos properties of the particle approximation models, one can prove that the above estimate converges to the true value, as $N \to \infty$, [1, 2]. More precisely, we have the almost sure convergence result

$$\mathbb{P}^N(T_A < T) = \prod_{p=1}^{m+1} \frac{N_p}{N} \xrightarrow[N\to\infty]{} \mathbb{P}(T_A < T) .$$

The genealogical tree based model associated with the above genetic-type algorithm represents the conditional distribution of the process evolving in the rare event regime. To be more precise, we let

$$\xi_{0,n}^i \leftarrow \xi_{1,n}^i \leftarrow \xi_{2,n}^i \leftarrow \cdots \leftarrow \xi_{n,n}^i$$

be the ancestral lines of the excursion-valued particles $(\xi_{n,n}^i)_{i\in I_n^N}$ having succeeded to reach the n-th level. For any bounded and measurable test function f defined on the excursion space, we have the almost sure result

$$1_{|I_n^N|>0} \times \frac{1}{|I_n^N|} \sum_{i\in I_n^N} f(\xi_{0,n}^i, \cdots, \xi_{n,n}^i) \xrightarrow[N\to\infty]{} \mathbb{E}\left(f(\mathcal{X}_0, \cdots, \mathcal{X}_n)|T_n < T\right) .$$

In the figure 8, we provide a schematic picture of a genealogical tree associated with $N = 4$ particles evolving on the lattice $\mathbb{Z}$ between a sequence of 3 upper-levels. Each particle starting at the origin, tends to move back to the set of non positive integer $O = -\mathbb{N}$. The prototype of this model is the simple random walk X_n on $\mathbb{Z}$ given by the transitions (1).

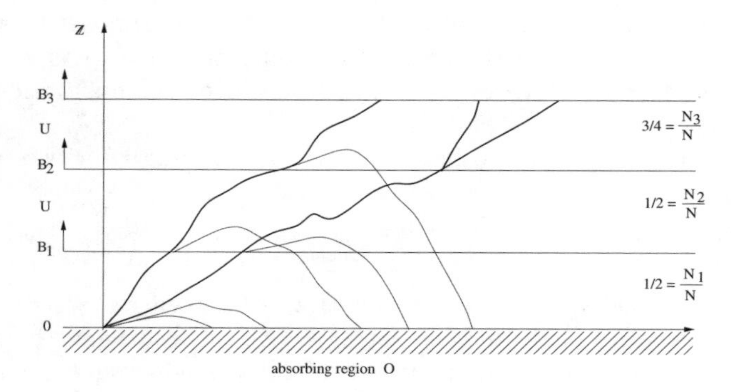

Fig. 8. Genealogical tree

3 Markov Chain and Random Excursion Models

Intuitively speaking, a sequence of random variables $(X_n)_{n\geq 0}$, taking values at each time n in some measurable state space $(E_n, \mathcal{E}_n)$, is said to be a Markov chain when its future and its past trajectories are independent, given the present state of the chain. This Markov property is an extension to a random phenomenon of the well-known property of a deterministic dynamical system, which basically says that the future position and velocity are uniquely defined as soon as they are known at a previous date.

A Markov chain is characterized by its Markov kernels $M_n(x_{n-1}, dx_n)$, which describe the conditional probability of the transition from the point $X_{n-1} = x_{n-1} \in E_{n-1}$ to the infinitesimal neighborhood dx_n of the point $x_n \in E_n$. More formally we have that,

$$\mathbb{P}(X_n \in dx_n | X_{n-1} = x_{n-1}) = M_n(x_{n-1}, dx_n).$$

3.1 Canonical Probability Space

Using the Markov dependence property, μ as the distribution of the initial random state X_0, we check that

$$\begin{aligned}\mathbb{P}_\mu((X_0,\cdots,X_n)\in d(x_0,\cdots,x_n))\\ &= M_n(x_{n-1},dx_n)\mathbb{P}_\mu\left((X_0,\cdots,X_{n-1})\in d(x_0,\cdots,x_{n-1})\right)\\ &= \mu(dx_0)M_1(x_0,dx_1)\cdots M_n(x_{n-1},dx_n)\end{aligned}$$

where $d(x_0,\cdots,x_n)$ stands for an infinitesimal neighborhood of the path point

$$(x_0,\cdots,x_n)\in E_0\times\cdots\times E_n\ .$$

We define the distribution $\mathbb{P}_{\mu,n}$of the canonical sequence $(X_0,\cdots,X_n)$ on $\Omega_n = \left(\prod_{p=0}^n E_p\right)$, equipped with the product σ-field $\mathcal{F}_n = \otimes_{p=0}^n \mathcal{E}_p$, by setting

$$\mathbb{P}_{\mu,n}(dx_{[0,n]}) = \mu(dx_0)M_1(x_0,dx_1)\cdots M_n(x_{n-1},dx_n).$$

By the consistency property of the collection $\mathbb{P}_{\mu,n}$, $n\in\mathbb{N}$, the Ionescu Tulcea's theorem ensures the existence of an overall distribution $\mathbb{P}_\mu$, on the whole path space $\Omega = \left(\prod_{n\geq 0} E_n\right)$, with finite-dimensional distributions $\mathbb{P}_{\mu,n}$. If we denote by $X_n, n\in\mathbb{N}$, the canonical projection mappings

$$X_n : \omega = (\omega_n)_{n\geq 0}\in\Omega \longrightarrow X_n(\omega) = \omega_n \in E_n\ ,$$

then for any $A_p\in\mathcal{E}_p, p\geq 0$, we have that

$$\begin{aligned}&\mathbb{P}_\mu((X_0,\cdots,X_n)\in(A_0\times\cdots\times A_n))\\ &= \int_{A_0\times\cdots\times A_n} \mu(dx_0)M_1(x_0,dx_1)\cdots M_n(x_{n-1},dx_n).\end{aligned}$$

For obvious reasons, the probability model defined in this way

$$(\Omega,\ \mathcal{F} = (\mathcal{F}_n)_{n\geq 0},\ X = (X_n)_{n\geq 0},\ \mathbb{P}_\mu), \tag{9}$$

is called the canonical realisation of the Markov chain, with transitions M_n, and initial distribution μ.

3.2 Path-Space Markov Models

As we mentionned in the introduction, the path space modelling is dictated by the excursions analysis of the process in the rare event regime.

Let $(E_n,\mathcal{E}_n)$ be an auxilairy collection of mesurable spaces, and let X_n be a nonanticipative sequence of E_n-valued random variables in the sense that the distribution of X_{n+1} on E_{n+1} only depends on the random states $(X_0,\cdots,X_n)$. By direct inspection, we notice that the path sequence

$$\mathcal{X}_n = X_{[0,n]} = (X_0,\cdots,X_n),$$

forms a nonhomogeneous Markov chain taking values in the product space

$$E_{[0,n]} = (E_0 \times \cdots \times E_n) \; ,$$

equipped with the product σ-field

$$\mathcal{F}_n = \mathcal{E}_0 \otimes \cdots \otimes \mathcal{E}_p \; .$$

In this situation, each point $x_{[0,n]} = (x_0, \cdots, x_n) \in E_{[0,n]}$ has to be thought of like a path from the origin up to time n.

When X_n is a Markov chain, with non necessarily time homogeneous transitions $M_n(x_{n-1}, dx_n)$ from E_{n-1} into E_n, the Markov chain $\mathcal{X}_n$ in path space is called the historical process, or the path process of the chain X_n.

The Markov transitions $\mathcal{M}_n$ of the chain $\mathcal{X}_n$ are connected to M_n by the formula

$$\mathcal{M}_{n+1}(x_{[0,n]}, dy_{[0,n+1]}) = \delta_{x_{[0,n]}}(dy_{[0,n]}) M_{n+1}(y_n, dy_{n+1}).$$

The motion of the path process $\mathcal{X}_n$ simply consists of extending each path of X_n with an elementary M_n-transition. In summary, we have the synthetic diagram

$$\mathcal{X}_{n-1} = X_{[0,n-1]} \longrightarrow \mathcal{X}_n = X_{[0,n]} = (X_{[0,n-1]}, X_n),$$

with the random state $X_n \sim M_{n+1}(X_{n-1}, \cdot)$.

4 Feynman-Kac Models

To describe precisely the Feynman-Kac models, we need to introduce some additional notation. Firstly, we denote by $\mathcal{B}_b(E)$ the set of bounded measurable functions on a given measurable space $(E, \mathcal{E})$. The expectation operators with respect to $\mathbb{P}_\mu$ and $\mathbb{P}_x$ are denoted by $\mathbb{E}_\mu$ and $\mathbb{E}_x$. For instance, for any $F_n \in \mathcal{B}_b(E_{[0,n]})$, we have

$$\mathbb{E}_\mu(F_n(X_{[0,n]})) = \int_{E_{[0,n]}} F_n(x_{[0,n]}) \mathbb{P}_{\mu,n}(dx_{[0,n]}).$$

We denote respectively by $\mathcal{M}(E_n)$, and $\mathcal{P}(E_n) \subset \mathcal{M}(E_n)$ the set of bounded and signed measures, and the subset of probability measures on the measurable space $(E_n, \mathcal{E}_n)$. We also recall that any Markov kernel M_n from E_{n-1} to E_n generates two operators: The first acting on $\mathcal{B}_b(E_n)$, taking value in $\mathcal{B}_b(E_{n-1})$, and defined by

$$\forall (x_{n-1}, f_n) \in (E_{n-1} \times \mathcal{B}_b(E_n)) \; ,$$

$$M_n(f_n)(x_{n-1}) = \int_{E_n} M_n(x_{n-1}, dx_n) f_n(x_n) \; .$$

The other one acting on measures $\mu_{n-1} \in \mathcal{P}(E_{n-1})$, taking values in $\mathcal{P}(E_n)$, and defined by

$$\forall(\mu_{n-1}, A_n) \in (\mathcal{P}(E_{n-1}) \times \mathcal{E}_n) ,$$
$$(\mu_{n-1}M_n)(A_n) = \int_{E_{n-1}} \mu_{n-1}(dx_{n-1})M_n(x_{n-1}, A_n) .$$

Finally, if $M_{n+1}(x_n, dx_{n+1})$ is a Markov transition from $(E_n, \mathcal{E}_n)$, to another measurable space $(E_{n+1}, \mathcal{E}_{n+1})$, then we denote by $M_nM_{n+1}$the composition operator,

$$M_nM_{n+1}(x_{n-1}, dx_{n+1}) = \int_{E_n} M_n(x_{n-1}, dx_n)M_{n+1}(x_n, dx_{n+1}).$$

Finally, for a given integral operator M from E_0 into E_1, for any $x \in E_0$ and $\varphi^1, \varphi^2 \in \mathcal{B}_b(E_1)$, we simplify notation, and we write

$$M[(\varphi^1 - M\varphi^1)(\varphi^2 - M\varphi^2)](x)$$

instead of

$$M[(\varphi^1 - M(\varphi^1)(x))(\varphi^2 - M(\varphi^2)(x))](x) = M(\varphi^1\varphi^2)(x) - M(\varphi^1)(x)M(\varphi^2)(x). \tag{10}$$

4.1 Description of the Models

We consider a given collection of bounded and $\mathcal{E}_n$-measurable nonnegative functions $G_n : E_n \to [0, \infty)$ such that for any $n \in \mathbb{N}$, we have

$$\mathbb{E}_\mu\Big(\prod_{p=0}^{n} G_p(X_p)\Big) > 0. \tag{11}$$

Definition 1. *The Feynman-Kac prediction and updated path models, associated with the pair (G_n, M_n) (and the initial distribution μ), are the sequence of measures on path space defined respectively, for any $n \in \mathbb{N}$, by the formulas*

$$\mathbb{Q}_{\mu,n}(dx_{[0,n]}) = \frac{1}{\mathcal{Z}_n}\Big\{\prod_{p=0}^{n-1} G_p(x_p)\Big\}\mathbb{P}_{\mu,n}(dx_{[0,n]}) ,$$
$$\widehat{\mathbb{Q}}_{\mu,n}(dx_{[0,n]}) = \frac{1}{\widehat{\mathcal{Z}}_n}\Big\{\prod_{p=0}^{n} G_p(x_p)\Big\}\mathbb{P}_{\mu,n}(dx_{[0,n]}) .$$

The normalizing constants

$$\mathcal{Z}_n = \mathbb{E}_\mu\Big(\prod_{p=0}^{n-1} G_p(x_p)\Big) \quad \text{and} \quad \widehat{\mathcal{Z}}_n = \mathcal{Z}_{n+1} = \mathbb{E}_\mu\Big(\prod_{p=0}^{n} G_p(x_p)\Big)$$

are also often called the partition functions.

Note that for any test function $F_n \in \mathcal{B}_b(E_{[0,n]})$, we have

$$\mathbb{Q}_{\mu,n}(F_n) = \frac{1}{\mathcal{Z}_n}\mathbb{E}_\mu\Big(F_n(X_0,\cdots,X_n)\prod_{p=0}^{n-1} G_p(X_p)\Big),$$

$$\widehat{\mathbb{Q}}_{\mu,n}(F_n) = \frac{1}{\widehat{\mathcal{Z}}_n}\mathbb{E}_\mu\Big(F_n(X_0,\cdots,X_n)\prod_{p=0}^{n} G_p(X_p)\Big).$$

The Feynman-Kac models have a particular dynamic structure. To describe precisely their evolution, it is convenient to introduce the flow of the time marginals.

Definition 2. *The sequence of bounded nonnegative measures γ_n, and $\widehat{\gamma}_n$, on E_n, and defined for any $f_n \in \mathcal{B}_b(E_n)$ by the formulas*

$$\gamma_n(f_n) = \mathbb{E}_\mu\Big(f_n(X_n)\prod_{p=0}^{n-1} G_p(X_p)\Big),$$

$$\widehat{\gamma}_n(f_n) = \mathbb{E}_\mu\Big(f_n(X_n)\prod_{p=0}^{n} G_p(X_p)\Big),$$

are called the unnormalized prediction, and updated, Feynman-Kac model associated with the pair (G_n, M_n). The sequence of distributions η_n, and $\widehat{\eta}_n$, on E_n, and defined for any $f_n \in \mathcal{B}_b(E_n)$ by

$$\eta_n(f_n) = \gamma_n(f_n)/\gamma_n(1) \quad \text{and} \quad \widehat{\eta}_n(f_n) = \widehat{\gamma}_n(f_n)/\widehat{\gamma}_n(1)$$

are called the normalized prediction, and updated, Feynman-Kac model associated with the pair (G_n, M_n).

To get one step further, we notice that

$$\gamma_n(f_n G_n) = \widehat{\gamma}_n(f_n), \quad \text{and} \quad \widehat{\eta}_n(f_n) = \frac{\gamma_n(f_n G_n)}{\gamma_n(G_n)} = \frac{\eta_n(f_n G_n)}{\eta_n(G_n)}. \tag{12}$$

An other key product formula that relates the "unnormalized models" $(\gamma_n, \widehat{\gamma}_n)$ with the Feynman-Kac distribution flow $(\eta_p)_{p\leq n}$, is given by

$$\gamma_n(f_n) = \eta_n(f_n)\prod_{p=0}^{n-1}\eta_p(G_p) \quad \text{and} \quad \widehat{\gamma}_n(f_n) = \widehat{\eta}_n(f_n)\prod_{p=0}^{n}\eta_p(G_p).$$

The identity (12) leads us to introduce the following transformation.

Definition 3. *The Boltzmann-Gibbs transformation associated with a potential function G_n on $(E_n, \mathcal{E}_n)$ is the mapping*

$$\Psi_n : \eta \in \mathcal{P}_n(E_n) \longrightarrow \Psi_n(\eta) \in \mathcal{P}_n(E_n)$$

from the subset $\mathcal{P}_n(E_n) = \{\eta \in \mathcal{P}(E_n) : \eta(G_n) > 0\}$ *into itself, and defined by*

$$\Psi_n(\eta)(dx_n) = \frac{1}{\eta(G_n)} G_n(x_n)\, dx_n \ .$$

In this notation, we see that

$$\widehat{\eta}_n = \Psi_n(\eta_n) \ , \qquad \text{and} \qquad \eta_n = \widehat{\eta}_{n-1} M_n \ . \tag{13}$$

The last identity comes from the following observation

$$\gamma_n(f_n) = \mathbb{E}_\mu\Big(M_n(f_n)(X_{n-1}) \prod_{p=0}^{n-1} G_p(X_p)\Big) = \widehat{\gamma}_{n-1}(M_n(f_n)) \ .$$

We conclude that, the Feynman-Kac flows $(\eta_n, \widehat{\eta}_n)$ are the solution of the nonlinear and measure-valued processes equations

$$\eta_n = \Phi_n(\eta_{n-1}) \ , \qquad \text{and } \widehat{\eta}_n = \widehat{\Phi}_n(\widehat{\eta}_{n-1}) \ , \tag{14}$$

with the one step mappings Φ_n, and $\widehat{\Phi}_n$, defined by

$$\Phi_n(\eta) = \Psi_{n-1}(\eta) M_n \ , \qquad \widehat{\Phi}_n = \Psi_n(\eta M_n) \ .$$

We emphasize that the above evolution analysis strongly relies on the fact that the potential functions $(G_n)_{n\geq 0}$ satisfy the regularity condition stated in (11). For instance, the measure-valued equations (14) may not be defined for any initial distribution η_0 or $\widehat{\eta}_0$, since it may be happen that $\eta_0(G_0) = 0$, or $\widehat{\eta}_0(G_0) = 0$. On the other hand, when the potential functions G_n are unbounded, the Boltzmann-Gibbs transformation Ψ_n are only defined on the set $\{\eta \in \mathcal{P}(E_n), 0 < \eta(G_n) < \infty\}$.

To solve these problems, we further require that the pairs (G_n, M_n) satisfy for any $x_n \in E_n$ the following condition:

$$0 < \widehat{G}_n(x_n) := M_{n+1}(G_{n+1})(x_n) \qquad \text{and} \qquad \sup_{x_n} |\widehat{G}_n(x_n)| = \|\widehat{G}_n\| < \infty \, . \tag{15}$$

In this situation, the integral operators

$$\widehat{M}_n(x_{n-1}, dx_n) = \frac{M_n(x_{n-1}, dx_n) G_n(x_n)}{M_n(G_n)(x_{n-1})}$$

are well-defined Markov-kernels from E_{n-1} to E_n . With this notation, the mapping $\widehat{\Phi}_n$ can be expressed as follows

$$\widehat{\Phi}_n = \widehat{\Psi}_{n-1}(\eta) \widehat{M}_n \ ,$$

where $\widehat{\Psi}_n$ is the Boltzmann-Gibbs transformation associated with the pair potential/kernel $(\widehat{G}_n, \widehat{M}_n)$ and the initial measure $\widehat{\eta}_0$. Thus the updated

Feynman-Kac models associated with the pair (G_n, M_n) and initial measure η_0 coincide with the prediction Feynman-Kac models associated with the pairs $(\widehat{G}_n, \widehat{M}_n)$ starting at $\widehat{\eta}_0$. As we mentionned above, the interpretation of the updated flow as a prediction flow associated with the pair $(\widehat{G}_n, \widehat{M}_n)$ is often more judicious. To illustrate this observation, we examine the situation where the potential function G_n may take some null values, and we set

$$\widehat{E}_n = \{x_n \in E_n : \quad G_n(x_n) > 0\} .$$

It may happen that $\widehat{E}_n$ is not M_n-accessible from any point in E_{n-1}. In this case, we may have $M_n(x_{n-1}, \widehat{E}_n) = 0$, for some $x_{n-1} \in E_{n-1}$, and therefore $M_n(G_n)(x_{n-1}) = 0$. In this situation, the condition (15) is clearly not met. So, we weaken it by considering the following condition

$$(\mathcal{A}) \qquad \forall x_n \in \widehat{E}_n,\ M_{n+1}(x_n, \widehat{E}_{n+1}) > 0, \text{ and } \eta_0(\widehat{E}_0) > 0 , \tag{16}$$

which says that the set $\widehat{E}_{n+1}$ is accessible from any point in $\widehat{E}_n$. This accessibility condition avoids some degenerate tunneling problems such as those represented in the figure 9.

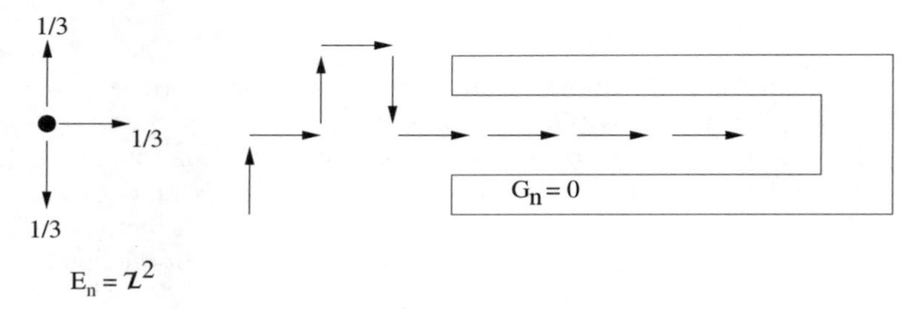

Fig. 9. Tunneling problem

Assuming the condition $(\mathcal{A})$, the condition (15) is only met for any $x_n \in \widehat{E}_n$, and the operators $\widehat{M}_n$ (defined for any $x_{n-1} \in \widehat{E}_{n-1}$) are well-defined Markov kernels from $\widehat{E}_{n-1}$ into $\widehat{E}_n$. Finally, we note that for any $\eta_0 \in \mathcal{P}(E_0)$, with $\eta_0(\widehat{E}_0) > 0$, the updated measure $\widehat{\eta}_0 = \Psi_0(\eta_0)$ is such that $\widehat{\eta}_0(\widehat{E}_0) = 1$.

Summarizing the discussion above, the updated Feynman-Kac measures $\widehat{\eta}_n \in \mathcal{P}(\widehat{E}_n)$ can be interpreted as the prediction models associated with the pair potential/kernel $(\widehat{G}_n, \widehat{M}_n)$ on the restricted state space $(\widehat{E}_n, \widehat{\mathcal{E}}_n)$, as soon as the accessibility condition $\mathcal{A}$ is met. We can also check that

$$\mathbb{E}_{\eta_0}\left(f_n(X_n)\prod_{p=0}^{n} G_p(X_p)\right) = \eta_0(G_0)\,\widehat{\mathbb{E}}_{\eta_0}\left(f_n(X_n)\prod_{p=0}^{n-1} \widehat{G}_p(X_p)\right) > 0 .$$

In particular, this shows that for any $n \in \mathbb{N}$, we have

$$\eta_n \in \mathcal{P}_n(E_n) = \{\eta \in \mathcal{P}(E_n) : \eta(G_n) > 0\} .$$

Therefore, the Feynman-Kac flow is a well-defined two-step updating/prediction model

$$\eta_n \in \mathcal{P}_n(E_n) \xrightarrow{\text{updating}} \widehat{\eta}_n \in \mathcal{P}_n(\widehat{E}_n) \xrightarrow{\text{prediction}} \eta_{n+1} \in \mathcal{P}_{n+1}(E_{n+1}) .$$

Finally, when the accessibility condition $(\mathcal{A})$ is not met, it may happen that

$$\widehat{\eta}_n M_{n+1}(G_{n+1}) = \eta_{n+1}(G_{n+1}) = 0 .$$

In this situation, the Feynman-Kac flow η_n is well-defined, up to the first time τ we have $\eta_\tau(G_\tau) = 0$. At time τ, the measure η_τ cannot be updated anymore. Recalling that $\eta_\tau(G_\tau) = \gamma_{\tau+1}(1)/\gamma_\tau(1)$, we also see that τ coincides with the first time that

$$\widehat{\gamma}_\tau(1) = \gamma_{\tau+1}(1) = \mathbb{E}_{\eta_0}\Big(\prod_{p=0}^{\tau} G_p(X_p)\Big) = 0 .$$

4.2 Physical Interpretations of the Feynman-Kac Models

We now provide different physical interpretations of the Feynman-Kac models. The first one is the traditional trapping interpretation, the second one is based on measure-valued, and interacting processes ideas, such as those arising in mathematical biology.

In the first part, we design a Feynman-Kac representation of distribution flows of a Markov particle evolving in an absorbing medium. As we mentionned in the introduction, these probabilistic models provide a physical interpretation of rare event probabilities in terms of absorption time distributions. In the second part, we set out an alternative representation in terms of non-linear and measure valued processes, the so-called McKean interpretation. The cornerstone of the particle interpretations, developped in this section, is the interpretation of the Feynman-Kac model as such the distribution of a non absorbed particle.

To clarify the presentation, we assume that the potential functions G_n are strictly positive. On the other hand, since the potential functions G_n are assumed to be bounded, we can replace in the definition of the normalized measures η_n, $\widehat{\eta}_n$, the functions G_n by $G_n/\|G_n\|$, without altering their nature. So, there is no loss of generality to assume that $0 < G_n(x_n) \leq 1$.

Killing Interpretation

Now, we identify the potential functions G_n with the multiplicative operator $\mathcal{G}_n$, acting on $\mathcal{B}_b(E_n)$, and defined by the formula

$$\mathcal{G}_n(f_n)(x_n) = G_n(x_n)\, f_n(x_n) .$$

We can alternatively see $\mathcal{G}_n$ as the integral operator on E_n defined by

$$\mathcal{G}_n(x_n, dy_n) = G_n(x_n)\delta_{x_n}(dy_n) .$$

In this connection, we note that $\mathcal{G}_n$ is a sub-Markovian kernel

$$\mathcal{G}_n(x_n, E_n) = G_n(x_n) \leq 1 .$$

The first way to turn the sub-Markovian kernels $\mathcal{G}_n$ into the Markov case consists in adding a cemetery point c to the state space E_n , and then extending the various quantities on the space $E_n^c = E_n \cup \{c\}$ as follows:

- The test functions f_n and the potential functions G_n are extended by setting $f_n(c) = 0 = G_n(c)$.
- The Markov transitions M_n are extended into transitions from E_{n-1}^c to E_n^c by setting $M_n^c(c, \cdot) = \delta_c$, and for each $x_{n-1} \in E_{n-1}$,

$$M_n^c(x_{n-1}, dx_n) = M_n(x_{n-1}, dx_n) .$$

- Finally, the Markov extension $\mathcal{G}_n^c$ of $\mathcal{G}_n$ is given by

$$\mathcal{G}_n^c(x_n, dy_n) = G_n(x_n)\delta_{x_n}(dy_n) + (1 - G_n(x_n))\delta_c(dy_n) .$$

The corresponding Markov chain

$$\left(\Omega^c = \prod_n E_n^c, \mathcal{F}^c = (\mathcal{F}_n^c)_{n\geq 0}, X = (X_n)_{n\geq 0}, \mathbb{P}_\mu^c\right) ,$$

with initial distribution $\mu \in \mathcal{P}(E_0)$ and elementary transitions

$$Q_{n+1}^c = \mathcal{G}_n^c M_{n+1}^c , \tag{17}$$

can be regarded as a Markov particle evolving in an environment, with absorbing obstacles related to potential functions G_n . In view of (17), we see that the motion is decomposed into two separate killing/exploration transitions,

$$X_n \xrightarrow{\text{killing}} \widehat{X}_n \xrightarrow{\text{exploration}} X_{n+1}$$

which are defined as follows:

- **Killing:** If $X_n = c$, then we set $\widehat{X}_n = c$. Otherwise the particle X_n is still alive. In this case, we perform the following random choice: With a probability $G(X_n)$, it remains in the same site and we set $\widehat{X}_n = X_n$; and with probability $1 - G_n(X_n)$, it is killed, and we set $\widehat{X}_n = c$.
- **Exploration:** Firstly, when the particle has been killed, we hace $\widehat{X}_n = c$, and we set $X_p = \widehat{X}_p = c$ for any $p > n$. Otherwise, the particle $\widehat{X}_n \in E_n$ evolves to a new location X_{n+1} in E_{n+1} , randomly chosen according to the distribution $M_{n+1}(\widehat{X}_n, \cdot)$.

In this physical interpretation, the Feynman-Kac flows $(\widehat{\eta}_n, \eta_n)$ represent the conditional distributions of a nonabsorbed Markov particle. To see this claim, we denote by T the time at which the particle has been killed

$$T = \inf\{n \geq 0 : \ \widehat{X}_n = c\} .$$

By construction, we have

$$\mathbb{P}^c_\mu(T > n) = \mathbb{P}^c_\mu(\widehat{X}_0 \in E_0, \cdots, \widehat{X}_n \in E_n) = \mathbb{E}_\mu\Big(\prod_{p=0}^n G_p(X_p)\Big) .$$

This shows that the normalized constants of $\widehat{\eta}_n$, and η_n , represent respectively the probability for the particle to be killed at a time strictly greater than or at least equal to n . That is, we have that

$$\widehat{\gamma}_n(1) = \mathbb{P}^c_\mu(T > n) \ \text{ and } \ \gamma_n(1) = \mathbb{P}^c_\mu(T \geq n) .$$

Similar arguments yield that

$$\widehat{\gamma}_n(f_n) = \mathbb{E}^c_\mu\left(f_n(X_n)1_{\{T>n\}}\right) \ \text{ and } \ \gamma_n(f_n) = \mathbb{E}^c_\mu\left(f_n(X_n)1_{\{T\geq n\}}\right) .$$

Finally, we conlude that

$$\widehat{\eta}_n(f_n) = \mathbb{E}^c_\mu(f_n(X_n)|T > n) \ \text{ and } \ \eta_n(f_n) = \mathbb{E}^c_\mu(f_n(X_n)|T \geq n) .$$

The subsets $G_n^{-1}((0,1))$ and $G_n^{-1}(0)$ are called respectively, the sets of soft and hard obstacles (at time n). A particle entering into a hard obstacle is instantly killed; whereas if it enters into a soft obstacle, its lifetime decreases. When the accessibility condition $(\mathcal{A})$ is met, we can replace the mathematical objects (η_0, E_n, G_n, M_n) by $(\widehat{\eta}_0, \widehat{E}_n, \widehat{G}_n, \widehat{M}_n)$. We define in this way a particle motion in an absorbing medium, with no hard obstacles. Loosely speaking, the hard obstacles have been replaced by repulsive obstacles. For instance, in the situation where $G_n = 1_{\widehat{E}_n}$, the Feynman-Kac model associated with (η_0, G_n, M_n) corresponds to a particle motion in an absorbing medium, with pure hard obstacle sets $\widehat{E}_n$; while the Feynman-Kac associated with $(\widehat{\eta}_0, \widehat{G}_n, \widehat{M}_n)$, corresponds to a particle motion in an absorbing medium, with only soft obstacles related to the potential functions $\widehat{G}_n$.

Interacting Process Interpretation

In interacting process literature, Feynman-Kac flows are alternatively interpreted as nonlinear measure-valued process. For instance, the distribution η_n in (14) is regarded as a solution of nonlinear recursive equations. This equation can be rewritten in the following form

$$\eta_{n+1} = \eta_n K_{n+1,\eta_n} , \tag{18}$$

where K_{n+1,η_n} is the collection of Markov kernels given by

$$K_{n+1,\eta_n}(x,dz) = S_{n,\eta_n}M_{n+1}(x,dz) = \int_{E_n} S_{n,\eta_n}(x,dy)M_{n+1}(y,dz) ,$$

with the selection type transitions

$$S_{n,\eta_n}(x,dy) = G_n(x)\delta_x(dy) + (1 - G_n(x))\Psi_n(\eta_n)(dy) .$$

Note that the corresponding evolution equation is now decomposed into two separate transitions

$$\eta_n \xrightarrow{S_{n,\eta_n}} \widehat{\eta}_n = \eta_n S_{n,\eta_n} \xrightarrow{M_{n+1}} \eta_{n+1} = \widehat{\eta}_n M_{n+1} , \tag{19}$$

In constrast with the killing interpretation, we have turned the sub-Markovian kernel $\mathcal{G}_n$ into the Markov case in a nonlinear way, by replacing the Dirac measure δ_c , by the Boltzmann-Gibbs jump distribution $\Psi_n(\eta_n)$.

The choice of $K_{n,\eta}$ is not unique. A collection of Markov kernels $K_{n,\eta}$, $\eta \in \mathcal{P}(E_n)$ satisfying the compatibility condition

$$\Phi_n(\eta) = \eta K_{n,\eta}$$

for any $\eta \in \mathcal{P}(E_n)$ is called a McKean interpretation of the flow η_n . In comparaison with (17), the motion of the canonical model $X_n \to X_{n+1}$ associated with the Markov kernels $(K_{n,\eta})_{\eta\in\mathcal{P}(E_n)}$ is the overlapping of an interacting jump, and an exploration transition

$$X_n \xrightarrow{\text{interacting jump}} \widehat{X}_n \xrightarrow{\text{exploration}} X_{n+1} .$$

These two mechanisms are defined as follows:

- **Interacting jump:** Given the position, and the distribution η_n at time n of the particle X_n , a jump is performed to a new site $\widehat{X}_n$, randomly chosen according to the distribution

$$S_{n,\eta_n}(X_n,\cdot) = G_n(X_n)\delta_{X_n} + (1 - G_n(X_n))\Psi_n(\eta_n) .$$

  In other words, with a probability $G_n(X_n)$ the particle remains in the same site, and we set $\widehat{X}_n = X_n$. Otherwise, it jumps to a new location, randomly chosen according to the Boltzmann-Gibbs distribution $\Psi_n(\eta_n)$. Notice that particles are attracted by regions with high potential values.
- **Exploration:** The exploration transition coincides with that of the killed particle model. During this stage, the particle evolves to a new site X_{n+1} , randomly chosen according to $M_{n+1}(\widehat{X}_n,\cdot)$.

5 Interacting Particle Systems

The basic idea behind the interacting particle systems is to associate to a given nonlinear dynamical structure, a sequence of E_n^N-valued Markov processes, in such a way that the configuration occupation measures converge, as $N \to \infty$, to the desired distribution. The parameter N represents the precision parameter, as well as the size of the systems. The state components of the E_n^N-valued Markov process are called particles.

5.1 Interacting Particle Interpretations

Hereafter, we suppose the potential functions G_n are bounded and strictly positive (the situation where G_n may take null values can be reduced to this situation, under appropriate accessibility conditions, by replacing η_n by $\widehat{\eta}_n$).

We recall that η_n satisfy the nonlinear recursive equation (18) where the kernels $K_{n,\eta}$ are a combination of a selection and mutation transition

$$K_{n+1,\eta} = S_{n,\eta} M_{n+1} \ . \tag{20}$$

The selection transition $S_{n,\eta}$ on E_n is given by

$$S_{n,\eta_n}(x, dy) = \varepsilon_n G_n(x) \delta_x(dy) + (1 - \varepsilon_n G_n(x)) \Psi_n(\eta_n)(dy) \ , \tag{21}$$

where ε_n stands for non negative number such that $\varepsilon_n G_n \leq 1$.

Definition 4. *The interacting particle model associated with a collection of Markov transitions $K_{n,\eta}, \eta \in \mathcal{P}(E_n), n \geq 1$, and with initial distribution η_0, is a sequence of nonhomogeneous Markov chains*

$$\left(\Omega^{(N)} = \prod_{n \geq 0} E_n^N, \ \mathcal{F}^N = (\mathcal{F}_n^N)_{n \geq 0}, \ \xi = (\xi_n)_{n \geq 0}, \ \mathbb{P}_{\eta_0}^N \right) ,$$

taking values at each time n in the product space E_n^N. That is, we have

$$\xi_n = (\xi_n^1, \cdots, \xi_n^N) \in E_n^N = \underbrace{E_n \times \cdots \times E_n}_{N times} .$$

The initial configuration ξ_0 consists of N independent, and identically distributed random variables, with common law η_0. Its elementary transitions from E_{n-1}^N into E_n^N are given by

$$\mathbb{P}_{\eta_0}^N \left(\xi_n \in dx_n | \xi_{n-1} \right) = \prod_{p=1}^N K_{n, m(\xi_{n-1})}(\xi_{n-1}^p, dx_n^p) ,$$

where

$$m(\xi_{n-1}) = \frac{1}{N} \sum_{i=1}^N \delta_{\xi_{n-1}^i}$$

is the empirical measure of the configuration ξ_{n-1} *of the system, and* $dx_n = dx_n^1 \times \cdots \times dx_n^N$ *is an infinitesimal neighborhood of a point* $x_n = (x_n^1, \cdots, x_n^N) \in E_n^N$.

The N-particle model, associated with the Markov transition $K_{n,\eta}$ given by (20), is the Markov chain ξ_n with elementary transitions

$$\mathbb{P}^N_{\eta_0}\Big(\xi_{n+1} \in dx_{n+1} | \xi_n\Big) = \int_{E_n^N} \mathcal{S}_n(\xi_n, dx_n) \mathcal{M}_{n+1}(x_n, dx_{n+1}) \ .$$

The Boltzmann-Gibbs transition $\mathcal{S}_n$, from E_n^N into itself, and the mutation transition $\mathcal{M}_{n+1}$, from E_n^N into E_{n+1}^N, are defined by the product formulas

$$\mathcal{S}_n(\xi_n, dx_n) = \prod_{p=1}^N S_{n,m(\xi_n)}(\xi_n^p, dx_n^p) \ ,$$

$$\mathcal{M}_{n+1}(x_n, dx_{n+1}) = \prod_{p=1}^N M_{n+1}(x_n^p, dx_{n+1}^p) \ .$$

This integral decomposition shows that (the deterministic) two-step updating/prediction transitions in (19) have been replaced by a two-step selection/mutation transitions (8)

$$\xi_n \in E_n^N \xrightarrow{\text{selection}} \widehat{\xi}_n \in E_n^N \xrightarrow{\text{mutation}} \xi_{n+1} \in E_{n+1}^N \ .$$

In more details, the motion of the particles is defined as follows:

- **Selection:** Given the configuration $\xi_n \in E_n^N$ of the system at time n, the selection transition consists in selecting randomly N particles $\widehat{\xi}_n^i$ with respective distribution $S_{n,m(\xi_n)}(\xi_n^i, \cdot)$. In other words, with a probability $\varepsilon_n G_n(\xi_n^i)$, we set $\widehat{\xi}_n^i = \xi_n^i$; otherwise, we select randomly a particle $\widetilde{\xi}_n^i$ with distribution

$$\Psi_n(m(\xi_n)) = \sum_{i=1}^N \frac{G_n(\xi_n^i)}{\sum_{j=1}^N G_n(\xi_n^j)} \delta_{\xi_n^i}, \text{ and we set } \widehat{\xi}_n^i = \widetilde{\xi}_n^i \ .$$

- **Mutation:** Given the selected configuration $\widehat{\xi}_n \in E_n^N$, the mutation transition consists in sampling randomly N independent particles ξ_{n+1}^i with respective distributions $M_{n+1}(\widehat{\xi}_n^i, \cdot)$.

5.2 Particle Models with Degenerate Potential

We now discuss the situation where G_n is not necessarily strictly positive. To avoid some complications, we suppose the accessibility condition $(\mathcal{A})$ is met.

Two strategies can be underlined. In view of the discussion given in Sect. 4.1, the first idea is to consider the N-particle approximation model associated with some McKean interpretation of the updated model $\widehat{\eta}_n = \Psi_n(\eta_n)$ which can be regarded as a sequence of measures on $\widehat{E}_n = G_n^{-1}(0,\infty)$. Furthermore, $\widehat{\eta}_n$ coincide with the prediction model starting at $\widehat{\eta}_0$ and associated with the pair of potentials/kernels $(\widehat{G}_n, \widehat{M}_n)$ on the state spaces $\widehat{E}_n$.

The potential function $\widehat{G}_n$ is now a strictly positive function on $\widehat{E}_n$ and the updated model $\widehat{\eta}_n$ satisfies the recursive equation

$$\widehat{\eta}_{n+1} = \widehat{\eta}_n \widehat{K}_{n+1,\eta_n} \quad \text{with} \quad \widehat{K}_{n+1,\eta} = \widehat{S}_{n,\eta} \widehat{M}_{n+1} .$$

The selection transitions are now Markov kernels, from $\widehat{E}_n$ into itself, and they are defined for any $x_n \in \widehat{E}_n$ by the formula

$$\widehat{S}_{n,\eta}(x_n, dy_n) = \varepsilon_n \widehat{G}_n(x_n) \delta_{x_n}(dy_n) + (1 - \varepsilon_n \widehat{G}_n(x_n)) \widehat{\Psi}_n(\eta)(dy_n) .$$

The Boltzmann-Gibbs transformation $\widehat{\Psi}_n$ is given by

$$\widehat{\Psi}_n(\eta)(dx_n) = \frac{1}{\eta(\widehat{G}_n)} \widehat{G}_n(x_n)\, \eta(dx_n) .$$

In this interpretation, the model $\widehat{\eta}_n$ satisfies the deterministic evolution equation

$$\widehat{\eta}_n \xrightarrow{\text{updating}} \widetilde{\eta}_n = \widehat{\eta}_n \widehat{S}_{n,\eta_n} \xrightarrow{\text{prediction}} \widehat{\eta}_{n+1} = \widetilde{\eta}_n \widehat{M}_{n+1} .$$

The N-particle associated with this McKean interpretation is defined as before.

The second strategy consists in still working with the McKean interpretation of the prediction flow associated with the collection of transitions $K_{n+1,\eta} = S_{n,\eta} M_{n+1}$ with $\eta \in \mathcal{P}_n(E_n)$. In this case the particle interpretation given in Definition 4 is not well-defined. Indeed, it may happen that the whole configuration ξ_n moves out of the set $\widehat{E}_n$. To describe rigorously the particle model we proceed as in Sect. 4.2. We add a cemetery point Δ to the product space E_n^N and we extend the test functions and the mutation/selection transitions $(\mathcal{S}_n, \mathcal{M}_n)$ on E_n^N to $E_n^N \cup \{\Delta\}$ as follows:

- The test functions $\varphi_n \in \mathcal{B}_b(E_n^N)$ are extended by setting $\varphi_n(\Delta) = 0$.
- The selection transitions $\mathcal{S}_n$, from E_n^N into itself, are extended into transitions on $E_n^N \cup \{\Delta\}$ by setting $\mathcal{S}_n(x, \cdot) = \delta_\Delta$, as soon as the empirical measure $m(x) \notin \mathcal{P}_n(E_n)$.
- The mutation transitions $\mathcal{M}_{n+1}$ are extended into transitions from $E_n^N \cup \{\Delta\}$ to $E_{n+1}^N \cup \{\Delta\}$ by setting $\mathcal{M}_{n+1}(\Delta, \cdot) = \delta_\Delta$.

The corresponding interacting particle model is a sequence of nonhomogeneous Markov chains, taking values at each time n in $E_n^N \cup \{\Delta\}$. It is defined by a two-step selection/mutation transition of the same nature as before:

$$\xi_n \in E_n^N \cup \{\Delta\} \xrightarrow{\text{selection}} \widehat{\xi}_n \in E_n^N \cup \{\Delta\} \xrightarrow{\text{mutation}} \xi_{n+1} \in E_{n+1}^N \cup \{\Delta\} .$$

The only difference is that the chain is killed at the first time n, we have $m(\xi_n) \notin \mathcal{P}_n(E_n)$. Let τ^N and τ be the dates at which respectively the chain and the Feynman-Kac model are killed:

$$\tau^N = \inf\{n \in \mathbb{N}; m(\xi_n)(G_n) = 0\}, \quad \text{and} \quad \tau = \inf\{n \in \mathbb{N}; \eta_n(G_n) = 0\} .$$

Then it is intuitively clear that $\tau^N \leq \tau$, and in Sect. 6.3 it will be proved that for any $n \leq \tau$ and $N \geq 1$ we have exponential estimate

$$\mathbb{P}^N_{\eta_0}(\tau^N \leq n) \leq a(n) \exp(-N/b(n)) .$$

In particular, this shows that $\lim_{N\to\infty} \mathbb{P}^N_{\eta_0}(\tau^N = \tau) = 1$.

5.3 Application to Particle Analysis of Rare Events

We use the notations and conventions as were introduced in Sects. 2.5 and 3. We recall that $X = (X_n)_{n\in\mathbb{N}}$ is a strong Markov chain taking values in some metric state space (S, d). The process X starts in some Borel set $O \subset S$ with a given probability distribution ν_0. We also consider a pair of Borel subsets (A, R), such that $A_0 \cap R = \emptyset = A \cap R$.

We associate with this pair, the first time T the process hits $A \cup R$, and we let T_R be the hitting time of the set R. We also assume that for any initial $x_0 \in O$, we have $\mathbb{P}_x(T < \infty) = 1$. One would like to estimate the quantities

$$\begin{aligned} \mathbb{P}(T < T_R) &= \mathbb{P}(X_T \in A) , \qquad (22) \\ \text{Law}(X_n; 0 \leq n \leq T | T < T_R) &= \text{Law}(X_n; 0 \leq n \leq T | X_T \in A) . \end{aligned}$$

It often happens that most of the realizations of X never reach the target set A, but are attracted, and absorbed by some non empty set R. These rare events are difficult to analyze numerically. One strategy to estimate these events is to consider the sequence of level-crossing excursions $\mathcal{X}_n$ associated with a splitting of the state space, namely

$$\mathcal{X}_0 = (0, X_0), \quad \text{and} \quad \mathcal{X}_n = (T_n, X_{[T_{n-1}, T_n]}) ,$$

with the entrance times $T_n = \inf\{n \geq 0 : X_n \in B_n \cup R\}$. This sequence forms a Markov chain taking value in the set of excursions $E = \cup_{p\geq 0}(\{p\} \times S^p)$.

One way to check whether or not a random path has succeeded to reach the desired n-th level is to consider the indicator potential functions $\mathcal{G}_n(q, x_{[p,q]}) = 1_{B_n}(x_q)$, with the convention $B_0 = O$. Using elementary calculations, we obtain the following Feynman-Kac representation of the desired quantities (22).

Proposition 1. *For any n and any $f_n \in \mathcal{B}_b(E)$, we have that*

$$\mathbb{E}\left(f_n(\mathcal{X}_0, \cdots, \mathcal{X}_n) \,;\, T_n < T_R\right) = \mathbb{E}\left(f_n(\mathcal{X}_0, \cdots, \mathcal{X}_n) \prod_{p=0}^{n} \mathcal{G}_p(\mathcal{X}_p)\right) .$$

The prediction Feynman-Kac model $\eta_n \in \mathcal{P}(E)$, defined by

$$\eta_n(f) = \gamma_n(f)/\gamma_n(1) \quad \text{with} \quad \gamma_n(f) = \mathbb{E}\left(f(\mathcal{X}_n)\prod_{p=0}^{n-1}\mathcal{G}_p(\mathcal{X}_p)\right),$$

satisfies the measure-valued dynamical system

$$\eta_{n+1} = \Phi_{n+1}(\eta_n) \quad \text{with} \quad \eta_0 = \delta_0 \otimes \nu_0 .$$

The mappings Φ_{n+1}, from $\mathcal{P}_n(E)$ into $\mathcal{P}(E)$, are defined by $\Phi_{n+1}(\eta) = \Psi_n(\eta)\mathcal{M}_{n+1}$, where the Markov kernels $\mathcal{M}_n(u, dv)$ represent the Markov transitions of the chain excursions $\mathcal{X}_n$. We have the following lemma

Lemma 1. *For any $n \geq 0$, we have*

$$\mathcal{P}(T_n < T_R) = \widehat{\gamma}_n(1) = \gamma_{n+1}(1) .$$

In addition, we have $\mathcal{P}(T_n < T_R | T_{n-1} < T_R) = \eta_n(G_n)$, and for any $f \in \mathcal{B}_b(E)$

$$\eta_n(f) = \mathbb{E}\left(f(T_n, X_{[T_{n-1},T_n]}) | T_{n-1} < T_R\right) ,$$
$$\widehat{\eta}_n(f) = \mathbb{E}\left(f(T_n, X_{[T_{n-1},T_n]}) | T_n < T_R\right) .$$

This lemma gives a Feynman-Kac interpretation of rare events probabilities. Since the potentials are indicator functions, it is more judicious to rewrite the Boltzmann-Gibbs transformations $\Psi_n(\eta) = \eta S_{n,\eta}$ in terms of the selection Markov transitions

$$S_{n,\eta}(u, dv) = (1 - 1_{\{\mathcal{G}_n^{-1}(1)\}}(u))\Psi_n(\eta)(dv) + 1_{\{\mathcal{G}_n^{-1}(1)\}}(u)\delta_u(dv) .$$

Note that $\mathcal{G}_n^{-1}(1)$ represents the collection of excursions in S entering the nth level B_n; that is, we have that

$$\mathcal{G}_n^{-1}(1) = \{u = (q, x_{[p,q]}) \in E;\ x_q \in B_n\} .$$

The particle interpretation of these discrete Feynman-Kac model is simply derived from Sect. 5.2. In this context, the particle model consists in evolving a collection of N-excursion valued particles

$$\xi_n^i = (T_n^i, X_{[T_{n-1}^i, T_n^i]}^i) \in E \cup \{\Delta\} ,$$
$$\widehat{\xi}_n^i = (\widehat{T}_n^i, \widehat{X}_{[T_{n-1}^i, T_n^i]}^i) \in E \cup \{\Delta\} .$$

The auxiliary point Δ stands for a cemetery point, the random time pairs (T_{n-1}^i, T_n^i) and $(\widehat{T}_{n-1}^i, \widehat{T}_n^i)$ represent the length of the corresponding excursions. At the time $n = 0$, the initial system consists of N independent, and identically distributed, S-valued random variables $\xi_0^i = (0, X_0^i)$, with common law $\eta_0 = \delta_0 \otimes \nu_0$. Since we have $\mathcal{G}_0(0,u) = 1$, there is no updating transition at time $n = 0$, and we set $\widehat{\xi}_0^i = \xi_0^i$, for each $1 \leq i \leq N$.

Mutation: The mutation stage $\widehat{\xi}_n \to \xi_{n+1}$ at time $n+1$ is defined as follows. If $\widehat{\xi}_n = \Delta$, we set $\xi_{n+1} = \Delta$. Otherwise, during the mutation, each selected excursion $\widehat{\xi}_n^i$ evolves randomly, and independently of each other, according to the Markov transition $\mathcal{M}_{n+1}$ of the chain $\mathcal{X}_n$. Thus, ξ_{n+1}^i is a random variable with distribution $\mathcal{M}_{n+1}(\widehat{\xi}_n^i, \cdot)$. More precisely, we set $T_n^i = \widehat{T}_n^i$, and the particle $\widehat{X}^i_{[T_{n-1}^i, T_n^i]}$ evolves randomly as a copy of the excursion process $(X_s)_{s \geq T_n^i}$ starting at $X_{T_n^i}$, and up to the first time T_{n+1}^i it visits B_{n+1}, or returns to R. The stopping time T_{n+1}^i represents the first time $t \geq T_n^i$ the ith excursion hits the set $B_{n+1} \cup R$.

Selection: The selection mechanism $\xi_{n+1} \to \widehat{\xi}_{n+1}$ is defined as follows. In the mutation stage, we have sampled N excursions ξ_{n+1}^i. Some of these particles have succeeded to reach the desired set B_{n+1}, and the other ones have entered into R. We denote by $I^N(n+1)$ the set of the labels of the particles having reached the $(n+1)$-th level, and we set $m(\xi_{n+1}) = N^{-1} \sum_{i=1}^N \delta_{(\xi_{n+1}^i)}$. Two situations may occur. If $I^N(n+1) = \emptyset$ then none of the particles have succeeded to hit the desired level. In this situation, we have $m(\xi_{n+1}) \notin \mathcal{P}_{n+1}(E)$, and the algorithm has to be stopped. In this case, we set $\widehat{\xi}_{n+1} = \Delta$. Otherwise, the selection transition is defined as follows. Each particle $\widehat{\xi}_{n+1}$ is sampled according to the selection distribution

$$
\begin{aligned}
&S_{n,m(\xi_{n+1})}(\xi_{n+1}^i, dv) \\
&= 1_{B_{n+1}}(X^i_{T_{n+1}^i})\delta_{\xi_{n+1}^i}(dv) + 1_{B_{n+1}^C}(X^i_{T_{n+1}^i})\Psi_n(m(\xi_{n+1}))(dv) \, .
\end{aligned}
$$

More precisely, if the i-th excursion has reached the desired level, then we set $\widehat{\xi}_{n+1}^i = \xi_{n+1}^i$. In the opposite case, the particle has not reached the $(n+1)$-th level, but it has visited the set R. In this case, $\widehat{\xi}_{n+1}^i$ is chosen randomly and uniformly in the set $\{\xi_{n+1}^j; j \in I^N(n+1)\}$ of all excursions having entered into B_{n+1}. In other words, each particle that doesn't enter into the $(n+1)$-th level is killed, and instantly a different particle in the B_{n+1} level splits into two offsprings.

For each time $n < \tau^N = \inf\{n \geq 0 \, : \, X^i_{T_n^i} \in R, 1 \leq i \leq n\}$, the N-particle approximation measures $(\gamma_n^N, \eta_n^N, \widehat{\eta}_n^N)$ associated with $(\gamma_n, \eta_n, \widehat{\eta}_n)$ are defined by

$$\widehat{\gamma}_n^N(1) = \gamma_n^N(\mathcal{G}_n) = N^{-n} \prod_{p=1}^{n} \mathrm{Card}(I^N(p)) \ ,$$

$$\eta_n^N = \frac{1}{N} \sum_{i=1}^{N} \delta_{\xi_n^i} \ ,$$

$$\widehat{\eta}_n^N = \Psi_n(\eta_n^N) = \frac{1}{\mathrm{Card}(I^N(n))} \sum_{i \in I^N(n)} \delta_{(T_n^i, X^i_{[T^i_{n-1}, T^i_n]})} \ .$$

Thus, $\widehat{\gamma}_n^N(1)$ is the proportion product of excursions having entered levels $B_1, \cdots, B_n$. Also notice that $\widehat{\eta}_n^N$ is the occupation measure of the excursions entering the nth level.

The asymptotic analysis of these particles measures will be discussed in the following sections. We will prove the following results (see notation (10)):

Theorem 1. *For any $n \geq 0$ and $N \geq 1$ we have*

$$\mathbb{P}(\tau^N \leq n) \leq a(n) \exp(-N/b(n)) \ .$$

The particle estimates are unbiased, $\mathbb{E}(\widehat{\gamma}_n^N(1) 1_{\{n<\tau^N\}}) = \mathbb{P}(T_n < T_R)$, and for any $p \geq 1$, and any $n \geq 0$, we have

$$\sqrt{N}\, \mathbb{E}\left(|\widehat{\gamma}_n^N(1) 1_{\{n<\tau^N\}} - \mathbb{P}(T_n < T_R)|^p\right)^{1/p} \leq a(p) b(n) \ ,$$

for some finite constants $a(p)$, $b(n) < \infty$ whose values only depend respectively on the parameters p and n.

In addition, for any $0 \leq n \leq m+1$, the sequence of random variables

$$W_{n+1}^N = \sqrt{N}(\gamma_n^N(1) 1_{\{\tau^N > n\}} - \mathbb{P}(T_n < T_R))$$

converges in law (as N tends to ∞) to a centered Gaussian random variable W_{n+1} with variance

$$\sigma_n^2 = \sum_{q=0}^{n+1} (\gamma_q(1))^2 \eta_{q-1}(K_{q,\eta_{q-1}}[Q_{q,n}(1) - K_{q,\eta_{q-1}} Q_{q,n}(1)]^2) \ .$$

The collection of functions $Q_{q,n+1}(1)$ on the excursion space E are defined for any $x = (x_n)_{s \leq n \leq t}$ by

$$Q_{q,n+1}(1)(t,x) = 1_{B_q}(x_t) \mathbb{P}(T_n < T_R | T_q = t, X_{T_q} = x_t) \ .$$

Example 1. When the set $S = \mathbb{R}^d$ is the Euclidean space, we can think of a sequence of centered decreasing balls with radius $1/(n+1)$

$$B_n = \mathcal{B}(0, \frac{1}{n+1}) \quad \text{and} \quad R = S \setminus \mathcal{B}(0, 1+\varepsilon)$$

for some $\varepsilon > 0$. Further assume that the process X exits the ball of radius $1+\varepsilon$ in finite time. In this situation, $\mathbb{P}(T < T_R)$ is the probability that X hits the smallest ball B_m, starting with $1/2 < |X_0| \leq 1$, and before exiting the ball of radius $1+\varepsilon$. The distribution (22) represents the conditional distribution of the process X in this ballistic regime (see Fig. 10).

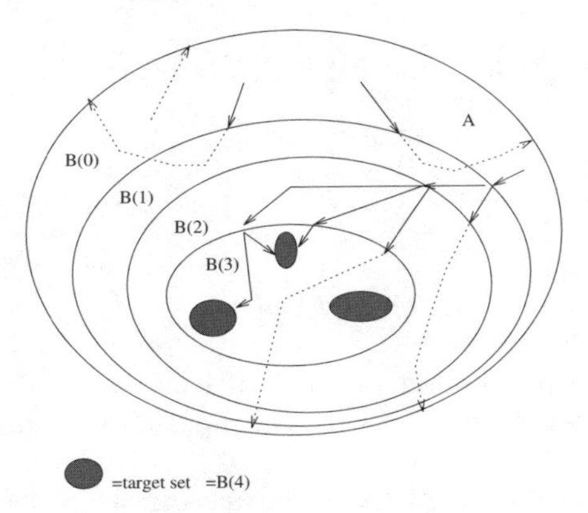

Fig. 10. Ballistic regime, target $B(4)$ with $N = 4$

6 Asymptotic Behavior

This section is concerned with the asymptotic behavior of particle approximation models, as the size of the systems tends to infinity. The principal convergence results are the following. Firstly, γ_n^N is an unbiaised estimator; that is, we have for any $f_n \in \mathcal{B}_b(E_n)$

$$\mathbb{E}_{\eta_0}^N(\gamma_n^N(f_n)1_{\{\tau_n^N \geq n\}}) = \gamma_n(f_n) .$$

Furthermore, we have the $\mathbb{L}_p$-estimates

$$\sqrt{N}\, \mathbb{E}_{\eta_0}^N[|\eta_n^N(f_n) - \eta_n(f_n)|^p]^{1/p} \leq a(p)b(n)\|f\| ,$$

which can be extended to a countable collection of uniformly bounded functions $\mathcal{F}_n \subset \mathcal{B}_b(E_n)$,

$$\sqrt{N}\, \mathbb{E}_{\eta_0}^N\left[\sup_{f_n \in \mathcal{F}_n} |\eta_n^N(f_n) - \eta_n(f_n)|^p\right]^{1/p} \leq a(p)b(n)I(\mathcal{F}_n) ,$$

for some finite constant $I(\mathcal{F}_n) < \infty$ that only depends on the class $\mathcal{F}_n$. Similar but exponential type estimates will be also covered. By instance, we have for any $\varepsilon > 0$ and N sufficiently large

$$\mathbb{P}_{\eta_0}^N\left[\sup_{f_n \in \mathcal{F}_n} |\eta_n^N(f_n) - \eta_n(f_n)| > \varepsilon\right] \leq d_n(\varepsilon, \mathcal{F}_n)\mathrm{e}^{-N\varepsilon^2/b(n)} ,$$

with a finite constant $d(\varepsilon, \mathcal{F}_n)$ depending on ε and the class $\mathcal{F}_n$. From these estimates and using the Borel-Cantelli lemma, we conclude the almost sure convergence result

$$\lim_{N\to\infty} \sup_{f_n \in \mathcal{F}_n} |\eta_n^N(f_n) - \eta_n(f_n)| = 0 .$$

The corresponding fluctuations and Central Limits Theorems will also be discussed in Sect. 6.5, in which the following result will be proved: For any $n \geq 0$, and $f \in \mathcal{B}_b(E_n)$, the sequence of random variables

$$W_n^N(f) = \sqrt{N}(\gamma_n^N(f_n)1_{\{\tau^N \geq n\}} - \gamma_n(f_n))$$

converges in law (as N tends to ∞) to a centered Gaussian random variable $W_n(f)$ with variance

$$\sigma_n^2(f) = \sum_{q=0}^{n} \gamma_q(1)^2 \eta_{q-1} \left(K_{q,\eta_{q-1}}[Q_{q,n}(f) - K_{q,\eta_{q-1}} Q_{q,n}(f)]^2\right) ,$$

where $Q_{p,n}(f)$ are some functions defined hereafter. We use the convention $\eta_{-1} = \eta_0 = K_{0,\eta_{-1}}$. Rephrasing these asymptotic results in the context of analysis of rare events leads to the Theorem 1.

6.1 Preliminaries

Feynman-Kac Semigroups

In this short section, we introduce the Feynman-Kac semigroups, $Q_{p,n}$ and $\Phi_{p,n}$, associated respectively with γ_n and η_n. They are defined by the formulas

$$Q_{p,n} = Q_{p+1} \cdots Q_{n-1} Q_n , \qquad \text{and} \qquad \Phi_{p,n} = \Phi_n \circ \Phi_{n-1} \circ \ldots \circ \Phi_{p+1} ,$$

with $Q_n(x_{n-1}, dx_n) = G_{n-1}(x_{n-1}) M_n(x_{n-1}, dx_n)$. We use the convention $Q_{n,n} = Id$ and $\Phi_{n,n} = Id$. These semigroups are alternatively defined by

$$Q_{p,n}(f_n)(x_p) = \mathbb{E}_{p,x_p} \left(f_n(X_n) \prod_{q=p}^{n-1} G_q(X_p) \right), \quad \Phi_{p,n}(\mu_p)(f_n) = \frac{\mu_p(Q_{p,n}(f_n))}{\mu_p(Q_{p,n}(1))} ,$$

where $\mathbb{E}_{p,x_p}$ is the expectation with respect the law of the shifted chain $(X_{p+n})_{n \geq 0}$. By definition of η_n and $Q_{p,n}$, we observe that

$$\eta_n(f_n) = \frac{\eta_p(Q_{p,n}(f_n))}{\eta_p(Q_{p,n}(1))} , \qquad \gamma_p(Q_{p,n}(1)) = \gamma_n(1) . \tag{23}$$

Now, introducing the pair potential/transition $(G_{p,n}, P_{p,n})$ defined by

$$G_{p,n} = Q_{p,n}(1) \quad \text{and} \quad P_{p,n}(f_n) = \frac{Q_{p,n}(f_n)}{Q_{p,n}(1)} ,$$

we deduce the following formula for the semigroup $\Phi_{p,n}$

$$\Phi_{p,n}(\mu_p) = \Psi_{p,n}(\mu_p) P_{p,n} ,$$

with the Boltzmann-Gibbs transformation, $\Psi_{p,n}$ from $\mathcal{E}_p$ into itself, defined by

$$\Psi_{p,n}(\mu_p)(f_p) = \mu_p(G_{p,n}(f_n))/\mu_p(G_{p,n}(1)) .$$

Some Inequalities for Independent Random Variables

In this section, we discuss some general inequalities for sequences of independent variables. These inequalities will be used in the following sections.

Let $(\mu_i)_{i\geq 1}$ be a sequence of probability measures on a given measurable state space $(E, \mathcal{E})$. We also consider a sequence of $\mathcal{E}$-measurable functions $(h_i)_{i\geq 1}$ such that $\mu_i(h_i) = 0$, for all $i \geq 1$. During the further development of this section we fix an integer $N \geq 1$. To clarify the presentation we slight abuse the notation and we denote respectively by

$$m(X) = \frac{1}{N}\sum_{i=1}^{N} \delta_{X^i} \quad \text{and} \quad \mu = \frac{1}{N}\sum_{i=1}^{N} \mu_i \ ,$$

the N-empirical measure associated to a collection of independent random variables $X = (X^i)_{i\geq 1}$, with respective distributions $(\mu_i)_{i\geq 1}$ and the N-averaged measure associated to the sequence of measures $(\mu_i)_{i\geq 1}$. When we are given N-sequences of points $x = (x^i)_{1\leq i\leq N} \in E^N$ and functions $(h_i)_{1\leq i\leq N} \in \mathcal{B}_b(E)^N$ we shall also use the following notations

$$m(x)(h) = \frac{1}{N}\sum_{i=1}^{N} h_i(x^i) \quad \text{and} \quad \sigma^2(h) = \frac{1}{N}\sum_{i=1}^{N} \mathrm{osc}^2(h_i) \ ,$$

where $\mathrm{osc}(h) = \sup\{|h(x) - h(y)|\}$ is the oscillation of the function h.

For any pair of integers (p, n), with $1 \leq p \leq n$, we denote by $(n)_p$ the quantity

$$(n)_p = \frac{n!}{(n-p)!} \ .$$

We have the following lemmas [2][§7.3]:

Lemma 2 (Chernov-Hoeffding).

$$\mathbb{P}\left(|m(X)(h)| \geq \varepsilon\right) \leq 2\mathrm{e}^{-2N\varepsilon^2/\sigma^2(h)} \ .$$

Lemma 3. *For any sequence of $\mathcal{E}$-measurable functions $(h_i)_{i\geq 1}$ such that $\mu_i(h_i) = 0$ and $\sigma(h) < \infty$ we have for any $p \geq 1$*

$$\sqrt{N}\ \mathbb{E}(|m(X)(h)|^p)^{\frac{1}{p}} \leq d(p)^{\frac{1}{p}}\ \sigma(h) \ , \tag{24}$$

with the sequence of finite constants $(d(n))_{n\geq 0}$ defined, for any $n \geq 1$, by the formulas

$$d(2n) = (2n)_n\ 2^{-n} \quad \text{and} \quad d(2n-1) = \frac{(2n-1)_n}{\sqrt{n-1/2}}\ 2^{-(n-1/2)} \ . \tag{25}$$

In addition we have for any $\varepsilon > 0$

$$\mathbb{E}(\exp(\varepsilon\sqrt{N}|m(X)(h)|)) \leq (1 + \varepsilon\sigma(h)/\sqrt{2})\ \exp(\varepsilon^2\sigma^2(h)/2) \ .$$

We now extend the previous results to the convergence of empirical processes with respect to some Zolotarev seminorm. Let $\mathcal{F}$ be a given collection of measurable functions $f : E \to \mathbb{R}$ such that $\|f\| = \sup_{x \in E} |f(x)| \leq 1$. We associate with $\mathcal{F}$ the Zolotarev seminorm on $\mathcal{P}(E)$ defined by

$$\|\mu - \nu\|_{\mathcal{F}} = \sup\{|\mu(f) - \nu(f)| \ : \ f \in \mathcal{F}\} .$$

No generality is lost and much convenience is gained by supposing that the unit constant function $f = 1 \in \mathcal{F}$. Furthermore, we shall suppose that $\mathcal{F}$ contains a countable and dense subset.

To measure the size of a given class $\mathcal{F}$, one considers the covering numbers $\mathcal{N}(\varepsilon, \mathcal{F}, L_p(\mu))$ defined as the minimal number of $L_p(\mu)$-balls of radius $\varepsilon > 0$ needed to cover $\mathcal{F}$. By $\mathcal{N}(\varepsilon, \mathcal{F})$ and by $I(\mathcal{F})$ we denote the uniform covering numbers and entropy integral given by

$$\mathcal{N}(\varepsilon, \mathcal{F}) = \sup\{\mathcal{N}(\varepsilon, \mathcal{F}, L_2(\eta)); \eta \in \mathcal{P}(E)\} ,$$
$$I(\mathcal{F}) = \int_0^1 \sqrt{\log(1 + \mathcal{N}(\varepsilon, \mathcal{F}))} d\varepsilon .$$

For more details and various examples the reader is invited to consult [14]. We have the following lemma [2][§7.3]:

Lemma 4. *For any $p \geq 1$, we have*

$$\sqrt{N}\, \mathbb{E}\left(\|m(X) - \mu\|_{\mathcal{F}}^p\right)^{1/p} \leq c \lfloor p/2 \rfloor !\, I(\mathcal{F}) ,$$

where c is a universal constant.

For any $\varepsilon > 0$ and $\sqrt{N} \geq 4\varepsilon^{-1}$, we have that

$$\mathbb{P}\left(\|m(X) - \mu\|_{\mathcal{F}} > 8\varepsilon\right) \leq 8 \mathcal{N}(\varepsilon, \mathcal{F}) e^{-N\varepsilon^2/2} .$$

6.2 Strong Law of Large Numbers

In the following picture, we have illustrated the random evolution of the N-particle approximation model:

$$\begin{array}{ccccccccc}
\eta_0 & \to & \eta_1 = \Phi_1(\eta_0) & \to & \eta_2 = \Phi_{0,2}(\eta_0) & \to & \cdots & \to & \eta_n = \Phi_{0,n}(\eta_0) \\
\Downarrow & & & & & & & & \\
\eta_0^N & \to & \Phi_1(\eta_0^N) & \to & \Phi_{0,2}(\eta_0^N) & \to & \cdots & \to & \Phi_{0,n}(\eta_0^N) \\
 & & \Downarrow & & & & & & \\
 & & \eta_1^N & \to & \Phi_2(\eta_0^N) & \to & \cdots & \to & \Phi_{1,n}(\eta_1^N) \\
 & & & & \Downarrow & & & & \\
 & & & & \eta_2^N & \to & \cdots & \to & \Phi_{2,n}(\eta_2^N) \\
 & & & & & & \Downarrow & & \vdots \\
 & & & & & & \eta_{n-1}^N & \to & \Phi_{n-1,n}(\eta_{n-1}^N) \\
 & & & & & & & & \Downarrow \\
 & & & & & & & & \eta_n^N
\end{array}$$

In this picture, the sampling errors are represented by the implication sign "$\Downarrow$". Using the identity $\Phi_{q-1,n}(\eta_{q-1}^N) = \Phi_{q,n}(\Phi_q(\eta_{q-1}^N))$, we observe that

$$\eta_n^N - \eta_n = \sum_{q=0}^{n} \left[\Phi_{q,n}(\eta_q^N) - \Phi_{q,n}(\Phi_q(\eta_{q-1}^N)) \right] , \tag{26}$$

with the convention $\Phi_0(\eta_{-1}^N) = \eta_0$. Note that each term on the r.h.s. represents the propagation of the pth sampling local error $\Phi_q(\eta_{q-1}^N) \Rightarrow \eta_q^N$. This pivotal formula will be of important use in the following. In addition, we have for each $\eta_1, \eta_2 \in \mathcal{P}(E_q)$ and $f \in \mathcal{B}_b(E_n)$

$$\begin{aligned}\Phi_{q,n}(\eta_1)(f) - \Phi_{q,n}(\eta_2)(f) = \frac{1}{\eta_2(G_{q,n})} &[(\eta_1(Q_{q,n}(f)) - \eta_2(Q_{q,n}(f))) \\ &+ \Phi_{q,n}(\eta_1)(f)(\eta_2(G_{q,n}) - \eta_1(G_{q,n}))] .\end{aligned}$$

We deduce the following formula which highlights the sampling errors:

$$\begin{aligned}\eta_n^N(f) - \eta_n(f) = \sum_{q=0}^{n} \frac{1}{\eta_{q-1}^N(G_{q,n})} &[(\eta_q^N(Q_{q,n}(f)) - \Phi_q(\eta_{q-1}^N)(Q_{q,n}(f))) \\ &+ \Phi_{q,n}(\eta_q^N)(f)(\Phi_q(\eta_{q-1}^N)(G_{q,n}) - \eta_q^N(G_{q,n}))] .\end{aligned} \tag{27}$$

6.3 Extinction Probabilities

The objective of this short section is to estimate the probability of extinction of a class of particle models, associated with bounded (by one) potential functions that may take null values. Let us recall that the limiting flow η_n is well-defined, only up to the first time τ we have $\eta_\tau(G_\tau) = 0$; that is

$$\tau = \inf\{n \in \mathbb{N} \, : \, \eta_n(G_n) = 0\} = \inf\{n \in \mathbb{N} \, : \, \gamma_{n+1} = 0\} .$$

In the same way, the N-interacting particle systems are only defined up to the time τ^N the whole configuration $\xi_n \in E_n^N$ first hits the hard obstacle set $(E_n \setminus \widehat{E}_n)^N$:

$$\tau^N = \inf\{n \in \mathbb{N} \, : \, \eta_n^N(G_n) = 0\}.$$

It follows the equivalence $(\tau^N \geq n) \Leftrightarrow (\xi_0 \in \widehat{E}_0, \cdots, \xi_{n-1} \in \widehat{E}_{n-1})$, which indicates that τ^N is a predictable Markov time with respect to the filtration $(\mathcal{F}_n^N)$, in the sense that $\{\tau^N \geq n\} \in \mathcal{F}_{n-1}^N$. We have the following rather crude but reassuring result [2][Theorem 7.4.1]

Theorem 2. *Suppose we have $\gamma_n(1) > 0$ for any $n \geq 0$. Then, for any $N \geq 1$ and $n \geq 0$, we have the estimate*

$$\mathbb{P}(\tau^N \leq n) \leq a(n) \mathrm{e}^{-N/b(n)} ,$$

for some constants $a(n)$ and $b(n)$ which depend only on n and $\gamma_{n+1}(1)$.

For a detailed proof, the reader is referred to [2][§7.4]. Its key idea is based on the following observation. Using formula (23), we obtain for any $p \leq n$,

$$\eta_n(G_n) = \frac{\eta_p(G_{p,n+1})}{\eta_p(G_{p,n})} = \frac{\gamma_{n+1}(1)}{\gamma_n(1)} \; .$$

Now, referring to the setting of Theorem 2, we obtain that $\eta_q(G_q) > 0$ for any $1 \leq q \leq n$, and therefore that $\tau > n$. In fact, assuming the condition $\gamma_n(1) > 0$ for all n, avoids the tunneling problems with probability one, so an exponential decrease of the extinction probabilities.

6.4 Convergence of Empirical Processes

This section provides precise estimates on the convergence of the particle density profiles when the size of the system tends to infinity. We start with the analysis of the unnormalized particles models and we show that this approximation particle has no bias. The central idea consists in expressing the difference between the particle measures and the limiting Feynman-Kac ones as such end values of martingale sequence.

We recall that a square integrable and $\mathcal{F}^N$-martingale $M^N = (M_n^N)_{n\geq 0}$ is an $\mathcal{F}^N$-adapted sequence such that $\mathbb{E}(M_n^N)^2 < \infty$ for all $n \geq 0$ and

$$\mathbb{E}(M_{n+1}^N | \mathcal{F}_n^N) = M_n^N \quad (\mathbb{P}^N - \text{a.s.}) \; .$$

The predictable quadratic characteristic of M^N is the sequence of random variables $\langle M^N \rangle = (\langle M^N \rangle_n)_{n\geq 0}$ defined by

$$\langle M^N \rangle_n = \sum_{p=0}^{n} \mathbb{E}((M_p^N - M_{p-1}^N)^2 | \mathcal{F}_{p-1}^N) \; ,$$

with the convention $\mathbb{E}((M_0^N - M_{-1}^N)^2 | \mathcal{F}_{-1}^N) = \mathbb{E}(M_0^N)^2$. The stochastic process $\langle M^N \rangle$ is also called the angle bracket of M^N and is the unique predictable increasing process such that the sequence $((M_n^N)^2 - \langle M^N \rangle_n)_{n\geq 0}$ is an $\mathcal{F}^N$-martingale.

In the following, we will use the simplified notation (10). For instance, if we consider the McKean model

$$K_{n,\eta}(x, \cdot) = G_{n-1}(x) M_n(x, \cdot) + (1 - G_{n-1}(x)) \Phi_n(\eta) \; , \tag{28}$$

we first observe that

$$K_{q,\eta}(\varphi - \Phi_q(\varphi)) = K_{q,\eta}(\varphi) - \Phi_q(\eta)(\varphi) = G_{q-1}(M_q(\varphi) - \Phi_q(\eta)(\varphi)) \; .$$

So, let $\tilde{\varphi}_q$ be the function defined by $\tilde{\varphi}_q = \varphi - \Phi_q(\eta)(\varphi)$. We obtain

$$\begin{aligned} K_{q,\eta}[\varphi - K_{q,\eta}(\varphi)]^2 &= K_{q,\eta}[\tilde{\varphi}_q - K_{q,\eta}(\tilde{\varphi}_q)]^2 \\ &= K_{q,\eta}(\tilde{\varphi}_q)^2 - (K_{q,\eta}(\tilde{\varphi}_q))^2 \\ &= K_{q,\eta}[\varphi - \Phi_q(\eta)(\varphi)]^2 - G_{q-1}^2[M_q(\varphi) - \Phi_q(\eta)(\varphi)]^2 . \end{aligned} \tag{29}$$

Furthermore, if we consider the McKean model

$$K_{n,\eta}(x,\cdot) = \Phi_n(\eta)(\cdot) , \tag{30}$$

we obtain

$$K_{q,\eta}[\varphi - K_{q,\eta}(\varphi)]^2 = \Phi_q(\eta)[\varphi - \Phi_q(\eta)(\varphi)]^2 . \tag{31}$$

These two formulas indicate that the particle model in the first case is more accurate than the other one.

Proposition 2. *For each $n \geq 0$ and $f_n \in \mathcal{B}_b(E_n)$, we let $\Gamma^N_{\cdot,n}(f_n)$ be the $\mathbb{R}$-valued process defined for any $p \in \{0, \cdots, n\}$ by*

$$\Gamma^N_{p,n}(f_n) = \gamma^N_p(Q_{p,n}f_n)1_{\{\tau^N \geq p\}} - \gamma_p(Q_{p,n}f_n) . \tag{32}$$

For any $p \leq n$, $\Gamma^N_{\cdot,n}(f_n)$ has the $\mathcal{F}^N$-martingale decomposition

$$\Gamma^N_{p,n}(f_n) = \sum_{q=0}^{p} \gamma^N_q(1)1_{\{\tau^N \geq p\}} \left[\eta^N_q(Q_{q,n}f_n) - \eta^N_{q-1}K_{q,\eta^N_{q-1}}(Q_{q,n}f_n)\right] , \tag{33}$$

and its bracket is given by

$$\langle \Gamma^N_{\cdot,n}(f_n)\rangle_p = \frac{1}{N}\sum_{q=0}^{p}(\gamma^N_q(1))^2\, 1_{\{\tau^N \geq p\}}\eta^N_{q-1}\left(K_{q,\eta^N_{q-1}}\left[Q_{q,n}f_n - K_{q,\eta^N_{q-1}}Q_{q,n}f_n\right]^2\right) ,$$

with the convention $\Phi_0(\eta^N_{-1}) = \eta_0 = K_{0,\eta^N_{-1}}$.

The first consequence of Proposition 2 is that γ^N_n is unbiased. More precisely, using the martingale decomposition (33) with $p = n$, we obtain for any $f \in \mathcal{F}_n$ the following identity

$$\mathbb{E}(\gamma^N_n(f)1_{\{\tau^N \geq p\}}) = \gamma_n(f) .$$

In fact, we have the more precise result [2][Theorem 7.4.2]

Theorem 3. *For each $p \geq 1, n \in \mathbb{N}$, and for any (separable) collection $\mathcal{F}_n$ of measurable functions $f : E_n \to \mathbb{R}$ such that $\|f\| \leq 1$ (and $1 \in \mathcal{F}_n$), we have for any $f \in \mathcal{F}_n$*

$$\mathbb{E}(\gamma^N_n(f)1_{\{\tau^N \geq p\}}) = \gamma_n(f) ,$$

and for any $r \leq n$

$$\sqrt{N}\,\mathbb{E}(\|1_{\{\tau^N\geq r\}}\gamma_r^N Q_{r,n}-\gamma_r Q_{r,n}\|_{\mathcal{F}_n}^p)^{1/p}\leq c(n+1)\lfloor p/2\rfloor!I(\mathcal{F}_n)\,.$$

In addition, for any $\varepsilon\geq 4/\sqrt{N}$ *, we have the exponential estimate*

$$\mathbb{P}\left(\|1_{\{\tau^N\geq r\}}\gamma_r^N Q_{r,n}-\gamma_r Q_{r,n}\|_{\mathcal{F}_n}>\varepsilon\right)\leq 8(n+1)\mathcal{N}(\varepsilon_n,\mathcal{F}_n)\mathrm{e}^{-N\varepsilon_n^2/2}\,,\quad (34)$$

with $\varepsilon_n=\varepsilon/(n+1)$ *.*

Applying the exponential estimate (34) with $r=n$ and $\varepsilon=\gamma_n(1)/2$, we obtain, for any pair (n,N) such that $\sqrt{N}\geq 8/\gamma_n(1)$, the following inequality

$$\mathbb{P}\left(1_{\{\tau^N\geq r\}}\gamma_n^N(1)\geq\gamma_n(1)/2\right)\geq 1-8(n+1)\mathcal{N}(\varepsilon_n,\mathcal{F}_n)\mathrm{e}^{-N\varepsilon_n^2/2}\,,$$

with $\varepsilon_n=\gamma_n(1)/(2(n+1))$. Now, to obtain some exponential estimate for the measure η_n^N, we use the following decomposition

$$(\eta_n^N(f)-\eta_n(f))1_{\{\tau^N\geq n\}}=\frac{\gamma_n(1)}{\gamma_n^N(1)}\,\gamma_n^N\left(\frac{1}{\gamma_n(1)}(f-\eta_n(f))\right)1_{\{\tau^N\geq n\}}\,.\quad (35)$$

If we set $f_n=\frac{1}{\gamma_n(1)}(f-\eta_n(f))$, then since $\gamma_n(f_n)=0$, (35) also reads

$$\begin{aligned}(\eta_n^N(f)-\eta_n(f))1_{\{\tau^N\geq n\}}&=\frac{\gamma_n(1)}{\gamma_n^N(1)}(\gamma_n^N(f_n)1_{\{\tau^N\geq n\}}-\gamma_n(f_n))\\&=\frac{\gamma_n(1)}{\gamma_n^N(1)}\varGamma_{n,n}^N(f_n)\,.\end{aligned}\quad (36)$$

Let $\varOmega_n^N$ be the set of events

$$\varOmega_n^N=\{\gamma_n^N(1)1_{\{\tau^N\geq n\}}\geq\gamma_n(1)/2\}\subset\{\tau^N\geq n\}\,.$$

Using Theorem 3, we have

$$\mathbb{P}(\varOmega_n^N)\geq 1-\frac{b(n)^2}{N}\,,$$

where $b(n)$ is a constant which depends on n only. If we combine this estimate with Theorem 3 and (36), we find that for any $f\in\mathcal{B}_b(E_n)$, with $\|f\|\leq 1$

$$\begin{aligned}|\mathbb{E}\left((\eta_n^N(f)-\eta_n(f))1_{\{\tau^N\geq n\}}\right)|&\leq|\mathbb{E}\left((\eta_n^N(f)-\eta_n(f))1_{\varOmega_n^N}\right)|+2\mathbb{P}((\varOmega_n^N)^2)\\&\leq\frac{b(n)^2}{N}\,,\end{aligned}$$

where $b(n)$ is a new constant which depends on n only. Finally by Theorem 2, we conclude that

$$|\mathbb{E}\left((\eta_n^N(f)1_{\{\tau^N\geq n\}}-\eta_n(f))\right)|\leq\frac{b(n)^2}{N}+a(n)\mathrm{e}^{-N/b(n)}\,.$$

A consequence of this result is the following extension of the Glivenko-Cantelli theorem to particle models.

Corollary 1. *Let $\mathcal{F}_n$ be a countable collection of functions f such that $\|f\| \leq 1$ and $\mathcal{N}(\varepsilon, \mathcal{F}_n) < \infty$ for any $\varepsilon > 0$. Then, for any time $n \geq 0$, $\|\eta_n^N(f)1_{\{\tau^N \geq n\}} - \eta_n(f)\|_{\mathcal{F}_n}$ converges almost surely to 0 as $N \to \infty$.*

Some time-uniform estimates can also be obtained when the pair (G_n, M_n) satisfies some regularity conditions. When these conditions are met the non-linear Feynman-Kac semigroup $\Phi_{p,n}$ has asymptotic stability properties which ensure that in some sense for each elementary term

$$[\Phi_{q,n}(\eta_n^N) - \Phi_{q,n}(\Phi_q(\eta_{q1}^N))] \to 0 \quad \text{as} \quad (n-q) \to \infty .$$

Consequently, according to (26), a uniform estimate of the sum of the "small errors" can be proved. The reader is invited to consult [2][§7.4] for more details about this subject.

6.5 Central Limit Theorems

Let us consider the particle approximation model $\xi_n = (\xi_n^i)_{1 \leq i \leq N}$ associated with a nonlinear measure-valued equation of the form

$$\eta_n = \eta_{n-1} K_{n,\eta_{n-1}} . \tag{37}$$

We will assume that $\gamma_n(1) > 0$ for all n. The n-th sampling error is the measure-valued random variable V_n^N defined by the formula

$$\eta_n^N = \eta_{n-1}^N K_{n,\eta_{n-1}}^N + V_n^N/\sqrt{N} . \tag{38}$$

Notice that V_n^N is itself the sum of the local errors induced by the random elementary transitions $\xi_{n-1}^i \rightsquigarrow \xi_n^i$ of the N particles; that is, we have

$$V_n^N = \sum_{i=1}^N \Delta_i V_n^N ,$$

with the "local" terms given for any $\varphi_n \in \mathcal{B}_b(E_n)$ by

$$\Delta_i V_n^N(\varphi_n) = \frac{1}{\sqrt{N}}[\varphi_n(\xi_n^i) - K_n, \eta_{n-1}^N(\varphi_n)(\xi_{n-1}^i)] .$$

By definition of the particle model, η_n^N is the empirical measure associated with a collection of conditionnaly independent random variables ξ_n^i with distributions $K_{n,\eta_{n-1}^N}(\xi_{n-1}^i, \cdot)$. From this we obtain that

$$\mathbb{E}_{\eta_0}^N[\eta_n^N(f_n)|\mathcal{F}_n^N] = \Phi_n(\eta_{n-1}^N)(f_n) = \eta_{n-1}^N K_{n,\eta_{n-1}^N} ,$$

where $\mathcal{F}_n^N = \sigma(\xi_0, \cdots, \xi_{n-1})$ is the σ-field asociated with the $\xi_0, \cdots, \xi_{n-1}$.

So we readily find that $\mathbb{E}(V_n^N(\varphi_n)) = 0$ and

$$\mathbb{E}(V_n^N(\varphi_n)^2) = \mathbb{E}(\eta_{n-1}^N(K_{n,\eta_{n-1}^N}[\varphi_n - K_{n,\eta_{n-1}^N}(\varphi_n)]^2)) .$$

In addition, for sufficiently regular McKean interpretation models, we have the asymptotic result

$$\lim_{N\to\infty} \mathbb{E}(V_n^N(\varphi_n)^2) = \eta_{n-1}(K_{n,\eta_{n-1}}[\varphi_n - K_{n,\eta_{n-1}}(\varphi_n)]^2) .$$

The formula (38) shows that the particle density η_n^N satisfy almost the same equation (37) as the limiting measures η_n. In fact [2][§9.3], $V_n^N(\varphi_n)$ converges in law to a Gaussian random variable $V_n(\varphi_n)$ such that

$$\mathbb{E}(V_n(\varphi_n)) = 0 \quad \text{and} \quad \mathbb{E}(V_n(\varphi_n)^2) = \eta_{n-1}(K_{n,\eta_{n-1}}[\varphi_n - K_{n,\eta_{n-1}}(\varphi_n)]^2) .$$

These elementary fluctuations give some insight on the asymptotic normal behavior of the local errors accumulated by the sampling scheme. Nevertheless, they do not give directly CLT result for the difference between the particle measures η_n^N or γ_n^N and the corresponding limiting measures η_n and γ_n.

Preliminaries

The key idea is to consider the one-dimensional $\mathcal{F}^N$-martingale

$$M_n^N(f) = \sqrt{N} \sum_{p=0}^{n} 1_{\{\tau^N \geq p\}}[\eta_p^N(f_p) - \Phi_p(\eta_{p-1}^N)(f_p)] ,$$

where f_p stands for some collection of measurable and bounded functions defined on E_p. The angle bracket of this martingale is given by the formula

$$\langle M^N(f)\rangle_n = \sum_{p=0}^{n} \eta_{p-1}^N[K_{p,\eta_{p-1}^N}((f_p - K_{p,\eta_{p-1}^N} f_p)^2)] .$$

Then [2][Theorem 9.3.1], for any sequence of bounded measurable functions f_p and $p \geq 0$, the $\mathcal{F}^N$-martingale $M_n^N(f)$ converges in law to a Gaussian martingale $M_n(f)$ such that for any $n \geq 0$

$$\langle M(f)\rangle_n = \sum_{p=0}^{n} \eta_{p-1}[K_{p,\eta_{p-1}}((f_p - K_{p,\eta_{p-1}} f_p)^2)] .$$

A first consequence of this result is the next corollary which expresses the fact that the local errors associated with the particle approximation sampling steps behave asymptotically as a sequence of independent and centered Gaussian random variables.

Corollary 2. *The sequence of random fields $\mathcal{V}_n^N = (V_p^N)_{0\leq p\leq n}$ converges in law, as $N \to \infty$, to a sequence $\mathcal{V}_n = (V_p)_{0\leq p\leq n}$ of $(n+1)$ independent and Gaussian random fields V_p with, for any $\varphi_p^1, \varphi_p^2 \in \mathcal{B}_p(E_p)$, $\mathbb{E}(V_p(\varphi_p^1)) = 0$ and*

$$\mathbb{E}(V_p(\varphi_p^1)V_p(\varphi_p^2)) = \eta_{p-1}(K_{p,\eta_{p-1}}[\varphi_p^1 - K_{p,\eta_{p-1}}(\varphi_p^1)][\varphi_p^2 - K_{p,\eta_{p-1}}(\varphi_p^2)]) .$$

We now are concerned with the fluctuations of the particle approximation measures γ_n^N nd η_n^N. Nevertheless, before we start, we recall some tools to transfer CLT such as the Slutsky's technique and the δ-method.Firstly, the Slutsky's theorem states that for any sequences of random variables $(X_n)_{n\geq 1}$ and $(Y_n)_{n\geq 1}$, taking value in some separable metric space (E,d), which are such that X_n converges in law, as $n\to\infty$, to some random variable X, and $d(X_n,Y_n)$ converges to 0 in probability, then Y_n converges in law, as $N\to\infty$, to X. We deduce of this theorem, that if X_n converges in law to some finite constant c (which implies the convergence in probability) and Y_n converges in law to some variable Y, then X_nY_n converges in law to cY.

The other tool, also known as the δ-method [2][§9.3], is the following lemma.

Lemma 5. *Let $(U_0^N,\cdots,U_n^N)_{N\geq 1}$ be a sequence of $\mathbb{R}^{n+1}$-valued random variables defined on some probability space and $(u_p)_{0\leq p\leq n}$ be a given point in $\mathbb{R}^{n+1}$. Suppose that*

$$\sqrt{N}\,(U_0^N-u_0,\cdots,U_n^N-u_n)$$

converges in law, as $N\to\infty$, to some random vector $(U_0,\cdots,U_n)$. Then, for any differentiable function $F_n:\mathbb{R}^{n+1}\to\mathbb{R}$ at the point $(u_p)_{0\leq p\leq n}$, the sequence

$$\sqrt{N}\,[F_n(U_0^N(\omega),\cdots,U_n^N(\omega))-F_n(u_0,\cdots,u_n)]$$

converges in law as $N\to\infty$ to the random variable $\sum_{p=0}^n \frac{\partial F_n}{\partial u_i}(u_0,\cdots,u_n)U_p$.

Unnormalized Measures

We consider the $\mathbb{R}$ valued process $\Gamma_{\cdot,n}^N(f_n)$ introduced in Proposition 2. As the reader may have certainly noticed, the martingale decomposition of $\Gamma_{\cdot,n}^N$, exhibited in Proposition, 2 is expressed in terms of the sequence of local errors V_n^N.

Let $\overline{\Gamma}_{\cdot,n}^N(f_n)$ be the random sequence defined as in (33) by replacing, in the summation, the terms $\gamma_q^N(1)1_{\{\tau^N\geq q\}}$ by their limiting values $\gamma_q(1)$. In order to combine the CLT stated in Corollary 2 with the δ-method, we rewrite the resulting random sequence as

$$\begin{aligned}\sqrt{N}\,\overline{\Gamma}_{n,n}^N(f_n) &= \sqrt{N}\sum_{q=0}^{p}\gamma_q(1)\left[\eta_q^N-\eta_{q-1}^N K_{q,\eta_{q-1}^N}\right](Q_{q,n}f_n)\\ &= \sqrt{N}\,F_n(U_{0,n}^N,\cdots,U_{n,n}^N)\,,\end{aligned}$$

with the random sequence $(U_{p,n}^N)_{0\leq p\leq n}$, and the function F_n given by

$$U_{p,n}^N=V_p^N(Q_{p,n}f_n)/\sqrt{N}\quad\text{and}\quad F_n(v_0,\cdots,v_n)=\sum_{q=0}^n\gamma_q(1)v_q\,.$$

Since for any $n \geq 0$ we have $\lim_{N\to\infty} \gamma_q^N(1)\ 1_{\{\tau^N \geq q\}} = \gamma_q(1)$ in probability, we easily deduce from Corollary 2, the Slutsky's theorem and the δ-method that the real-valued random variable $\sqrt{N}\ (\gamma_n^N(f_n)1_{\{\tau^N\geq n\}} - \gamma_n(f_n))$ converges in law to the centered Gaussian random variable $W_n^\gamma(f_n) = \sum_{q=0}^n \gamma_q(1)V_p(Q_{p,n}f_n)$ with variance

$$\sigma_n^2(f) = \sum_{q=0}^n (\gamma_q(1))^2 \eta_{q-1}\left(K_{q,\eta_{q-1}}[Q_{p,n}f_n - K_{q,\eta_{q-1}}Q_{p,n}f_n]^2\right) .$$

With the McKean model (28), the formula (29) gives the following new expression for the variance

$$\begin{aligned} \sigma_n^2(f) = &\sum_{q=0}^n (\gamma_q(1))^2 \eta_q((Q_{q,n}f - \eta_q(Q_{q,n}f))^2) \\ &- \sum_{q=1}^n (\gamma_q(1))^2 \eta_{q-1}\left(G_{q-1}^2(M_q Q_{q,n}f - \eta_q(Q_{q,n}f))^2\right). \end{aligned} \tag{39}$$

Normalized Measures

Using formula (35) and the Slutsky's theorem, we obtain that the sequence of real-valued random variables

$$W_n^{\eta,N}(f) = \sqrt{N}\ (\eta_n^N(f) - \eta_n(f))1_{\{\tau^N\geq n\}}$$

converges to the Gaussian random variable W_n^η given by

$$W_n^\eta(f) = W_n^\gamma\left(\frac{1}{\gamma_n(1)}(f - \eta_n(f))\right) .$$

Now, let the semigroups $\overline{Q}_{p,n}$ and the functions $f_{p,n}$ be respectively defined by

$$\overline{Q}_{p,n} = \frac{\gamma_p(1)}{\gamma_n(1)} Q_{p,n} , \quad \text{and} \quad f_{p,n} = \overline{Q}_{p,n}(f - \eta_n f) . \tag{40}$$

Then, the variance of the Gaussian random variable $W_n^\eta(f)$ is given by the formula

$$\mathbb{E}(W_n^\eta(f)^2) = \sum_{p=0}^n \eta_{p-1}\left(K_{p,\eta_{p-1}}[f_{p,n} - K_{p,\eta_{p-1}}f_{p,n}]^2\right) . \tag{41}$$

Killing Interpretations and Related Comparisons

One of the best ways to interpret the fluctuations variances developed previously is to use the Feynman-Kac killing interpretations provided in Sect. 4.2.

In this context, X_n is regarded as a Markov particle evolving in an absorbing medium with obstacles related to $[0,1]$-valued potentials. Using the same notation and terminology as was used in Sect. 4.2, the Feynman-Kac semigroup $Q_{p,n}$ has the following interpretation

$$\begin{aligned} Q_{p,n}(x_p, dx_n) &= \int \left\{ \prod_{q=p}^{n-1} G_q(x_q) \right\} M_{p+1}(x_p, dx_{p+1} \cdots M_n(x_{n-1}, dx_n) \\ &= \mathbb{P}^c_{p,x_p}(X_n \in dx_n, T \geq n) \; , \end{aligned}$$

where $\mathbb{P}^c_{p,x_p}$ represents the distribution of the absorbed particle evolution model starting at $X_p = x_p$ at time p. In this context, the variance of the fluctuation variable $W_n^\gamma(1)$, associated with the McKean interpretation model (30), is given by

$$\begin{aligned} \mathbb{E}(W_n^\gamma(1)^2) &= \gamma_n(1)^2 \sum_{p=0}^{n} \eta_p \left([1 - G_{p,n}/\eta_p(G_{p,n})]^2 \right) \\ &= \mathbb{P}^c(T \geq n)^2 \sum_{p=0}^{n} \int_{E_p} \mathbb{P}^c(X_p \in dx_p | T \geq p) \left[\frac{\mathbb{P}^c_{p,x_p}(T \geq n)}{\mathbb{P}^c(T \geq n | T \geq p)} - 1 \right]^2 . \end{aligned}$$

We further assume that for any $n \geq p$ and η_p-a.e. $x_p, y_p \in \widehat{E}_p$, we have

$$\mathbb{P}^c_{p,x_p}(T \geq n) \geq \delta \mathbb{P}^c_{p,y_p}(T \geq n) \; , \tag{42}$$

for some $\delta > 0$ (see [2][Proposition 4.3.3] for sufficient conditions to obtain the condition (42)). In this case we have

$$\mathbb{E}(W_n^\gamma(1)^2) \leq b(\delta)(n+1)\mathbb{P}^c(T \geq n)^2,$$

for some finite constant $b(\delta)$.

The killing interpretation also suggests another evolution model based on N independent and identically distributed copies X^i of the absorbed particle evolution model. The Monte Carlo approximation is now given by $N^{-1} \sum_{i=1}^N 1_{\{T^i \geq n\}}$, where T^i represents the absorption time of the i-th particle. It is well known that the fluctuation variance $\sigma_n^{MC}(1)^2$ of this scheme is given by

$$\sigma_n^{MC}(1)^2 = \mathbb{P}^c(T \geq n)(1 - \mathbb{P}^c(T \geq n)) \; .$$

From previous considerations we find that

$$\frac{\sigma_n^{MC}(1)^2}{\mathbb{E}(W_n^\gamma(1)^2)} \geq \frac{1}{b(\delta)(n+1)} \frac{1 - \mathbb{P}^c(T \geq n)}{\mathbb{P}^c(T \geq n)} \to \infty \; ,$$

as soon as $\mathbb{P}^c(T \geq n) = o(1/n)$.

In addition, according to the formulas (41) and (31), and the observation that $\eta_q(f_{q,n}) = 0$, the variance of the random field W_n^η can also be described for any $f \in \mathcal{B}_b(E_n)$ as

$$\mathbb{E}(W_n^\eta(f)^2) = \sum_{p=0}^{n} \eta_p(f_{p,n}^2) \;.$$

If we choose the McKean model (28) then, according to the formula (29), we conclude that the variance of the random field W_n^η is defined for any $f \in \mathcal{B}_b(E_n)$ by the formula

$$\mathbb{E}(W_n^\eta(f)^2) = \sum_{p=0}^{n} \eta_p(f_{p,n}^2) - \sum_{p=1}^{n} \eta_{p-1}[(G_{p-1}M_p(f_{p,n}))^2] \;.$$

Then, we readily see that the variance of the corresponding CLT is strictly smaller than the one associated with the McKean interpretation $K_{n,\eta}(x_{n-1},\cdot) = \Phi_n(\eta)$.

Application to Rare Event Analysis

We use the same notation and conventions as introduced in Sect. 5.3. Using the fluctuation analysis stated in the Sect. 6.5, we have the following theorem

Theorem 4. *For any $0 \leq n \leq m+1$, the sequence of random variables*

$$W_{n+1}^N = \sqrt{N}\;(1_{\{\tau^N > n\}}\gamma_{n+1}^N(1) - \mathbb{P}(T_n < T_R))$$

converges in law (as N tends to ∞) to a Gaussian random variable W_{n+1} with mean 0 and variance

$$\sigma_n^2 = \sum_{q=0}^{n+1} (\gamma_q(1))^2 \eta_{q-1}\left(K_{q,\eta_{q-1}}[Q_{q,n+1}(1) - K_{q,\eta_{q-1}}Q_{q,n+1}(1)]^2\right) \;.$$

The collection of functions $Q_{q,n+1}(1)$ on the excursion space E are defined for any $x = (x_n)_{s \leq n \leq t}$ by

$$Q_{q,n+1}(1)(t,x) = 1_{B_q}(x_t)\mathbb{P}(T_n < T_R | T_q = t, X_{T_q} = x_t) \;.$$

Explicit calculations of σ_n are in general difficult to obtain since they rely on an explicit knowledge of the semigroup $Q_{q,n}$. Nevertheless, in the context of rare event analysis, an alternative can be provided. Firstly, according to the formula (39), the variance σ_n^2 takes the form

$$\sigma_n^2 = \mathbb{P}(T_n < T_R)^2 (a_n - b_n) \;,$$

with

$$a_n = \frac{1}{\gamma_{n+1}(1)^2} \sum_{q=0}^{n+1} (\gamma_q(1))^2 \eta_q((Q_{q,n+1}(1) - \eta_q(Q_{q,n+1}(1)))^2)$$

$$b_n = \frac{1}{\gamma_{n+1}(1)^2} \sum_{q=1}^{n+1} (\gamma_q(1))^2 \eta_{q-1}\left(G_{q-1}^2 (M_q Q_{q,n+1}(1) - \eta_q(Q_{q,n+1}(1)))^2\right) \;.$$

Then we observe that $\gamma_p(1) = \mathbb{P}(T_{p-1} < T_R)$ and

$$\eta_q Q_{q,n+1}(1) = \gamma_{n+1}(1)/\gamma_p(1) = \mathbb{P}(T_n < T_R | T_{p-1} < T_R) ,$$

from which we conclude that

$$a_n = \sum_{q=0}^{n+1} \mathbb{E}\left([\Delta^n_{q-1,q}(T_q, X_{T_q}) 1_{\{T_q < T_R\}} - 1]^2 | T_{q-1} < T_R \right) ,$$

where

$$\Delta^n_{p,q}(t,x) = \mathbb{P}(T_n < T_R | T_q = t, X_{T_q} = x) / \mathbb{P}(T_n < T_R | T_p < T_R) .$$

In much the same way, we find

$$\begin{aligned} b_n &= \sum_{q=0}^{n} \mathbb{E}(1_{\{T_q < T_R\}} [\Delta^n_{q,q}(T_q, X_{T_q}) - 1]^2 | T_{q-1} < T_R) \\ &= \sum_{q=0}^{n} \mathbb{P}(T_q < T_R | T_{q-1} < T_R) \mathbb{E}[\Delta^n_{q,q}(T_q, X_{T_q}) - 1]^2 | T_q < T_R) . \end{aligned}$$

7 Conclusion

In this paper, we presented a new technique for reducing the number of trials in Monte Carlo simulation of rare events based on interacting particle algorithms. The main idea behind these models is to decompose the state space into a judicious choice of threshold subsets related to the system evolution. A natural probabilistic interpretation of this approach has been provided by Multilevel Feynman-Kac excursion models. After having introduced these models, we stated asymptotic results as the number of particles tends to infinity.

Nevertheless, asymptotic reassuring results are not satisfactory if one wants to achieve bounds that are useful for a fixed number of particles. For instance, the analysis of the stopping time of the algorithm is far from complete and many questions remain to be answered. Other interesting questions are left such on the optimal selection of the splitting levels, and how the number of the levels, the number of particles and the number of independent simulations influence the accuracy.

Further work will concentrate on practicable estimates of the size of the particle models that ensures a given precision, or in refinements to these interacting particle algorithms.

References

1. F. Cérou, P. Del Moral, F. LeGland, and P. Lezaud. Genetic Genealogical Models in Rare Events Analysis. *Publication du Laboratoire de Statistique et Probabilités, Toulouse III*, 2002.

2. P. Del Moral. *Feynman-Kac Formulae: Genealogical and Interacting Particle Systems with Applications.* Springer-Verlag, 2004.
3. P. Del Moral and J. Jacod. Interacting Particle Filtering With Discrete Observations. In A. Doucet, N. de Freitas, and N. Gordon, editors, *Sequential Monte Carlo Methods in Practice.* Springer Verlag, 2001.
4. P. Del Moral and L. Miclo. Branching and interacting particle systems approximations of Feynman-Kac formulae with applications to non-linear filtering. In J. Azma, M. Émery, M. Ledoux, and M. Yor, editors, *Séminaire de Probabilités XXXIV*, volume 1729 of *Lecture Notes in Mathematics*, pages 1–14. Springer, 2000.
5. P. Del Moral and L. Miclo. Particle Approximations of Lyapunov Exponents Connected to Schrödinger Operators and Feynman-Kac Semigroups. *ESAIM: Probability and Statistics*, 7:171–208, 2003.
6. A. Doucet, N. de Freitas, and N. Gordon. An Introduction to Sequential Monte Carlo Methods. In A. Doucet, N. de Freitas, and N. Gordon, editors, *Sequential Monte Carlo Methods in Practice.* Springer Verlag, 2001.
7. P. Embrechts, C. Klüppelberg, and T. Mikosch. *Modelling Extremal Events for Insurance and Finance.* Springer, 2003.
8. T.E. Harris and H. Kahn. Estimation of particle transmission by random sampling. *Natl. Bur. Stand. Appl. Math. Ser.*, 12:27–30, 1951.
9. J. Krystul and H. Blom. Sequential Monte Carlo Simulation of Rare Event Probability. *Hybridge report D9.3, http://www.nlr.nl/public/hosted-sites/hybridge/*, 2004.
10. A. Lagnoux. Rare Events Simulation. *Publication du Laboratoire de Statistique et Probabilités, Toulouse III*, 2003.
11. R. D. Reiss and M. Thomas. *Statistical Analysis of Extreme Values.* Birkhäuser, 1997.
12. M.N. Rosenbluth and A.W. Rosenbluth. Monte-Carlo calculations of the average extension of macromolecular chains. *J. Chem. Phys.*, 23:356–359, 1995.
13. A. Sznitman. *Brownian Motion Obstacles and Random Media.* Springer-Verlag, 1998.
14. A.N. Van der Vaart and J.A. Wellner. *Weak Convergence and Empirical Processes with Applications to Statistics.* Series in Statistics. Springer, 1996.
15. M. Villén-Altamirano and J. Villén-Altamirano. A method for accelerating rare event simulations. In *13th Int. Teletraffic Congress,ITC 13 (Queueing, Performance and Control in ATM)*, pages 71–76, Copenhagen, Denmark, 1991.
16. M. Villén-Altamirano and J. Villén-Altamirano. RESTART: An efficient and general method for fast simulation of rare events. Technical Report 7, Departamento de Mætmatica Aplicada, E.U. Informática, Universidad Politécnica de Madrid, 1997.

Compositional Specification of a Multi-agent System by Stochastically and Dynamically Coloured Petri Nets

Mariken H.C. Everdij[1], Margriet B. Klompstra[1], Henk A.P. Blom[1], and Bart Klein Obbink[1]

National Aerospace Laboratory NLR, P.O. Box 90502, 1006 BM Amsterdam, The Netherlands, everdij@nlr.nl, klompstr@nlr.nl, blom@nlr.nl, bklein@nlr.nl

Summary. For safety-critical operations in the nuclear and chemical industries, Petri nets have proven to be useful for the compositional specification of appropriate accident risk assessment models. For air traffic operations, the development of such model is more challenging due to the high distribution of and complex interactions between the multiple agents involved. The specific problems are: A) Need for a hierarchy from low level Petri nets to the complete Petri net; B) Duplication of arcs and transitions within a low level Petri net; C) Cluttering of interconnections. The chapter develops adequate solutions for each of these problems. The solution approaches are first explained graphically, and next formally. The approach developed is illustrated for an air traffic operation example.

1 Introduction

The aim of this chapter is to extend the compositional specification power of Petri nets for application to a multi-agent hybrid system. The motivating type of application is accident risk assessment of safety-critical operations in general, and of air traffic operations in particular. For safety-critical operations in e.g. nuclear and chemical industries, it is common practice that accident risk assessment models are being developed to provide valuable feedback to the process of design and certification of a change (e.g. [16], [18]). Accident risk assessment could play a similar valuable role in the design of novel air traffic operations.

By the very nature of air traffic management, the various decision-makers are highly distributed: per aircraft there is a crew of pilots, and per air traffic control centre there are many human operators. In addition, the safety related decision-making process involves interactions of these humans with each other and with:

- a random and often unpredictable environment, e.g. varying wind, thunderstorms, etc.,

H.A.P. Blom, J. Lygeros (Eds.): Stochastic Hybrid Systems, LNCIS 337, pp. 325–350, 2006.

- a large set of procedural rules and guidelines,
- many technical and automation support systems,
- decision-makers at airline operation centres.

These aspects make accident risk assessment for air traffic operations a very challenging application area, the decision making process of which is significantly more complex than it is of operations in other safety-critical industries as is illustrated in Figure 1 of the Introduction of this book. This makes the specification of an unambiguous mathematical model of air traffic operations a very challenging task.

The most advanced approaches that have been developed in literature to model accident risk of safety-critical operations in nuclear and chemical industries make use of the compositional specification power of Petri nets to instantiate a model, and subsequently use stochastic analysis and Monte Carlo simulation (e.g. [16]) to evaluate the model. Since their introduction in the 1960s, Petri nets have shown their usefulness for many practical applications in different industries (e.g. [4]). Various Petri net extensions and generalisations, new analysis techniques, and numerous supporting computer tools have been developed, which further increased their modelling opportunities, though falling short for air traffic operations. In order to capture the characteristics of air traffic operations through a Petri net, [7] extended Dynamically Coloured Petri Net (DCPN) of [6], [5], [8] to Stochastically and Dynamically Coloured Petri Net (SDCPN) and proved that there exists a close relationship with the larger class of Generalised Stochastic Hybrid Processes (GSHP) needed to model air traffic operations. Basically, a DCPN is an extension of Coloured Stochastic Petri Net (e.g. [12]), in the sense that in DCPN the token colours evolve in time (dynamically) as solutions of differential equations while the tokens reside in their places. In an SDCPN, differential equations are replaced by stochastic differential equations. The DCPN formalism has been successfully used in practical air traffic applications, (e.g. [2], [3]). However, it was found that when being used for modelling more and more complex multi-agent hybrid systems, the compositional specification power of Petri nets reaches its limitations. More specifically, the following problems were identified:

A. Need for a hierarchy from low level Petri nets to the complete Petri net. For the modelling of a complete Petri net for complex systems, a hierarchical approach is necessary in order to be able to separate local modelling issues from global or interaction modelling issues.
B. Duplication of arcs and transitions within a low level Petri net. Often the addition of an interconnection between two low-level Petri nets leads to a duplication of transition and arcs in the receiving Petri net.
C. Cluttering of interconnections. The number of interconnections between the different low level Petri nets tends to grow quadratically with the size of the Petri net.

2 Compositional Specification Challenge

In literature, approaches have been developed to address problem A. These approaches are outlined below.

Ref. [14] introduced Hierarchical Coloured Petri Nets. These Hierarchical CPNs allow a set of subnets, called pages, to be related to each other, in such a way that together they constitute a single model. The pages interact with each other in a well-defined way. A page can also be substituted by a place or a transition, in order to show its role in the larger model, or to postpone its detailed modelling until later. In addition to these substitution transitions and places, Hierarchical CPN allow invocation transitions (CPN is temporarily extended with a new instance of an invocation subpage), place fusion (a set of places is folded into a single place) and transition fusion (a set of transitions is folded into a single transition). The pages that interact solve problem A.

More recent approaches also address problem A; they consider elementary Petri nets that have input (or entry) and output (or exit) places through which these Petri nets are coupled with other Petri nets. One example approach is $B(PN)^2$ (Basic Petri Net Programming Notation), introduced by Best and Hopkins (see e.g. [10]). The compositional denotational semantics of $B(PN)^2$ programs can be given in terms of M-nets (modular multilabelled nets), which form an algebra of composable high-level Petri nets. These Petri net components have at least one entry place and at least one exit place. Several composition operations (e.g. parallel composition, sequential composition) are defined to couple the Petri nets. Communication is performed by transition synchronisation. Another example approach is by [15], who introduced the concept of Petri net Components and showed how systems can be composed from components. These components have input and output places and components can be connected at these input and output places. Ref. [15] also provides the compositional semantics.

Also addressing problem A are [9] and [11], who consider sub-Petri nets that model parallel systems, and draw these sub-Petri nets in separate boxes. Places and transitions in different sub-Petri nets are coupled by arcs to model interactions. Ref. [9] uses Synchronous Interpreted Petri Nets (SIPN) as basis and shows how the interactions can be used to model synchronisation or priority of the parallel systems. Ref. [9] also allows hierarchy: a macroplace can be exploded (or imploded) to form (or hide) a complete sub-Petri net. Ref. [11] uses Generalised Stochastic Petri nets (GSPNs), refers to the sub-Petri nets as modules, and adopts the requirement that there should be exactly one token in each module; transitions in a module are not allowed to consume a token from another module without returning one immediately. Therefore, [11] introduced three module coupling mechanisms: 1) marking tests; 2) common transitions; 3) interconnection blocks. In addition, in order to improve the compactness of the module, [11] recommends two rules, called optimisation rules: 1) avoidance of immediate internal transitions; 2) module folding using memories.

For addressing problem B, some ideas from literature are useful. In order to avoid the duplication of transitions, one might apply transition fusion as proposed by [14], or module folding of [11].

The aim of this chapter is to combine and adopt the approaches from literature that solve problem A and to develop approaches to solve problems B and C, and to organise these new developments into a compositional specification approach for SDCPN. In addition, also the effectiveness of the approach is illustrated for the modelling of an air traffic example. The relations between our approach and those found in literature are explained in Sections 4 and 5, and are summarised below.

To solve problem A, the compositional specification of a SDCPN for a complex process or operation starts with developing a Local Petri Net (LPN) for each agent that exists in the process or operation (e.g. air traffic controller, pilot, navigation and surveillance equipment). Counterparts of LPNs in literature are the modules of [11], the pages of [14] and the components of [15]. An essential difference is that our LPNs (and [11]s modules) are connected with each other such that the number of tokens residing in an LPN is not influenced by these interconnections, while [14] and [15] do not pose this restriction.

We use two types of interconnections between nodes and arcs in different LPNs:

- Enabling arc (or inhibitor arc) from one place in one LPN to one transition in another LPN. These types of arcs have been used widely in Petri net literature, including [11] for inhibitor arcs and [9] for both types.
- Interaction Petri Net (IPN) from one (or more) transition(s) in one LPN to one (or more) transition(s) in another LPN. These IPNs are similar to the interconnection blocks of [11]. If an IPN consists of one place only, then the connection of two LPNs through an IPN also has some similarity with place fusion, see e.g. [14] or [15], except that our IPN will not change the number of tokens in its connecting LPNs.

Each LPN is surrounded by a box. This boxing idea has also been used by e.g. [9] or [15]. Next, to solve problems B and C, we identify additional interconnections between LPNs that allow, with well-defined meanings, arcs to initiate and/or to end on the edge of the box surrounding an LPN. To the authors knowledge this element has no counterpart in Petri net literature; however, it is based on how [13] composes statecharts. The meaning of these interconnections from or to an edge of a box allows several arcs or transitions to be represented by only one arc or transition. In that sense, there is a relation with transition fusion used by [14] and with module folding used by [11].

This chapter is organised as follows: Section 3 briefly outlines SDCPN; for a more complete definition we refer to [7]. Section 4 outlines how a SDCPN can be specified in a logical sequence for each entity of an agent, and explains how the entities of agents are connected without changing the structure of low level entities. This solves problem A above. Section 5 defines some new Petri net clustering types which avoid the internal duplication problem (problem B) and the problem of cluttering interconnections (problem C). It is noted that these clustering types can also be applied to other Petri net extensions than SDCPN. Section 6 extends the SDCPN definition of Section 4 to include these new clustering types. In Section 7 the approach developed is illustrated for an air traffic operation example. Finally, Section 8 gives concluding remarks.

3 Stochastically and Dynamically Coloured Petri Net

This section gives a definition of SDCPN. Subsection 3.1 describes the SDCPN elements, while Subsection 3.2 describes the SDCPN evolution rules. These elements and evolution rules together form the SDCPN definition. For a more formal SDCPN definition and a simple SDCPN example we refer to [5] and to [7].

3.1 SDCPN Elements

A Stochastically and Dynamically Coloured Petri Net [7] is given by the following tuple: SDCPN = ($\mathcal{P}$, $\mathcal{T}$, $\mathcal{A}$, $\mathcal{N}$, $\mathcal{S}$, $\mathcal{C}$, $\mathcal{V}$, $\mathcal{W}$, $\mathcal{G}$, $\mathcal{D}$, $\mathcal{F}$, $\mathcal{I}$), where:

$\mathcal{P}$ is a set of places

$\mathcal{T}$ is a set of transitions which consists of a set of guard transitions ($\mathcal{T}_G$), a set of delay transitions ($\mathcal{T}_D$) and a set of immediate transitions ($\mathcal{T}_I$).

$\mathcal{A}$ is a set of arcs which consists of a set of ordinary arcs ($\mathcal{A}_o$), a set of enabling arcs ($\mathcal{A}_e$) and a set of inhibitor arcs ($\mathcal{A}_i$).

$\mathcal{N}$ is a node function which maps each arc to an ordered pair of one transition and one place; multiple arcs between the same place and transition are allowed.

$\mathcal{S}$ is a set of colour types for the tokens occurring in the net (a colour is the value of an object or process in Petri net terminology). Each colour type is to be of the form $\mathbb{R}^n$.

$\mathcal{C}$ is a colour function which maps each place to a colour type in $\mathcal{S}$.

$\mathcal{V}$ and $\mathcal{W}$ are sets of place-specific colour functions which together describe the behaviour of the colour of a token while it resides in its place. For each place, these elements determine a stochastic differential equation, which is locally Lipschitz continuous.

$\mathcal{G}$ is a set of Boolean-valued transition guards associating each transition in $\mathcal{T}_G$ with a guard function. This guard function is continuously evaluated when the transition has a token in each of its input places, i.e., when there is at least one token per input arc of the transition present. The guard function must evaluate to True before the transition is allowed to fire (i.e. remove and produce tokens). This happens when the colours of the input tokens of the transition (which can change value through time) reach particular transition-specific value combinations.

$\mathcal{D}$ is a set of transition delays associating each transition in $\mathcal{T}_D$ with a delay function. This delay function is continuously evaluated when the transition has a token in each of its input places. The delay function determines for how long the transition must wait before it is allowed to fire (i.e. remove and produce tokens). The firing rate depends on the colours of the input tokens of the transition (which can change value through time) and is determined by a Poisson point process.

$\mathcal{F}$ is a set of (probabilistic) firing functions. For each transition it describes the quantity and colours of the tokens produced by the transition at its firing. A transition produces 0 or 1 token per outgoing arc; this quantity and the colour of the produced tokens is according to a transition-specific probabilistic mapping rule that may depend on the colours of the input tokens.

$\mathcal{I}$ is an initial marking which defines the set of tokens initially present, i.e., it specifies in which places they initially reside, and the colours they initially have.

The set of places $\mathcal{P}$, the set of transitions $\mathcal{T}$, the set of arcs $\mathcal{A}$ and the node function $\mathcal{N}$ are defined in a Petri net graph. Figure 1 shows the graphical representation of the elements in $\mathcal{P}$, $\mathcal{T}$ and $\mathcal{A}$. The node function $\mathcal{N}$ describes how these components are connected into a Petri net graph.

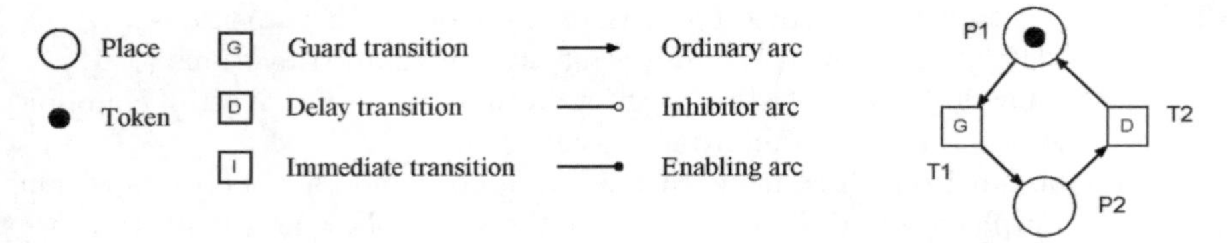

Fig. 1. Notation for places, tokens, transitions and arcs in Petri net graphs. On the right-hand- side there is a simple example Petri net graph, with two places and two transitions and a token in place P1

3.2 SDCPN Evolution

Tokens and the associated colour values in a SDCPN evolve over time quite similar as in a Coloured Stochastic Petri Nets (e.g. [12]). The main additions are that the colour of a token while it is residing in a place is an element of $\mathbb{R}^n$ (where n is place-specific) and may evolve according to a differential equation that is governed by the colour functions $\mathcal{V}$ and $\mathcal{W}$ of the specific place where the token resides, and that guard and delay transitions take the evolving colour values into account. For DCPN, this differential equation is an ordinary differential equation, completely described by $\mathcal{V}$ only. For SDCPN, this differential equation is a stochastic differential equation, described by $\mathcal{V}$ (for the flow part) and $\mathcal{W}$ (for the Brownian motion term).

Tokens can be removed from places by transitions that are connected to these places by incoming ordinary arcs. A transition can only remove tokens if two conditions are both satisfied. If this is the case, the transition is said to be enabled. The first condition is that the transition must have at least one token per ordinary arc and one token per enabling arc in each of its input places and have no token in the input places to which it is connected by an inhibitor arc. When the first condition holds, the transition is said to be pre-enabled. The second condition differs per type of transition. For immediate transitions the second condition is automatically satisfied if the transition is pre-enabled. For guard transitions the second condition is specified by the set of transition guards $\mathcal{G}$ and for delay transitions it is specified by the set of transition delays $\mathcal{D}$, see their description in Subsection 3.1.

When these two conditions are satisfied, the transition removes the tokens from the input places by which it is connected through an ordinary arc. It does not remove the tokens from places by which it is connected through an enabling arc. Subsequently, the transition produces a token for some or all of its output places, specified by the firing function $\mathcal{F}$. The colour of a produced token (which must be of the correct type, indicated by what $\mathcal{C}$ defines for the output place), and the place for which it is produced is also specified by the firing function $\mathcal{F}$. The evaluation of $\mathcal{G}$, $\mathcal{D}$ and $\mathcal{F}$ may be dependent on the colours of the input tokens of the corresponding transition.

In order to avoid ambiguity, for a DCPN the following rules apply when two transitions are enabled simultaneously:

- R_0 The firing of an immediate transition has priority over the firing of a guard or a delay transition.
- R_1 If one transition becomes enabled by two or more disjoint sets of input tokens at exactly the same time, then it will fire these sets of tokens independently, at the same time.
- R_2 If one transition becomes enabled by two or more non-disjoint sets of input tokens at exactly the same time, and the firing of one set disables the other, then the set that is fired is selected randomly.
- R_3 If two or more transitions become enabled at exactly the same moment by disjoint sets of input tokens, then they will fire at the same time.
- R_4 If two or more transitions become enabled at exactly the same moment by non-disjoint sets of input tokens, then the transition that will fire is selected randomly, with the same probability for each transition.

Note that in Rule R_2 there is a conflict between tokens fighting for the same transition and in Rule R_4 there is a conflict between transitions fighting for the same set of tokens; these rules settle these fights by appointing a random winner. In Rules R_1 and R_3 there is no conflict between tokens fighting for the same transition or for transitions fighting for the same tokens; these firings can occur independently and these rules only take care of the timing of these individual firings. For a motivation of why these rules are chosen like this, see [5].

4 Local Petri Nets-Based Specification of an SDCPN

The compositional specification of a Stochastically and Dynamically Coloured Petri Net for a complex process with many different interacting agents such as exist in air traffic operations (e.g. air traffic controllers, pilots, navigation and surveillance equipment), is a bottom-up process (note that for DCPN this process is the same, except that no Brownian motion terms need to be defined). Prior to starting this compositional process, per agent the relevant low level functional entities have to be identified based on expert domain knowledge of that agent. For this chapter we assume that these low-level functional entities are given per agent. The compositional specification idea is then first to specify one small Petri net per functional entity of an agent, and refer to this as a Local Petri Net (LPN). Next, the interactions between these LPNs are specified. Note that our LPN definition has similar counterparts in Petri net literature. For example, [11] considers Generalised Stochastic Petri Nets to be composed of Modules in a similar way as we propose for SDCPN to be composed of LPNs. Ref. [14] proposes Hierarchical Coloured Petri Nets to be composed of Pages, while [15] considers the composition of Petri Net Components. An essential difference is that our LPNs (and Fotas modules) are connected with each other such that the number of tokens residing in an LPN (or module) is not influenced by these interconnections, while [14] and [15] do not pose this restriction.

The specification of the various elements of one LPN is explained in Subsection 4.1; this has to be accomplished for all LPNs. Subsection 4.2 describes how the interconnections between these LPNs are established.

4.1 Specification of Local Petri Net

Specification of elements $\mathcal{P}$, $\mathcal{T}$, $\mathcal{A}$, $\mathcal{N}$

First, places (drawn as circles) are identified for the LPN. These places may represent operational or physical conditions (nominal modes and non-nominal modes). Next, the transitions are identified: If between two places, say P_1 and P_2, a switch might occur, one transition (rectangle) is drawn, with two arcs (arrows) connecting the places with the transition. The places are gathered in the set of places $\mathcal{P}$, the transitions are gathered in the set of transitions $\mathcal{T}$, and the arcs are gathered in the set of arcs $\mathcal{A}$. The node function $\mathcal{N}$ describes for each arc which place and transition it connects.

Specification of elements $\mathcal{S}$, $\mathcal{C}$, $\mathcal{V}$ and $\mathcal{W}$

A complex stochastic dynamic process such as in air traffic operations cannot be described by places and transitions alone. Usually, some (piecewise) continuous valued timed processes are identified, which can be influenced by and which can influence the LPN places and transitions. In SDCPN, continuous valued processes are associated with tokens that reside in the places. In general, if a token resides in a particular place, its value changes according to a differential equation that is associated with that place. In this step, all places are checked on whether a continuous valued process can be associated with it. Note that on the other hand, it may happen that identified continuous valued processes lead to a necessary introduction of new places. For example: one continuous-valued process appears to alternately follow two different differential equations; in that case two places need to be introduced, each associated with one of these differential equations. All continuous valued process types are collected in the set $\mathcal{S}$. The mapping of each place of the Petri net to one of these types is described by $\mathcal{C}$. The sets of place-specific colour functions $\mathcal{V}$ and $\mathcal{W}$ describe how the colour of a token changes while it is staying in a place. For each place, $\mathcal{V}$ and $\mathcal{W}$ specify the coefficients of a differential equation which describes the rate of change of the token colour: if $\mathcal{V}_P$ and $\mathcal{W}_P$ define the token colour functions for place P that has colour type $\mathcal{C}_P$, then the colour $c_t^P \in \mathcal{C}_P$ of a token in place P at time t satisfies: $dc_t^P = \mathcal{V}_P(c_t^P)dt + \mathcal{W}_P(c_t^P)dw_t$, where $\{w_t\}$ is Brownian motion.

Specification of elements $\mathcal{G}$, $\mathcal{D}$, $\mathcal{F}$ and $\mathcal{I}$ in Local PN terms

Next, for each transition, one should determine whether it is a guard transition, a delay transition or an immediate transition. A guard transition fires based on the combined colours of its input tokens reaching some value. A delay transition models a duration, e.g. of an action. An immediate transition fires without delay. The guard transitions are collected in the set $\mathcal{T}_G$, the delay transitions are collected in $\mathcal{T}_D$ and the immediate transitions are collected in the set $\mathcal{T}_I$. Subsequently, the guards $\mathcal{G}$ and the delays $\mathcal{D}$ are specified in detail. The firing function $\mathcal{F}$ describes the colours of the tokens fired by a transition into its output places, given the colours of the tokens in the input places. Finally, the initial marking $\mathcal{I}$ describes which place(s) of the LPN initially contain one or more token(s) and describes the initial colour values of these tokens, hence it describes the initial state of the process modelled by the SDCPN.

4.2 Interconnections Between LPNs

The interconnections between the LPNs have to be specified in a way that allows to start at the lowest level and then step by step going up to the highest level, and such that an interconnection at a higher level does not imply a significant change at a lower level. The typical exception on this is caused by non-local influences on $\mathcal{G}$, $\mathcal{D}$ and $\mathcal{F}$. In order to improve into this desired direction, in this subsection some specific types of interconnections are identified.

Following [11] one step in enabling a systematic bottom-up specification of a Petri net is to ensure that each LPN always contains exactly one token. For air traffic types of applications it often is useful to allow multiple tokens to be within one LPN, e.g. one for each aircraft. Hence we relaxed the Fota-principle to the following requirement: all interconnections between LPNs shall be such that the number of tokens in an LPN is not directly influenced by these interconnections. Subsequently we identified two types of interconnections that satisfied our above requirement:

- Enabling arc (or inhibitor arc) from one place in one LPN to one transition in another LPN.
- Interaction Petri Net (IPN) from one (or more) transition(s) in one LPN to one (or more) transition(s) in another LPN.

Enabling and inhibitor interconnections are illustrated in Figures 2 and 3, respectively. Note that in these figures, each LPN is surrounded by a box. This boxing idea has also been used by e.g. [9] or [15].

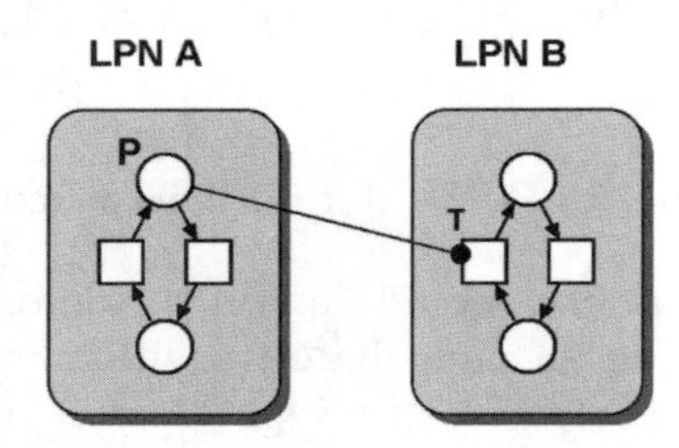

Fig. 2. Illustration of an enabling arc from one place in LPN A to one transition in LPN B

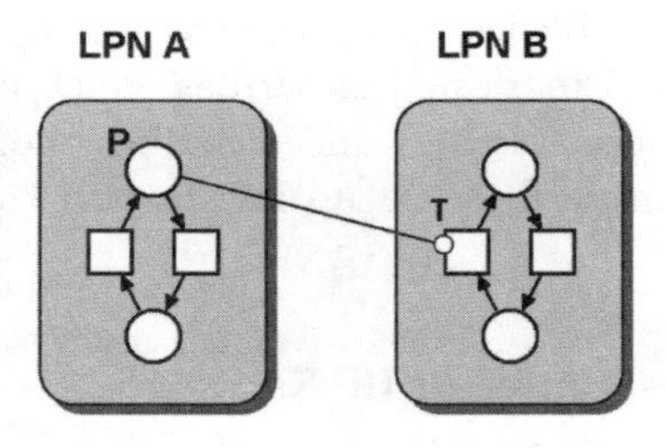

Fig. 3. Illustration of an inhibitor arc from one place in LPN A to one transition in LPN B

Enabling and inhibitor arcs are used to describe how agents modelled by individual LPNs influence each other. The transition at the tip of the arc (i.e. transition T in LPN B in Figures 2 and 3) can only fire if the process modelled by LPN A is in a particular state or marking, and when it fires, it may use the information existing in this marking of LPN A. For example, it may appear that the guard or delay of transition T is dependent of the colour of the token in place P. In those cases, the Petri net graph needs to be extended with (enabling) arcs from place P to transition T in order to get access to this information (and the guard or delay of the transition, which in the previous subsection has only be defined locally, needs to be adapted). Since tokens are not consumed through enabling arcs at a transition firing, the state of LPN A is not changed through this firing. Ref. [9] uses enabling arcs like this to model synchronisation. [11] uses GSPN which do not support enabling arcs although they do support inhibitor arcs; however, [11] does allow tokens of other modules be consumed and immediately placed back, which is similar to using an enabling arc.

An Interaction Petri Net (IPN) consists of at least one place, and zero or more transitions. It connects, by means of ordinary arcs, one or more transition(s) in one LPN with one or more transition(s) in another LPN. If there are transitions in the IPN, and if these transitions are connected with other LPNs, then only enabling or inhibitor arcs can be used for the connections of these transitions with other LPNs. An example of an IPN is illustrated in Figure 4. It can be easily verified that an IPN does not influence the number of tokens in the LPNs it connects.

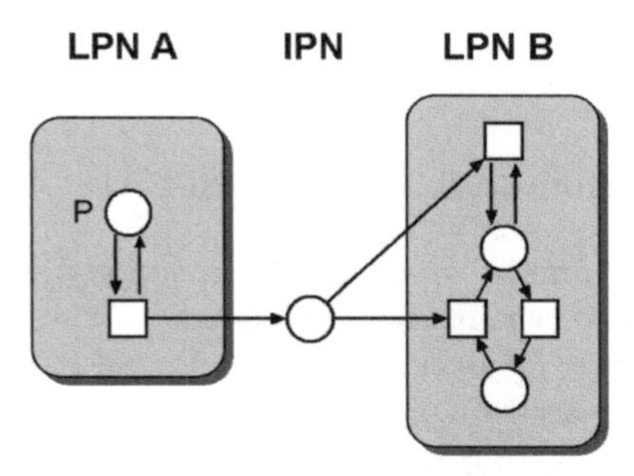

Fig. 4. Illustration of an Interaction Petri Net from one transition in LPN A to two transitions in LPN B

Interaction Petri Nets are used when enabling or inhibitor arcs are insufficient to model the interconnection between two agents. For example, it can hold on to state information from its input LPN (i.e. LPN A in Figure 4) while the state of LPN A itself evolves further. Also, IPNs can be used to connect two transitions, while enabling or inhibitor arcs always connect a place with a transition. Note that our IPNs are similar to the Interconnection blocks of [11]. The connection of two LPNs through an Interaction Petri Net also has some similarity with Place Fusion, see e.g. [14] or [15], except that our Interaction Petri Net will not change the number of tokens in its connecting LPNs.

5 Extension with Interconnection Mapping Types

Interconnections between LPNs through enabling (or inhibitor) arcs and IPNs might lead to a combinatorial growth of the number of interconnections with the size of the Petri net. To avoid this combinatorial growth as much as possible, in this section hierarchical clustering and interconnection mapping approaches are graphically developed, based on how [13] composes statecharts:

1. Interconnection mapping types I and II are defined to avoid possible duplication of transitions and arcs within LPNs caused by specifying interconnections between LPNs.
2. Interconnection mapping types III, IV and V are defined to avoid cluttering of interconnections between places and transitions of different LPNs.
3. Interconnection mapping types VI and VII define interconnections from or to hierarchical clusters of LPNs, which reduce the cluttering of interconnections.
4. Combinations of interconnection mapping types, and an additional interconnection mapping type VIII that avoids a duplication of transitions and arcs within an LPN and duplication of arcs between LPNs.

In Section 6, the SDCPN definition is extended to include these interconnection mapping types.

5.1 Avoid Duplication of Transitions and Arcs Within an LPN

Figure 5 shows an example where interconnections between LPNs lead to duplication of transitions and arcs within one of these LPNs. A transition from place P3 to place P4 occurs if either P1 or P2 contains a token. To model this, it is necessary to use two transitions T1 and T2 between P3 and P4. The use of only one transition between P3 and P4 would model an and relation (i.e. both P1 *and* P2 contain a token) instead of an or relation.

In most cases, the duplicated transitions and arcs do not have an essentially different meaning, and they are mostly introduced to be able to make use of colours of tokens residing in other LPNs. In particular, these duplicated transitions have the same guard or delay and the same firing function. This makes that duplication leads to reduced readability. This subsection presents some interconnection mapping types to avoid such duplication.

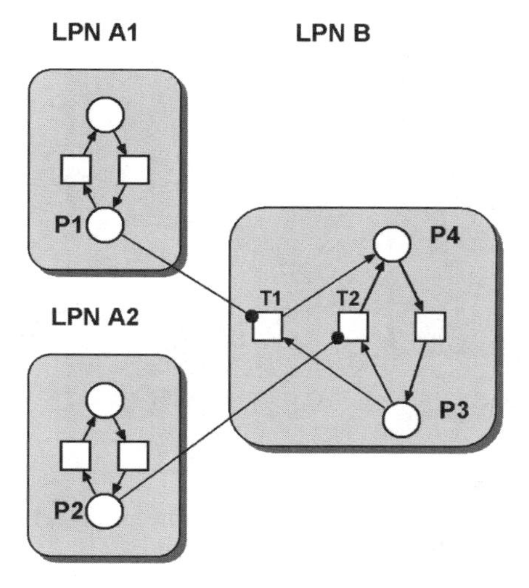

Fig. 5. Illustration of duplication of transitions within an LPN

LPN interconnection mapping type I

A set of s enabling arcs initiating on s places, merging into one arc and ending on one transition, means that this transition is duplicated s times, and that s enabling arcs are drawn between the s places and the s resulting transitions. This type of arc is called merging arc. The transition at the end of the merging arc should be in a different LPN than the s places that are at the beginning of the arc. Figure 6 shows an example of this interconnection mapping type. Note that in order to avoid confusion when using this interconnection mapping type, the s duplicated transitions should have the same guard or delay function and the same firing function and their input places should have the same colour type. Interconnection mapping type I is not defined with inhibitor or ordinary arcs instead of enabling arcs.

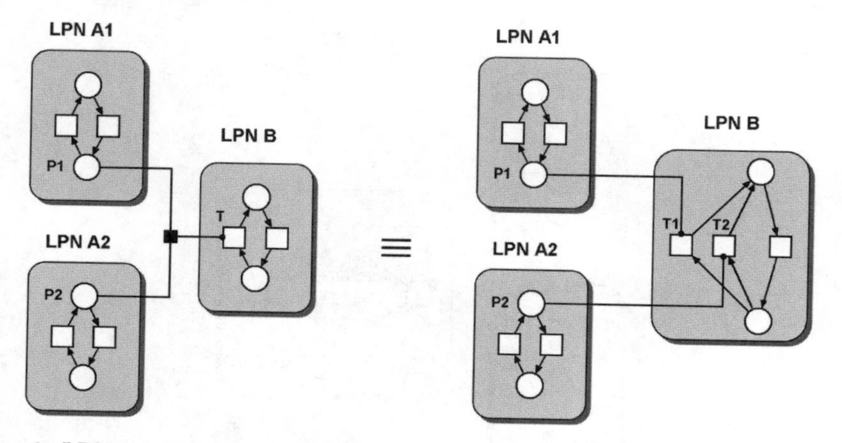

Fig. 6. LPN interconnection mapping type I. The point where several arcs merge into one arc is represented by a small black square

LPN interconnection mapping type II

An enabling arc initiating on the edge of an LPN box and ending on a transition in another LPN box, means that enabling arcs initiate from all places in the first LPN and end on duplications of this transition in the second LPN. Figure 7 shows an example of this interconnection mapping type. The duplicated transitions should have the same guard or delay function and the same firing function and their input places should have the same colour type. Interconnection mapping type II is not defined with inhibitor or ordinary arcs instead of enabling arcs.

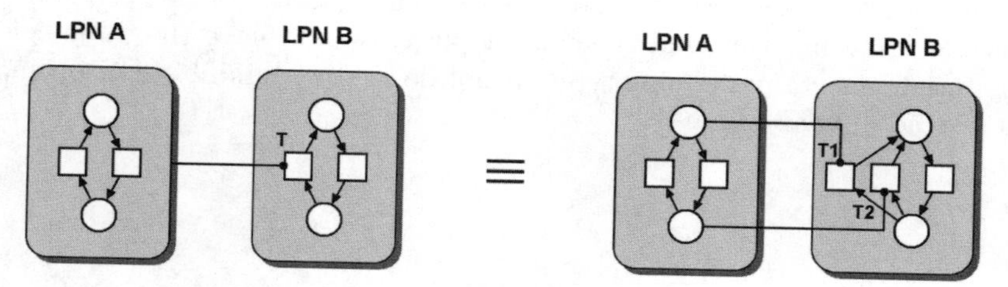

Fig. 7. LPN interconnection mapping type II

5.2 Avoid Cluttering of Interconnections Between LPNs

The interconnection mapping types in the previous subsection avoid the duplication problem, but not the cluttering due to the many enabling arcs and IPNs between places and transitions of different LPNs. If several LPNs are interconnected in one graph the result becomes unreadable. This subsection presents some interconnection mapping types to avoid this:

- Interconnection mapping type III can be applied to avoid enabling arcs cluttering.
- Interconnection mapping types IV and V can be applied to avoid IPNs cluttering.

LPN interconnection mapping type III

An enabling arc ending on the edge of an LPN box, means that enabling arcs end on each transition in this LPN. Figure 8 shows an example of this interconnection mapping type. Interconnection mapping type III can also be used with inhibitor arcs instead of enabling arcs. It cannot be used with ordinary arcs, due to the restriction that the number of tokens in an LPN should remain the same.

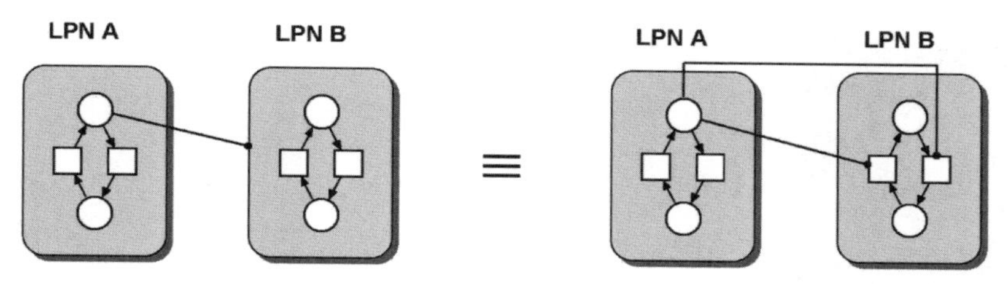

Fig. 8. LPN interconnection mapping type III

LPN interconnection mapping type IV

An ordinary arc initiating on the edge of an LPN box and ending on a place within an IPN means that ordinary arcs initiate from all transitions in this LPN. Figure 9 shows an example of this interconnection mapping type. Interconnection mapping type IV is not defined with enabling or inhibitor arcs instead of ordinary arcs.

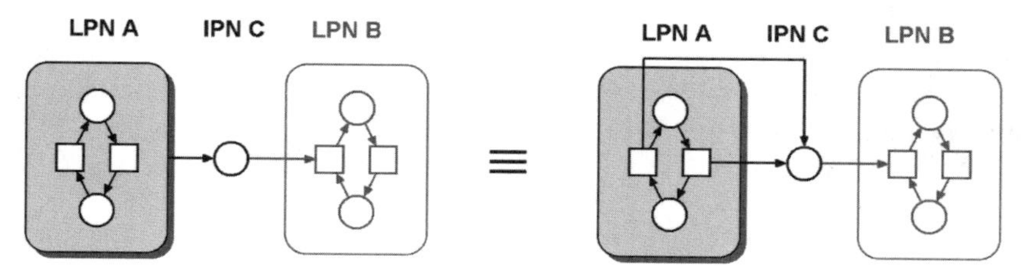

Fig. 9. LPN interconnection mapping type IV

LPN interconnection mapping type V

An ordinary arc ending on the edge of an LPN box and starting from a place within an IPN means that ordinary arcs end on each transition in this LPN. Figure 10 shows an example of this interconnection mapping type. Interconnection mapping type IV is not defined with enabling or inhibitor arcs instead of ordinary arcs.

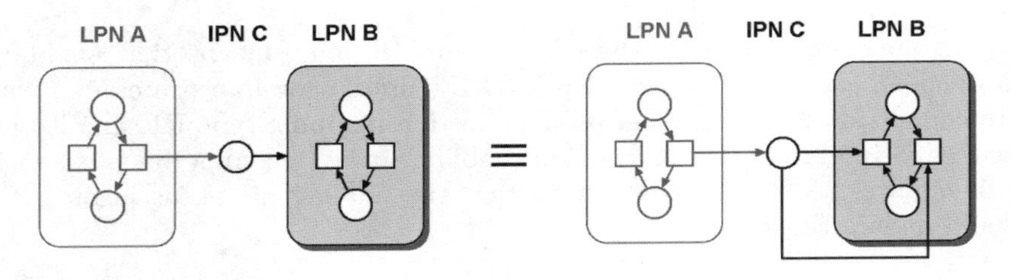

Fig. 10. LPN interconnection mapping type V

5.3 Clustering of LPNs

In this subsection, we define enabling arcs that go from or to a cluster of LPNs. This is done following the next two interconnection mapping types. Figures 11 and 12 show examples of these interconnection mapping types.

LPN interconnection mapping type VI

Suppose there is one LPN A and a set of n LPNs B_i (i=1,2, ..., n) which is enclosed by a large box. An enabling arc initiating on the edge of LPN A and ending on the edge of the large box with the set of LPNs B_i, means that the enabling arc represents n actual enabling arcs, initiating on the edge of LPN A and ending on the edge of each LPN B_i. Interconnection mapping type VI can also be defined from a place to a large box of LPNs (by means of an enabling arc), or from a place within an IPN to a large box of LPNs (by means of an ordinary arc). It is not defined with inhibitor arcs instead of enabling arcs. Note that the right hand side of Figure 11 makes use of a combination of interconnection mapping types II and III. For more examples of such combinations, see Subsection 5.4.

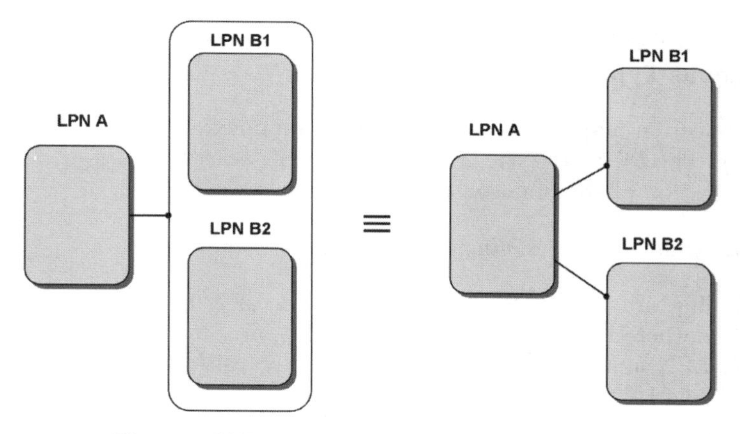

Fig. 11. LPN interconnection mapping type VI

LPN interconnection mapping type VII

Suppose there is a set of n LPNs A_i (i=1,2, ..., n) which is enclosed by a large box and one LPN B. An enabling arc initiating on the edge of the large box with the set of LPNs A_i (i=1,2, ..., n) and ending on the edge of LPN B, means that the enabling arc represents n actual enabling arcs, initiating on the edge of each LPN A_i and ending on the edge of LPN B. Interconnection mapping type VII can also be defined from a large box to a transition. It is not defined with ordinary or inhibitor arcs instead of enabling arcs.

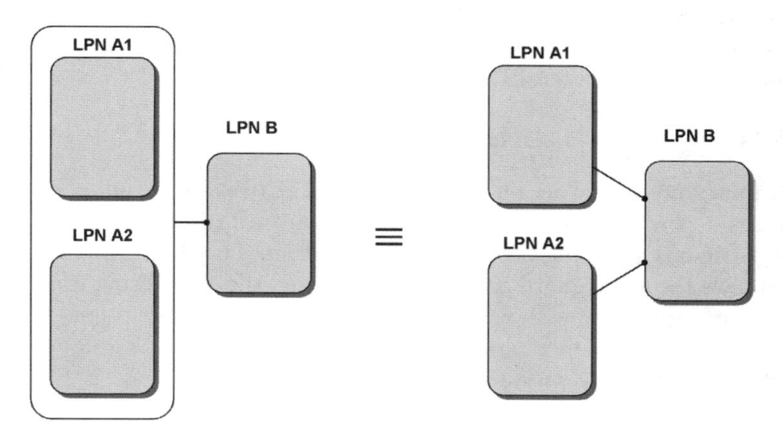

Fig. 12. LPN interconnection mapping type VII

5.4 Combinations of Interconnection Mapping Types and Additional Type

Interconnection mapping types can also be combined, such as interconnection mapping type I with II, type II with III, type IV with V, or type VI with VII. An illustration of combination of II with III is given below.

LPN interconnection mapping types II and III combined

An enabling arc initiating on the edge of an LPN box and ending on the edge of another LPN box, means that enabling arcs initiate from all places in the first LPN and end on duplications of all transitions in the second LPN. Figure 16-fig:LPN IMP II and III shows an example of this combination of interconnection mapping types II and III.

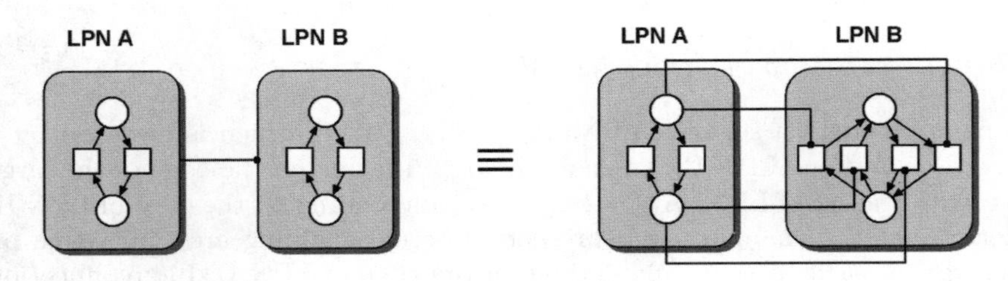

Fig. 13. LPN interconnection mapping types II and III combined

Finally, we introduce an additional interconnection mapping type which avoids duplication of transitions and arcs within an LPN, and consequently cluttering of arcs between LPNs:

LPN interconnection mapping type VIII

An ordinary arc initiating on the edge of an LPN box and ending on a transition inside the same box, means that ordinary arcs initiate from all places in the LPN box to duplications of this transition. The duplicated transitions should have the same guard or delay function and the same firing function and their set of input places should have the same set of colour types. Figure 14 illustrates how this avoids both the duplication of transitions and arcs within an LPN, and the duplication of arcs between LPNs.

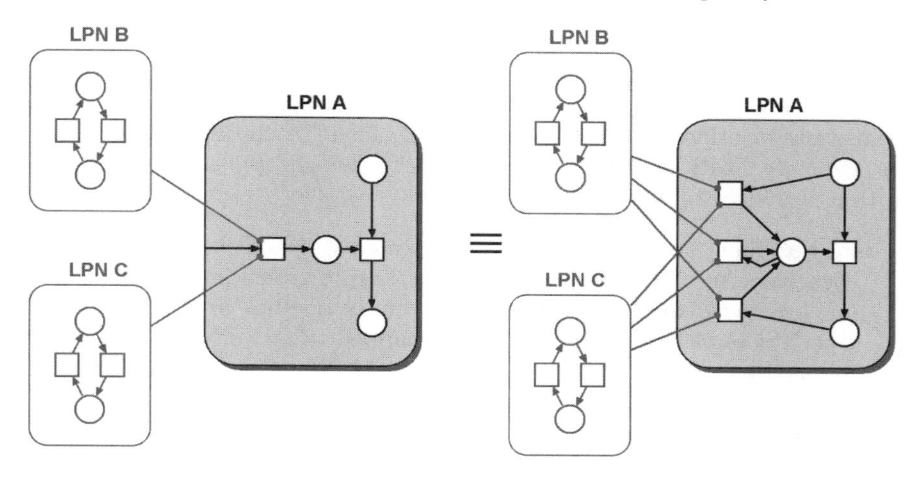

Fig. 14. LPN interconnection mapping type VIII, which avoids duplication of arcs and transitions within an LPN and duplication of arcs between LPNs

Remark: The interconnection mapping types introduced in this section could also be used for other types of Petri nets than SDCPN, provided that these other types of Petri nets support the same graphical elements as SDCPN, such as enabling arcs. If this is not the case, the interconnection mapping types might still be used, but then the restriction that the number of tokens in an LPN cannot be changed by the interconnections must be removed.

6 Extension of SDCPN with Interconnection Mapping Types I Through VIII

This subsection extends the SDCPN definition of [7] to include the interconnection mapping types identified in Section 5. The extension is referred to as SDCPN^{imt}.

SDCPN^{imt} is a tuple ($\mathcal{P}$, $\mathcal{T}$, $\mathcal{B}$, $\mathcal{A}^{imt}$, $\mathcal{L}$, $\mathcal{N}^{imt}$, $\mathcal{S}$, $\mathcal{C}$, $\mathcal{V}$, $\mathcal{W}$, $\mathcal{G}$, $\mathcal{D}$, $\mathcal{F}$, $\mathcal{I}$), where $\mathcal{P}$, $\mathcal{T}$, $\mathcal{S}$, $\mathcal{C}$, $\mathcal{V}$, $\mathcal{W}$, $\mathcal{G}$, $\mathcal{D}$, $\mathcal{F}$, $\mathcal{I}$ are as in the definition of SDCPN (Section 3 or Ref. [7]), and the other elements are outlined below:

$\mathcal{A}^{imt}$ is the set of arcs in the SDCPN^{imt}. It equals the set of arcs $\mathcal{A} = \mathcal{A}_o \cup \mathcal{A}_e \cup \mathcal{A}_i$ as defined in [7], but extended with a set of merging arcs ($\mathcal{A}_m$). In other words, $\mathcal{A}^{imt} = \mathcal{A} \cup \mathcal{A}_m$.

A merging arc is a set of $s \geq 2$ enabling arcs merging into one arc, where s can be different for each merging arc. The merging point is denoted by a small black square.

$\mathcal{B}$ is a set of boxes which consists of a set $\mathcal{B}_L$ of LPN-boxes and a set $\mathcal{B}_C$ of Cluster-boxes.

Each box in $\mathcal{B}$ is drawn as a rectangle with rounded corners. Note that at this definition level, each element of $\mathcal{B}$ is just an empty box. The box function $\mathcal{L}$ (see definition next) will specify the actual contents (i.e. the places and transitions or other boxes) of each box in $\mathcal{B}$. There, each LPN-box in $\mathcal{B}_L$ will be associated with an LPN, and each cluster-box in $\mathcal{B}_C$ will be associated with a cluster of LPNs.

$\mathcal{L}$ is a box function which specifies the contents of each box in $\mathcal{B}$: $\mathcal{L}$ maps each place in $\mathcal{P}$ to zero or one box in $\mathcal{B}$, each transition in $\mathcal{T}$ to zero or one box in $\mathcal{B}$, and each box in $\mathcal{B}_L$ to zero or one box in $\mathcal{B}_C$. Places (and transitions) that form IPNs are not mapped to an LPN-box but can be mapped to a cluster-box, and at least two LPN-boxes should be mapped to each cluster-box.

For each LPN-box in $\mathcal{B}_L$, the box function specifies which places in $\mathcal{P}$ and which transitions in $\mathcal{T}$ are drawn in it to form an LPN; for each cluster-box in $\mathcal{B}_C$ it specifies which (at least two) LPN-boxes in $\mathcal{B}_L$ are drawn in it to form a cluster of LPNs. Some places (and transitions) are not inside any LPN-box; these form the IPNs. It is however possible that IPNs are part of a cluster-box (although they are not part of an LPN-box). Similarly, not all LPN-boxes need to be inside a cluster-box.

$\mathcal{N}^{imt}$ is a node function which maps each arc in $\mathcal{A}_m$ to an ordered pair of which the first component is a set of places (but not in IPNs) or boxes, and the second component is a transition. Furthermore, $\mathcal{N}^{imt}$ maps each arc in $\mathcal{A} = \mathcal{A}_o \cup \mathcal{A}_e \cup \mathcal{A}_i$, to an ordered pair of nodes, where a node is a place, a transition, an LPN-box or a cluster-box. Multiple arcs between the same pair of nodes are allowed (but not both an inhibitor arc and another type of arc). There are the following restrictions:

- Ordinary arcs can only be drawn from a place to a transition within the same LPN- box, from a transition to a place within the same LPN-box, from a place in an IPN to a transition, from a transition to a place in an IPN, from a place in an IPN to an LPN-box, from an LPN-box to a place in an IPN, from a place in an IPN to a cluster-box, or from an LPN box to a transition in the same LPN box.
- Enabling arcs can only be drawn from a place to a transition within the same LPN- box, from a place to a transition in a different LPN-box or in an IPN, from a place (but not in an IPN) to an LPN- or cluster-box, from an LPN- or cluster-box to a transition, or between two boxes (i.e. LPN-LPN, LPN-cluster, cluster-LPN or cluster-cluster).
- Inhibitor arcs can only be drawn from a place to a transition within the same LPN- box, from a place to a transition in a different LPN-box or in an IPN, or from a place (but not in an IPN) to an LPN box.
- Merging arcs can only be drawn from a set of places (but not in IPNs) or boxes, to a transition that is in another LPN than these places or boxes. The input places of a merging arc should be of the same type.

Note that the guards $\mathcal{G}$, delays $\mathcal{D}$, and firing functions $\mathcal{F}$ defined for DCPN^{imt} are equal to those defined for SDCPN. However, since the elements $\mathcal{G}$, $\mathcal{D}$, and $\mathcal{F}$ use the colours of the transition input tokens as input, their evaluation is a little more complicated in the sense that from the SDCPN^{imt} graph it is not immediately obvious which places are the input and output places of the transitions. These input and output places become clear if the SDCPN^{imt} graph is extended to a SDCPN graph, i.e. the cluttered one without the interconnection mapping types. Some rules that avoid this for the most-often used interconnection mapping types are given below. Here, only the between-LPN interconnections are considered. The pre-enabling, enabling or firing of each transition is also dependent on the colours of the input tokens along the within-LPN connections, but to keep the description brief, these are not considered here.

- If a transition has an incoming merging arc (see e.g. interconnection mapping type I), it is pre-enabled if it has a token in at least one of the places also connected to this merging arc. The transition is enabled if it is enabled by this input token as described in [7] (e.g. its guard evaluates to true, or its delay has passed). If there are tokens in several of these input places, the transition guard or delay function uses their colours in parallel for its evaluation.
- If a transition has an (enabling) incoming arc connected with an LPN-box (see e.g. interconnection mapping type II), then it is pre-enabled if there is at least one token somewhere in this input LPN-box (and this is usually the case). It is enabled if it is enabled by this input token as described in [7].
- If the LPN-box in which a transition resides has an input place (see e.g. interconnection mapping type III), then the transition is pre-enabled if there is a token in this input place, and its guard or delay uses the colour of this token for its evaluation as described in [7].
- If the LPN-box in which a transition resides has an input LPN-box (see e.g. interconnection mapping type II combined with III), then the transition is pre-enabled if there is at least one token somewhere in the input LPN-box (and this is usually the case) and its guard or delay uses the colour of this token for its evaluation as described in [7].

7 Free Flight Air Traffic Example

The compositional specification approach described has been used to specify an initial SDCPN^{imt} for a risk assessment of the Free Flight based air traffic operation adopted in [17]. Free Flight -sometimes referred to as Self Separation Assurance- is a concept where pilots are allowed to select their trajectory freely at real time, at the cost of acquiring responsibility for conflict prevention. It changes ATM in a fundamental way: the centralised control becomes a distributed one, responsibilities and tasks transfer from ground to air, ATC sectorization and routes are removed and new technologies are brought in. Before such concept can be implemented, it is necessary to determine its level of safety. The aim of this section is to illustrate the SDCPN^{imt} specification developed in [1] for a collision risk model for Free Flight.

7.1 LPNs of the Free Flight Air Traffic Example

In the Free Flight air traffic example, the airspace is an En-Route Airspace without fixed routes or an active ATC specifying routes. All aircraft flying in this airspace are assumed to be properly equipped and enabled for Free Flight: the pilots can try to optimise their trajectory, due to the enlarged freedom to choose path and flight level. The pilots are only limited by their responsibility to maintain airborne separation, in which they are assisted by a system called ASAS (Airborne Separation Assistance System). This can be considered as a system processing the information flows from the data-communication links between aircraft, the navigation systems and the aircraft guidance and control systems. ASAS detects conflicts, determines conflict resolution manoeuvres and presents the relevant information to the aircrew.

The number of agents involved in the Free Flight operation is huge and ranges from the Control Flow Management Unit to flight attendants. In the setting chosen for an initial risk assessment, the following agents are taken into account:

- A Pilot-Flying in each aircraft,
- A Pilot-Non-Flying in each aircraft,
- A number of systems and entities per aircraft, like the aircrafts position evolution and the Conflict Management Support systems,
- A number of global systems and entities, like the communication frequencies and the satellite system.

As explained in the beginning of Section 4, LPNs are specified for each relevant functional entity of each agent. It was judged sufficient to specify the following number of LPNs for the agents:

- 6 LPNs for each Pilot-Flying,
- 2 LPNs for each Pilot-Non-Flying,
- 36 LPNs for the systems and entities of each aircraft,
- 7 LPNs for the environment.

The actual number of LPNs in the whole model then depends on the number N of aircraft involved, and equals $7 + N \times (6 + 2 + 36)$.

7.2 Interconnected LPNs of Pilot Flying

This subsection illustrates, for the specific Free Flight air traffic example, a Petri Net model for the Pilot Flying as agent. A graphical representation of all LPNs the Pilot-Flying consists of, is given in Figure 15.

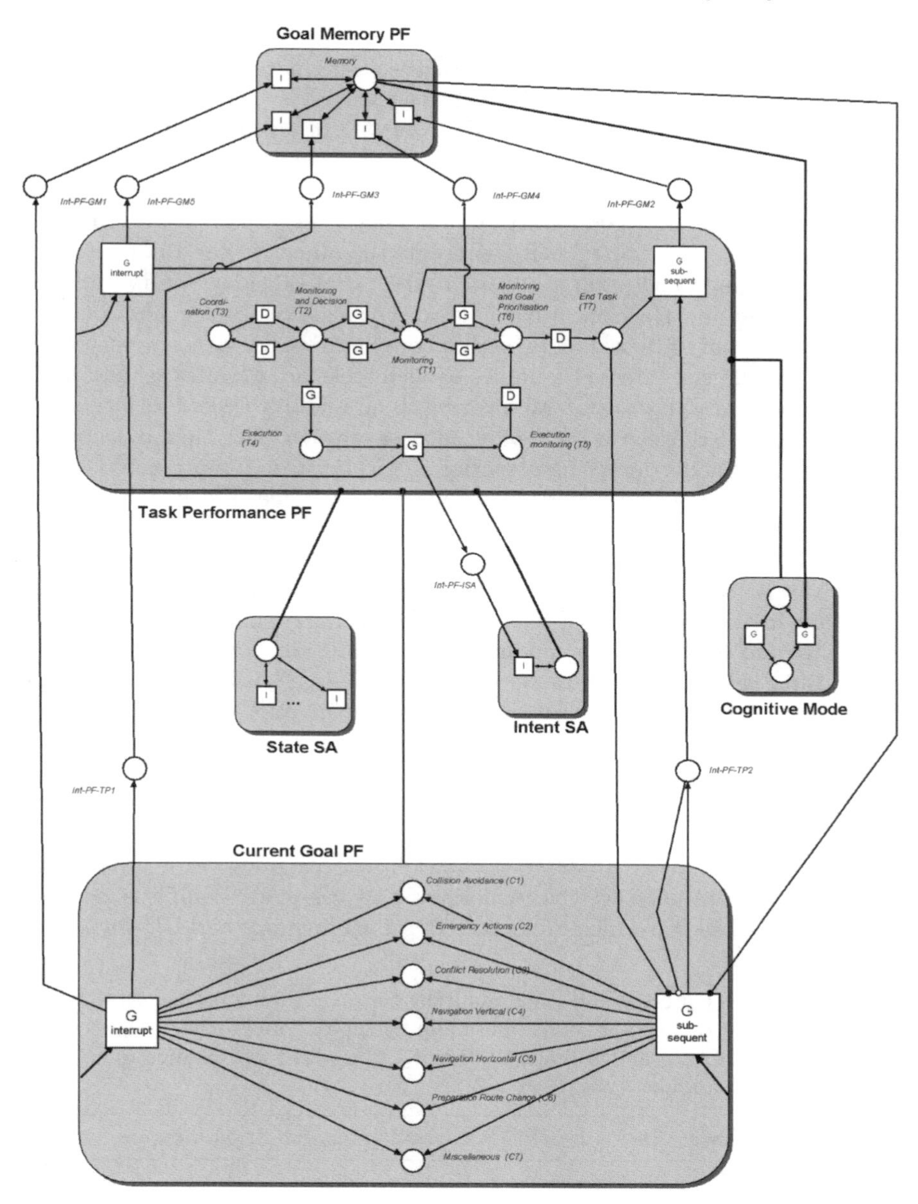

Fig. 15. The agent Pilot-Flying in Free Flight is modelled by 6 different LPNs, and a number of ordinary and enabling arcs and some IPNs, consisting of one place and input and output arcs

The Human-Machine-Interface where sound or visual clues might indicate that attention should be paid to a particular issue, is represented by a LPN that does not belong to the Pilot-Flying as agent and is therefore not depicted in the Figure. Similarly, the arcs to or from any other agent are not shown in Figure 15. Because of the very nature of Petri Nets, these arcs can easily be added during the follow-up specification cycle. To get an understanding of the different LPNs, a good starting point might be the LPN Current Goal (at the bottom of the figure) as it represents the objective the Pilot-Flying is currently working on. Examples of such goals are Collision Avoidance, Conflict Resolution and Horizontal Navigation. For each of these goals, the pilot executes a number of tasks in a prescribed or conditional order, represented in the LPN Task Performance. Examples of such tasks are Monitoring and Decision, Execution and Execution Monitoring. If all relevant tasks for the current goal are considered executed, the pilot chooses another goal, thereby using his memory (where goals deserving attention might be stored, represented by the LPN Goal Memory) and the Human-Machine-Interface. His memory where goals deserving attention might be stored is represented as the LPN Goal Memory in Figure 15.

So, the LPNs Current Goal, Task Performance, and Goal Memory are important in the modelling of which task the Pilot-Flying is executing. The other three LPNs are important in the modelling on how the Pilot-Flying is executing the tasks. The LPN State SA, where SA stands for Situation Awareness, represents the relevant perception of the pilot about the states of elements in his environment, e.g. whether he is aware of an engine failure. The LPN Intent SA represents the intent, e.g. whether he needs to leave the Free Flight Airspace. The LPN Cognitive mode represents whether the pilot is in an opportunistic mode, leading to a high but error-prone throughput, or in a tactical mode, leading to a moderate throughput with a low error probability.

There are many interactions (which, in some cases, are complex) between these individual LPNs, which are depicted as enabling arcs and IPNs with one

Table 1. Numbers of interconnection mapping types and Petri net elements before and after application of interconnection mapping types. The number of places (i.e. 19 places within LPNs and 8 places between LPNs) does not change due to the interconnection mapping types

Number of elements	In Figure 15	Without interconnection mapping types
Within LPNs	27 transitions	279 transitions
	66 arcs	642 arcs
Between LPNs	16 ordinary arcs	293 ordinary arcs
	7 enabling arcs	1023 enabling arcs
	1 inhibitor arc	7 inhibitor arcs
Total	117	2244

place only. The use of the new interconnection mapping types makes that the figure is still readable. Interconnection mapping types I, IV, V, VI and VII have not been used, while type II has been used 2×, type III 4×, and type VIII 3×. Table 1 shows that without the use of these interconnection mapping types the figure really would be cluttered with duplicated transitions and arcs within LPNs, and with connections drawn between LPNs.

8 Concluding Remarks

For the compositional specification of a multi-agent hybrid system this chapter has introduced a hierarchical extension of the compositional specification power of Petri nets, which avoids the need for all kinds of low level changes once making connections at a higher model level. Moreover the problem of combinatorial growth of the number of interconnections with the size of the Petri net is remedied. The effectiveness of the SDCPN based compositional specification is illustrated for an air traffic example of a taxiing aircraft crossing an active runway.

After a SDCPN^{imt} has been specified for a particular application, the next step is to analyse it, and to investigate and assess particular characteristics of the application. This can be done in various ways; for example, the DCPN^{imt} can be directly used as basis of a computer (e.g. Monte Carlo) simulation. Instead of this, or in addition to this, a particular SDCPN property can be used: In [7] it has been proven that, under a few conditions, a SDCPN is equivalent to a particular powerful subclass of hybrid state Markov process, named Generalised Stochastic Hybrid Process (GSHP), see 16-BujorianuLygerosGloverPola2003. Due to this equivalence, typical GSHP properties can be used to analyse the SDCPN^{imt}, even without elaborating the particular transformation from SDCPN^{imt} to GSHP for the application considered, and Monte Carlo simulations can be run which make use of GSHP mathematical properties. By those means, a collision risk model for Free Flight is instantiated using Petri-nets and the new interconnection mapping types. The usage of the new interconnection mapping types improves simplicity, readability and resilience against modelling errors.

References

1. G.J. Bakker, B. Klein Obbink, M.B. Klompstra, and H.A.P. Blom. DCPN specification of a free flight air traffic operation, working document. Technical report, National Aerospace Laboratory NLR, August 2004. HYBRIDGE WP9.2, Version 0.3, http://www.nlr.nl/public/hosted-sites/hybridge/.
2. H.A.P. Blom, M.B. Klompstra, and G.J. Bakker. Accident risk assessment of simultaneous converging instrument approaches. *Air Traffic Control Quarterly*, 11(2):123–155, 2003.

3. H.A.P. Blom, S.H. Stroeve, M.H.C. Everdij, and M.N.J. Van der Park. Human cognition performance model to evaluate safe spacing in air traffic. *Human Factors and Aerospace Safety*, 2:59–82, 2003.
4. R. David and H. Alla. Petri nets for the modeling of dynamic systems - a survey. *Automatica*, 30(2):175–202, 1994.
5. M.H.C. Everdij and H.A.P. Blom. Petri nets and hybrid state Markov processes in a power-hierarchy of dependability models. *Proc. IFAC Conference on Analysis and Design of Hybrid System, Saint-Malo, Brittany, France*, pages 355–360, June 2003.
6. M.H.C. Everdij and H.A.P. Blom. Piecewise deterministic Markov processes represented by dynamically coloured Petri nets. *Stochastics*, 77(1):1–29, February 2005.
7. M.H.C. Everdij and H.A.P. Blom. Representing Generalised Stochastic Hybrid Processes through Stochastically and Dynamically Coloured Petri Nets. In H.A.P. Blom and J. Lygeros, editors, *Stochastic Hybrid Systems: Theory and Applications.* HYBRIDGE final report PD23 for European Commission, 2005.
8. M.H.C. Everdij, H.A.P. Blom, and M.B. Klompstra. Dynamically Coloured Petri Nets for air traffic management safety purposes. *Proc. 8th IFAC Symposium on Transportation Systems, Chania, Greece*, pages 184–189, 1997.
9. M. Fernandes, M. Adamski, and A.J. Proença. Vhdl generation from hierarchical Petri net specifications of parallel controller. *IEE Proceedings: Computers and Digital Techniques*, 144:127–137, March 1997.
10. H. Fleischhack and B. Grahlmann. A Petri net semantics for $B(PN)^2$ with procedures. *Parallel and Distributed Software Engineering*, 1997.
11. N.O. Fota, M. Kaaniche, and K. Kanoun. A modular and incremental approach for building complex stochastic Petri net models. *Proc. First Int. Conf. on Mathematical Methods in Reliability*, 1997.
12. P.J. Haas. *Stochastic Petri Nets, Modelling, Stability, Simulation.* Springer-Verlag, New York, 2002.
13. D. Harel. Statecharts: a visual formalism for complex systems. *Science of Computer Programming*, 8:231–274, 1987.
14. P. Huber, K. Jensen, and R.M. Shapiro. Hierarchies in Coloured Petri Nets. In G. Rozenberg, editor, *Advances in Petri nets 1990. Lecture notes in Computer Science*, volume 483, pages 313–341. Springer, Berlin Heidelberg New York, 1990. Also Chapter 7 in: K. Jensen and G. Rozenberg (eds), High-level Petri Nets; Theory and Application, Springer-Verlag, 1991.
15. E. Kindler. A compositional partial order semantics for Petri net components. In Pierre Azema and Gianfranco Balbo, editors, *18th International Conference on Application and Theory of Petri nets, LNCS, 1248.* Springer-Verlag, June 1997.
16. P.E. Labeau, C. Smidts, and S. Swaminathan. Dynamic reliability: towards an integrated platform for probabilistic risk assessment. *Reliability Engineering and Systems Safety*, 68:219–254, 2000.
17. B. Klein Obbink. Description of advanced operation: Free flight. Technical report, National Aerospace Laboratory NLR, 2002, 2004. HYBRIDGE WP9.1 report, http://www.nlr.nl/public/hosted-sites/hybridge/.
18. Royal Society. *Risk assessment, Report of a Royal Society Study Group.* Royal Society, London, 1983.

A Sequential Particle Algorithm that Keeps the Particle System Alive

François LeGland[1] and Nadia Oudjane[2,3]

[1] IRISA / INRIA, Campus de Beaulieu, 35042 RENNES Cédex, France, legland@irisa.fr
[2] EDF, Division R&D, 92141 CLAMART Cédex, France, nadia.oudjane@edf.fr
[3] Département de Mathématiques, Université Paris XIII, 93430 VILLETANEUSE, France, n_oudjane@math.univ-paris13.fr

Summary. A sequential particle algorithm proposed by Oudjane (2000) is studied here, which uses an adaptive random number of particles at each generation and guarantees that the particle system never dies out. This algorithm is especially useful for approximating a nonlinear (normalized) Feynman–Kac flow, in the special case where the selection functions can take the zero value, e.g. in the simulation of a rare event using an *importance splitting* approach. Among other results, a central limit theorem is proved by induction, based on the result of Rényi (1957) for sums of a random number of independent random variables. An alternate proof is also given, based on an original central limit theorem for triangular arrays of martingale increments spread across generations with different random sizes.

1 Introduction

The problem considered here is the particle approximation of the linear (unnormalized) flow and of the associated nonlinear (normalized) flow, defined by

$$\langle \gamma_n, \phi \rangle = \mathbb{E}[\phi(X_n) \prod_{k=0}^{n} g_k(X_k)] \qquad \text{and} \qquad \langle \mu_n, \phi \rangle = \frac{\langle \gamma_n, \phi \rangle}{\langle \gamma_n, 1 \rangle} \ ,$$

for any bounded measurable function ϕ, where $\{X_k \, , \, k = 0, 1, \cdots, n\}$ is a Markov chain with initial probability distribution η_0 and transition kernels $\{Q_k \, , \, k = 1, \cdots, n\}$, and where $\{g_k \, , \, k = 0, 1, \cdots, n\}$ are given bounded measurable nonnegative functions, known as selection or fitness functions. This problem has been widely studied [3] and the focus here is on the special case where the selection functions can possibly take the zero value. Clearly

$$\gamma_k = g_k \, (\gamma_{k-1} \, Q_k) = g_k \, (\mu_{k-1} \, Q_k) \, \langle \gamma_{k-1}, 1 \rangle \qquad \text{and} \qquad \gamma_0 = g_0 \, \eta_0 \ , \qquad (1)$$

or equivalently $\gamma_k = \gamma_{k-1} \, R_k$ where the nonnegative (unnormalized) kernel R_k is defined by $R_k(x, dx') = Q_k(x, dx') \, g_k(x')$, hence

H.A.P. Blom, J. Lygeros (Eds.): Stochastic Hybrid Systems, LNCIS 337, pp. 351–389, 2006.

$$\langle \gamma_k, 1 \rangle = \langle \mu_{k-1}\, Q_k, g_k \rangle \, \langle \gamma_{k-1}, 1 \rangle \qquad \text{and} \qquad \langle \gamma_0, 1 \rangle = \langle \eta_0, g_0 \rangle \;, \tag{2}$$

and the minimal assumption made throughout this paper is that $\langle \gamma_n, 1 \rangle > 0$, or equivalently that $\langle \eta_0, g_0 \rangle > 0$ and $\langle \mu_{k-1}\, Q_k, g_k \rangle > 0$ for any $k = 1, \cdots, n$, otherwise the problem is not well defined. There are many practical situations where the selection functions can possibly take the zero value

- simulation of a rare event using an *importance splitting* approach [3, Section 12.2], [1, 6],
- simulation of a Markov chain conditionned or constrained to visit a given sequence of subspaces of the state space (this includes tracking a mobile in the presence of obstacles : when the mobile is hidden behind an obstacle, occlusion occurs and no observation is available at all, however this information can still be used, with a selection function equal to the indicator function of the region hidden by the obstacle),
- simulation of a r.v. in the tail of a given probability distribution,
- nonlinear filtering with bounded observation noise,
- implementation of a *robustification* approach in nonlinear filtering, using truncation of the likelihood function [10, 16],
- algorithms of approximate nonlinear filtering, where hidden state and observation are simulated jointly, and where the simulated observation is validated against the actual observation [4, 5, 18, 19], e.g. when there is no explicit expression available for the likelihood function, or when a likelihood function does not even exist (nonadditive observation noise, noise–free observations, etc.).

This work has been announced in [12], and it is organized as follows. In Section 2, the (usual) nonsequential particle algorithm is presented, and the potential difficulty that arises if the selection functions can possibly take the zero value, i.e. the possible extinction of the particle system, is addressed. Refined $\mathbb{L}^1$ error estimates are stated in Theorem 1, for the purpose of comparison with the sequential particle algorithm, and the central limit theorem proved in [3, Section 9.4] is recalled. In Section 3, the sequential particle algorithm already proposed in [15, 11] is introduced, which uses an adaptive random number of particles at each generation and automatically keeps the particle system alive, i.e. which ensures its non–extinction. The main contributions of this work are $\mathbb{L}^1$ error estimates stated in Theorem 3 and a central limit theorem stated in Theorem 4. An interesting feature of the sequential particle algorithm is that a fixed performance can be guaranteed in advance, at the expense of a random computational effort : this could be seen as an adaptive rule to automatically choose the number of particles. To get a fair comparison of the nonsequential and sequential particle algorithms, the time–averaged random number of simulated particles, which is an indicator of how much computational effort has been used, is used as a normalizing factor. The different behaviour of the two particle algorithms is illustrated on the simple example of binary selection functions (taking only the value 0 or 1). The proof of Theorem 4 relies on results stated in Section 4 for sums of a random number of i.i.d. random variables, especially when this random number is a stopping time. A conditional version of the central limit theorem, known in sequential analysis as the Anscombe theorem and proved in [17], is stated in Theorem 6, and a central limit theorem for triangular arrays of martingale increments spread across generations with different random sizes, is stated in Theorem 7.

The remaining part of this work is devoted to proofs of the main results. The central limit theorem for the sequential particle algorithm, stated in Theorem 4, is proved in Section 5 by induction, based on the central limit theorem stated in Theorem 6 for sums of a random number of i.i.d. random variables, and an alternate proof is given in Section 6, based on the central limit theorem stated in Theorem 7 for triangular arrays of martingale increments spread across generations with different random sizes. Finally, Theorems 6 and 7 are proved in Appendices A and B respectively. Further details, including the proofs of Theorems 1 and 3 can be found in [13, Sections 5 and 6].

2 Nonsequential Particle Algorithm

The evolution of the normalized (nonlinear) flow $\{\mu_k \, , \, k = 0, 1, \cdots, n\}$ is described by the following diagram

$$\mu_{k-1} \longrightarrow \eta_k = \mu_{k-1} \, Q_k \longrightarrow \mu_k = g_k \cdot \eta_k \ ,$$

with initial condition $\mu_0 = g_0 \cdot \eta_0$, where the notation $\cdot$ denotes the projective product. It follows from (2) and from the definition that

$$\mathbb{E}[\prod_{k=0}^n g_k(X_k)\,] = \langle \gamma_n, 1 \rangle = \prod_{k=0}^n \langle \eta_k, g_k \rangle \ ,$$

i.e. the expectation of a product is replaced by the product of expectations. Notice that the ratio

$$\rho_k = \frac{\sup_{x \in E} g_k(x)}{\langle \eta_k, g_k \rangle}$$

is an indicator of how difficult a given problem is : indeed, a large value of ρ_k means that regions where the selection function g_k is large have a small probability under η_k. The idea behind the particle approach is to look for an approximation

$$\mu_k \approx \mu_k^N = \sum_{i=1}^N w_k^i \, \delta_{\xi_k^i} \ ,$$

in the form of the weighted empirical probability distribution associated with the particle system $(\xi_k^i, w_k^i \, , \, i = 1, \cdots, N)$, where N denotes the number of particles. The weights and positions of the particles are chosen is such a way that the evolution of the approximate sequence $\{\mu_k^N \,,\, k=0,1,\cdots,n\}$ is described by the following diagram

$$\mu_{k-1}^N \longrightarrow \eta_k^N = S^N(\mu_{k-1}^N\, Q_k) \longrightarrow \mu_k^N = g_k \cdot \eta_k^N \;,$$

with initial condition defined by $\mu_0^N = g_0 \cdot \eta_0^N$ and $\eta_0^N = S^N(\eta_0)$, where the notation $S^N(\mu)$ denotes the empirical probability distribution associated with an N–sample with common probability distribution μ. In practice, particles

- are selected according to their respective weights $(w_{k-1}^i \,,\, i=1,\cdots,N)$ (selection step),
- move according to the Markov kernel Q_k (mutation step),
- are weighted by evaluating the fitness function g_k (weighting step).

Starting from (1) and introducing the particle approximation

$$\gamma_k^N = g_k\; S^N(\mu_{k-1}^N\, Q_k)\; \langle \gamma_{k-1}^N, 1\rangle = g_k\, \eta_k^N\; \langle \gamma_{k-1}^N, 1\rangle \;,$$

and

$$\gamma_0^N = g_0\; S^N(\eta_0) = g_0\, \eta_0^N \;,$$

for the unnormalized (linear) flow, it is easily seen that

$$\langle \gamma_k^N \,, 1\rangle = \langle \eta_k^N \,, g_k\rangle\; \langle \gamma_{k-1}^N, 1\rangle \qquad \text{and} \qquad \langle \gamma_0^N \,, 1\rangle = \langle \eta_0^N \,, g_0\rangle \;, \tag{3}$$

hence

$$\frac{\gamma_k^N}{\langle \gamma_k^N \,, 1\rangle} = g_k \cdot \eta_k^N = \mu_k^N \qquad \text{and} \qquad \frac{\gamma_0^N}{\langle \gamma_0^N \,, 1\rangle} = g_0 \cdot \eta_0^N = \mu_0^N \;.$$

However, if the function g_k can possibly take the zero value, and even if $\langle \eta_k, g_k\rangle > 0$, it can happen that $\langle \eta_k^N \,, g_k\rangle = 0$, i.e. it can happen that the evaluation of the function g_k returns the zero value for all the particles generated at the end of the mutation step : in such a situation, the particle systems dies out and the algorithm cannot continue. A reinitialization procedure has been proposed and studied in [5], in which the particle system is generated afresh from an arbitrary restarting probability distribution ν. Alternatively, one could be interested by the behavior of the algorithm until the extinction time of the particle system, defined by

$$\tau^N = \inf\{k \geq 0 \;:\; \langle \eta_k^N \,, g_k\rangle = 0\} \;.$$

Under the assumption that $\langle \gamma_n, 1\rangle > 0$, the probability $\mathbb{P}[\tau^N \leq n]$ that the algorithm cannot continue up to the time instant n goes to zero with exponential rate [3, Theorem 7.4.1].

Example 1 (Binary selection). In the special case of binary selection functions (taking only the value 0 or 1), i.e. indicator functions $g_k = 1_{A_k}$ of Borel subsets for any $k=0,1,\cdots,n$, it holds

$$p_0 = \langle \eta_0, g_0 \rangle = \mathbb{P}[X_0 \in A_0] \ ,$$

and

$$p_k = \langle \eta_k, g_k \rangle = \mathbb{P}[X_k \in A_k \mid X_0 \in A_0, \cdots, X_{k-1} \in A_{k-1}] \ ,$$

for any $k = 1, \cdots, n$, and it follows from (2) that

$$P_n = \mathbb{P}[X_0 \in A_0, \cdots, X_n \in A_n] = \langle \gamma_n, 1 \rangle = \prod_{k=0}^{n} \langle \eta_k, g_k \rangle \ .$$

On the *good set* $\{\tau^N > n\}$, the nonsequential particle algorithm results in the following approximations

$$p_k \approx p_k^N = \langle \eta_k^N, g_k \rangle = \frac{|I_k^N|}{N} \qquad \text{where} \qquad I_k^N = \{i = 1, \cdots, N \ : \ \xi_k^i \in A_k\} \ ,$$

denotes the set of successful particles within an N–sample with common probability distribution η_0 (for $k = 0$) and $\mu_{k-1}^N Q_k$ (for $k = 1, \cdots, n$), and it follows from (3) that

$$P_n \approx P_n^N = \langle \gamma_n^N, 1 \rangle = \prod_{k=0}^{n} \langle \eta_k^N, g_k \rangle = \prod_{k=0}^{n} \frac{|I_k^N|}{N} \ .$$

In other words, the probability P_n of a successful sequence is approximated as the product of the fraction of successful particles at each generation, and each transition probability p_k separately is approximated as the fraction of successful particles at the corresponding generation. Notice that the computational effort, i.e. the number N of simulated particles at each generation, is *fixed* in advance, whereas the number $|I_k^N|$ of successful particles at the k–th generation is *random*, and could even be zero.

The following results have been obtained for the nonsequential particle algorithm with a constant number N of particles : a nonasymptotic estimate [3, Theorem 7.4.3]

$$\sup_{\phi \,:\, \|\phi\|=1} \mathbb{E}|\, 1_{\{\tau^N > n\}} \, \langle \mu_n^N, \phi \rangle - \langle \mu_n, \phi \rangle \,| \le \frac{c_n^0}{\sqrt{N}} + \mathbb{P}[\tau^N \le n] \ ,$$

and a central limit theorem (see [3, Section 9.4] for a slightly different algorithm)

$$\sqrt{N}\, [\, 1_{\{\tau^N > n\}} \, \langle \mu_n^N, \phi \rangle - \langle \mu_n, \phi \rangle \,] \Longrightarrow \mathcal{N}(0, v_n^0(\phi)) \ ,$$

in distribution as $N \uparrow \infty$, with an explicit expression for the asymptotic variance. In the simple case where the fitness functions are positive, i.e. cannot take the zero value, these results are well–known and can be found in [7, Proposition 2.9, Corollary 2.20], where the proof relies on a central limit theorem for triangular arrays of martingale increments, or in [9, Theorem 4], where the same central limit theorem is obtained by induction.

For the purpose of comparison with the sequential particle algorithm, the following nonasymptotic error estimates are proved in [13, Section 5].

Theorem 1. *With the extinction time τ^N defined by*

$$\tau^N = \inf\{k \geq 0 \, : \, \langle \eta_k^N , g_k \rangle = 0\} \; ,$$

it holds

$$\mathbb{E} | \, 1_{\{\tau^N \, > \, n\}} \, \frac{\langle \gamma_n^N , 1 \rangle}{\langle \gamma_n , 1 \rangle} - 1 \, | \leq z_n^N + \mathbb{P}[\tau^N \leq n] \; , \tag{4}$$

and

$$\sup_{\phi \, : \, \|\phi\|=1} \mathbb{E} | \, 1_{\{\tau^N \, > \, n\}} \, \langle \mu_n^N , \phi \rangle - \langle \mu_n , \phi \rangle \, | \leq 2 \, z_n^N + \mathbb{P}[\tau^N \leq n] \; , \tag{5}$$

where the sequence $\{z_k^N \, , \, k = 0, 1, \cdots , n\}$ satisfies the linear recursion

$$z_k^N \leq \rho_k \, (1 + \frac{\sqrt{\rho_k}}{\sqrt{N}}) \, z_{k-1}^N + \frac{\sqrt{\rho_k}}{\sqrt{N}} \qquad \text{and} \qquad z_0^N \leq \frac{\sqrt{\rho_0}}{\sqrt{N}} \; . \tag{6}$$

Remark 1. The forcing term in (6) is $\sqrt{\rho_k}/\sqrt{N}$, and

$$\limsup_{N \uparrow \infty} \, [\sqrt{N} \, z_k^N] \leq \rho_k \, \limsup_{N \uparrow \infty} \, [\sqrt{N} \, z_{k-1}^N] + \sqrt{\rho_k} \; ,$$

and

$$\limsup_{N \uparrow \infty} \, [\sqrt{N} \, z_0^N] \leq \sqrt{\rho_0} \; .$$

Notice that with a fixed number N of simulated particles, the performance is $\sqrt{\rho_k}/\sqrt{N}$ and depends on ρ_k : as a result, it is not possible to guarantee in advance a fixed performance, since ρ_k is not known.

For completeness, the central limit theorem obtained in [3, Section 9.4] for a slightly different algorithm is recalled below.

Theorem 2 (Del Moral). *With the extinction time τ^N defined by*

$$\tau^N = \inf\{k \geq 0 \, : \, \langle \eta_k^N , g_k \rangle = 0\} \; ,$$

it holds

$$\sqrt{N} \, [\, 1_{\{\tau^N \, > \, n\}} \, \frac{\langle \gamma_n^N , 1 \rangle}{\langle \gamma_n , 1 \rangle} - 1 \,] \Longrightarrow \mathcal{N}(0, V_n^0) \; ,$$

and

$$\sqrt{N} \, [\, 1_{\{\tau^N \, > \, n\}} \, \langle \mu_n^N , \phi \rangle - \langle \mu_n , \phi \rangle \,] \Longrightarrow \mathcal{N}(0, v_n^0(\phi)) \; ,$$

in distribution as $N \uparrow \infty$, for any bounded measurable function ϕ, with the asymptotic variance

$$V_n^0 = \sum_{k=0}^n \frac{\mathrm{var}(g_k \, R_{k+1:n} \, 1, \eta_k)}{\langle \eta_k , g_k \, R_{k+1:n} \, 1 \rangle^2} \; ,$$

and

$$v_n^0(\phi) = \sum_{k=0}^{n} \frac{\mathrm{var}(g_k\, R_{k+1:n}\, (\phi - \langle \mu_n, \phi \rangle), \eta_k)}{\langle \eta_k, g_k\, R_{k+1:n}\, 1 \rangle^2} \ ,$$

respectively, where

$$R_{k+1:n}\, \phi(x) = R_{k+1} \cdots R_n\, \phi(x) = \mathbb{E}[\phi(X_n) \prod_{p=k+1}^{n} g_p(X_p) \mid X_k = x] \ ,$$

for any $k = 0, 1, \cdots, n$, *with the convention* $R_{n+1:n}\, \phi(x) = \phi(x)$, *for any* $x \in E$.

Remark 2. Notice that

$$\langle \eta_0, g_0\, R_{1:n}\, (\phi - \langle \mu_n, \phi \rangle)\, \rangle = \langle \gamma_0\, R_{1:n}, \phi - \langle \mu_n, \phi \rangle\, \rangle = \langle \gamma_n, \phi - \langle \mu_n, \phi \rangle\, \rangle = 0 \ ,$$

and

$$\langle \eta_k, g_k\, R_{k+1:n}\, (\phi - \langle \mu_n, \phi \rangle)\, \rangle = \langle \mu_{k-1}\, R_{k:n}, \phi - \langle \mu_n, \phi \rangle\, \rangle = \frac{\langle \gamma_n, \phi - \langle \mu_n, \phi \rangle\, \rangle}{\langle \gamma_{k-1}, 1 \rangle} = 0 \, ,$$

for any $k = 1, \cdots, n$, hence the following equivalent expression holds for the asymptotic variance

$$v_n^0(\phi) = \sum_{k=0}^{n} \frac{\langle \eta_k, |g_k\, R_{k+1:n}\, [\phi - \langle \mu_n, \phi \rangle]\, |^2\, \rangle}{\langle \eta_k, g_k\, R_{k+1:n}\, 1 \rangle^2} \ .$$

Example 2 (Binary selection). In the special case of binary selection functions, it holds

$$R_{k+1:n}\, 1(x) = \mathbb{P}[X_{k+1} \in A_{k+1}, \cdots, X_n \in A_n \mid X_k = x] \ ,$$

for any $k = 0, 1, \cdots, n$, with the convention $R_{n+1:n}\, 1(x) = 1$, for any $x \in E$, and it follows from Theorem 2 that

$$\sqrt{N}\, [\, 1_{\{\tau^N > n\}}\, \frac{P_n^N}{P_n} - 1] \Longrightarrow \mathcal{N}(0, V_n^0) \ ,$$

in distribution as $N \uparrow \infty$, with the asymptotic variance

$$V_n^0 = \sum_{k=0}^{n} (\frac{1}{p_k} - 1) + \sum_{k=0}^{n} \frac{1}{p_k}\, \frac{\mathrm{var}(R_{k+1:n}\, 1, \mu_k)}{\langle \mu_k, R_{k+1:n}\, 1 \rangle^2} \ .$$

Indeed, since $g_k^2 = g_k$, it holds

$$\begin{aligned}\frac{\mathrm{var}(g_k\, R_{k+1:n}\, 1, \eta_k)}{\langle \eta_k, g_k\, R_{k+1:n}\, 1\rangle^2} &= \frac{\langle \eta_k, g_k\, |R_{k+1:n}\, 1|^2\rangle}{\langle \eta_k, g_k\, R_{k+1:n}\, 1\rangle^2} - 1 \\ &= \frac{1}{p_k}\, \frac{\langle \mu_k, |R_{k+1:n}\, 1|^2\rangle}{\langle \mu_k, R_{k+1:n}\, 1\rangle^2} - 1 \\ &= (\frac{1}{p_k} - 1) + \frac{1}{p_k}\, [\, \frac{\langle \mu_k, |R_{k+1:n}\, 1|^2\rangle}{\langle \mu_k, R_{k+1:n}\, 1\rangle^2} - 1]\ ,\end{aligned}$$

for any $k = 0, 1, \cdots, n$.

3 Sequential Particle Algorithm

The purpose of this work is to study a sequential particle algorithm, already proposed in [15, 11], which automatically keeps the particle system alive, i.e. which ensures its non–extinction. For any level $H > 0$, and for any $k = 0, 1, \cdots, n$, define the random number of particles

$$N_k^H = \inf\{N \geq 1\, :\, \sum_{i=1}^{N} g_k(\xi_k^i) \geq H\, \sup_{x \in E} g_k(x)\}\ ,$$

where the random variables $\xi_0^1, \cdots, \xi_0^i, \cdots$ are i.i.d. with common probability distribution η_0 (for $k = 0$), and where, conditionally w.r.t. the σ–algebra $\mathcal{H}_{k-1}^H$ generated by the particle system until the $(k-1)$–th generation, the random variables $\xi_k^1, \cdots, \xi_k^i, \cdots$ are i.i.d. with common probability distribution $\mu_{k-1}^H\, Q_k$ (for $k = 1, \cdots, n$). The particle approximation $\{\mu_k^H\, ,\, k = 0, 1, \cdots, n\}$ is now parameterized by the level $H > 0$, and its evolution is described by the following diagram

$$\mu_{k-1}^H \xrightarrow{\hspace{2cm}} \eta_k^H = S^{N_k^H}(\mu_{k-1}^H\, Q_k) \xrightarrow{\hspace{2cm}} \mu_k^H = g_k \cdot \eta_k^H\ ,$$

with initial condition defined by $\mu_0^H = g_0 \cdot \eta_0^H$ and $\eta_0^H = S^{N_0^H}(\eta_0)$. Starting from (1) and introducing the particle approximation

$$\gamma_k^H = g_k\, S^{N_k^H}(\mu_{k-1}^H\, Q_k)\, \langle \gamma_{k-1}^H, 1\rangle = g_k\, \eta_k^H\, \langle \gamma_{k-1}^H, 1\rangle\ ,$$

and

$$\gamma_0^H = g_0\, S^{N_0^H}(\eta_0) = g_0\, \eta_0^H\ ,$$

for the unnormalized (linear) flow, it is easily seen that

$$\langle \gamma_k^H, 1\rangle = \langle \eta_k^H, g_k\rangle\, \langle \gamma_{k-1}^H, 1\rangle \qquad \text{and} \qquad \langle \gamma_0^H, 1\rangle = \langle \eta_0^H, g_0\rangle\ , \tag{7}$$

hence

$$\frac{\gamma_k^H}{\langle \gamma_k^H, 1 \rangle} = g_k \cdot \eta_k^H = \mu_k^H \qquad \text{and} \qquad \frac{\gamma_0^H}{\langle \gamma_0^H, 1 \rangle} = g_0 \cdot \eta_0^H = \mu_0^H \ .$$

Clearly, $N_k^H \geq H$ and if $\langle \mu_{k-1}^H \, Q_k, g_k \rangle > 0$ — a sufficient condition for which is

$$\widehat{g}_k(x) = Q_k \, g_k(x) = \mathbb{E}[g_k(X_k) \mid X_{k-1} = x] > 0 \ ,$$

for any x in the support of μ_{k-1}^H — then the random number N_k^H of particles is a.s. finite, see Section 4 below. Moreover

$$\langle \eta_0^H, g_0 \rangle = \langle S^{N_0^H}(\eta_0), g_0 \rangle = \frac{1}{N_0^H} \sum_{i=1}^{N_0^H} g_0(\xi_0^i) \geq \frac{H}{N_0^H} \sup_{x \in E} g_0(x) > 0 \ ,$$

and

$$\langle \eta_k^H, g_k \rangle = \langle S^{N_k^H}(\mu_{k-1}^H \, Q_k), g_k \rangle = \frac{1}{N_k^H} \sum_{i=1}^{N_k^H} g_k(\xi_k^i) \geq \frac{H}{N_k^H} \sup_{x \in E} g_k(x) > 0 \ ,$$

for any $k = 1, \cdots, n$, i.e. the particle system never dies out and the algorithm can always continue, by construction.

Remark 3. It follows from Lemma 3 below that $\dfrac{N_0^H}{H} \longrightarrow \rho_0$ in probability, and in view of Remark 8 (ii) below, if $\langle \mu_{k-1}^H \, Q_k, g_k \rangle > 0$ then $\dfrac{N_k^H}{H \, \rho_k^H} \longrightarrow 1$ in probability as $H \uparrow \infty$, where

$$\rho_k^H = \frac{\sup_{x \in E} g_k(x)}{\langle \mu_{k-1}^H \, Q_k, g_k \rangle} \ ,$$

for any $k = 1, \cdots, n$.

Remark 4. For any $k = 0, 1, \cdots, n$ and any integer $i \geq 1$, let $\mathcal{F}_{k,i}^H = \mathcal{F}_{k,0}^H \vee \sigma(\xi_k^1, \cdots, \xi_k^i)$, where $\mathcal{F}_{0,0}^H = \{\emptyset, \Omega\}$ (for $k = 0$) and $\mathcal{F}_{k,0}^H = \mathcal{H}_{k-1}^H$ (for $k = 1, \cdots, n$) by convention. The random number N_k^H is a stopping time w.r.t. $\mathcal{F}_k^H = \{\mathcal{F}_{k,i}^H \, , \, i \geq 0\}$, which allows to define the σ–algebra $\mathcal{F}_{k,N_k^H}^H = \mathcal{H}_k^H$: clearly N_k^H is measurable w.r.t. $\mathcal{H}_k^H$, and therefore the random variable

$$\sigma_k^H = N_0^H + \cdots + N_k^H \ ,$$

is measurable w.r.t. $\mathcal{H}_k^H$.

Example 3 (Binary selection). In the special case of binary selection functions, the sequential particle algorithm results in the following approximations

$$p_k \approx p_k^H = \langle \eta_k^H, g_k \rangle = \frac{H}{N_k^H} \qquad \text{where} \qquad N_k^H = \inf\{N \geq 1 \, : \, |I_k^N| = H\} \ ,$$

for any integer $H \geq 1$, and where for any integer $N \geq 1$

$$I_k^N = \{i = 1, \cdots, N \; : \; \xi_k^i \in A_k\} \; ,$$

denotes the set of successful particles within an N–sample with common probability distribution η_0 (for $k = 0$) and $\mu_{k-1}^H Q_k$ (for $k = 1, \cdots, n$), and it follows from (7) that

$$P_n \approx P_n^H = \langle \gamma_n^H, 1 \rangle = \prod_{k=0}^{n} \langle \eta_k^H, g_k \rangle = \prod_{k=0}^{n} \frac{H}{N_k^H} \; .$$

Notice that the approximation $\mu_k^H = g_k \cdot \eta_k^H$ obtained here is exactly the empirical probability distribution associated with an H–sample that would be obtained using the rejection method, with common probability distribution $g_0 \cdot \eta_0$ (for $k = 0$) and $g_k \cdot (\mu_{k-1}^H Q_k)$ (for $k = 1, \cdots, n$). Here again, the probability P_n of a successful sequence is approximated as the product of the fraction of successful particles at each generation, and each transition probability p_k separately is approximated as the fraction of successful particles at the corresponding generation. In opposition to the nonsequential particle algorithm, notice that the number H of successful particles at each generation is *fixed* in advance, whereas the computational effort, i.e. the number N_k^H of simulated particles needed to get H successful particles exactly at the k–th generation, is *random*.

The main contributions of this paper are the following results for the sequential particle algorithm with a random number of particles, defined by the level $H > 0$: a nonasymptotic estimate (which was already obtained in [11, Theorem 5.4] in a different context), see Theorem 3 below

$$\sup_{\phi \, : \, \|\phi\|=1} \mathbb{E}| \, \langle \mu_n^H - \mu_n, \phi \rangle \, | \leq \frac{c_n}{\sqrt{H}} \; ,$$

and a central limit theorem, see Theorem 4 below

$$\sqrt{H} \, \langle \mu_n^H - \mu_n, \phi \rangle \Longrightarrow \mathcal{N}(0, v_n(\phi)) \; ,$$

in distribution as $H \uparrow \infty$, with an explicit expression for the asymptotic variance.

Theorem 3. *If $\langle \mu_{k-1}^H Q_k, g_k \rangle > 0$ for any $k = 1, \cdots, n$ — a sufficient condition for which is*

$$\widehat{g}_k(x) = Q_k \, g_k(x) = \mathbb{E}[g_k(X_k) \mid X_{k-1} = x] > 0 \; ,$$

for any x in the support of μ_{k-1}^H — then

$$\mathbb{E}| \, \frac{\langle \gamma_n^H, 1 \rangle}{\langle \gamma_n, 1 \rangle} - 1 \, | \leq z_n^H \qquad \text{and} \qquad \sup_{\phi \, : \, \|\phi\|=1} \mathbb{E}| \, \langle \mu_n^H - \mu_n, \phi \rangle \, | \leq 2 \, z_n^H \; , \tag{8}$$

where the sequence $\{z_k^H \, , \, k = 0, 1, \cdots, n\}$ *satisfies the linear recursion*

$$z_k^H \leq \rho_k \, (1 + \omega_H + \omega_H^2) \, z_{k-1}^H + \omega_H \, (1 + \omega_H \, \rho_k) \ , \tag{9}$$

and

$$z_0^H \leq \omega_H \, (1 + \omega_H \, \rho_0) \ ,$$

where $\omega_H = \dfrac{1}{H} \, \sqrt{H+1}$ *is of order* $1/\sqrt{H}$.

The proof of Theorem 3 can be found in [13, Section 6].

Remark 5. Up to higher order terms, the forcing term in (9) is $1/\sqrt{H}$ exactly (which is equivalent to $\sqrt{\rho_k^H} / \sqrt{N_k^H}$, in view of Remark 3), and

$$\limsup_{H \uparrow \infty} \, [\sqrt{H} \, z_k^H] \leq \rho_k \, \limsup_{H \uparrow \infty} \, [\sqrt{H} \, z_{k-1}^H] + 1 \qquad \text{and} \qquad \limsup_{H \uparrow \infty} \, [\sqrt{H} \, z_0^H] \leq 1 \, .$$

In opposition to the nonsequential particle algorithm, notice that it is possible here to guarantee in advance a fixed performance of $1/\sqrt{H}$ exactly, without any knowledge of ρ_k, at the expense of using an adaptive random number N_k^H of simulated particles : this could be seen as an adaptive rule to automatically choose the number of particles.

Remark 6. It follows from Theorem 3 that $\langle \mu_{k-1}^H \, Q_k, g_k \rangle \to \langle \eta_k, g_k \rangle$ in probability, hence $\rho_k^H \to \rho_k$ in probability, for any $k = 1, \cdots, n$, and it follows from Remark 3 that $\dfrac{N_k^H}{H} \to \rho_k$ in probability as $H \uparrow \infty$, for any $k = 0, 1, \cdots, n$.

Theorem 4. *If* $\langle \mu_{k-1}^H \, Q_k, g_k \rangle > 0$ *for any* $k = 1, \cdots, n$ *— a sufficient condition for which is*

$$\widehat{g}_k(x) = Q_k \, g_k(x) = \mathbb{E}[g_k(X_k) \mid X_{k-1} = x] > 0 \ ,$$

for any x *in the support of* μ_{k-1}^H *— then*

$$\sqrt{H} \, [\, \frac{\langle \gamma_n^H, 1 \rangle}{\langle \gamma_n, 1 \rangle} - 1 \,] \Longrightarrow \mathcal{N}(0, V_n) \ , \tag{10}$$

and

$$\sqrt{H} \, \langle \mu_n^H - \mu_n, \phi \rangle \Longrightarrow \mathcal{N}(0, v_n(\phi)) \ , \tag{11}$$

in distribution as $H \uparrow \infty$, *for any bounded measurable function* ϕ, *with the asymptotic variance*

$$V_n = \sum_{k=0}^{n} \frac{\mathrm{var}(g_k \, R_{k+1:n} \, 1, \eta_k)}{\langle \eta_k, g_k \, R_{k+1:n} \, 1 \rangle^2} \, \frac{1}{\rho_k} \ ,$$

and

$$v_n(\phi) = \sum_{k=0}^{n} \frac{\mathrm{var}(g_k\, R_{k+1:n}\, (\phi - \langle \mu_n, \phi \rangle), \eta_k)}{\langle \eta_k, g_k\, R_{k+1:n}\, 1 \rangle^2} \, \frac{1}{\rho_k} \; ,$$

respectively, where

$$R_{k+1:n}\, \phi(x) = R_{k+1} \cdots R_n\, \phi(x) = \mathbb{E}[\phi(X_n) \prod_{p=k+1}^{n} g_p(X_p) \mid X_k = x] \; ,$$

for any $k = 0, 1, \cdots, n$, *with the convention* $R_{n+1:n}\, \phi(x) = \phi(x)$, *for any* $x \in E$.

In view of Remark 2, the following equivalent expression holds for the asymptotic variance

$$v_n(\phi) = \sum_{k=0}^{n} \frac{\langle \eta_k, |g_k\, R_{k+1:n}\, [\phi - \langle \mu_n, \phi \rangle]\, |^2 \rangle}{\langle \eta_k, g_k\, R_{k+1:n}\, 1 \rangle^2} \, \frac{1}{\rho_k} \; .$$

Remark 7. To prove Theorem 4, it is enough to prove that

$$\sqrt{H}\, \frac{\langle \gamma_n^H - \gamma_n, \phi \rangle}{\langle \gamma_n, 1 \rangle} \Longrightarrow \mathcal{N}(0, V_n(\phi)) \; , \tag{12}$$

for any bounded measurable function ϕ, where the asymptotic variance $V_n(\phi)$ is defined by

$$V_n(\phi)\, \langle \gamma_n, 1 \rangle^2$$

$$= \mathrm{var}(g_0\, R_{1:n}\, \phi, \eta_0)\, \frac{1}{\rho_0} + \sum_{k=1}^{n} \mathrm{var}(g_k\, R_{k+1:n}\, \phi, \eta_k)\, \frac{\langle \gamma_{k-1}, 1 \rangle^2}{\rho_k} \; , \tag{13}$$

or equivalently by

$$V_n(\phi) = \sum_{k=0}^{n} \frac{\mathrm{var}(g_k\, R_{k+1:n}\, \phi, \eta_k)}{\langle \eta_k, g_k\, R_{k+1:n}\, 1 \rangle^2} \, \frac{1}{\rho_k} \; ,$$

since

$$\langle \gamma_n, 1 \rangle = \langle \gamma_0\, R_{1:n}, 1 \rangle = \langle \eta_0, g_0\, R_{1:n}\, 1 \rangle \; ,$$

and since

$$\langle \gamma_n, 1 \rangle = \langle \gamma_{k-1}\, R_{k:n}, 1 \rangle = \langle \gamma_{k-1}, 1 \rangle\, \langle \eta_k, g_k\, R_{k+1:n}\, 1 \rangle \; ,$$

for any $k = 1, \cdots, n$. Indeed, notice that

$$\langle \mu_n^H - \mu_n, \phi \rangle = \langle \frac{\gamma_n^H}{\langle \gamma_n^H, 1 \rangle}, \phi - \langle \mu_n, \phi \rangle \, \rangle = \frac{\langle \gamma_n, 1 \rangle}{\langle \gamma_n^H, 1 \rangle}\, \langle \frac{\gamma_n^H - \gamma_n}{\langle \gamma_n, 1 \rangle}, \phi - \langle \mu_n, \phi \rangle \, \rangle \; ,$$

for any bounded measurable function ϕ, and it follows from Theorem 3 that $\langle \gamma_n^H, 1 \rangle \to \langle \gamma_n, 1 \rangle$ in probability as $H \uparrow \infty$, hence (10) and (11) follow from (12) and from the Slutsky lemma, with $V_n = V_n(1)$ and $v_n(\phi) = V_n(\phi - \langle \mu_n, \phi \rangle)$, respectively.

Moreover, if (12) holds, then

$$(\sqrt{H}\,[\,\frac{\langle \gamma_n^H, 1 \rangle}{\langle \gamma_n, 1 \rangle} - 1\,], \sqrt{H}\,\langle \mu_n^H - \mu_n, \phi_1 \rangle, \cdots, \sqrt{H}\,\langle \mu_n^H - \mu_n, \phi_d \rangle)\ ,$$

converge jointly in distribution as $H \uparrow \infty$ to a Gaussian limit, for any bounded measurable functions $\phi_1, \cdots, \phi_d$, using the Cramér–Wold device.

Two different proofs of (12) are given in Sections 5 and 6, respectively. A first proof follows the approach of [9, Theorem 4] by induction, and relies on an extension of a central limit theorem for sums of a random number of i.i.d. random variables, see Theorem 6 below. An alternate proof follows the approach of [3, Chapter 9], see also [7, Proposition 2.9, Corollary 2.20], and relies on an original central limit theorem for triangular arrays of martingale increments spread across generations with different random sizes, see Theorem 7 below.

To get a fair comparison of the nonsequential and sequential particle algorithms, the time–averaged random number of simulated particles, which is an indicator of how much computational effort has been used, can be used as a normalizing factor instead of the level $H > 0$. It follows from Remark 6 that

$$\frac{1}{H}\,[\,\frac{1}{n+1}\sum_{k=0}^{n} N_k^H\,] \longrightarrow \frac{1}{n+1}\sum_{k=0}^{n} \rho_k\ ,$$

in probability as $H \uparrow \infty$, hence under the assumptions of Theorem 4, and using the Slutsky lemma

$$[\,\frac{1}{n+1}\sum_{k=0}^{n} N_k^H\,]^{1/2}\,\frac{\langle \gamma_n^H - \gamma_n, 1 \rangle}{\langle \gamma_n, 1 \rangle} \Longrightarrow \mathcal{N}(0, V_n^*)\ ,$$

and

$$[\,\frac{1}{n+1}\sum_{k=0}^{n} N_k^H\,]^{1/2}\,\langle \mu_n^H - \mu_n, \phi \rangle \Longrightarrow \mathcal{N}(0, v_n^*(\phi))\ ,$$

in distribution as $H \uparrow \infty$, with the asymptotic variance

$$V_n^* = [\,\frac{1}{n+1}\sum_{k=0}^{n} \rho_k\,]\,V_n \qquad \text{and} \qquad v_n^*(\phi) = [\,\frac{1}{n+1}\sum_{k=0}^{n} \rho_k\,]\,v_n(\phi)\ ,$$

respectively, where V_n and $v_n(\phi)$ are defined in Theorem 4. Notice that the asymptotic variances V_n^0 and $v_n^0(\phi)$ defined in Theorem 2 for the nonsequential particle algorithm coincide with the asymptotic variances V_n^* and $v_n^*(\phi)$ for the renormalized sequential particle algorithm respectively, in the special case where $\rho_0 = \rho_1 = \cdots = \rho_n$.

Example 4 (Binary selection). In the special case of binary selection functions, the support of μ_{k-1}^H is contained in A_{k-1}, and if

$$Q_k(x, A_k) = \mathbb{P}[X_k \in A_k \mid X_{k-1} = x] > 0 \ ,$$

for any $x \in A_{k-1}$, then it follows from Theorem 4 that

$$\sqrt{H}\,(\frac{P_n^H}{P_n} - 1) \Longrightarrow \mathcal{N}(0, V_n) \ ,$$

in distribution as $H \uparrow \infty$, with the asymptotic variance

$$V_n = \sum_{k=0}^{n} (1 - p_k) + \sum_{k=0}^{n} \frac{\operatorname{var}(R_{k+1:n}\, 1, \mu_k)}{\langle \mu_k, R_{k+1:n}\, 1\rangle^2} \ ,$$

since $1/\rho_k = \langle \eta_k, g_k \rangle = p_k$ for any $k = 0, 1, \cdots, n$.

4 Limit Theorems in Sequential Analysis

In this section, some basic properties are proved for sums of a random number of i.i.d. random variables, especially when this random number is a stopping time. Let $\xi_1, \cdots, \xi_i, \cdots$ be i.i.d. random variables with common probability distribution μ, and let Λ be a nonnegative bounded measurable function, possibly taking the zero value. For any $H > 0$, consider the stopping time

$$N_H = \inf\{N \geq 1 : \sum_{i=1}^{N} \Lambda(\xi_i) \geq H\,\lambda\} \qquad \text{where} \qquad \lambda = \sup_{x \in E} \Lambda(x) \ .$$

Lemma 1. *If $\langle \mu, \Lambda \rangle > 0$, then the stopping time N_H is a.s. finite and integrable.*

PROOF. By the strong law of large numbers, it follows that

$$\frac{1}{N} \sum_{i=1}^{N} \Lambda(\xi_i) \longrightarrow \langle \mu, \Lambda \rangle \ ,$$

a.s. as $N \uparrow \infty$, and if $\langle \mu, \Lambda \rangle > 0$, then

$$\sum_{i=1}^{N} \Lambda(\xi_i) \longrightarrow \infty \ ,$$

and the finite level $H\,\lambda$ is reached after a finite number of steps, i.e. the stopping time N_H is a.s. finite. In addition, for any $a > 0$

$$\mathbb{P}[N_H > N] = \mathbb{P}[\sum_{i=1}^{N} \Lambda(\xi_i) < H\,\lambda] \;=\; \mathbb{P}[\exp\{-a \sum_{i=1}^{N} \Lambda(\xi_i)\} > e^{-a\,H\,\lambda}]$$

$$\leq\; e^{a\,H\,\lambda}\, r^N \;,$$

by independence, where

$$r = \mathbb{E}[\exp\{-a\,\Lambda(\xi)\}] = \int_E e^{-a\,\Lambda(x)}\,\mu(dx) = \langle \mu, e^{-a\,\Lambda} \rangle \;,$$

and $r < 1$ if and only if $\langle \mu, \Lambda \rangle > 0$. This proves that the stopping time N_H is integrable, and the estimate

$$\mathbb{E}[N_H] = \sum_{N=0}^{\infty} \mathbb{P}[N_H > N] \leq e^{a\,H\,\lambda} \sum_{N=0}^{\infty} r^N \leq \frac{e^{a\,H\,\lambda}}{1-r} < \infty \;,$$

holds. □

Lemma 2. *If* $\langle \mu, \Lambda \rangle > 0$, *then the rough estimate*

$$\sup_{\phi\,:\,\|\phi\|=1} \{ \mathbb{E}\,|\,\langle S^{N_H}(\mu) - \mu, \Lambda\,\phi \rangle\,|^2 \}^{1/2} \leq \omega_H\,\lambda \;,$$

and the refined estimate

$$\sup_{\phi\,:\,\|\phi\|=1} \mathbb{E}\,|\,\langle S^{N_H}(\mu) - \mu, \Lambda\,\phi \rangle\,| \leq \omega_H\,[\,\langle \mu, \Lambda \rangle + \omega_H\,\lambda\,] \;,$$

hold, where $\omega_H = \dfrac{1}{H}\,\sqrt{H+1}$ *is of order* $1/\sqrt{H}$.

PROOF. Let

$$\delta_H = \Lambda\,(S^{N_H}(\mu) - \mu) \qquad \text{and} \qquad \delta'_H = \frac{\Lambda\,(S^{N_H}(\mu) - \mu)}{\langle S^{N_H}(\mu), \Lambda \rangle} \;.$$

Notice that

$$\delta_H = \delta'_H\,\langle S^{N_H}(\mu), \Lambda \rangle \;=\; \delta'_H\,[\,\langle \mu, \Lambda \rangle + \langle \delta_H, 1 \rangle\,]$$

$$=\; \delta'_H\,[\,\langle \mu, \Lambda \rangle + \langle \delta'_H, 1 \rangle\,\langle S^{N_H}(\mu), \Lambda \rangle\,] \;,$$

hence

$$|\,\langle \delta_H, \phi \rangle\,| \leq |\,\langle \delta'_H, \phi \rangle\,|\,\lambda \;,$$

and

$$|\,\langle \delta_H, \phi \rangle\,| \leq |\,\langle \delta'_H, \phi \rangle\,|\,[\,\langle \mu, \Lambda \rangle + |\,\langle \delta'_H, 1 \rangle\,|\,\lambda\,] \;,$$

for any bounded measurable function ϕ. It follows from (the proof of) Lemma 5.4 in [11] that

$$\sup_{\phi \,:\, \|\phi\|=1} \{ \mathbb{E} \,|\, \langle \delta'_H, \phi \rangle \,|^2 \,\}^{1/2} \leq \frac{1}{H} \sqrt{H+1} = \omega_H \ ,$$

which immediately proves the rough estimate, and using the Cauchy–Schwartz inequality and the Minkowski triangle inequality yields

$$\begin{aligned} \mathbb{E} \,|\, \langle \delta_H, \phi \rangle \,| \quad & \leq \quad \mathbb{E}[\,|\, \langle \delta'_H, \phi \rangle \,|\, [\, \langle \mu, \Lambda \rangle + |\, \langle \delta'_H, 1 \rangle \,|\, \lambda \,]\,] \\ & \leq \quad \{ \mathbb{E} \,|\, \langle \delta'_H, \phi \rangle \,|^2 \,\}^{1/2} \,[\, \langle \mu, \Lambda \rangle + \{ \mathbb{E} \,|\, \langle \delta'_H, 1 \rangle \,|^2 \,\}^{1/2} \, \lambda \,] \\ & \leq \quad \omega_H \, [\, \langle \mu, \Lambda \rangle + \omega_H \, \lambda \,] \, \|\phi\| \ , \end{aligned}$$

which proves the refined estimate. □

Lemma 3. *If* $\langle \mu, \Lambda \rangle > 0$, *then* $\dfrac{N_H}{H\,\rho} \to 1$ *and* $\dfrac{H}{N_H} \to \dfrac{1}{\rho}$ *in* $\mathbb{L}^2$ *as* $H \uparrow \infty$, *with rate* $1/\sqrt{H}$, *where* $\rho = \dfrac{\lambda}{\langle \mu, \Lambda \rangle}$.

PROOF. For any $N \geq 1$, define

$$D_N = \sum_{i=1}^N \Lambda(\xi_i) \qquad \text{and} \qquad M_N = \sum_{i=1}^N [\Lambda(\xi_i) - \langle \mu, \Lambda \rangle\,] = D_N - N\,\langle \mu, \Lambda \rangle \ .$$

By definition of the stopping time N_H, it holds

$$H\,\lambda \leq D_{N_H} = D_{N_H - 1} + \Lambda(\xi_{N_H}) \leq (H+1)\,\lambda \ ,$$

hence, upon subtracting $H\,\lambda$ throughout

$$0 \leq D_{N_H} - H\,\lambda \leq \lambda \ .$$

Using the decomposition

$$N_H\,\langle \mu, \Lambda \rangle - H\,\lambda = D_{N_H} - H\,\lambda - M_{N_H} \ ,$$

and the triangle inequality yields

$$|N_H\,\langle \mu, \Lambda \rangle - H\,\lambda| \leq |D_{N_H} - H\,\lambda| + |M_{N_H}| \leq \lambda + |M_{N_H}| \ .$$

Since $\langle \mu, \Lambda \rangle > 0$, it follows from Lemma 1 that the stopping time N_H is integrable, and it follows from the Wald identity, see e.g. [14, Proposition IV–4–21], that

$$\mathbb{E}[D_{N_H}] = \mathbb{E}[N_H]\,\langle \mu, \Lambda \rangle \qquad \text{and} \qquad \mathbb{E}|M_{N_H}|^2 = \mathbb{E}[N_H]\,\mathrm{var}(\Lambda, \mu) \ ,$$

hence

$$\mathbb{E}|M_{N_H}|^2 = \frac{\text{var}(\Lambda,\mu)}{\langle \mu, \Lambda \rangle}\, \mathbb{E}[D_{N_H}] \leq (H+1)\, \lambda^2 \ ,$$

since $\text{var}(\Lambda,\mu) = \langle \mu, \Lambda^2 \rangle - \langle \mu, \Lambda \rangle^2 \leq \langle \mu, \Lambda^2 \rangle \leq \lambda \, \langle \mu, \Lambda \rangle$, and since $D_{N_H} \leq (H+1)\, \lambda$. Using the Minkowski triangle inequality yields

$$\{\mathbb{E}|N_H \, \langle \mu, \Lambda \rangle - H\, \lambda|^2\}^{1/2} \leq \lambda + \{\mathbb{E}|M_{N_H}|^2\}^{1/2} \leq (\sqrt{H+1} + 1)\, \lambda \ ,$$

and, upon dividing by $H\, \lambda$ throughout

$$\{\mathbb{E}|\frac{N_H}{H\, \rho} - 1|^2\}^{1/2} \leq \frac{1}{H}\, (\sqrt{H+1} + 1) = \omega_H + \frac{1}{H} \ ,$$

where ω_H is of order $1/\sqrt{H}$. Since $N_H \geq H$, it holds

$$|\frac{H}{N_H} - \frac{1}{\rho}| \leq \frac{N_H}{H}\, |\frac{H}{N_H} - \frac{1}{\rho}| = \frac{1}{H\, \lambda}\, |N_H \, \langle \mu, \Lambda \rangle - H\, \lambda| \ ,$$

hence

$$\{\mathbb{E}|\frac{H}{N_H} - \frac{1}{\rho}|^2\}^{1/2} \leq \frac{1}{H}\, (\sqrt{H+1} + 1) = \omega_H + \frac{1}{H} \ . \qquad \square$$

Remark 8. A direct look into the proofs of Lemma 2 and Lemma 3 shows that a conditional version of the same results holds under the following assumptions. For any $H > 0$, let $\xi_1^H, \cdots, \xi_i^H, \cdots$ be i.i.d. random variables conditionally w.r.t. the σ–algebra $\mathcal{F}^H$, with common conditional probability distribution μ_H, let Λ be a nonnegative bounded measurable function, possibly taking the zero value, and consider the stopping time

$$N_H = \inf\{N \geq 1 \, : \, \sum_{i=1}^{N} \Lambda(\xi_i) \geq H\, \lambda\} \qquad \text{where} \qquad \lambda = \sup_{x \in E} \Lambda(x) \ .$$

If $\langle \mu_H, \Lambda \rangle > 0$, then (i) the rough estimate

$$\sup_{\phi \, : \, \|\phi\|=1} \{\, \mathbb{E}[\, | \, \langle S^{N_H}(\mu_H) - \mu_H, \Lambda\, \phi \rangle \, |^2 \mid \mathcal{F}^H \,] \, \}^{1/2} \leq \omega_H\, \lambda \ ,$$

and the refined estimate

$$\sup_{\phi \, : \, \|\phi\|=1} \mathbb{E}[\, | \, \langle S^{N_H}(\mu_H) - \mu_H, \Lambda\, \phi \rangle \, | \mid \mathcal{F}^H] \leq \omega_H \, [\, \langle \mu_H, \Lambda \rangle + \omega_H\, \lambda \,] \ ,$$

hold, where ω_H of order $1/\sqrt{H}$, and (ii)

$$\{\mathbb{E}[\, |\frac{N_H}{H \rho_H} - 1|^2 \mid \mathcal{F}^H]\}^{1/2} \leq \omega_H + \frac{1}{H} \ ,$$

and

$$\{\mathbb{E}[\, |\frac{H}{N_H} - \frac{1}{\rho_H}|^2 \mid \mathcal{F}^H]\}^{1/2} \leq \omega_H + \frac{1}{H} \ ,$$

with $\rho_H = \dfrac{\lambda}{\langle \mu_H, \Lambda \rangle}$.

The following central limit theorem, known in sequential analysis as the Anscombe theorem, has been proved in [17] for sums of a random number of i.i.d. random variables, see also [8, Theorem I.3.1] or [20, Theorem 2.40].

Theorem 5 (Anscombe). *For any $H > 0$, let $\rho_H > 0$ be a deterministic constant, and let $X_1^H, \cdots, X_i^H, \cdots$ be i.i.d. random variables with zero mean and variance σ_H^2. If $r_H = \lfloor H \rho_H \rfloor \to \infty$, if*

$$\frac{N_H}{H \rho_H} \longrightarrow 1 ,$$

in probability, and if the Lindeberg condition

$$\mathbb{E}[1_{\{|\frac{X_i^H}{\sigma_H}| \geq c \sqrt{r_H}\}} |\frac{X_i^H}{\sigma_H}|^2] \longrightarrow 0 ,$$

holds for any $c > 0$, then

$$\frac{1}{\sqrt{N_H}} \sum_{i=1}^{N_H} \frac{X_i^H}{\sigma_H} \Longrightarrow \mathcal{N}(0,1) \qquad \text{and} \qquad \frac{1}{\sqrt{H \rho_H}} \sum_{i=1}^{N_H} \frac{X_i^H}{\sigma_H} \Longrightarrow \mathcal{N}(0,1) ,$$

in distribution as $H \uparrow \infty$.

Remark 9. Using the Slutsky lemma, and since $\dfrac{\sqrt{N_H}}{\sqrt{H \rho_H}} \to 1$ in probability as $H \uparrow \infty$, the two convergence results are indeed equivalent.

The next theorem provides a stronger result, with a precise statement on the convergence of conditional characteristic functions, in a special case where both σ_H^2 and ρ_H are random variables. It is used in an essential way in Section 5, in the proof of Theorem 4 by induction.

Theorem 6. *For any $H > 0$, let $X_1^H, \cdots, X_i^H, \cdots$ be i.i.d. random variables conditionally w.r.t. the σ–algebra $\mathcal{F}^H$, with zero conditional mean and conditional variance σ_H^2, and let $\rho_H > 0$ be a $\mathcal{F}^H$–measurable r.v. If $r_H = \lfloor H \rho_H \rfloor \to \infty$ in probability, if*

$$F_H(d) = \mathbb{P}[\,|\frac{N_H}{H \rho_H} - 1| > d \mid \mathcal{F}^H\,] \longrightarrow 0 ,$$

in probability for any $d > 0$, and if the conditional Lindeberg condition

$$R_H(c) = \mathbb{E}[1_{\{|\frac{X_i^H}{\sigma_H}| \geq c \sqrt{r_H}\}} |\frac{X_i^H}{\sigma_H}|^2 \mid \mathcal{F}^H] \longrightarrow 0 ,$$

holds in probability for any $c > 0$, then for any fixed real number u

$$\mathbb{E}[\exp\{i\,\frac{u}{\sqrt{H\rho_H}}\sum_{j=1}^{N_H}\frac{X_j^H}{\sigma_H}\}\mid \mathcal{F}^H] \longrightarrow \exp\{-\tfrac{1}{2}\,u^2\}\ , \tag{14}$$

in $\mathbb{L}^1$ *as* $H \uparrow \infty$. *If in addition* $\dfrac{\sigma_H}{\sqrt{\rho_H}} \to \dfrac{\sigma}{\sqrt{\rho}}$ *in probability, then for any fixed real number* u

$$\mathbb{E}[\exp\{i\,u\,\sqrt{H}\,\frac{1}{N_H}\sum_{j=1}^{N_H} X_j^H\}\mid \mathcal{F}^H] \longrightarrow \exp\{-\tfrac{1}{2}\,\frac{u^2\,\sigma^2}{\rho}\}\ , \tag{15}$$

in $\mathbb{L}^1$ *as* $H \uparrow \infty$.

Using the Lebesgue dominated convergence theorem, it is sufficient to prove that (14) and (15) hold in probability. The proof of Theorem 6 is postponed to Appendix A.

Remark 10. If (14) holds, then in particular

$$Z_H = \frac{1}{\sqrt{H\rho_H}}\sum_{j=1}^{N_H}\frac{X_j^H}{\sigma_H} \Longrightarrow \mathcal{N}(0,1)\ ,$$

in distribution as $H \uparrow \infty$.

Remark 11. If $F_H(d) \to 0$ in probability for any $d > 0$, (or equivalently in $\mathbb{L}^1$ using the Lebesgue dominated convergence theorem), then equivalently $\mathbb{E}[F_H(d)] \to 0$ for any $d > 0$, since these r.v.'s are nonnegative, which means that $\dfrac{N_H}{H\rho_H} \to 1$ in probability as $H \uparrow \infty$.

The last result of this section is a central limit theorem for triangular arrays of martingale increments spread across generations with random sizes. It is used in an essential way in Section 6, in an alternate proof of Theorem 4.

Theorem 7. *For any* $k = 0, 1, \cdots, n$, *let* $\mathcal{F}_k^H = \{\mathcal{F}_{k,i}^H\,,\, i \geq 0\}$ *be an increasing sequence of* σ*–algebras, let* N_k^H *be a stopping time w.r.t.* $\mathcal{F}_k^H$, *which allows to define the* σ*–algebra* $\mathcal{H}_k^H = \mathcal{F}_{k,N_k^H}^H$, *assume that* $\mathcal{F}_{0,0}^H = \{\emptyset, \Omega\}$ *(for* $k = 0$*) and* $\mathcal{F}_{k,0}^H = \mathcal{H}_{k-1}^H$ *(for* $k = 1, \cdots, n$*), and let* $\{X_{k,i}^H\,,\, i \geq 1\}$ *be a sequence of square integrable random variables adapted to* $\mathcal{F}_k^H$, *such that*

$$\mathbb{E}[X_{k,i}^H \mid \mathcal{F}_{k,i-1}^H] = 0\ , \tag{16}$$

$$\mathbb{E}[\,|X_{k,i}^H|^2 \mid \mathcal{F}_{k,i-1}^H] = V_{k,0}^H\ , \tag{17}$$

and

$$\mathbb{E}[\,|X_{k,i}^H|^2\, 1_{\{|X_{k,i}^H| > \varepsilon\}} \mid \mathcal{F}_{k,i-1}^H] \leq Y_{k,0}^{H,\varepsilon}\ , \tag{18}$$

for any $i \geq 1$, *where* $V_{k,0}^H$ *and* $Y_{k,0}^{H,\varepsilon}$ *are measurable w.r.t.* $\mathcal{F}_{k,0}^H$. *If for any* $\varepsilon > 0$

$$\sum_{k=0}^{n} N_k^H \, V_{k,0}^H \longrightarrow W_n \qquad \textit{and} \qquad \sum_{k=0}^{n} N_k^H \, Y_{k,0}^{H,\varepsilon} \longrightarrow 0 \; , \tag{19}$$

in probability, then

$$S_n^H = \sum_{k=0}^{n} \sum_{i=1}^{N_k^H} X_{k,i}^H \Longrightarrow \mathcal{N}(0, W_n) \; ,$$

in distribution as $H \uparrow \infty$.

The proof of Theorem 7 is postponed to Appendix B. The idea is to rewrite S_n^H as a single sum across all generations, and to use a central limit theorem for triangular arrays of martingale increments [2, Theorem 2.8.42].

5 Proof of Theorem 4 by induction

In view of Remark 7 above, the problem reduces to prove (12), i.e. to prove asymptotic normality for the unnormalized linear flow. The proof given below follows the approach of [9, Theorem 4] by induction.

PROOF OF THEOREM 4. Notice first that

$$\langle \gamma_0^H - \gamma_0, \phi \rangle = \langle S^{N_0^H}(\eta_0) - \eta_0, g_0 \, \phi \rangle = \frac{1}{N_0^H} \sum_{j=1}^{N_0^H} X_{0,j}^H(\phi) \; ,$$

where

$$X_{0,j}^H(\phi) = g_0(\xi_0^j) \, \phi(\xi_0^j) - \langle \eta_0, g_0 \, \phi \rangle \; ,$$

for any $j = 1, \cdots, N_0^H$, and where $\xi_0^1, \cdots, \xi_0^j, \cdots$ are i.i.d. random variables with common probability distribution η_0, hence the random variables $X_{0,1}^H(\phi), \cdots, X_{0,j}^H(\phi), \cdots$ are i.i.d. with zero mean and with variance $\mathrm{var}(g_0 \, \phi, \eta_0)$ independent of $H > 0$. It follows from Lemma 3 that $\dfrac{N_0^H}{H} \to \rho_0$ in probability as $H \uparrow \infty$, hence the assumptions of Theorem 5 are satisfied, and the induction assumption (12) holds at step 0, with

$$V_0(\phi) \, \langle \gamma_0, 1 \rangle^2 = \mathrm{var}(g_0 \, \phi, \eta_0) \, \frac{1}{\rho_0} \; .$$

Assume now that the induction assumption (12) holds at step $(k-1)$. Notice that

$$\gamma_k^H - \gamma_k = \gamma_k^H - \gamma_{k-1}^H \, R_k + (\gamma_{k-1}^H - \gamma_{k-1}) \, R_k \; ,$$

hence

$$\langle \gamma_k^H - \gamma_k, \phi \rangle = \langle \gamma_k^H - \gamma_{k-1}^H \, R_k, \phi \rangle + \langle \gamma_{k-1}^H - \gamma_{k-1}, R_k \, \phi \rangle \; ,$$

for any bounded measurable function ϕ, and the last term goes to zero in probability as $H \uparrow \infty$. Notice also that

$$\begin{aligned} \langle \gamma_k^H - \gamma_{k-1}^H \, R_k, \phi \rangle &= \langle S^{N_k^H}(\mu_{k-1}^H \, Q_k) - \mu_{k-1}^H \, Q_k, g_k \, \phi \rangle \; \langle \gamma_{k-1}^H, 1 \rangle \\ &= \frac{1}{N_k^H} \sum_{j=1}^{N_k^H} X_{k,j}^H(\phi) \; , \end{aligned}$$

where

$$X_{k,j}^H(\phi) = [\, g_k(\xi_k^j) \, \phi(\xi_k^j) - \langle \mu_{k-1}^H \, Q_k, g_k \, \phi \rangle \,] \; \langle \gamma_{k-1}^H, 1 \rangle \; ,$$

for any $j = 1, \cdots, N_k^H$, and where, conditionally w.r.t. the σ–algebra $\mathcal{H}_{k-1}^H$ generated by the particle system up to the $(k-1)$–th generation, the random variables $\xi_k^1, \cdots, \xi_k^j, \cdots$ are i.i.d. with common probability distribution $\mu_{k-1}^H \, Q_k$, hence the random variables $X_{k,1}^H(\phi), \cdots, X_{k,j}^H(\phi), \cdots$ are i.i.d. with zero conditional mean and with conditional variance

$$(\sigma_k^H(\phi))^2 = \mathrm{var}(g_k \, \phi, \mu_{k-1}^H \, Q_k) \; \langle \gamma_{k-1}^H, 1 \rangle^2 \; .$$

In view of Remark 8 (ii)

$$F_H(d) = \mathbb{P}[\, | \frac{N_k^H}{H \rho_k^H} - 1 | > d \mid \mathcal{H}_{k-1}^H \,] \longrightarrow 0 \qquad \text{with} \qquad \rho_k^H = \frac{\sup_{x \in E} g_k(x)}{\langle \mu_{k-1}^H \, Q_k, g_k \rangle} \; ,$$

in probability for any $d > 0$, as $H \uparrow \infty$. It follows from Theorem 3 that $\langle \gamma_{k-1}^H, 1 \rangle \to \langle \gamma_{k-1}, 1 \rangle$, $\langle \mu_{k-1}^H \, Q_k, g_k \rangle \to \langle \eta_k, g_k \rangle$ and $\mathrm{var}(g_k \, \phi, \mu_{k-1}^H \, Q_k) \to \mathrm{var}(g_k \, \phi, \eta_k)$ in probability, hence $\rho_k^H \to \rho_k$ and $\sigma_k^H(\phi) \to \sigma_k(\phi)$ in probability as $H \uparrow \infty$, with

$$\sigma_k^2(\phi) = \mathrm{var}(g_k \, \phi, \mu_{k-1} \, Q_k) \; \langle \gamma_{k-1}, 1 \rangle^2 \; .$$

Therefore, the assumptions of Theorem 6 are satisfied, and for any fixed real number u, it holds

$$\begin{aligned} &\mathbb{E}[\, \exp\{ i \, u \, \sqrt{H} \; \langle \gamma_k^H - \gamma_{k-1}^H \, R_k, \phi \rangle \} \mid \mathcal{H}_{k-1}^H] \\ &= \mathbb{E}[\, \exp\{ i \, u \, \sqrt{H} \, \frac{1}{N_k^H} \sum_{j=1}^{N_k^H} X_{k,j}^H(\phi) \} \mid \mathcal{H}_{k-1}^H] \\ &\longrightarrow \exp\{ -\tfrac{1}{2} \, u^2 \, \frac{\sigma_k^2(\phi)}{\rho_k} \} \; , \end{aligned} \tag{20}$$

in $\mathbb{L}^1$ as $H \uparrow \infty$. Notice that

$$\mathbb{E}[\exp\{i\, u\, \sqrt{H}\, \langle \gamma_k^H - \gamma_k, \phi \rangle\}]$$

$$- \exp\{-\tfrac{1}{2}\, u^2\, \frac{\sigma_k^2(\phi)}{\rho_k} - \tfrac{1}{2}\, u^2\, V_{k-1}(R_k\, \phi)\, \langle \gamma_{k-1}, 1 \rangle^2\}$$

$$= \mathbb{E}[\,[\mathbb{E}[\exp\{i\, u\, \sqrt{H}\, \langle \gamma_k^H - \gamma_{k-1}^H\, R_k, \phi \rangle\} \mid \mathcal{H}_{k-1}^H] - \exp\{-\tfrac{1}{2}\, u^2\, \frac{\sigma_k^2(\phi)}{\rho_k}\}\,]$$

$$\exp\{i\, u\, \sqrt{H}\, \langle \gamma_{k-1}^H - \gamma_{k-1}, R_k\, \phi \rangle\}\,]$$

$$+ \exp\{-\tfrac{1}{2}\, u^2\, \frac{\sigma_k^2(\phi)}{\rho_k}\}\; \mathbb{E}[\exp\{i\, u\, \sqrt{H}\, \langle \gamma_{k-1}^H - \gamma_{k-1}, R_k\, \phi \rangle\}]$$

$$- \exp\{-\tfrac{1}{2}\, u^2\, \frac{\sigma_k^2(\phi)}{\rho_k}\}\; \exp\{-\tfrac{1}{2}\, u^2\, V_{k-1}(R_k\, \phi)\, \langle \gamma_{k-1}, 1 \rangle^2\}\;,$$

and the triangle inequality yields

$$|\, \mathbb{E}[\exp\{i\, u\, \sqrt{H}\, \langle \gamma_k^H - \gamma_k, \phi \rangle\}]$$

$$- \exp\{-\tfrac{1}{2}\, u^2\, \frac{\sigma_k^2(\phi)}{\rho_k} - \tfrac{1}{2}\, u^2\, V_{k-1}(R_k\, \phi)\, \langle \gamma_{k-1}, 1 \rangle^2\}\,|$$

$$\leq \mathbb{E}|\, \mathbb{E}[\exp\{i\, u\, \sqrt{H}\, \langle \gamma_k^H - \gamma_{k-1}^H\, R_k, \phi \rangle\} \mid \mathcal{H}_{k-1}^H] - \exp\{-\tfrac{1}{2}\, u^2\, \frac{\sigma_k^2(\phi)}{\rho_k}\}\,|$$

$$+ \,|\, \mathbb{E}[\exp\{i\, u\, \sqrt{H}\, \langle \gamma_{k-1}^H - \gamma_{k-1}, R_k\, \phi \rangle\}]$$

$$- \exp\{-\tfrac{1}{2}\, u^2\, V_{k-1}(R_k\, \phi)\, \langle \gamma_{k-1}, 1 \rangle^2\}\,|\;,$$

where the first term goes to zero using (20), and the second term goes to zero since the induction assumption (12) holds at step $(k-1)$, as $H \uparrow \infty$. Therefore, the induction assumption (12) holds at step k, with

$$V_k(\phi)\, \langle \gamma_k, 1 \rangle^2 = \frac{\sigma_k^2(\phi)}{\rho_k} + V_{k-1}(R_k\, \phi)\, \langle \gamma_{k-1}, 1 \rangle^2\;,$$

and iterating the above relation yields

$$\begin{aligned} V_n(\phi)\, \langle \gamma_n, 1 \rangle^2 \;&=\; V_0(R_{1:n}\, \phi)\, \langle \gamma_0, 1 \rangle^2 + \sum_{k=1}^{n} \sigma_k^2(R_{k+1:n}\, \phi)\, \frac{1}{\rho_k} \\ &=\; \mathrm{var}(g_0\, R_{1:n}\, \phi, \eta_0)\, \frac{1}{\rho_0} + \sum_{k=1}^{n} \mathrm{var}(g_k\, R_{k+1:n}\, \phi, \eta_k)\, \frac{\langle \gamma_{k-1}, 1 \rangle^2}{\rho_k}\;, \end{aligned}$$

which is the expression given in (13) for the asymptotic variance. In view of Remark 7, this finishes the proof of Theorem 4. □

6 Alternate Proof of Theorem 4

The alternate proof given below follows the approach of [3, Chapter 9], see also [7, Proposition 2.9, Corollary 2.20], and relies on an approximate decomposition of $\sqrt{H}\,\langle \gamma_n^H - \gamma_n, \phi \rangle$ in terms of a triangular array of martingale increments, with a different random number $\sigma_n^H = N_0^H + \cdots + N_n^H$ of such increments on each different row of the array. This requires a specific central limit theorem, see Theorem 8 below, which is of independent interest.

Let $f = (f_0, f_1, \cdots, f_n)$ be an arbitrary collection of bounded measurable functions. For any $i = 1, \cdots, N_0^H$, the random variable

$$X_{0,i}^H(f) = \frac{1}{\rho_0 \sqrt{H}}\, [f_0(\xi_0^i) - \langle \eta_0, f_0 \rangle] \qquad \text{where} \qquad \rho_0 = \frac{\sup_{x \in E} g_0(x)}{\langle \eta_0, g_0 \rangle} \;,$$

is measurable w.r.t. $\mathcal{F}_{0,i}^H$, where $\xi_0^1, \cdots, \xi_0^i, \cdots$ are i.i.d. random variables with common probability distribution η_0. Moreover

$$\mathbb{E}[X_{0,i}^H(f) \mid \mathcal{F}_{0,i-1}^H] = 0 \;, \tag{21}$$

and

$$\mathbb{E}[\,|X_{0,i}^H(f)|^2 \mid \mathcal{F}_{0,i-1}^H] = \frac{1}{(\rho_0 \sqrt{H})^2}\, \mathrm{var}(f_0, \eta_0) = V_{0,0}^H(f) \;, \tag{22}$$

with the convention $\mathcal{F}_{0,0}^H = \{\emptyset, \Omega\}$. Notice that

$$|X_{0,i}^H(f)| \le \frac{1}{\rho_0 \sqrt{H}}\, 2\, \|f_0\| \qquad \text{and} \qquad \rho_0 \ge 1 \;,$$

hence for any $\varepsilon > 0$

$$\begin{aligned} &\mathbb{E}[\,|X_{0,i}^H(f)|^2\, 1_{\{|X_{0,i}^H(f)| > \varepsilon\}} \mid \mathcal{F}_{0,i-1}^H] \\ &\qquad \le \frac{1}{\rho_0\, H}\, (2\, \|f_0\|)^2\, 1_{\{\frac{1}{\sqrt{H}}\, 2\, \|f_0\| > \varepsilon\}} = Y_{0,0}^{H,\varepsilon}(f) \;. \end{aligned} \tag{23}$$

For any $k = 1, \cdots, n$, the random variable

$$\rho_k^H = \frac{\sup_{x \in E} g_k(x)}{\langle \mu_{k-1}^H\, Q_k, g_k \rangle} \;,$$

is measurable w.r.t. $\mathcal{H}_{k-1}^H = \mathcal{F}_{k,0}^H$, and for any $i = 1, \cdots, N_k^H$, the random variable

$$X_{k,i}^H(f) = \frac{\langle \gamma_{k-1}^H, 1 \rangle}{\rho_k^H \sqrt{H}}\, [f_k(\xi_k^i) - \langle \mu_{k-1}^H\, Q_k, f_k \rangle] \;,$$

is measurable w.r.t. $\mathcal{F}^H_{k,i}$, where, conditionally w.r.t. the σ–algebra $\mathcal{H}^H_{k-1}$ generated by the particle system up to the $(k-1)$–th generation, the random variables $\xi^1_k, \cdots, \xi^i_k, \cdots$ are i.i.d. with common probability distribution $\mu^H_{k-1}\, Q_k$. Moreover

$$\mathbb{E}[X^H_{k,i}(f) \mid \mathcal{F}^H_{k,i-1}] = 0 \ , \tag{24}$$

and

$$\mathbb{E}[\, |X^H_{k,i}(f)|^2 \mid \mathcal{F}^H_{k,i-1}] = \frac{\langle \gamma^H_{k-1}, 1\rangle^2}{(\rho^H_k \sqrt{H})^2} \, \mathrm{var}(f_k, \mu^H_{k-1}\, Q_k) = V^H_{k,0}(f) \ , \tag{25}$$

where the random variable $V^H_{k,0}(f)$ is measurable w.r.t. $\mathcal{H}^H_{k-1} = \mathcal{F}^H_{k,0}$. Notice that

$$|X^H_{k,i}(f)| \leq \frac{\langle \gamma^H_{k-1}, 1\rangle}{\rho^H_k \sqrt{H}} \, 2 \, \|f_k\| \qquad \text{and} \qquad \rho^H_k \geq 1 \ ,$$

hence for any $\varepsilon > 0$

$$\begin{aligned} &\mathbb{E}[\, |X^H_{k,i}(f)|^2 \, 1_{\{|X^H_{k,i}(f)| \, > \, \varepsilon\}} \mid \mathcal{F}^H_{k,i-1}] \\ &\leq \frac{\langle \gamma^H_{k-1}, 1\rangle^2}{\rho^H_k \, H} \, (2 \, \|f_k\|)^2 \, 1_{\{\frac{\langle \gamma^H_{k-1}, 1\rangle}{\sqrt{H}} \, 2 \, \|f_k\| \, > \, \varepsilon\}} = Y^{H,\varepsilon}_{k,0}(f) \ , \end{aligned} \tag{26}$$

where the random variable $Y^{H,\varepsilon}_{k,0}(f)$ is measurable w.r.t. $\mathcal{H}^H_{k-1} = \mathcal{F}^H_{k,0}$.

Theorem 8. *For any collection $f = (f_0, f_1, \cdots, f_n)$ of bounded measurable functions*

$$S^H_n(f) = \sum_{k=0}^{n} \sum_{i=1}^{N^H_k} X^H_{k,i}(f) \Longrightarrow \mathcal{N}(0, W_n(f)) \ ,$$

in distribution as $H \uparrow \infty$, with asymptotic variance

$$W_n(f) = \frac{1}{\rho_0} \, \mathrm{var}(f_0, \eta_0) + \sum_{k=1}^{n} \frac{\langle \gamma_{k-1}, 1\rangle^2}{\rho_k} \, \mathrm{var}(f_k, \eta_k) \ .$$

Remark 12. Since the mapping $f \longmapsto S^H_n(f)$ is linear (which incidentally implies that the mapping $f \longmapsto W_n(f)$ is quadratic), the result of Theorem 8 is easily extended to any collection $f = (f_0, f_1, \cdots, f_n)$ of d–dimensional bounded measurable functions, using the Cramér–Wold device, and it follows from the structure of the asymptotic variance that the random variables

$$(\sum_{i=1}^{N^H_0} X^H_{0,i}(f), \cdots, \sum_{i=1}^{N^H_k} X^H_{k,i}(f), \cdots, \sum_{i=1}^{N^H_n} X^H_{n,i}(f)) \ ,$$

are mutually independent, asymptotically as $H \uparrow \infty$.

PROOF OF THEOREM 8. It follows from (21) and (24), from (22) and (25), and from (23) and (26), that the assumptions (16), (17) and (18) of Theorem 7 are satisfied, respectively. It follows from Theorem 3 that $\langle \gamma_{k-1}^H, 1 \rangle \to \langle \gamma_{k-1}, 1 \rangle$ and $\mathrm{var}(f_k, \mu_{k-1}^H \, Q_k) \to \mathrm{var}(f_k, \eta_k)$ in probability as $H \uparrow \infty$, for any $k = 1, \cdots, n$. Therefore, it follows from Remark 3 that

$$\begin{aligned}
&\sum_{k=0}^{n} N_k^H \, V_{k,0}^H(f) \\
&\qquad = \frac{N_0^H}{H \, \rho_0} \, \frac{1}{\rho_0} \, \mathrm{var}(f_0, \eta_0) + \sum_{k=1}^{n} \frac{N_k^H}{H \, \rho_k^H} \, \frac{\langle \gamma_{k-1}^H, 1 \rangle^2}{\rho_k^H} \, \mathrm{var}(f_k, \mu_{k-1}^H \, Q_k) \\
&\qquad \longrightarrow W_n(f) = \frac{1}{\rho_0} \, \mathrm{var}(f_0, \eta_0) + \sum_{k=1}^{n} \frac{\langle \gamma_{k-1}, 1 \rangle^2}{\rho_k} \, \mathrm{var}(f_k, \eta_k) \; ,
\end{aligned}$$

and

$$\begin{aligned}
&\sum_{k=0}^{n} N_k^H \, Y_{k,0}^{H,\varepsilon}(f) \\
&\qquad \leq \frac{N_0^H}{\rho_0 \, H} \, (2 \, \|f_0\|)^2 \, 1_{\{\frac{1}{\sqrt{H}} \, 2 \, \|f_0\| > \varepsilon\}} \\
&\qquad\quad + \sum_{k=1}^{n} \frac{N_k^H}{\rho_k^H \, H} \, \langle \gamma_{k-1}^H, 1 \rangle^2 \, (2 \, \|f_k\|)^2 \, 1_{\{\frac{\langle \gamma_{k-1}^H, 1 \rangle}{\sqrt{H}} \, 2 \, \|f_k\| > \varepsilon\}} \\
&\qquad \longrightarrow 0 \; ,
\end{aligned}$$

in probability as $H \uparrow \infty$, and the proof follows from Theorem 7. □

PROOF OF THEOREM 4. For any bounded measurable function ϕ, the following decomposition holds

$$\begin{aligned}
\langle \gamma_n^H - \gamma_n, \phi \rangle &= \sum_{k=1}^{n} \langle \gamma_k^H - \gamma_{k-1}^H \, R_k, R_{k+1:n} \, \phi \rangle + \langle \gamma_0^H - \gamma_0, R_{1:n} \, \phi \rangle \\
&= \sum_{k=1}^{n} \langle \gamma_{k-1}^H, 1 \rangle \, \langle g_k \, (\eta_k^H - \mu_{k-1}^H \, Q_k), R_{k+1:n} \, \phi \rangle \\
&\qquad + \langle g_0 \, (\eta_0^H - \eta_0), R_{1:n} \, \phi \rangle \\
&= \sum_{k=1}^{n} \langle \gamma_{k-1}^H, 1 \rangle \, \langle \eta_k^H - \mu_{k-1}^H \, Q_k, f_k \rangle + \langle \eta_0^H - \eta_0, f_0 \rangle \; ,
\end{aligned}$$

where the collection $f = (f_0, f_1, \cdots, f_n)$ of bounded measurable functions is defined by

$$f_k(x) = g_k(x)\, R_{k+1:n}\, \phi(x) \; ,$$

for any $k = 0, 1, \cdots, n$, with the convention $R_{n+1:n}\, \phi(x) = \phi(x)$, for any $x \in E$. Notice that

$$\begin{aligned} \langle \eta_0^H - \eta_0, f_0 \rangle &= \frac{1}{N_0^H} \sum_{i=1}^{N_0^H} [f_0(\xi_0^i) - \langle \eta_0, f_0 \rangle\,] \\ &= \frac{1}{\sqrt{H}} \sum_{i=1}^{N_0^H} X_{0,i}^H(f) + (1 - \frac{N_0^H}{H\, \rho_0})\, \langle \eta_0^H - \eta_0, f_0 \rangle \; , \end{aligned}$$

and

$$\begin{aligned} &\langle \gamma_{k-1}^H, 1 \rangle\, \langle \eta_k^H - \mu_{k-1}^H\, Q_k, f_k \rangle \\ &\quad = \frac{\langle \gamma_{k-1}^H, 1 \rangle}{N_k^H} \sum_{i=1}^{N_k^H} [f_k(\xi_k^i) - \langle \mu_{k-1}^H\, Q_k, f_k \rangle\,] \\ &\quad = \frac{1}{\sqrt{H}} \sum_{i=1}^{N_k^H} X_{k,i}^H(f) + \langle \gamma_{k-1}^H, 1 \rangle\, (1 - \frac{N_k^H}{H\, \rho_k^H})\, \langle \eta_k^H - \mu_{k-1}^H\, Q_k, f_k \rangle \; , \end{aligned}$$

for any $k = 1, \cdots, n$. Taking the sum of both sides for $k = 0, 1, \cdots, n$ yields

$$\sqrt{H}\, \langle \gamma_n^H - \gamma_n, \phi \rangle = S_n^H(f) + \varepsilon_0^H(f) + \sum_{k=1}^{n} \langle \gamma_{k-1}^H, 1 \rangle\, \varepsilon_k^H(f) \; ,$$

where

$$\varepsilon_0^H(f) = \sqrt{H}\, (1 - \frac{N_0^H}{H\, \rho_0})\, \langle \eta_0^H - \eta_0, f_0 \rangle \; ,$$

and where

$$\varepsilon_k^H(f) = \sqrt{H}\, (1 - \frac{N_k^H}{H\, \rho_k^H})\, \langle \eta_k^H - \mu_{k-1}^H\, Q_k, f_k \rangle \; ,$$

for any $k = 1, \cdots, n$. Using the Cauchy–Schwartz inequality, it follows from Lemma 3 and from the rough estimate of Lemma 2 that

$$\begin{aligned} \mathbb{E}|\varepsilon_0^H(f)| &\leq \sqrt{H}\, \{\mathbb{E}|1 - \frac{N_0^H}{H\, \rho_0}|^2\, \}^{1/2}\, \{\mathbb{E}\, |\langle \eta_0^H - \eta_0, f_0 \rangle|^2\, \}^{1/2} \\ &\leq \sqrt{H}\, (\omega_H + \frac{1}{H})\, \omega_H\, \|g_0\|\, \|R_{1:n}\, \phi\| \; , \end{aligned}$$

and in view of Remark 8 (i) and (ii)

$$\begin{aligned}\mathbb{E}[\,|\varepsilon_k^H(f)|\mid \mathcal{H}_{k-1}^H] &\leq \sqrt{H}\,\{\mathbb{E}[\,|1-\frac{N_k^H}{H\,\rho_k^H}|^2 \mid \mathcal{H}_{k-1}^H]\,\}^{1/2} \\ &\qquad \{\mathbb{E}[\,|\langle \eta_k^H - \mu_{k-1}^H\,Q_k, f_k\rangle|^2 \mid \mathcal{H}_{k-1}^H]\,\}^{1/2} \\ &\leq \sqrt{H}\,(\omega_H + \frac{1}{H})\,\omega_H\,\|g_k\|\,\|R_{k+1:n}\,\phi\|\ ,\end{aligned}$$

for any $k = 1, \cdots, n$, hence

$$\begin{aligned}&\mathbb{E}\,|\sqrt{H}\,\langle \gamma_n^H - \gamma_n, \phi\rangle - S_n^H(f)\,| \\ &\leq \omega'_H\,[\,\|g_0\|\,\|R_{1:n}\,\phi\| + \sum_{k=1}^n \mathbb{E}[\,\langle \gamma_{k-1}^H, 1\rangle\,]\,\|g_k\|\,\|R_{k+1:n}\,\phi\|\,]\ ,\end{aligned}$$

where $\omega'_H = \sqrt{H}\,(\omega_H + \frac{1}{H})\,\omega_H$ is of order $1/\sqrt{H}$. It follows from the above discussion that

$$\sqrt{H}\,\langle \gamma_n^H - \gamma_n, \phi\rangle - S_n^H(f) \longrightarrow 0\ ,$$

in probability, and it follows from Theorem 7 that $\sqrt{H}\,\langle \gamma_n^H - \gamma_n, \phi\rangle$ converges in distribution as $H \uparrow \infty$ to a Gaussian random variable with zero mean and with variance

$$W_n(f) = \frac{1}{\rho_0}\,\mathrm{var}(g_0\,R_{1,n}\,\phi, \eta_0) + \sum_{k=1}^n \frac{\langle \gamma_{k-1}, 1\rangle^2}{\rho_k}\,\mathrm{var}(g_k\,R_{k+1,n}\,\phi, \eta_k)\ ,$$

which proves (12) with the expression given in (13) for the asymptotic variance. In view of Remark 7, this finishes the proof of Theorem 4. □

A Proof of Theorem 6

Proof of Theorem 6. For any $H > 0$, notice that

$$\begin{aligned}Z_H &= \frac{1}{\sqrt{H \rho_H}} \sum_{i=1}^{N_H} \frac{X_i^H}{\sigma_H} \\ &= \frac{\sqrt{r_H}}{\sqrt{H \rho_H}}\,[\,\frac{1}{\sqrt{r_H}} \sum_{i=1}^{r_H} \frac{X_i^H}{\sigma_H} + \frac{1}{\sqrt{r_H}}\,(\sum_{i=1}^{N_H} \frac{X_i^H}{\sigma_H} - \sum_{i=1}^{r_H} \frac{X_i^H}{\sigma_H})\,] \qquad (27) \\ &= a_H\,(S_H + O_H)\ ,\end{aligned}$$

with

$$a_H = \frac{\sqrt{r_H}}{\sqrt{H \rho_H}} \leq 1\ , \qquad O_H = \frac{1}{\sqrt{r_H}}\,(\sum_{i=1}^{N_H} \frac{X_i^H}{\sigma_H} - \sum_{i=1}^{r_H} \frac{X_i^H}{\sigma_H})\ ,$$

and

$$S_H = \frac{1}{\sqrt{r_H}} \sum_{i=1}^{r_H} \frac{X_i^H}{\sigma_H} \; .$$

Notice that

$$0 \leq 1 - a_H = 1 - \frac{\sqrt{r_H}}{\sqrt{H \rho_H}} \leq 1 - \frac{\sqrt{r_H}}{\sqrt{r_H + 1}} \leq \frac{1}{2\, r_H + 1} = \varepsilon(r_H) \; ,$$

where the straightforward estimate $1 - \dfrac{\sqrt{x}}{\sqrt{x+1}} \leq \dfrac{1}{2\,x+1}$ holds for any nonnegative real number $x \geq 0$, and

$$\mathbb{E}[\, |S_H| \mid \mathcal{F}^H] \leq 1 \; ,$$

using the Cauchy–Schwartz inequality. Therefore, for any fixed real number u

$$\begin{aligned}
&\mathbb{E}[\exp\{i \, \frac{u}{\sqrt{H \rho_H}} \sum_{j=1}^{N_H} \frac{X_j^H}{\sigma_H}\} \mid \mathcal{F}^H] - \exp\{-\tfrac{1}{2}\, u^2\} \\
&= \mathbb{E}[\exp\{i\, u\, Z_H\} \mid \mathcal{F}^H] - \exp\{-\tfrac{1}{2}\, u^2\} \\
&= \mathbb{E}[\exp\{i\, u\, Z_H\} - \exp\{i\, u\, S_H\} \mid \mathcal{F}^H] \\
&\quad + \mathbb{E}[\exp\{i \, \frac{u}{\sqrt{r_H}} \sum_{j=1}^{r_H} \frac{X_j^H}{\sigma_H}\} \mid \mathcal{F}^H] - \exp\{-\tfrac{1}{2}\, u^2\} \; .
\end{aligned}$$

The last term can be controlled easily, using classical estimates in the central limit theorem for sums of i.i.d. random variables. Moreover, for any $B > 0$ and any $0 < d < 1$

$$\begin{aligned}
&\mathbb{E}[\, | \exp\{i\, u\, Z_H\} - \exp\{i\, u\, S_H\} | \mid \mathcal{F}^H] \\
&\leq \mathbb{E}[1_{\{|O_H| \,\leq\, B\}} \, |\exp\{i\, u\, Z_H\} - \exp\{i\, u\, S_H\}| \mid \mathcal{F}^H] \\
&\quad + \mathbb{E}[1_{\{|O_H| \,>\, B\}} \, |\exp\{i\, u\, Z_H\} - \exp\{i\, u\, S_H\}| \mid \mathcal{F}^H] \\
&\leq |u| \; ((1 - a_H)\, \mathbb{E}[\, |S_H| \mid \mathcal{F}^H] + a_H\, B) + 2\, \mathbb{P}[\, |O_H| > B \mid \mathcal{F}^H] \\
&\leq |u| \; (\varepsilon(r_H) + B) + 2\, \mathbb{P}[\, |O_H| > B \, , \, |\frac{N_H}{H \rho_H} - 1| \leq d \mid \mathcal{F}^H] \\
&\quad + 2\, \mathbb{P}[\, |\frac{N_H}{H \rho_H} - 1| > d \mid \mathcal{F}^H] \; ,
\end{aligned}$$

and the second term can be controlled using the Kolmogorov maximal inequality, along the lines of the proof of [8, Theorem I.3.1].

► For any fixed real number u

$$\mathbb{E}[\exp\{i\,u\,S_H\} \mid \mathcal{F}^H] = \mathbb{E}[\exp\{i\,\frac{u}{\sqrt{r_H}} \sum_{j=1}^{r_H} \frac{X_j^H}{\sigma_H}\} \mid \mathcal{F}^H] = (\Phi_N(u))^{r_H} \;,$$

by independence, where

$$\Phi_H(u) = \mathbb{E}[\exp\{i\,\frac{u}{\sqrt{r_H}}\,\frac{X_j^H}{\sigma_H}\} \mid \mathcal{F}^H] \;,$$

does not depend on $j = 1, \cdots, r_H$, and it follows from [13, Lemma C.3] that

$$|\Phi_H(u) - (1 - \tfrac{1}{2}\,\frac{u^2}{r_H})| \leq \tfrac{1}{6}\,c\,\frac{|u|^3}{r_H} + R_H(c)\,\frac{u^2}{r_H} \;,$$

for any $c > 0$. Using the straightforward estimate $|x^r - y^r| \leq r\,|x - y|$, which holds for any integer r and for any complex numbers x, y such that $|x| \leq 1$ and $|y| \leq 1$, yields

$$|(\Phi_H(u))^{r_H} - (1 - \tfrac{1}{2}\,\frac{u^2}{r_H})^{r_H}| \leq r_H\,|\Phi_H(u) - (1 - \tfrac{1}{2}\,\frac{u^2}{r_H})| \leq \tfrac{1}{6}\,c\,|u|^3 + R_H(c)\,u^2 \;.$$

Using the same estimate again and the straightforward estimate $|e^{-x} - (1 - x)| \leq \frac{1}{2}\,x^2$, which holds for any nonnegative real number $x \geq 0$, yields

$$|\exp\{-\tfrac{1}{2}\,u^2\} - (1 - \tfrac{1}{2}\,\frac{u^2}{r_H})^{r_H}| \leq r_H\,|\exp\{-\tfrac{1}{2}\,\frac{u^2}{r_H}\} - (1 - \tfrac{1}{2}\,\frac{u^2}{r_H})| \leq \tfrac{1}{8}\,\frac{u^4}{r_H} \;.$$

Combining the above estimates together and using the triangle inequality, yields

$$|\mathbb{E}[\exp\{i\,u\,S_H\} \mid \mathcal{F}^H] - \exp\{-\tfrac{1}{2}\,u^2\}| \leq \tfrac{1}{6}\,c\,|u|^3 + R_H(c)\,u^2 + \tfrac{1}{8}\,\frac{u^4}{r_H} \;, \qquad (28)$$

and on the *good set* $\{r_H > r\}$

$$|\mathbb{E}[\exp\{i\,u\,S_H\} \mid \mathcal{F}^H] - \exp\{-\tfrac{1}{2}\,u^2\}| \leq \tfrac{1}{6}\,c\,|u|^3 + R_H(c)\,u^2 + \tfrac{1}{8}\,\frac{u^4}{r} \;.$$

► Notice that

$$|N_H - r_H| \leq |N_H - H\rho_H| + 1 \leq |\frac{N_H}{H\rho_H} - 1|\;H\rho_H + 1 \;,$$

hence if $|\dfrac{N_H}{H\rho_H} - 1| \leq d$, then $|N_H - r_H| \leq d\,(r_H + 1) + 1 = d_H$, and either $r_H - \lceil d_H \rceil \leq N_H \leq r_H$ or $r_H \leq N_H \leq r_H + \lceil d_H \rceil$. Therefore, for any $B > 0$ and any $0 < d < 1$

$$\mathbb{P}[\,|O_H| > B\,,\, |\frac{N_H}{H\rho_H} - 1| \leq d \mid \mathfrak{F}^H]$$

$$\leq \mathbb{P}[\,|\sum_{i=1}^{N_H} \frac{X_i^H}{\sigma_H} - \sum_{i=1}^{r_H} \frac{X_i^H}{\sigma_H}| > B\,\sqrt{r_H}\,,\, |N_H - r_H| \leq \lceil d_H \rceil \mid \mathfrak{F}^H]$$

$$\leq \mathbb{P}[\max_{r_H - \lceil d_H \rceil \leq N \leq r_H} |\sum_{i=1}^{r_H} \frac{X_i^H}{\sigma_H} - \sum_{i=1}^{N} \frac{X_i^H}{\sigma_H}| > B\,\sqrt{r_H} \mid \mathfrak{F}^H]$$

$$+ \mathbb{P}[\max_{r_H \leq N \leq r_H + \lceil d_H \rceil} |\sum_{i=1}^{N} \frac{X_i^H}{\sigma_H} - \sum_{i=1}^{r_H} \frac{X_i^H}{\sigma_H}| > B\,\sqrt{r_H} \mid \mathfrak{F}^H]$$

$$\leq 2\,\frac{1}{B^2\,r_H}\,\lceil d_H \rceil$$

$$\leq 2\,\frac{d\,(r_H + 1) + 2}{B^2\,r_H}\;,$$

using the Kolmogorov maximal inequality, and on the *good set* $\{r_H > r\} \in \mathfrak{F}^H$

$$\mathbb{P}[\,|O_H| > B\,,\, |\frac{N_H}{H\rho_H} - 1| \leq d \mid \mathfrak{F}^H] \leq 2\,\frac{d\,(r+1) + 2}{B^2\,r}\;.$$

► Combining the above estimates, using the triangle inequality and taking $B = d^{1/3}$, yields

$$|\,\mathbb{E}[\exp\{i\,\frac{u}{\sqrt{H\rho_H}}\,\sum_{j=1}^{N_H} \frac{X_j^H}{\sigma_H}\} \mid \mathfrak{F}^H] - \exp\{-\tfrac{1}{2}\,u^2\}\,|$$

$$\leq 2 \cdot 1_{\{r_H \leq r\}} + R_H(c)\,u^2 + 2\,F_H(d)$$

$$+ \tfrac{1}{6}\,c\,|u|^3 + \tfrac{1}{8}\,\frac{u^4}{r} + |u|\,(\varepsilon(r) + d^{1/3}) + 4\,\frac{d\,(r+1) + 2}{d^{2/3}\,r}\;.$$

Taking d so that $d \to 0$ and $d^{2/3}\,r \to \infty$ when $r \uparrow \infty$, it is possible for any $a > 0$, to find $r > 0$ large enough, $c > 0$ small enough, such that

$$\tfrac{1}{6}\,c\,|u|^3 + \tfrac{1}{8}\,\frac{u^4}{r} + |u|\,(\varepsilon(r) + d^{1/3}) + 4\,\frac{d\,(r+1) + 2}{d^{2/3}\,r} < \tfrac{1}{2}\,a\;,$$

in which case

$$\mathbb{P}[\,|\,\mathbb{E}[\exp\{i\,\frac{u}{\sqrt{H\rho_H}}\,\sum_{j=1}^{N_H} \frac{X_j^H}{\sigma_H}\} \mid \mathfrak{F}^H] - \exp\{-\tfrac{1}{2}\,u^2\}\,| > a]$$

$$\leq \mathbb{P}[2 \cdot 1_{\{r_H \leq r\}} + R_H(c)\,u^2 + 2\,F_H(d) > \tfrac{1}{2}\,a]\;,$$

which goes to zero as $H \uparrow \infty$: this terminates the proof of (14).

To prove (15), notice that

$$\sqrt{H}\,\frac{1}{N_H}\sum_{j=1}^{N_H} X_j^H = \frac{H\,\rho_H}{N_H}\,\frac{\sigma_H}{\sqrt{\rho_H}}\,\frac{1}{\sqrt{H\,\rho_H}}\sum_{j=1}^{N_H}\frac{X_j^H}{\sigma_H} = c_H\,Z_H\ ,$$

where

$$c_H = \frac{H\,\rho_H}{N_H}\,\frac{\sigma_H}{\sqrt{\rho_H}} \qquad \text{and} \qquad c = \frac{\sigma}{\sqrt{\rho}}\ ,$$

and where Z_H is defined in (27), hence

$$\mathbb{E}[\exp\{i\,u\,\sqrt{H}\,\frac{1}{N_H}\sum_{j=1}^{N_H} X_j^H\} \mid \mathcal{F}^H] - \exp\{-\tfrac{1}{2}\,\frac{u^2\,\sigma^2}{\rho}\}$$

$$= \mathbb{E}[\exp\{i\,u\,c_H\,Z_H\}] - \exp\{-\tfrac{1}{2}\,u^2\,c^2\}$$

$$= \mathbb{E}[\exp\{i\,u\,c_H\,Z_H\} - \exp\{i\,u\,c\,Z_H\} \mid \mathcal{F}^H]$$

$$+\,\mathbb{E}[\exp\{i\,\frac{u\,c}{\sqrt{H\,\rho_H}}\sum_{j=1}^{N_H}\frac{X_j^H}{\sigma_H}\} \mid \mathcal{F}^H] - \exp\{-\tfrac{1}{2}\,u^2\,c^2\}\ .$$

The last term goes to zero in $\mathbb{L}^1$ as $H \uparrow \infty$, using (14). Moreover, for any $b > 0$ and any $0 < d < 1$

$$\mathbb{E}|\,\mathbb{E}[\exp\{i\,u\,c_H\,Z_H\} - \exp\{i\,u\,c\,Z_H\} \mid \mathcal{F}^H]\,|$$

$$\leq \mathbb{E}|\,\exp\{i\,u\,c_H\,Z_H\} - \exp\{i\,u\,c\,Z_H\}\,|$$

$$\leq \mathbb{E}[\,1_{\{|\frac{\sigma_H}{\sqrt{\rho_H}} - \frac{\sigma}{\sqrt{\rho}}| \leq b\,,\ |\frac{N_H}{H\,\rho_H} - 1| \leq d\}}$$

$$|\,\exp\{i\,u\,c_H\,Z_H\} - \exp\{i\,u\,c\,Z_H\}\,|\,]$$

$$+\,2\,\mathbb{P}[\,|\frac{N_H}{H\,\rho_H} - 1| > d] + 2\,\mathbb{P}[\,|\frac{\sigma_H}{\sqrt{\rho_H}} - \frac{\sigma}{\sqrt{\rho}}| > b]\ .$$

The last two terms go to zero as $H \uparrow \infty$, by assumption and in view of Remark 11. Next, if $|\dfrac{N_H}{H\,\rho_H} - 1| \leq d$, then clearly $|\dfrac{H\,\rho_H}{N_H} - 1| \leq \dfrac{d}{1-d}$, and since

$$c_H - c = \frac{H\,\rho_H}{N_H}\,\frac{\sigma_H}{\sqrt{\rho_H}} - \frac{\sigma}{\sqrt{\rho}} = (\frac{H\,\rho_H}{N_H} - 1)\,\frac{\sigma_H}{\sqrt{\rho_H}} + (\frac{\sigma_H}{\sqrt{\rho_H}} - \frac{\sigma}{\sqrt{\rho}})\ ,$$

then it holds

$$1_{\{|\frac{\sigma_H}{\sqrt{\rho_H}} - \frac{\sigma}{\sqrt{\rho}}| \leq b\,,\, |\frac{N_H}{H\rho_H} - 1| \leq d\}} |c_H - c|$$

$$\leq \frac{d}{1-d} (\frac{\sigma}{\sqrt{\rho}} + b) + b = \frac{d\,\frac{\sigma}{\sqrt{\rho}} + b}{1-d} \; .$$

Therefore, using the straightforward estimate $|e^{i\,x} - e^{i\,x'}| \leq \min(|x - x'|, 2)$ which holds for any real numbers x, x', yields

$$\mathbb{E}[\, 1_{\{|\frac{\sigma_H}{\sqrt{\rho_H}} - \frac{\sigma}{\sqrt{\rho}}| \leq b\,,\, |\frac{N_H}{H\rho_H} - 1| \leq d\}} \, |\exp\{i\,u\,c_H\,Z_H\} - \exp\{i\,u\,c\,Z_H\}\,|\,]$$

$$\leq \mathbb{E}[\, 1_{\{|\frac{\sigma_H}{\sqrt{\rho_H}} - \frac{\sigma}{\sqrt{\rho}}| \leq b\,,\, |\frac{N_H}{H\rho_H} - 1| \leq d\}} \min(|u|\,|c_H - c|\,|Z_H|, 2)\,]$$

$$\leq \mathbb{E}\, \min(|u|\, \frac{d\,\frac{\sigma}{\sqrt{\rho}} + b}{1-d}\, |Z_H|, 2) \; ,$$

hence using (14) yields

$$\limsup_{H \uparrow \infty} \mathbb{E}|\, \mathbb{E}[\, \exp\{i\,u\,c_H\,Z_H\} - \exp\{i\,u\,c\,Z_H\} \mid \mathcal{F}^H]\,|$$

$$\leq \limsup_{H \uparrow \infty} \mathbb{E}[\, 1_{\{|\frac{\sigma_H}{\sqrt{\rho_H}} - \frac{\sigma}{\sqrt{\rho}}| \leq b\,,\, |\frac{N_H}{H\rho_H} - 1| \leq d\}} \, |\exp\{i\,u\,c_H\,Z_H\} - \exp\{i\,u\,c\,Z_H\}\,|\,]$$

$$\leq \mathbb{E}\, \min(|u|\, \frac{d\,\frac{\sigma}{\sqrt{\rho}} + b}{1-d}\, |Z|, 2) \; ,$$

in view of Remark 10, where Z is a standard Gaussian r.v. (with zero mean and unit variance). Finally, using the Lebesgue dominated convergence theorem, it follows that

$$\mathbb{E}[\, \exp\{i\,u\,c_H\,Z_H\} - \exp\{i\,u\,c\,Z_H\} \mid \mathcal{F}^H] \longrightarrow 0 \; ,$$

in $\mathbb{L}^1$ as $H \uparrow \infty$, since $b > 0$ and $0 < d < 1$ are arbitrary : this terminates the proof of (15).

□

B Proof of Theorem 7

By definition, S_n^H is written as a double sum over generations with different random sizes : the idea is to rewrite this as a single sum across all generations, and to use a central limit theorem for triangular arrays of martingale increments [2, Theorem 2.8.42]. Notice that the i–th particle within the k–th generation can be associated in a unique way with an integer p between 1 and $\sigma_n^H = N_0^H + \cdots + N_n^H$: clearly $p_{k,i} = \sigma_{k-1}^H + i$, and conversely the random integers k_p and i_p are defined by

$$k_p = \inf\{k \geq 0 \,:\, \sigma_k^H \geq p\} \qquad \text{and} \qquad i_p = p - \sigma_{k_p-1}^H \ ,$$

with the convention $\sigma_{-1}^H = 0$, or in other words $k_p = k$ and $i_p = i$ if and only if

$$\sigma_{k-1}^H + 1 \leq p = \sigma_{k-1}^H + i \leq \sigma_k^H \ ,$$

with $1 \leq i \leq N_k^H$, see Figure 1.

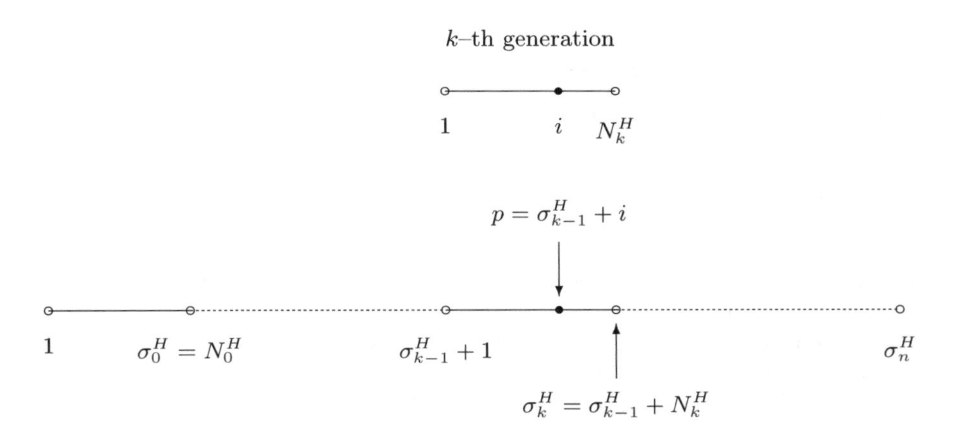

Fig. 1. The i–th particle within the k–th generation (above), seen as the p–th particle across all generations (below)

For any $k = 0, 1, \cdots, n$ and any integer $i \geq 1$

$$\{k_p = k, i_p = i\} = \{p = \sigma_{k-1}^H + i, i \leq N_k^H\} \in \mathcal{F}_{k,i-1}^H \subset \mathcal{F}_{k,i}^H \ ,$$

since $\{p = \sigma_{k-1}^H + i\} \in \mathcal{H}_{k-1}^H$ and $\{i \leq N_k^H\} \in \mathcal{F}_{k,i-1}^H$, which allows to define the σ–algebra $\mathcal{G}_p^H = \mathcal{F}_{k_p,i_p}^H$ in the usual way : by definition, $A \in \mathcal{G}_p^H$ if and only if $A \cap \{k_p = k, i_p = i\} \in \mathcal{F}_{k,i}^H$ for any $k = 0, 1, \cdots, n$ and any integer $i \geq 1$. Using this new labeling of the particle system yields

$$S_n^H = \sum_{k=0}^{n} \sum_{i=1}^{N_k^H} X_{k,i}^H = \sum_{p=1}^{\sigma_n^H} U_p^H \; ,$$

where the time changed random variable $U_p^H = X_{k_p,i_p}^H$ is measurable w.r.t. $\mathcal{G}_p^H$, for any $p = 1, \cdots, \sigma_n^H$: indeed for any Borel subset B, any $k = 0, 1, \cdots, n$ and any integer $i \geq 1$

$$\{U_p^H \in B\} \cap \{k_p = k, i_p = i\} = \{X_{k,i}^H \in B\} \cap \{k_p = k, i_p = i\} \; ,$$

hence $\{U_p^H \in B\} \in \mathcal{G}_p^H$, since $\{X_{k,i}^H \in B\} \in \mathcal{F}_{k,i}^H$ and $\{k_p = k, i_p = i\} \in \mathcal{F}_{k,i-1}^H$. Moreover, the random variable σ_n^H is a stopping time w.r.t. $\mathcal{G}^H = \{\mathcal{G}_p^H \; , \; p \geq 1\}$: indeed, for any integer $p \geq 1$, any $k = 0, 1, \cdots, n$ and any integer $i \geq 1$

$$\begin{aligned} \{\sigma_n^H = p\} \cap \{k_p = k, i_p = i\} &= \{\sigma_n^H = p\} \cap \{p = \sigma_{k-1}^H + i, 1 \leq i \leq N_k^H\} \\ &= \begin{cases} \emptyset \; , & \text{if } k \neq n, \\ \{p = \sigma_{k-1}^H + i\} \cap \{N_k^H = i\} \; , & \text{if } k = n, \end{cases} \end{aligned}$$

hence $\{\sigma_n^H = p\} \in \mathcal{G}_p^H$ since $\{p = \sigma_{k-1}^H + i\} \in \mathcal{H}_{k-1}^H$ and $\{N_k^H = i\} \in \mathcal{F}_{k,i}^H$.

PROOF OF THEOREM 7. To apply Theorem 2.8.42 in [2] to

$$S_n^H = \sum_{p=1}^{\sigma_n^H} U_p^H \; ,$$

where σ_n^H is a stopping time w.r.t. $\mathcal{G}^H = \{\mathcal{G}_p^H \; , \; p \geq 1\}$, and where the random variable U_p^H is measurable w.r.t. $\mathcal{G}_p^H$ for any $p = 1, \cdots, \sigma_n^H$, the following three conditions have to be checked : a martingale increment property, the convergence of conditional variances, and a conditional Lindeberg condition. These three conditions follow immediately from (19) and from

$$\sum_{p=1}^{\sigma_n^H} \mathbb{E}[U_p^H \mid \mathcal{G}_{p-1}^H] = 0 \; ,$$

$$\sum_{p=1}^{\sigma_n^H} \mathbb{E}[\, |U_p^H|^2 \mid \mathcal{G}_{p-1}^H] = \sum_{k=0}^{n} N_k^H \, V_{k,0}^H \; ,$$

and

$$\sum_{p=1}^{\sigma_n^H} \mathbb{E}[\, |U_p^H|^2 \, 1_{\{|U_p^H| \, > \, \varepsilon\}} \mid \mathcal{G}_{p-1}^H] = \sum_{k=0}^{n} N_k^H \, Y_{k,0}^{H,\varepsilon} \; ,$$

which follow from (16), (17) and (18) respectively, using properties of the past σ–algebra $\mathcal{G}_{p-1}^H$, see [13, Lemmas B.1 and B.2], and using the preservation of the martingale property under time change, see Lemma 4 below. Indeed, it follows from Lemma 4 and from (16), that

$$\mathbb{E}[U_p^H \mid \mathcal{G}_{p-1}^H] = 0 \ ,$$

for any integer $p \geq 1$, hence

$$\sum_{p=1}^{\sigma_n^H} \mathbb{E}[U_p^H \mid \mathcal{G}_{p-1}^H] = 0 \ .$$

Similarly, it follows from Corollary 1 and from (17), that

$$\mathbb{E}[\,|U_p^H|^2 \mid \mathcal{G}_{p-1}^H] = V_{k_p,0}^H \ ,$$

for any integer $p \geq 1$, hence

$$\sum_{p=1}^{\sigma_n^H} \mathbb{E}[\,|U_p^H|^2 \mid \mathcal{G}_{p-1}^H]$$

$$= \sum_{p=1}^{\sigma_n^H} V_{k_p,0}^H = \sum_{k=0}^{n} \sum_{p=\sigma_{k-1}^H+1}^{\sigma_k^H} V_{k_p,0}^H = \sum_{k=0}^{n} N_k^H \, V_{k,0}^H \ ,$$

since $k_p = k$ if $\sigma_{k-1}^H + 1 \leq p \leq \sigma_k^H$. Finally, it follows from Corollary 2 and from (18), that

$$\mathbb{E}[\,|U_p^H|^2 \, 1_{\{|U_p^H| \, > \, \varepsilon\}} \mid \mathcal{G}_{p-1}^H] \leq Y_{k_p,0}^{H,\varepsilon} \ ,$$

for any integer $p \geq 1$ and any $\varepsilon > 0$, hence

$$\sum_{p=1}^{\sigma_n^H} \mathbb{E}[\,|U_p^H|^2 \, 1_{\{|U_p^H| \, > \, \varepsilon\}} \mid \mathcal{G}_{p-1}^H]$$

$$\leq \sum_{p=1}^{\sigma_n^H} Y_{k_p,0}^{H,\varepsilon} = \sum_{k=0}^{n} \sum_{p=\sigma_{k-1}^H+1}^{\sigma_k^H} Y_{k_p,0}^{H,\varepsilon} = \sum_{k=0}^{n} N_k^H \, Y_{k,0}^{H,\varepsilon} \ ,$$

since $k_p = k$ if $\sigma_{k-1}^H + 1 \leq p \leq \sigma_k^H$. □

Lemma 4. *If for any $k = 0, 1, \cdots, n$ and any integer $i \geq 1$*

$$\mathbb{E}[F_{k,i} \mid \mathcal{F}_{k,i-1}^H] = 0 \ ,$$

then for any integer $p \geq 1$, the time changed random variable $G_p = F_{k_p, i_p}$ satisfies

$$\mathbb{E}[G_p \mid \mathcal{G}_{p-1}^H] = 0 \ .$$

Corollary 1. *If for any $k = 0, 1, \cdots, n$ and any integer $i \geq 1$*

$$\mathbb{E}[F_{k,i} \mid \mathcal{F}^H_{k,i-1}] = \widehat{F}^H_k \ ,$$

where the random variable $\widehat{F}^H_k$ is measurable w.r.t. $\mathcal{F}^H_{k,0}$, then for any integer $p \geq 1$, the time changed random variables $G_p = F_{k_p,i_p}$ and $\widehat{G}^H_p = \widehat{F}^H_{k_p}$ satisfy

$$\mathbb{E}[G_p \mid \mathcal{G}^H_{p-1}] = \widehat{G}^H_p \ .$$

PROOF. Notice that

$$\mathbb{E}[F_{k,i} - \widehat{F}^H_k \mid \mathcal{F}^H_{k,i-1}] = 0 \ ,$$

under the assumption, and it follows from Lemma 4 above that

$$\mathbb{E}[G_p - \widehat{G}^H_p \mid \mathcal{G}^H_{p-1}] = 0 \ ,$$

or equivalently

$$\mathbb{E}[G_p \mid \mathcal{G}^H_{p-1}] = \widehat{G}^H_p \ ,$$

since $\widehat{G}^H_p$ is measurable w.r.t. $\mathcal{G}^H_{p-1}$, in view of [13, Lemma B.2]. □

Corollary 2. *If for any $k = 0, 1, \cdots, n$ and any integer $i \geq 1$*

$$F_{k,i} \leq F^*_k \ ,$$

where the random variable F^*_k is measurable w.r.t. $\mathcal{F}^H_{k,0}$, then for any integer $p \geq 1$, the time changed random variables $G_p = F_{k_p,i_p}$ and $G^*_p = F^*_{k_p}$ satisfy

$$\mathbb{E}[G_p \mid \mathcal{G}^H_{p-1}] \leq G^*_p \ .$$

PROOF. Notice that

$$\mathbb{E}[\max(F_{k,i} - F^*_k, 0) \mid \mathcal{F}^H_{k,i-1}] = 0 \ ,$$

since $\max(F_{k,i} - F^*_k, 0) = 0$ under the assumption, and it follows from Lemma 4 above that

$$\mathbb{E}[\max(G_p - G^*_p, 0) \mid \mathcal{G}^H_{p-1}] = 0 \ ,$$

hence using the Jensen inequality yields

$$\max(\mathbb{E}[G_p - G^*_p \mid \mathcal{G}^H_{p-1}], 0) = 0 \qquad \text{i.e.} \qquad \mathbb{E}[G_p - G^*_p \mid \mathcal{G}^H_{p-1}] \leq 0 \ ,$$

or equivalently

$$\mathbb{E}[G_p \mid \mathcal{G}^H_{p-1}] \leq G^*_p \ ,$$

since G^*_p is measurable w.r.t. $\mathcal{G}^H_{p-1}$, in view of [13, Lemma B.2]. □

PROOF OF LEMMA 4. First, recall the following identity

$$\sum_{i=1}^{N \wedge M} 1_{\{I=i\}} = \sum_{i=1}^{M} 1_{\{N \geq i\}} \, 1_{\{I=i\}} \; ,$$

which is easily obtained using summation by parts. For any $A \in \mathcal{G}_{p-1}^H$, and any integer $M \geq 1$

$$\mathbb{E}[G_p \, 1_A \sum_{k=0}^{n} 1_{\{k_p = k\}} \sum_{i=1}^{N_k^H \wedge M} 1_{\{i_p = i\}}]$$

$$= \mathbb{E}[G_p \, 1_A \sum_{k=0}^{n} 1_{\{k_p = k\}} \sum_{i=1}^{M} 1_{\{N_k^H \geq i\}} \, 1_{\{i_p = i\}}]$$

$$= \sum_{k=0}^{n} \sum_{i=1}^{M} \mathbb{E}[G_p \, 1_{A \,\cap\, \{k_p = k, i_p = i\}} \, 1_{\{N_k^H \geq i\}}]$$

$$= \sum_{k=0}^{n} \sum_{i=1}^{M} \mathbb{E}[F_{k,i} \, 1_{A \,\cap\, \{k_p = k, i_p = i\}} \, 1_{\{N_k^H \geq i\}}] \; .$$

Notice that $A \cap \{k_p = k, i_p = i\} \in \mathcal{F}_{k,i-1}^H$ in view of [13, Lemma B.1] and $\{N_k^H \geq i\} \in \mathcal{F}_{k,i-1}^H$, hence

$$\mathbb{E}[F_{k,i} \, 1_{A \,\cap\, \{k_p = k, i_p = i\}} \, 1_{\{N_k^H \geq i\}}]$$

$$= \mathbb{E}[\, \mathbb{E}[F_{k,i} \mid \mathcal{F}_{k,i-1}^H] \, 1_{A \,\cap\, \{k_p = k, i_p = i\}} \, 1_{\{N_k^H \geq i\}}] = 0 \; ,$$

under the assumption, and

$$\mathbb{E}[G_p \, 1_A \sum_{k=0}^{n} 1_{\{k_p = k\}} \sum_{i=1}^{N_k^H \wedge M} 1_{\{i_p = i\}}] = 0 \; ,$$

or equivalently

$$\mathbb{E}[\, G_p^+ \, 1_A \sum_{k=0}^{n} 1_{\{k_p = k\}} \sum_{i=1}^{N_k^H \wedge M} 1_{\{i_p = i\}}]$$

$$= \mathbb{E}[\, G_p^- \, 1_A \sum_{k=0}^{n} 1_{\{k_p = k\}} \sum_{i=1}^{N_k^H \wedge M} 1_{\{i_p = i\}}] \; .$$

where $G_p^+ = \max(G_p, 0)$ and $G_p^- = \max(-G_p, 0)$. Finally, using the monotone convergence theorem yields

$$\begin{aligned}
\mathbb{E}[G_p^+ \, 1_A] &= \mathbb{E}[G_p^+ \, 1_A \sum_{k=0}^{n} 1_{\{k_p = k\}} \sum_{i=1}^{N_k^H} 1_{\{i_p = i\}}] \\
&= \lim_{M \uparrow \infty} \mathbb{E}[G_p^+ \, 1_A \sum_{k=0}^{n} 1_{\{k_p = k\}} \sum_{i=1}^{N_k^H \wedge M} 1_{\{i_p = i\}}] \\
&= \lim_{M \uparrow \infty} \mathbb{E}[G_p^- \, 1_A \sum_{k=0}^{n} 1_{\{k_p = k\}} \sum_{i=1}^{N_k^H \wedge M} 1_{\{i_p = i\}}] \\
&= \mathbb{E}[G_p^- \, 1_A \sum_{k=0}^{n} 1_{\{k_p = k\}} \sum_{i=1}^{N_k^H} 1_{\{i_p = i\}}] = \mathbb{E}[G_p^- \, 1_A] \; ,
\end{aligned}$$

or equivalently $\mathbb{E}[G_p \, 1_A] = 0$, hence $\mathbb{E}[G_p \mid \mathcal{G}_{p-1}^H] = 0$, since $A \in \mathcal{G}_{p-1}^H$ is arbitrary. □

Acknowledgments

The first author gratefully thanks Natacha Caylus for her careful reading of an earlier version of this work, and for suggesting the proof given in Appendix A for the second part of Theorem 6. We both thank Pierre Del Moral for his warm support, and for suggesting the alternate approach, based on a central limit theorem for triangular arrays of martingale increments, which is followed in Section 6 for the proof of Theorem 4.

References

1. Frédéric Cérou, Pierre Del Moral, François Le Gland, and Pascal Lezaud. Limit theorems for the multilevel splitting algorithm in the simulation of rare events. In Michael E. Kuhl, Natalie M. Steiger, Frank B. Armstrong, and Jeffrey A. Joines, editors, *Proceedings of the 2005 Winter Simulation Conference, Orlando 2005*, December 2005.
2. Didier Dacunha-Castelle and Marie Duflo. *Probability and Statistics II.* Springer–Verlag, Berlin, 1986.
3. Pierre Del Moral. *Feynman–Kac Formulae. Genealogical and Interacting Particle Systems with Applications.* Probability and its Applications. Springer–Verlag, New York, 2004.
4. Pierre Del Moral and Jean Jacod. Interacting particle filtering with discrete observations. In Arnaud Doucet, Nando de Freitas, and Neil Gordon, editors, *Sequential Monte Carlo Methods in Practice*, Statistics for Engineering and Information Science, chapter 3, pages 43–75. Springer–Verlag, New York, 2001.
5. Pierre Del Moral, Jean Jacod, and Philip Protter. The Monte Carlo method for filtering with discrete–time observations. *Probability Theory and Related Fields*, 120(3):346–368, 2001.

6. Pierre Del Moral and Pascal Lezaud. Branching and interacting particle interpretation of rare event probabilities. In Henk Blom and John Lygeros, editors, *Stochastic Hybrid Systems : Theory and Safety Critical Applications*, Lecture Notes in Control and Information Sciences, chapter 14. Springer–Verlag, Berlin. To appear.
7. Pierre Del Moral and Laurent Miclo. Branching and interacting particle systems approximations of Feynman–Kac formulae with applications to nonlinear filtering. In Jacques Azéma, Michel Émery, Michel Ledoux, and Marc Yor, editors, *Séminaire de Probabilités XXXIV*, volume 1729 of *Lecture Notes in Mathematics*, pages 1–145. Springer–Verlag, Berlin, 2000.
8. Allan Gut. *Stopped Random Walks. Limit Theorems and Applications.* Applied Probability (A Series of the Applied Probability Trust). Springer–Verlag, New York, 1988.
9. Hans R. Künsch. Recursive Monte Carlo filters : Algorithms and theoretical analysis. *The Annals of Statistics*, 33(5):1983–2021, October 2005.
10. François Le Gland and Nadia Oudjane. A robustification approach to stability and to uniform particle approximation of nonlinear filters : the example of pseudo-mixing signals. *Stochastic Processes and their Applications*, 106(2):279–316, August 2003.
11. François Le Gland and Nadia Oudjane. Stability and uniform approximation of nonlinear filters using the Hilbert metric, and application to particle filters. *The Annals of Applied Probability*, 14(1):144–187, February 2004.
12. François Le Gland and Nadia Oudjane. A sequential particle algorithm that keeps the particle system alive. In *Proceedings of the 13th European Signal Processing Conference, Antalya 2005*. EURASIP, September 2005.
13. François Le Gland and Nadia Oudjane. A sequential algorithm that keeps the particle system alive. Rapport de Recherche 5826, INRIA, February 2006. `ftp://ftp.inria.fr/INRIA/publication/publi-pdf/RR/RR-5826.pdf`.
14. Jacques Neveu. *Discrete–Parameter Martingales*, volume 10 of *North–Holland Mathematical Library*. North–Holland, Amsterdam, 1975.
15. Nadia Oudjane. *Stabilité et Approximations Particulaires en Filtrage Non–Linéaire — Application au Pistage.* Thèse de Doctorat, Université de Rennes 1, Rennes, December 2000. `ftp://ftp.irisa.fr/techreports/theses/2000/oudjane.pdf`.
16. Nadia Oudjane and Sylvain Rubenthaler. Stability and uniform particle approximation of nonlinear filters in case of non ergodic signal. *Stochastic Analysis and Applications*, 23(3):421–448, May 2005.
17. Alfréd Rényi. On the asymptotic distribution of the sum of a random number of independent random variables. *Acta Mathematica Academiae Scientiarum Hungaricae*, 8:193–199, 1957.
18. Vivien Rossi. *Filtrage Non–Linéaire par Noyaux de Convolution — Application à un Procédé de Dépollution Biologique.* Thèse de Doctorat, École Nationale Supérieure Agronomique de Montpellier, Montpellier, December 2004.
19. Vivien Rossi and Jean-Pierre Vila. Nonlinear filtering in discrete time : A particle convolution approach. *Statistical Inference for Stochastic Processes.* To appear.
20. David Siegmund. *Sequential Analysis. Tests and Confidence Intervals.* Springer Series in Statistics. Springer–Verlag, New York, 1985.

Index

Lecture Notes in Control and Information Sciences

Edited by M. Thoma and M. Morari

Further volumes of this series can be found on our homepage: springer.com

Vol. 311: Lamnabhi-Lagarrigue, F.; Loría, A.; Panteley, E. (Eds.)
Advanced Topics in Control Systems Theory
294 p. 2005 [1-85233-923-3]

Vol. 310: Janczak, A.
Identification of Nonlinear Systems Using Neural Networks and Polynomial Models
197 p. 2005 [3-540-23185-4]

Vol. 309: Kumar, V.; Leonard, N.; Morse, A.S. (Eds.)
Cooperative Control
301 p. 2005 [3-540-22861-6]

Vol. 308: Tarbouriech, S.; Abdallah, C.T.; Chiasson, J. (Eds.)
Advances in Communication Control Networks
358 p. 2005 [3-540-22819-5]

Vol. 307: Kwon, S.J.; Chung, W.K.
Perturbation Compensator based Robust Tracking Control and State Estimation of Mechanical Systems
158 p. 2004 [3-540-22077-1]

Vol. 306: Bien, Z.Z.; Stefanov, D. (Eds.)
Advances in Rehabilitation
472 p. 2004 [3-540-21986-2]

Vol. 305: Nebylov, A.
Ensuring Control Accuracy
256 p. 2004 [3-540-21876-9]

Vol. 304: Margaris, N.I.
Theory of the Non-linear Analog Phase Locked Loop
303 p. 2004 [3-540-21339-2]

Vol. 303: Mahmoud, M.S.
Resilient Control of Uncertain Dynamical Systems
278 p. 2004 [3-540-21351-1]

Vol. 302: Filatov, N.M.; Unbehauen, H.
Adaptive Dual Control: Theory and Applications
237 p. 2004 [3-540-21373-2]

Vol. 301: de Queiroz, M.; Malisoff, M.; Wolenski, P. (Eds.)
Optimal Control, Stabilization and Nonsmooth Analysis
373 p. 2004 [3-540-21330-9]

Vol. 300: Nakamura, M.; Goto, S.; Kyura, N.; Zhang, T.
Mechatronic Servo System Control
Problems in Industries and their Theoretical Solutions
212 p. 2004 [3-540-21096-2]

Vol. 299: Tarn, T.-J.; Chen, S.-B.; Zhou, C. (Eds.)
Robotic Welding, Intelligence and Automation
214 p. 2004 [3-540-20804-6]

Vol. 298: Choi, Y.; Chung, W.K.
PID Trajectory Tracking Control for Mechanical Systems
127 p. 2004 [3-540-20567-5]

Vol. 297: Damm, T.
Rational Matrix Equations in Stochastic Control
219 p. 2004 [3-540-20516-0]

Vol. 296: Matsuo, T.; Hasegawa, Y.
Realization Theory of Discrete-Time Dynamical Systems
235 p. 2003 [3-540-40675-1]

Vol. 295: Kang, W.; Xiao, M.; Borges, C. (Eds)
New Trends in Nonlinear Dynamics and Control, and their Applications
365 p. 2003 [3-540-10474-0]

Vol. 294: Benvenuti, L.; De Santis, A.; Farina, L. (Eds)
Positive Systems: Theory and Applications (POSTA 2003)
414 p. 2003 [3-540-40342-6]

Vol. 293: Chen, G. and Hill, D.J.
Bifurcation Control
320 p. 2003 [3-540-40341-8]

Vol. 292: Chen, G. and Yu, X.
Chaos Control
380 p. 2003 [3-540-40405-8]

Vol. 291: Xu, J.-X. and Tan, Y.
Linear and Nonlinear Iterative Learning Control
189 p. 2003 [3-540-40173-3]

Vol. 290: Borrelli, F.
Constrained Optimal Control of Linear and Hybrid Systems
237 p. 2003 [3-540-00257-X]

Vol. 289: Giarré, L. and Bamieh, B.
Multidisciplinary Research in Control
237 p. 2003 [3-540-00917-5]

Vol. 288: Taware, A. and Tao, G.
Control of Sandwich Nonlinear Systems
393 p. 2003 [3-540-44115-8]

Vol. 287: Mahmoud, M.M.; Jiang, J.; Zhang, Y.
Active Fault Tolerant Control Systems
239 p. 2003 [3-540-00318-5]

Vol. 286: Rantzer, A. and Byrnes C.I. (Eds)
Directions in Mathematical Systems Theory and Optimization
399 p. 2003 [3-540-00065-8]

Vol. 285: Wang, Q.-G.
Decoupling Control
373 p. 2003 [3-540-44128-X]

Vol. 284: Johansson, M.
Piecewise Linear Control Systems
216 p. 2003 [3-540-44124-7]

Vol. 283: Fielding, Ch. et al. (Eds)
Advanced Techniques for Clearance of Flight Control Laws
480 p. 2003 [3-540-44054-2]

Vol. 282: Schröder, J.
Modelling, State Observation and Diagnosis of Quantised Systems
368 p. 2003 [3-540-44075-5]

Vol. 281: Zinober A.; Owens D. (Eds)
Nonlinear and Adaptive Control
416 p. 2002 [3-540-43240-X]

Vol. 280: Pasik-Duncan, B. (Ed)
Stochastic Theory and Control
564 p. 2002 [3-540-43777-0]

Vol. 279: Engell, S.; Frehse, G.; Schnieder, E. (Eds)
Modelling, Analysis, and Design of Hybrid Systems
516 p. 2002 [3-540-43812-2]

Vol. 278: Chunling D. and Lihua X. (Eds)
H_∞ Control and Filtering of Two-dimensional Systems
161 p. 2002 [3-540-43329-5]

Printing: Krips bv, Meppel
Binding: Stürtz, Würzbu